Chemical
Oceanography

VOLUME 2
2ND EDITION

Chemical Oceanography

Edited by

J. P. RILEY

Department of Oceanography,
The University of Liverpool, England

and

G. SKIRROW

Department of Inorganic, Physical and Industrial Chemistry,
The University of Liverpool, England

VOLUME 2

2ND EDITION

1975

ACADEMIC PRESS

LONDON   NEW YORK   SAN FRANCISCO

A Subsidiary of Harcourt Brace Jovanovich, Publishers

ACADEMIC PRESS INC. (LONDON) LTD.
24/28 Oval Road,
London NW1

United States Edition published by
ACADEMIC PRESS INC.
111 Fifth Avenue
New York, New York 10003

Library of Congress Catalog Card Number: 74-5679
ISBN: 0-12-588602-0

Printed in Great Britain by
PAGE BROS (NORWICH) LTD
NORWICH

Contributors to Volume 2

G. E. FOGG, *Marine Science Laboratories, University College of North Wales, Menai Bridge, Anglesey, Wales*

K. GRASSHOFF, *Institut für Meereskunde, Kiel, Federal German Republic*

PETER S. LISS, *School of Environmental Sciences, University of East Anglia, Norwich, England*

T. R. PARSONS, *Institute of Oceanography, The University of British Columbia, Vancouver, B.C., Canada*

GEOFFREY SKIRROW, *Donnan Laboratories, The University of Liverpool, Liverpool, England*

C. P. SPENCER, *Marine Science Laboratories, University College of North Wales, Menai Bridge, Anglesey, Wales*

P. J. LE B. WILLIAMS, *Department of Oceanography, University of Southampton, Southampton, England*

Preface to the Second Edition

Rapid progress has occurred in all branches of Chemical Oceanography since the publication of the first edition of this book a decade ago. Particularly noteworthy has been the tendency to treat the subject in a much more quantitative fashion; this has become possible because of our much improved understanding of the physical chemistry of sea water systems in terms of ionic and molecular theories. For these reasons chapters dealing with sea water as an electrolyte system, with speciation and with aspects of colloid chemistry are now to be considered as essential in any up-to-date treatment of the subject. Fields of research which were little more than embryonic only ten years ago, for example sea surface chemistry, have now expanded so much that they merit separate consideration. Since the previous edition, there has arisen a general awareness of the potential threat to the sea caused by man's activities, in particular its use as a "rubbish bin" and a receptacle for toxic wastes. Although it was inevitable that there should be some over-reaction to this, there is real cause for concern. Clearly, it is desirable to have available reasoned discussions of this topic and also an examination of the role of the sea as a potential source of raw materials in view of the imminent exhaustion of many high grade ores; these subjects are treated in the second, third and forth volumes.

Most branches of marine chemistry make use of analytical techniques; the number and range of these has increased dramatically over recent years. Consequently, it has been necessary to expand greatly and restructure the sections dealing with analytical methodology. These developments are extending increasingly into the very important and rapidly developing area of organic chemistry.

Rapid advances which have taken place in geochemistry, particularly those that have stemmed from the Deep Sea Drilling Project, have made it necessary to devote a whole volume to topics in sedimentary geochemistry.

Both the range and accuracy of the physical constants available have increased since the first edition and a selection of tabulated values of these constants are to be found at the end of each of the first four volumes.

No attempt has been made to discuss Physical Oceanography except where a grasp of the physical concepts is necessary for a better understanding of the chemistry. For a treatment of the physical processes occurring in the sea the reader is referred to the numerous excellent texts now available on physical oceanography. Likewise, since the distribution of salinity in the sea is of greater relevance to the physical oceanographer and is well discussed in these texts, it will not be considered in the present volumes.

This series is not intended to serve as a practical handbook of Marine Chemistry, and if practical details are required the original references given in the text should be consulted. In passing, it should be mentioned that, although those practical aspects of sea water chemistry which are of interest to biologists are reasonably adequately covered in the "Manual of Sea Water Analysis" by Strickland and Parsons, there is an urgent need for a more general laboratory manual.

The editors are most grateful to the various authors for their helpful co-operation which has greatly facilitated the preparation of this book. They would particularly like to thank Messrs R. F. C. Mantoura and A. Dickson for their willing assistance with the arduous task of proof reading; without their aid many errors would have escaped detection. They would also like to acknowledge the courtesy of the various copyright holders, both authors and publishers, for permission to use tables, figures, and photographs. In conclusion, they wish to thank Academic Press, and in particular Mr. E. A. S. Cotton, for their efficiency and ready co-operation which has much lightened the task of preparing this book for publication.

Liverpool   J. P. RILEY

November, 1974   G. SKIRROW

CONTENTS

Chapter 9 by GEOFFREY SKIRROW

The Dissolved Gases—Carbon Dioxide

Chapter 10 by PETER S. LISS

Chemistry of the Sea Surface Microlayer

Chapter 11 *by* C. P. SPENCER

The Micronutrient Elements

Chapter 12 *by* P. J. le B. WILLIAMS

Biological and Chemical Aspects of Dissolved Organic Material in Sea Water

Chapter 13 *by* T. R. PARSONS

Particulate Organic Carbon in the Sea

Contents of Volume 1

Chapter 1 *by* K. F. BOWDEN
Oceanic and Estuarine Mixing Processes
Chapter 2 *by* M. WHITFIELD
Sea Water as an Electrolyte Solution
Chapter 3 *by* WERNER STUMM and PHYLLIS A. BRAUNER
Chemical Speciation
Chapter 4 *by* GEORGE A. PARKS
Adsorption in the Marine Environment
Chapter 5 *by* FRED T. MACKENZIE
Sedimentary Cycling and the Evolution of Sea Water
Chapter 6 *by* T. R. S. WILSON
Salinity and the Major Elements of Sea Water
Chapter 7 *by* PETER G. BREWER
Minor Elements in Sea Water
Chapter 8 *by* DANA R. KESTER
Dissolved Gases Other Than CO_2
Tables of Chemical and Physical Constants
Subject Index

Contents of Volume 3

Contents of Volume 4

Symbols and units used in the text

A list of the more important symbols used in the text is given below. It is not exhaustive and inevitably there is some duplication of usage since some symbols have different accepted usages in two or more disciplines. The generally accepted symbols have been altered only when there is a possibility of ambiguity.

Concentration. There are several systems in common use for expressing concentration. The more important of these are the molarity scale (g molecules l^{-1} of solution $=$ mol l^{-1}) usually designated by c_i, the molality scale (g molecules kg^{-1} of solvent* $=$ mol kg^{-1}) designated by m_i, and the mole fraction scale usually denoted by x_i, which is of more fundamental significance in physical chemistry. In each instance the subscript i indicates the solute species; when i is an ion the charge is not included in the subscript unless confusion is likely to arise. Some other means of indicating the concentration are also to be found in the text, these include: g or mg kg^{-1} of solution (for major components), μg or ng l^{-1} or kg^{-1} of solution (for trace elements and nutrients) and μg-at l^{-1} of solution (for nutrients). Factors for conversion of μg to μg-at are to be found in Appendix Tables 4 and 5.

Activity. When an activity or activity coefficient is associated with a species the symbols a_i and γ_i are used respectively regardless of the method of expressing concentration, where the subscript i has the significance indicated above. Further qualifying symbols may be added as superscripts and/or subscripts as circumstances demand. It is important to realize that the numerical values of the activity and activity coefficient depend on the standard state chosen. It should also be noted that since activity is a relative quantity it is dimensionless.

UNITS

Where practicable SI units (and the associated notations) have been adopted in the text except where their usage goes contrary to established oceanographic practice.

---

* A common practice is to regard sea water as the solvent for minor constituents.

LENGTH

Å	= Ångstrom unit	$= 10^{-10}$ m
nm	= nanometre	$= 10^{-9}$ m
μm	= micrometre	$= 10^{-6}$ m
mm	= millimetre	$= 10^{-3}$ m
cm	= centimetre	$= 10^{-2}$ m
m	= metre	
km	= kilometre	$= 10^{3}$ m
mi	= nautical mile (6080 ft)	$= 1.85$ km

WEIGHT

pg	= picogram	$= 10^{-12}$ g
ng	= nanogram	$= 10^{-9}$ g
μg	= microgram	$= 10^{-6}$ g
mg	= milligram	$= 10^{-3}$ g
g	= gram	
kg	= kilogram	$= 10^{3}$ g
ton	= metric ton	$= 10^{6}$ g

VOLUME

μl	= microlitre	$= 10^{-6}$ l
ml	= millilitre	$= 10^{-3}$ l
l	= litre	
dm^3	= litre	

CONCENTRATION

ppm	= parts per million (μg g^{-1} or mg l^{-1})
ppb	= parts per billion (ng g^{-1} or μg l^{-1})
μg-at l^{-1}	= μg atoms l^{-1} = (μg/atomic weight) l^{-1}

ELECTRICAL

V	= volt
A	= ampere
Ω	= ohm

TIME

s	= second
min	= minute
h	= hour
d	= day
yr	= year

ENERGY AND FORCE

J	= Joule	= 0·2390 cal
N	= Newton	= 10^5 dynes
W	= Watt	

LIGHT FLUX

klux	= kilolux

GENERAL SYMBOLS

A	alkalinity (meq l^{-1})
A	absorbance = optical density (log I_0/I)
a_i	activity of component i
AOU	apparent oxygen utilization (see Vol. 1 p. 535)
B.O.D.	biochemical oxygen demand
C	heat capacity
c_G	molar concentration of gas G, at saturation
c_i	molar concentration of species i; charges of the species i are omitted where these are obvious
$c_{\Sigma B}$	total molar concentration of boric acid
CA	carbonate alkalinity (meq l^{-1})
Cl‰	chlorinity (g kg^{-1} = ‰)
D_g	molecular diffusion coefficient of gas g in the gas phase
D_l	molecular diffusion coefficient of a gas in the liquid phase
d.p.m.	radioactive disintegrations per minute
E	the E.M.F. of a cell
$E°$	standard potential
E_G	exchange coefficient of a gas, G

E_h redox potential

E_l liquid junction potential

E_P EMF at a pressure P

E_1 EMF at one atmosphere pressure

e_g exit coefficient of a gas (see Vol. 1 p. 512)

F Faraday equivalent of electric charge

G Gibbs free energy

H enthalpy

h Planck's constant

h relative humidity (%)

I ionic strength and attenuated light intensity

I_k light intensity for saturation (p. 390)

I_{mpc} light intensity of most penetrating component

I_o light intensity at surface or initially

I_z light intensity at depth z

K equilibrium constant

K^* stoichiometric equilibrium constant

K_a generalized acid ionization constant

K_{arag} thermodynamic (activity) solubility product of aragonite

K'_{arag} $=[c_{Ca(T)} \cdot C_{CO_3(T)}]_{arag} =$ apparent stoichiometric solubility product of aragonite

K'_B apparent first ionization constant of boric acid.

K_{calc} thermodynamic (activity) solubility product of calcite

K'_{calc} $=[C_{Ca(T)} \cdot C_{CO_3(T)}]_{calc} =$ apparent stoichiometric solubility product of calcite

K_G Henry's Law constant for gas G.

K_x coefficient of eddy diffusion in the x direction (similarly K_y and K_z)

$K_{0(TD)}$ thermodynamic equilibrium constant for hydration of carbon dioxide

$K_{1(TD)}$ thermodynamic first ionization constant of carbonic acid (infinitely dilute solution in pure water activity scale)

K'_1 apparent first acid ionization constant

K'_{1b} first apparent ionization constant of carbonic acid according to Buch (see p. 40)

K'_{1l} first apparent ionization constant of carbonic acid according to Lyman (see p. 40)

K'_{1h} first apparent ionization constant of carbonic acid according to Hansson (see p. 40)

$K_{2(TD)}$ thermodynamic second ionization constant of carbonic acid (infinitely dilute solution in pure water scale)

K_2'	second apparent ionization constant of carbonic acid according to Buch and Lyman
K_{2h}'	second apparent ionization constant of carbonic acid according to Hansson
k	rate constant
k	relative growth constant of algae
k_d	rate coefficient for gas transfer defined according to equation 9.142 (p. 135)
k_e	vertical extinction coefficient
k_{ob}	relative growth constant of algae
m_i	molarity of component, i
N_a	Avogadro's number
n_i	number of moles of component i in a mixture
P	primary production (mg C m^{-2} day^{-1})
P	total phytoplankton population per unit area
P_G	partial pressure of gas G in solution
P_h	photosynthetic quotient
P_i	polarizability of an ion i
P_{max}	maximum rate of photosynthesis
P_T	total pressure
p_G	partial pressure of gas G in the atmosphere
p_o	saturated vapour pressure of pure water
p_s	saturated vapour pressure of sea water or an aqueous solution
pH_1	pH at one atmosphere pressure
pH_p	pH at pressure p
Q	composite rate coefficient for gas transfer through gas-liquid interface (p. 134)
R	rate of production or loss of material
R	gas constant
R	relative photosynthesis (see p. 423)
S	entropy
S	salinity (g kg^{-1} = ‰)
s	solubility coefficient defined according to equation 9.135 (see p. 134)
T	temperature in K
t	temperature in °C
t	time
U	energy
u	component of velocity of water in x-direction
$\overline{V}$	volume
V_i	partial molar volume of species i

$\bar{V}_i^0$ partial molar volume species i at infinite dilution in pure water (p. 70)

$\bar{V}_i^{*0}$ partial molar volume of species i at infinite dilution in sea water (p. 70)

v, w components of velocity of water in y and z directions respectively.

Y see Vol. 1 p. 60

Z atomic number

z depth

z charge on an ion

Greek Symbols

α generalized solubility coefficient of a gas in a liquid (mol 1^{-1} atm^{-1})

α_f isotopic fractionation factor

β Bunsen coefficient of a gas (see Vol. 1 p. 501)

β van Slyke buffer capacity

γ_i activity coefficient of species i

Δ change of (as in ΔG)

$\Delta\,^{14}C$ see p. 9

$\delta^{13}C$ permillage enrichment of ^{13}C relative to a given standard (see p. 8)

$\delta\,^{14}C$ permillage enrichment of ^{14}C relative to a given standard (see p. 8)

δ_g thickness of gas phase boundary layer

δ_l thickness of liquid phase boundary layer

ε buffer capacity for CO_2 uptake (see p. 120)

ε molar absorptivity

λ radio-nuclide decay constant

μ ionic strength (more correctly I)

μ relative growth constant of plankton

$\boldsymbol{\mu}_i$ chemical potential of component i

v stoichiometric coefficient

v frequency

ρ density

ΣCO_2 total carbon dioxide content of a solution = $c_{H_2CO_3} + c_{CO_{2(aq)}} + c_{HCO_{3(T)}} + c_{CO_{3(T)}}$

σ_t specific gravity of sea water at $t°C$

τ residence time of an element in sea water

ϕ osmotic coefficient

Ω_{calc} degree of calcium carbonate saturation

Superscripts

$\ominus$ standard state

$-$ partial molar quantity (e.g. $\bar{V}$ etc.)

$'$ apparent constant

Chapter 9

The Dissolved Gases—Carbon Dioxide

GEOFFREY SKIRROW
Donnan Laboratories, The University, Liverpool, England

9.1. INTRODUCTION

Dissolved CO_2 is only one component of a complex cyclic system some of the more important reservoirs and pathways of which are indicated in Fig. 9.1 (see also Fig. 9.12 and Table 9.1). Not all aspects of this system are fully understood, but before discussing those which have been at least semi-quanti-

1

tatively characterized it is convenient to examine some features of the individual reservoirs in order to place the subject in perspective.

The concentration of CO_2 in the atmosphere is about 320 p.p.m., the observed value depending on locality (including height above ground level). Often, diurnal or annual variations occur, and superimposed on these short-term effects are longer term changes of possible natural origin and a progressive increase attributable to the use of fossil fuels. Contributions to the

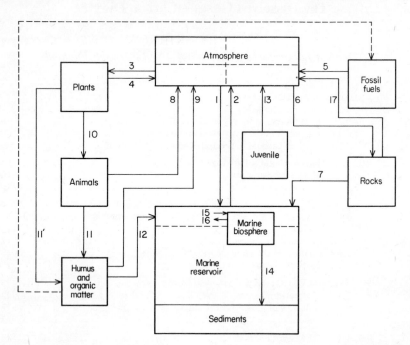

FIG. 9.1. The carbon cycle. Only the main reservoirs and transfer routes of CO_2 are shown. The relative (areal) sizes of the reservoirs shown in the diagram are of no significance. Dotted lines within the atmospheric and marine reservoirs indicate mixing barriers (atmosphere, troposphere-stratosphere and interhemisphere mixing; marine reservoir, thermocline indicated, but barriers exist between major water masses). CO_2 transfers: 1, 2, atmosphere–sea interchange; 3, 4, photo-synthesis and respiration; 5, burning of coal and oil; 6, 7, weathering of rocks by, e.g.

$$CaCO_3 + H_2O + CO_2 \rightarrow Ca^{2+} + 2HCO_3^- ;$$

8, animal respiration; 9, oxidation of humus and organic debris; 10, consumption of plants by animals; 11, 11', humus formation; 12 transfer of humic material to ocean; 13, juvenile CO_2; 14, decay of organic matter and precipitation of $CaCO_3$ debris from the marine biosphere leading to the generation of CO_2 (HCO_3^- and CO_3^{2-}) (some of the organic matter reaches the sediments and some $CaCO_3$ dissolves in the water column); 15, 16, photosynthesis and respiration in the marine biosphere; 17, generation of CO_2 by lime burning, cement manufacture etc. Dotted route indicates potential conversion of organic debris (e.g. to peat). Remobilization of carbon in sediments not indicated. (See also Fig. 9.12).

atmospheric carbon dioxide total are made by juvenile CO_2, the biosphere, the burning of fossil fuels, industrial processes such as cement manufacture, forest fires, agricultural activity, respiratory processes and passage from the oceans across the air–sea interface.

Atmospheric CO_2 is utilized by the biosphere and in the weathering of rocks, particularly carbonate ones. This weathering leads ultimately to the addition of inorganic carbon to the oceanic reservoir. The partial pressure of CO_2 in the atmosphere (p_{CO_2}) and that of the dissolved CO_2 of the sea (P_{CO_2}) are, at best, in approximate equilibrium. Noticeable local and regional disequilibria occur and lead to a net transfer across the interface in one direction or the other (Sections 9.4 and 9.5). The rate of this transfer is of particular interest when considering the ability of the oceans to buffer any large and long-term industrially derived input of carbon dioxide to the atmosphere. The interest is not simply an academic one since there are possible undesirable climatological consequences if the oceans are not able to accommodate such an increase. Present indications are that the mean residence time of CO_2 molecules in the atmosphere is several years—much longer than those for water or for more soluble gases such as SO_2, but shorter than for O_2 and N_2.

The problem of the uptake of CO_2 by the sea is complicated by factors other than the direct rate of transfer across the interface. The total capacity of the sea to accommodate, by uptake, an atmospheric CO_2 increase is considerable. However, the sea is not a single well-mixed reservoir, but is, as it were, a layered system. Various models for this system have been proposed (Section 9.4), and according to one of the simplest of these the atmosphere has direct access only to the upper layer above the thermocline. Turn-over of the sea is slow, and even if equilibrium between the atmosphere and the upper layer were to be achieved rapidly, equilibrium with the total sea would take many centuries. In addition, CO_2 in the sea participates in a complex equilibrium which may be summarized by

$$H_2O + CO_2(aq) \rightleftharpoons H_2CO_3 \rightleftharpoons H^+ + HCO_3^- \rightleftharpoons 2H^+ + CO_3^{2-}$$

and addition of further CO_2 induces responses in this system which lead to a pH depression and an opposition to further CO_2 entry. Clearly, the process of equilibration of atmospheric CO_2 with that in the sea is a complex business involving a number of processes some of which—the mixing of different water masses—are mainly physical, readjustment of chemical equilibrium being rapid. Other processes, notably those at the interface, involve both slow physical steps (diffusion through a boundary layer) and chemical steps of which some (the hydration and dehydration reactions of CO_2) may also be slow and possibly rate determining.

The simplified scheme given above illustrates a particularly important property of the sea water carbonate system, i.e. its pH buffering ability. This buffering is largely responsible for oceanic pH control when the time scales being considered are hundreds or thousands of years; however, for longer time scales additional buffering is likely to arise from the interaction of sea water with the sediments. For a quantitative interpretation of the carbonate buffer effect it is necessary to know the concentrations of the individual components of the carbonate system viz, $c_{HCO_3(T)}$, $c_{CO_3(T)}$ (these terms including not only the amounts of these ions as such, but also the amounts of them bound as ion pairs such as $CaCO_3^0$, $MgCO_3^0$, and $NaCO_3^-$, see Section 9.2.3.7) and $c_{CO_2(T)}$, the dissolved CO_2 (really $c_{H_2CO_3} + c_{CO_2(aq)}$). In addition, the related quantities ΣCO_2 ($= c_{HCO_3(T)} + c_{CO_3(T)} + c_{CO_2(T)}$) and P_{CO_2} are often required. It is a matter of oceanographical, geochemical and ecological interest to be able to obtain values for these quantities on the basis of routine shipboard measurements using a minimum of experimental determinations. Each step in the equilibrium sequence can be characterized by an equilibrium constant, and knowledge of these constants as functions of salinity, temperature and pressure enables values for the individual components to be calculated from suitable pairs of experimental determinations such as alkalinity and pH or P_{CO_2} and ΣCO_2. Considerable effort has been devoted to the determination of these constants (Section 9.2.3.4) and it turns out that, for several reasons, it is often more convenient to use apparent rather than thermodynamic ones. By combining these with suitable experimental determinations it is possible to show that $HCO_3^-(T)$ and, more particularly, $CO_3^{2-}(T)$ and dissolved carbon dioxide show considerable concentration changes from place to place, these changes often being related to physical, biological or chemical events in the water.

Thus, CO_2 is removed from solution in surface waters when photosynthesis occurs and by evasion from the surface when P_{CO_2} is high relative to p_{CO_2} as, for example, as a result of diurnal solar heating. It is generated by the oxidation of decaying organic matter in sub-surface waters, this production being accompanied by a pH fall, the loss of dissolved oxygen and the liberation of nutrients from the organic material. At this oxidation stage the carbon cycle overlaps with those of oxygen and the various nutrients. Changes in the equilibrium position of the carbonate system also occur as a result of temperature changes when a mass of water moves from a warmer to a cooler region, or in response to the pressure dependence of the equilibrium constants when a mass of water changes depth. An increase in the total inorganic carbon and in the alkalinity accompanies the dissolution of calcium carbonate.

Problems associated with the dissolution or precipitation of calcium carbonate are intimately linked with the sea water carbonate system. The surface waters of much of the Worlds Ocean are supersaturated with respect to calcium carbonate; calcium is delivered to the surface by upwelling of deeper water and also in terrestrial run-off. The failure of $CaCO_3$ to precipitate from ocean surface waters seems to be a consequence of the slow kinetics of nucleation or of nucleus growth as a result of the retarding effect exerted by the presence of certain ions (e.g. Mg^{2+}) and also possibly by organic material in the water. However, this is a difficult subject to study even in simple dilute solutions under laboratory controlled conditions. Even when precipitation does occur, as expected, the metastable aragonite first appears rather than calcite; (vaterite formation under present oceanic conditions is not sufficiently important for consideration here). Surface water calcium is used by shell-building organisms. Following their deaths, their calcareous tests fall through the water column and as they do so they experience a changing environment. However, below a certain depth—the saturation depth—the water is no longer saturated with respect to calcium carbonate, partly as a result of the pressure dependence of the relevant equilibria. Despite this, $CaCO_3$ still occurs in sediments below the saturation depth and is relatively abundant down to the compensation depth below which there is a dramatic fall-off in the proportion of $CaCO_3$ in the sediments. The origin of the compensation depth is a matter of controversy, but in one currently urged view it is a kinetic phenomenon reflecting a state of balance between the rate of sedimentation and the (slow) rate of dissolution. Factors controlling the rate of solution of natural calcareous particles are not fully understood, but the process may be subject to slow, diffusion controlled, steps. The return of deep water to the surface is accompanied by a change of the pressure and temperature and a readjustment of the equilibrium distribution of the carbonate system occurs; under these new conditions (increased $c_{CO_3(T)}$) the water is now supersaturated.

Approximate time scales for many of the transfer processes within the carbon system can be obtained by making use of radiocarbon data combined with information on the total carbon content of the various sub-reservoirs. Table 9.1 lists some of the suggested values for the carbon contents of natural exchange reservoirs and the fluxes associated with them. Some of these values are not very precisely known; in particular, those for the marine biosphere, the terrestrial biosphere and humus are uncertain. The division of the sea into a two-layered structure at the thermocline is a gross oversimplification of a complex of interacting reservoirs. More sophisticated models are to be found in the literature and are indicated in Section 9.4.

Carbon Isotopes and Isotope Fractionation
Natural atmospheric carbon dioxide comprises $^{12}CO_2$, $^{13}CO_2$ and $^{14}CO_2$,

TABLE 9.1

Properties of the natural carbon reservoirs

(i) *Carbon contents of the natural reservoirs* (10^{20} g C)
Atmosphere

pre-industrial	0·00616 ($\equiv$ 293 p.p.m.)	(a)
ca. 1973	0·00675 ($\equiv$ 323 p.p.m.)	
Ocean (inorganic C)		
above thermocline	0·0075	
deep sea	0·373	
total inorganic C	0·38	(b)
Ocean (organic C)	0·01	(c)
Terrestrial biosphere	0·0164	(d)
Calcareous tests plus detrital carbonates	0·0129	(e)
Organic carbon in sedimentary rocks	68·2	(f)
Carbonate rocks	183	(g)

(a) Keeling (1973a), based on Bray (1959).
(b) The total inorganic carbon concentration is depth dependent and generally tends to be higher in deeper water. Surface water ΣCO_2 values are usually close to 2·1 mmol l^{-1} (2·52 mg Cl^{-1}) except near regions of upwelling. Below ca. 1000 m, values close to 2·32 mmol l^{-1} (2·8 mg Cl^{-1}) are commonly observed. The inorganic carbon is distributed between $CO_2(aq)$, H_2CO_3, HCO_3^- and CO_3^{2-}. The proportions depend on temperature, depth, pH etc., but the bulk of ΣCO_2 is HCO_3^-. Ion pairing of HCO_3^- and CO_3^{2-} with, e.g. Ca^{2+} and Mg^{2+} occurs.
The thickness selected for the surface layer is, within limits, arbitrary, but usually is between 50 and 250 m.
(c) Approximate value. The distribution of organic carbon is not uniform, but depends on geographical area and depth. Also DOC > POC > C in living marine biosphere (see Chapter 12).
(d) Keeling (1973a). The value applies to the vegetative biosphere. Reported estimates differ widely, and the absolute size and trend with time are uncertain. Keeling (1973a) has made estimates of the long-lived and short-lived fractions of this biomass.
(e) Pytkowicz (1973b). Assumes that only the upper 10 cm of the sediments are in exchange contact with the overlying water.
(f) Pytkowicz (1973b).
(g) Pytkowicz (1973b).

(ii) *Fluxes of carbon between natural reservoirs* (10^{14} g C yr^{-1})

Juvenile CO_2	0·082	(a)
Atmosphere–sea	700	(b)
Industrial CO_2 (ca. 1970)	40	(c)
Photosynthesis on land	500	(d)
Photosynthesis in the sea	500	(e)
Weathering of carbonate rocks	4·92	(f)
Oxidation in the sea	ca. 450	(g)

TABLE 9.1—*contd.*

Properties of the natural carbon reservoirs

(a) Pytkowicz (1973b).
(b) Depends on value chosen for the mean residence time of CO_2 in the atmosphere with respect to transfer to the sea. This is probably somewhat less than a decade. In pre-industrial times the reverse flux was probably slightly greater than the air–sea flux. (Pytkowicz, 1973b).
(c) Value increasing progressively with time.
(d) Approximate value. Estimates differ somewhat. Keeling (1973b) gives the sum of the rates of production of short- and long-lived C on land as $560 \times 10^{14} \text{g C yr}^{-1}$.
(e) Approximate value. Estimates differ widely. $(200–1400 \times 10^{14} \text{g C yr}^{-1})$ (see Chapter 14).
(f) Pytkowicz (1973b). The value is twice that for the loss of carbon in the rock which is weathered since the weathering solution contributes an equal amount of carbon.
(g) Value chosen to match the photosynthesis on the assumption that although most of the organic matter is reconverted to CO_2 some is incorporated in the sediments. Much of the oxidation takes place in the upper layers of the water column. (See Pytkowicz (1973b) for discussion).

(iii) $\delta^{13}C$ *values* (‰ Chicago PDB standard)

Atmospheric CO_2	-7
Terrestrial biosphere and humus	-25
Planktonic organic material	-20 to -30
Dissolved organic carbon	-21 to -24
Inorganic carbon in the sea	$+2$ to $-1\cdot0$
Calcareous material	$+2$

Values for calcareous material are variable (see Kroopnick 1974). The $\delta^{13}C$ of the inorganic carbon of the sea depends on depth since additions to this carbon are derived from both the oxidation of organic debris and the dissolution of calcareous material (see also Deuser and Hunt, 1969 and Chapter 12).

although $^{14}CO_2$, which is radioactive (β-decay) with a half-life of 5720 ± 30 years (Olsson and Karlen, 1963), forms only some $10^{-10}\%$ of the total. The main source of natural carbon-14 is the upper atmosphere where it is formed by the interactions of cosmic ray neutrons with nitrogen ($^{14}N(n, p)^{14}C$) at a rate of about $2\cdot5$ atoms $^{14}C \text{ cm}^{-2} \text{ s}^{-1}$ although some variations with sunspot activity occurs (Lingfelter, 1963—see also Baxter and Walton, 1971). The ^{14}C so formed reacts readily with oxygen or ozone (possibly via CO—see MacKay *et al.*, 1963) to give CO_2 which mixes relatively quickly with the atmosphere (Section 9.3). Despite its low abundance, $^{14}CO_2$ is of value as a natural tracer (in, for example, the determination of CO_2 transfer rates between natural reservoirs and in carbon-14 dating work) although many factors complicate the interpretation of observed activities. It now appears that the atmospheric concentration of ^{14}C has not been as steady over past millenia as was previously supposed (Dyck, 1965; Stuiver, 1965; Suess, H. E., 1973). In addition, the output of inactive CO_2 derived from the burning of fossil fuels and industrial activity over the last century has tended to dilute the natural level (Suess effect—Suess, 1955; Baxter and Walton, 1970, 1971), and recent injections of artifical (nuclear bomb produced) carbon-14 have had the opposite effect of enhancing the level (see e.g. Baxter and Walton, 1971). Even in the absence of these last two disturbing effects, a number of routine precautions are necessary in the interpretation of radio-carbon data.

Thus, for example, because of the natural decay of carbon-14 and the finite time taken for inter-reservoir transfer, true isotopic equilibrium is not attained in the establishment of chemical equilibrium between the atmosphere and an aqueous solution. Instead, if the above disturbing effects are ignored, something approximating to a steady-state is reached in which the rate of entry of active carbon to a reservoir is balanced by its rate of exit plus its rate of decay in the reservoir. This is the basis underlying certain methods for the determination of transfer rates. However, even if true isotopic equilibrium between the CO_2 of the atmosphere and that in an aqueous solution were to be attained in the sense that loss by decay could be ignored (infinite transfer rates), it would be found that the relative abundances of the three isotopes in the two phases would differ. This is because the equilibria and kinetics of the transfer process depend on isotopic mass; consequently, fractionation occurs during the transfer. The enrichment of radio-carbon in such a transfer can be calculated from a knowledge of the simultaneous enrichment of the stable carbon-13.

The fractionation factor for carbon-13 is defined by

$$\alpha_f = R_t/R_s$$

where R_t and R_s are the $^{13}C/^{12}C$ ratios in the two compounds t and s respectively (Craig, 1954). In order to provide a working scale of reference, R_s is a standard of known isotopic composition (Craig, 1957) and the composition of the test sample is expressed as the enrichment per mille ($\delta^{13}C$) relative to the standard, i.e.,

$$\delta^{13}C = \left[\frac{R_t - R_s}{R_s} \right] \times 1000\%_0.$$

In a carbon isotope fractionation process, the carbon-14 fractionation factor should be the square of the corresponding factor for carbon-13. Both factors are close to unity and, consequently, the permillage enrichment of ^{14}C should be twice that for ^{13}C for the same source material. A $\delta^{14}C$ value can be defined in terms of the sample and a standard (e.g. the NBS oxalic acid standard), and combination of the known $\delta^{13}C$ and $\delta^{14}C$ values by means of the equation

$$\Delta^{14}C = \delta^{14}C - (2\delta^{13}C + 50)\left[\frac{1000 + \delta^{14}C}{1000}\right]\%_0$$

gives $\Delta^{14}C$ values. This scale provides a convenient framework within which radiocarbon activities can be discussed (Broecker and Olson, 1961; Baxter and Walton, 1971).

9.2. THE SEA WATER CARBONATE SYSTEM

9.2.1. GENERAL COMMENTS

Aqueous solutions containing salts of weak acids or bases show a resistance to pH change (buffering) on the addition of acids or bases. The effect is, of course, well understood, and may be briefly summarized as follows. In a solution of a weak acid HA in the presence of the anion A^- the ionization equilibrium

$$HA \rightleftharpoons H^+ + A^- \qquad \qquad (9.1)$$

will result in a solution pH given by

$$pH = pK_a + \log_{10}\frac{c_A}{c_{HA}} + \log_{10}\frac{\gamma_A}{\gamma_{HA}} \qquad (9.2)$$

$$= pK_a' + \log_{10}\frac{c_A}{c_{HA}}$$

where K_a is the acid ionization constant (thermodynamic) of the acid, K_a' is the corresponding apparent constant, and γ_A and γ_{HA} are the activity co-efficients of the anion and the unionized acid respectively. If K_a is small, c_A and c_{HA} will be approximately equal to the stoichiometric concentrations of the salt and acid respectively. Small additions of H^+ or OH^- ions will alter only slightly the second term on the right hand side of (9.2) and will have little effect on the last term. For an increment Δc_H of hydrogen ion, the anion concentration will become $c_A - \Delta c_H$ and the acid concentration will become $c_{HA} + \Delta c_H$. Thus, provided that Δc_H is not too large, the pH changes only slightly. It should be noted that since the activity coefficients, particularly that of A^-, depend on the nature of the electrolyte present, the pH of the buffer solution is a function not only of c_{HA} and c_A, but also of the electrolyte composition.

A measure of the ability of a solution to accommodate additions of acid or base without appreciable pH change is the buffer capacity (van Slyke, 1922) defined by

$$\beta = \frac{dc_b}{dpH} \tag{9.3}$$

where dc_b is an increment of strong base. For the above system β can be shown to be given by

$$\beta = 2 \cdot 303 \left[\frac{K'_a \cdot (c_A + c_{HA}) \cdot c_H}{(K'_a + c_H)^2} + c_H + c_{OH} \right] \tag{9.4a}$$

The last two terms in the bracket of this expression determine the buffer capacity at high and low pH values respectively. At intermediate pH values they can be neglected and it can be seen that the buffer capacity is then proportional to the total concentration of A-containing species, $(c_A + c_{HA})$. For a dilute solution of a weak dibasic acid H_2A which is capable of ionizing to give HA^- and A^{2-}, the buffer capacity is easily shown to be given by

$$\beta = 2 \cdot 303 \left[\frac{c_T \cdot K'_1 \cdot c_H [c_H^2 + K'_1 \cdot K'_2 + 4K'_2 \cdot c_H]}{[c_H^2 + K'_1 \cdot c_H + K'_1 \cdot K'_2]^2} + c_H + c_{OH} \right] \tag{9.4b}$$

where c_T $(= c_{H_2A} + c_{HA} + c_A)$ is the total concentration of A-containing species and K'_1 and K'_2 are respectively the first and second apparent ionization constants of the acid. From this expression it is possible to show that, in addition to the high buffer capacity expected at extreme pH values, there are now two regions of maximum buffer capacity in the intermediate pH range.

In sea water remote from contaminated or anoxic regions, the pH is mainly controlled by the $CO_2/HCO_3^-/CO_3^{2-}$ system. Other weak electrolytes slightly augment this effect (e.g. borate, phosphate, silicate and arsenate), and over long periods of time, interaction with the sediments is likely to be of importance (Sillén, 1961; Garrels, 1965; Pytkowicz, 1968, 1973b. See also Chapter 5). However, for short periods of time and for open ocean waters, the low concentration of other potential buffers (Table 9.2) means that only carbonate and borate need to be considered. The high precision which can now be attained in the determination of alkalinity and pH means that contributions from biosynthetic changes in nitrate (Gunderson and Mountain, 1973; Dyrssen and Gunderson, 1974) and from phosphate (Dyrssen and Gunderson, 1974; Almgren et al., 1974b) can now be considered. Table 9.2 shows that in recent marine sediments and in anoxic waters the concentrations of protolytic materials, particularly sulphide, may rise to levels at which it is no longer strictly correct to regard the carbonate and borate systems as the only ones which contribute to the alkalinity, the buffer capacity and pH control (see Section 9.6 and Richards, 1965; Richards et al., 1965; Knull and Richards, 1969; Berner et al., 1970; Gaines and Pilson, 1972; Ben-Yakov, 1973).

If influences on pH other than those arising from the carbonate system are disregarded, the appropriate form of (9.4b) can be derived for the sea water system by making use of equation (9.85) (Section 9.2.3.5) and remembering that $c_T \equiv \Sigma CO_2$ and that $dc_b = dCA$. It should be noted that the maximum buffer capacity ($\beta_{max} = 2\cdot303 \ \Sigma CO_2/4$) develops at pH values close to 6·0 and 9·0. Natural sea water of pH about 8·0 is not in the region of maximum buffer capacity; at 25°C, β is about 4×10^{-4} mol base pH^{-1}.

The buffer capacity implied in equations (9.4a and b) is equivalent to the strong base (or strong acid) increment necessary to cause unit pH increase (or decrease). It is, of course, inversely related to the slope of the pH-strong acid curve (as normally plotted) noted during the titration of sea water. Under natural circumstances, for open ocean water the principal pH influencing substance which is likely to be added or withdrawn is carbon dioxide itself, and in this respect the quantity of interest is the buffer capacity with respect to CO_2 addition, i.e., $d\Sigma CO_2/dpH$ where ΣCO_2 is the total inorganic carbon $(= c_{CO_2(aq)} + c_{H_2CO_3(T)} + c_{HCO_3(T)} + c_{CO_3(T)})$. From expression (9.85) it is easy to show that if CA changes negligibly during the CO_2 exchange

$$\beta' = \frac{d\Sigma CO_2}{dpH} = -\frac{2\cdot303 . CA . c_H}{K_1'}\left[\frac{c_H^2 + K_1' . K_2' + 4K_2' . c_H}{[c_H + 2K_2']^2}\right]$$

and in most surface ocean water β' differs little from β. In fact, the ratio of these two capacities is

$$\frac{\beta'}{\beta} = -\frac{CA}{\Sigma CO_2}\left[\frac{K_1' . c_H + K_1' . K_2' + c_H^2}{K_1'[c_H + 2K_2']}\right]^2 = -\frac{\Sigma CO_2}{CA}$$

and since for most sea water samples ΣCO_2 and CA do not differ by more than a few percent, the ratio of the two capacities is close to -1.

Recently, the possibility has been considered that the sea water buffer system is influenced by the presence of ion pairs, and it has been suggested that changes in the proportions of the major ions in sea water should lead to a change in the amount of ion pairing and thereby to a change of pH (Wangersky, 1972a, b). Wangersky's comments have been criticized by Pytkowicz (1972a), and Whitfield (1974a) has given detailed calculations based on an ion pairing model which show that the pH stability is not so much a consequence of a buffering effect arising from the presence of ion pairs, but rather a result of the fact that, under natural circumstances, the metal-carbonate ion pairs form only a small proportion of the total cation concentration. Both Pytkowicz and Whitfield insist that, although pairing influences the effective concentrations of HCO_3^- and CO_3^{2-} and also the pH, the primary buffering is realized through the process

$$2HCO_3^- \rightleftharpoons CO_2\,(aq) + H_2O + CO_3^{2-}$$

For other comments on pH buffering in natural aqueous systems see Stumm and Morgan (1970).

The main steps in the carbonate equilibria can be summarized by

$$CO_2(g) \rightleftharpoons CO_2(aq) \tag{9.5}$$

$$CO_2(aq) + H_2O \rightleftharpoons H_2CO_3 \tag{9.6}$$

$$H_2CO_3 \rightleftharpoons H^+ + HCO_3^- \tag{9.7}$$

$$HCO_3^- \rightleftharpoons H^+ + CO_3^{2-} \tag{9.8}$$

$$H_2O \rightleftharpoons H^+ + OH^- \tag{9.9}$$

TABLE 9.2

Approximate concentrations of protolytic materials in sea water and sediment pore water.

Sea water (pH ca. 8·3 at 10°C and 35‰ S)	$mol\,l^{-1}$
bicarbonate	ca. 20×10^{-4}
carbonate	ca. $2\cdot0 \times 10^{-4}$
total boron	$4\cdot5 \times 10^{-4}$
borate	ca. $1\cdot0 \times 10^{-4}$
total phosphate	1×10^{-6}
total arsenite	4×10^{-8}
sulphide	2×10^{-5}
organic acids	traces only

Pore water from sediment core

depth below sediment surface (cm)	pH	ΣCO_2	Sulphide	NH_3	PO_4^{3-}
			$mol\,l^{-1} \times 10^3$		
0–15	7·6	2·7	0·0	0·15	0·042
40–50	8·0	13·6	3·8	0·53	0·035
85–100	7·9	23·7	5·0	2·05	0·155
135–150	8·0	29·7	0·6	3·5	0·073
175–185	8·0	36·2	1·2	4·26	0·122

From core taken in Saanich Inlet (Nissenbaum *et al.*, 1972) and quoted by Ben-Yaakov, 1973.

the last of these reactions can probably be more correctly written as in (9.22) below. Explicit recognition can be given to the solvolysis of weak acid anions by writing for example,

$$CO_3^{2-} + H_2O \rightleftharpoons HCO_3^- + OH^- \qquad (9.10)$$

although this reaction is already accounted for by (9.8) and (9.9). A more complete description considers ion pair formation, viz,

$$Ca^{2+} + CO_3^{2-} \rightleftharpoons CaCO_3^0 \qquad (9.11)$$

$$Mg^{2+} + CO_3^{2-} \rightleftharpoons MgCO_3^0 \qquad (9.12)$$

$$Na^+ + CO_3^{2-} \rightleftharpoons NaCO_3^- \qquad (9.13)$$

$$Ca^{2+} + HCO_3^- \rightleftharpoons CaHCO_3^+ \qquad (9.14)$$

$$Mg^{2+} + HCO_3^- \rightleftharpoons MgHCO_3^+ \qquad (9.15)$$

$$Na^+ + HCO_3^- \rightleftharpoons NaHCO_3^0 \qquad (9.16)$$

and, in addition, reactions involving solid phases (solution and precipitation of $CaCO_3$, attack on sediments) should also be taken into account. However, solid phase reactions will not be considered in this Section.

Reactions (9.5) to (9.16) have been written as equilibria. For equilibrium to be maintained, the forward and reverse steps of each reaction should not be too slow. Immediately below, available kinetic evidence is examined briefly to see whether kinetic factors may cause disequilibrium in the carbonate system. The discussion then passes to the core of this Section and examines the equilibria themselves, their implications and their uses in oceanographical investigations.

9.2.2. RATE FACTORS IN THE CARBONATE SYSTEM

Of the reactions contributing to the proposed equilibria, only those for the passage of CO_2 across the phase boundary and the carbon dioxide hydration and dehydration reactions, viz,

$$CO_2(aq) + H_2O \rightarrow H_2CO_3 \qquad (9.17)$$

and

$$H_2CO_3 \rightarrow H_2O + CO_2(aq) \qquad (9.18)$$

have rate coefficients sufficiently small to make kinetic factors of possible importance. However, the reaction

B

$$CO_2(aq) + OH^- \rightarrow HCO_3^-, \qquad (9.19)$$

which can occur directly, and its reverse, viz.,

$$HCO_3^- \rightarrow CO_2(aq) + OH^-, \qquad (9.20)$$

not included explicitly in the original scheme should also be considered.

9.2.2.1. *Rates of uptake and loss of CO_2 by aqueous solutions*

The uptake and loss of CO_2 by aqueous solutions is considered in Section 9.5, and the following remarks are confined to the more essential points. In the most widely used model for the transport of gas across a gas–liquid boundary, the limiting rate factor is passage through the pair of boundary layers assumed to exist, one on each side of the interface. Of these two layers, that on the liquid side is the more important for most atmospheric gases. Transport through these layers is by molecular diffusion, and under quiescent conditions when the boundary layers are relatively thick, equilibrium will be reached only slowly. Under natural conditions, turbulence in the bulk phases reduces each boundary layer thickness and so increases the transport rate. For gases such as CO_2 which are able to react with water, the transfer rate may be augmented relative to that expected in the absence of reaction because the consequent reduction in its concentration in solution (accompanied by the generation of ionic species) leads to an enhanced driving force behind the uptake process. The amount of augmentation depends on the reaction rate; that is, for CO_2 it depends on the rates of (9.17) and (9.18) (see below). However, even at its most rapid, phase transfer is slow compared with simple ionization steps, and precise equilibrium between the atmospheric and marine concentrations of carbon dioxide is hardly to be expected, and in fact, is rarely found (Section 9.6).

9.2.2.2. *Rates of reactions involving dissolved CO_2*

The rate coefficients of the two components comprising the equilibrium (9.6), i.e. (9.17) and (9.18), are sufficiently small to merit consideration here. The physiological and industrial importance of these steps (and (9.19) and (9.20)) has stimulated much interest (Faurholt, 1924; Roughton and Booth, 1937; Mills and Urey, 1940; Roughton, 1941; Berg and Patterson, 1953; Wissburn *et al.*, 1954; Pinsent *et al.*, 1956; Ljunggren and Lamm, 1958; Sirs, 1958; Kern, 1960; Eigen *et al.*, 1961; Danckwerts and Melkersson, 1962; Ho and Sturtevant, 1963; Gibbons and Edsall, 1963; Sharma and Dankwerts, 1963).

The interconversion routes between dissolved CO_2, H_2CO_3 and HCO_3^- are probably best depicted by

$$\text{(a)} \quad H^+ + HCO_3^- \underset{k_{ba}}{\overset{k_{ab}}{\rightleftharpoons}} H_2CO_3 \quad \text{(b)}$$

$$k_{ca} \Big\Vert k_{ac} \qquad\qquad k_{cb} \Big\Vert k_{bc}$$

$$CO_2 + H_2O \quad \text{(c)}$$

rather than by

$$H^+ + HCO_3^- \rightleftharpoons H_2CO_3 \rightleftharpoons CO_2 + H_2O$$

(Eigen *et al.*, 1961; Ho and Sturtevant, 1963; Gibbons and Edsall, 1963), and it is convenient to recognize net rate coefficients for the generation of bicarbonate from CO_2 (k'_{ca}) and for the reverse process (k'_{ac}) such that, for example, $k'_{ac} = k_{ac} + k_{bc}/K_1$, when K_1 is the first ionization constant of H_2CO_3. The hydration process is first order with respect to CO_2, and at 25°C the first order coefficient is about 0.037 s^{-1} (Table 9.3) indicating a half-life for dissolved CO_2 of several minutes. The reverse process has a second order rate coefficient at 25°C of between 5×10^4 and about 10×10^4 $1 \text{ mol}^{-1} \text{ sec}^{-1}$ (Table 9.3), the exact value probably depending on experimental conditions such as the solution composition. The activation energy for the hydration process is about 80 kJ mol^{-1}, and, contrary to earlier reports, is almost independent of temperature between 0 and 40°C (Gibbons and Edsall, 1963).

Reaction (9.19) has a second order rate coefficient of about $8500 \text{ l mol}^{-1} \text{ s}^{-1}$ at 25°C. Since the rate expression for this process is

$$\frac{dc_{CO_2}}{dt} = -k_{19}c_{CO2(aq)} \cdot c_{OH} \qquad (9.21)$$

it follows that below pH 9 (i.e., for most natural waters) the rate of loss of $CO_2(aq)$ by (9.19) is negligible compared with that by the $CO_2(aq) + H_2O$ reaction. The activation energy for this process is reported to be about 55 kJ mol^{-1}. The first order rate coefficient for (9.20) at 25°C is about $2 \times 10^{-4} \text{ s}^{-1}$.

These rate coefficients might be expected to change slightly with depth because of the effect of pressure. However, the effect is probably not very important over the range of pressures which apply in the oceans (Pytkowicz, 1968).

All the rate coefficients cited above are likely to be lower limiting values since they have been obtained from measurements using systems free from catalytic material. It is known that many substances enhance the rates of the hydration and dehydration reactions (see above references). Such materials include phosphate, borate and arsenate ions, and particularly the enzyme carbonic anhydrase. Little is known about the kinetics of the hydration and

TABLE 9.3

Rate coefficients for the reactions of carbon dioxide in aqueous solution.

k'_{ac}	$5{\cdot}5 \times 10^4\,\mathrm{l\,mol^{-1}\,s^{-1}}$	Gibbons and Edsall (1963)
	$10{\cdot}2 \times 10^4\,\mathrm{l\,mol^{-1}\,s^{-1}}$	Ho and Sturtevant (1963)
k'_{ca}	$0{\cdot}0375\,\mathrm{s^{-1}}$	Gibbons and Edsall (1963)
	$0{\cdot}0358\,\mathrm{s^{-1}}$	Ho and Sturtevant (1963)
	$0{\cdot}037\,\mathrm{s^{-1}}$	Roughton (1941)
k_{19}	$8500\,\mathrm{l\,mol^{-1}\,s^{-1}}$	Pinsent *et al.* (1956)
k_{20}	$2 \times 10^{-4}\,\mathrm{s^{-1}}$	

dehydration reactions in natural sea water, and it would be interesting to know how the high ionic strength and the organic content influence the rate coefficients. Investigations by Berger and Libby (1969) suggest that these coefficients may vary with locality. Samples from 200 ft depth equilibrated more rapidly with atmospheric CO_2 than did surface water, possibly as a consequence of the presence in deep water of enzymatic material which is destroyed near the surface.

So far as is known, all other homogeneous reactions of the sea water carbonate system are sufficiently rapid to justify the assumption that they reach equilibrium.

9.2.3. EQUILIBRIA IN THE SEA WATER CARBONATE SYSTEM

Discussions of the sea water carbonate system invariably make use of a number of concepts such as alkalinity and carbonate alkalinity which have rather speciailized meanings together with other, more elusive, concepts (such as activity and pH). Sections 9.2.3.1 and 9.2.3.2 below deal with these. The main features of the equilibria are considered in subsequent Sections.

9.2.3.1. *Ionization of water and the pH concept*

Water is a weak electrolyte having more complex properties than its apparent stoichiometric simplicity might suggest (see Chapter 2). It is strongly associated because of pronounced hydrogen bonding, and it shows considerable structure which is influenced by temperature and the presence of dissolved ions. The high dielectric constant and high (exothermic) enthalpy of solvation of ions by water make it a good solvent for electrolytes. Pure water ionizes in accordance with (9.9) which is more correctly written

$$H_2O(l) \rightleftharpoons H^+(aq) + OH^-(aq) \tag{9.22}$$

where $H^+(aq)$ signifies all species such as H_3O^+, $H_5O_2^+$, etc known to exist in aqueous solution. For simplicity, in the following account these various species are denoted by H^+, and for all dissolved ions the suffix (aq) is omitted except when its omission might cause confusion.

The thermodynamic ionization constant of (9.22) is given by

$$K_{H_2O(TD)} = \frac{a_H \cdot a_{OH}}{a_{H_2O}} \tag{9.23}$$

$$= \frac{c_H \cdot c_{OH}}{a_{H_2O}} \cdot \gamma_H \cdot \gamma_{OH} \tag{9.24}$$

$$= K_w \frac{\gamma_H \cdot \gamma_{OH}}{a_{H_2O}} \tag{9.25}$$

where, for a component i, c_i signifies concentration, a_i the thermodynamic activity and γ_i the activity coefficient. K_w is the concentration ionic product of water. For dilute aqueous solutions the activity convention usually adopted is one in which the activity coefficients tend to unity at infinite dilution. On this scale, for pure water, the small degree of ionization justifies setting γ_H, γ_{OH} and a_{H_2O} equal to unity. K_w is thus equal to $K_{H_2O(TD)}$ (and also to the activity ionic product, $a_H \cdot a_{OH}$). At 25°C, K_w is ca. 10^{-14}, corresponding to a value for the pH (tentatively defined as $-\log_{10} c_H$) of 7. The concentration ionic product is increased by a rise in temperature (in accordance with the high endothermicity—ca. 230 kJ mol^{-1}—of the ionization) and also by increased pressure (Table 9.4).

TABLE 9.4

Effect of pressure on the ionic product of water.
($I = 0.1$ M; 15°C)

Pressure (atm)	$K_{w(P)}/K_{w(1)}$	
	(a)	(b)
200	1·20	1·20
400	1·41	1·42
600	1·62	1·67
800	1·91	1·94
1000	2·19	2·24
1200	2·51	2·56
1400	2·88	2·90
1600	3·31	3·26

(a) Measured values (Whitfield, 1972).
(b) Calculated values (Millero *et al.*, 1972).

In the presence of electrolytes the activity coefficients of H^+ and OH^- defined on the above scale are no longer unity. At constant temperature and pressure, $K_{H_2O(TD)}$ is, of course, independent of the salt concentration, and changes in the activity coefficients of H^+, OH^- and H_2O are accompanied by corresponding changes in K_w which, for moderate salt concentrations, increases (decrease of pK_w) with increasing ionic strength (Harned and Owen, 1958). Knowledge of K_w for sea water is important when investigations are made of the ionizations of protolytes having very high (>9) pK values and for the quantitative investigation of hydroxy complexes. In this connection Culberson and Pytkowicz (1973) have recently determined pK_w for sea water at 13·02°C and 25°C. The values they obtained are summarized in Table 9.5 where a comparison with the values obtained by Dyrssen and Hansson (1973) and Hansson (1972) is given. The latter's original figures have been adjusted by Culberson and Pytkowicz to make the concentration scales for the two sets compatible. Dyrssen and Hansson (1973) also showed not only that K_w increased with increasing ionic strength but that it was also a function of ionic composition since values found for synthetic sea water are noticeably larger than those for NaCl solutions of comparable ionic strength. This is because in sea water, the presence of SO_4^{2-}, Mg^{2+} and Ca^{2+} enables HSO_4^-, $MgOH^+$ and $CaOH^+$ to form. The procedure used to determine the c_H and c_{OH} values used for the sea water K_w estimates paid no regard to the ion pairs involving H^+ and OH^-. That is,

$$c_{H(T)} = c_{H(f)} + c_{HSO_4} \tag{9.26}$$

and

$$c_{OH(T)} = c_{OH(f)} + c_{MgOH} + c_{CaOH} \tag{9.27}$$

where $c_{H(f)}$ and $c_{OH(f)}$ are the "true" (free) concentrations of these species, and $c_{H(T)}$ and $c_{OH(T)}$ are the corresponding total concentrations.

The pH concept. pH is an operational concept defined in terms of two EMF measurements made on cells of the type

glass electrode | solution (X) ‖ KCl(aq) | reference electrode

in which in the first measurement X is a standard pH buffer and in the second it is the test solution. Denoting the two EMF readings by E_s and E_x respectively, the relationship between them and the two pH values (pH_x and pH_s) is

$$pH_x - pH_s = \frac{E_x - E_s}{2·303RT/F} \tag{9.28}$$

A pH standard adopted by the National Bureau of Standards and the

TABLE 9.5

Comparison of measured values of pK_w^{sw}.

Temperature (°C)	Salinity (‰)	pK_w^{sw} (molar)		Difference
		Hansson 1972*	Culberson and Pytkowicz (1973)	
25·0	19·90	13·304	13·325	+0·021
	26·87	13·244	13·253	+0·009
	34·82	13·176	13·199	+0·023
13·02	34·82	13·654	13·681	+0·027

* Interpolated from Hansson's (1972) data by linear least squares. His values of K_w^{sw} were multiplied by the density of sea water squared to convert them to molar concentrations. From Culberson and Pytkowicz (1973).

British Standard Specifications is a 0·05 M solution of potassium hydrogen phthalate, the pH of which is defined as

$$pH_s = 4·00 + \tfrac{1}{2}\left[\frac{t - 15}{100}\right]^2 \qquad (9.29)$$

where t is the temperature in °C, although other buffers having pH values closer to that of sea water are often used (see Section 20.3.1.1.).

On this basis, an agreed scale of acidity is thereby established which allows at least intercomparison of different solutions, and often this is all that is required. However, interpretation of measured values in terms of fundamental thermodynamic properties is difficult. Thus, since for pH cells of the type indicated above,

$$E_x - E_s = \frac{2·303\,RT}{F}\left[\log_{10} a_{H(s)} - \log_{10} a_{H(x)}\right] + E_{l(x)} - E_{l(s)} \qquad (9.30)$$

where $E_{l(x)}$ and $E_{l(s)}$ are the appropriate liquid junction potentials, then from (9.28) and (9.30)

$$pH_x - pH_s = \log_{10} a_{H(s)} - \log_{10} a_{H(x)} + \frac{E_{l(x)} - E_{l(s)}}{2·303\,RT/F}$$

It is clear that only if it were possible to assign a value to the activity of the hydrogen ion of the standard and at the same time to known the difference in the liquid junction potentials of the two solutions could the activity of the hydrogen ions in the test solution be determined. Notwithstanding the limitations associated with the concept of individual ionic activities, attempts have been made to standardize reference buffers, although inevitably the methods used to assign $a_{H(s)}$ rest on extra-thermodynamic assumptions. The liquid junction potential is a serious source of uncertainty in the theoretical interpretation of measured pH, particularly when the difference between the pH values of the test and standard solutions is large and when the two solutions differ considerably in ionic strength. These considerations mean that the often quoted relationship

$$pH = -\log_{10}(a_H) \qquad (9.31)$$

is not generally valid, and is acceptable only under a very restricted range of conditions. In the view of Bates (1973), rigorous theoretical interpretation is possible only for dilute aqueous solutions of buffers and simple salt solutions having ionic strengths no greater than 0·1. Suspensions and colloidal solutions are excluded. Clearly, these requirements are not met by ocean waters, for which the ionic strength is close to 0·7, and by many estuarine waters. This does not mean that pH values cannot be obtained for these media. By observing proper precautions it is possible to make such pH measurements, often with a high degree of reproducibility, and these values may be of use for record or comparison purposes or for use in combination with suitably determined apparent-acidity constants (Section 9.2.3.5). However, they cannot be interpreted in terms of (9.31) without straining the basic theoretical framework. However, Hawley and Pytkowicz (1973) have attempted to show that the residual potential which arises when electrodes are transfered from dilute buffer solutions to 0·725 M NaCl solution is about 0·32 mV, and they consider that the value for transfer to sea water is similar.

It should be noted that many earlier pH determinations made use of the Sørensen pH scale (psH). This is related to the more modern scale by

$$pH = psH + X \qquad (9.32)$$

where X has the values 0·034, 0·036, 0·034, 0·028 and 0·023 at 12, 20, 25, 30 and 35°C respectively (Bates, 1973).

Procedures have been suggested for avoiding some of the difficulties associated with the conventional method of pH measurement for oceanographical purposes. It will be appreciated that some of the difficulties stem from the fact that the reference solution is dilute whereas the test solution has a relatively high ionic strength. The basis of the proposed method is to make use of organic buffers (e.g. *tris*(hydroxymethyl)aminomethane plus its hydrochloride) made up, not in distilled water, but in synthetic sea water of known salinity (Hansson, 1973b). Use of this buffer in a slightly different context had previously been explored by Smith and Hood (1964). The activity scale employed in Hansson's work was based not on the infinitely dilute convention normally employed in physico-chemical investigations, but on a scale such that $\gamma_H (=a_H/c_{H(T)})$ tended to unity as $c_{H(T)}$ tended to zero in the medium used, for the concentration range of c_H for which $c_{H(T)}/c_{H(f)}$ is constant. As noted in the previous Section, $c_{H(T)} = c_{H(f)} + c_{HSO_4}$ where $c_{H(f)}$ is the "true" hydrogen ion concentration (i.e. $c_{H_3O} + c_{H_2O_5} +$ etc.).

This activity scale is such that for the purpose for which it was designed, viz., the redetermination of the acidity constants of carbonic and boric acids in sea water (Section 9.2.3.4), $c_{H(T)}$ is effectively the same as a_H over the range of H^+ concentrations used, and for this reason pH values on the Hansson scale (see below) have a more distinct meaning than do those determined on, say, the N.B.S. scale. By means of carefully designed titration procedures, Hansson was able to establish a self-consistent tabulation of pH values as a function of salinity and temperature for synthetic sea water buffer systems prepared as indicated in Table 9.6 (see also Table A9.1 of the Appendix to this Chapter and Section Vol. 3 20.3.1).

Although this buffer system is rather temperature sensitive (Almgren *et al.*, 1974a) and suffers from the disadvantage that the salinity of the buffer should match that of the test solution, its use in place of the usual dilute solution buffers means that difficulties and uncertainties associated with the liquid junction potential problem, which always arise when a test solution is compared with a buffer, are much reduced, and the precision associated with the assignment of a pH value to the sea water sample is improved. However, it should be understood that the direct comparison of pH measurements made on the basis of the sea water scale with determinations made on other scales involves some uncertainty since no clearly defined fundamental relationship exists between the various scales. Accordingly, when pH measurements are used to enable calculations on sea water samples to be made (as for example when estimating the equilibrium concentrations of the various components of the carbonate system from measurements of the alkalinity and pH—see Section 9.2.3.5), it is vitally important that the equilibrium constants used in the calculation should be based on the same pH scale as that used for the determination of the pH of the sample. Sea going experience with the Hansson type buffer scales has been obtained by Almgren *et al.* (1974a, b) and some details of this are to be found in Section 20.3.1.1.

Measurement of pH (see also Section 20.3.1.1). Reliable and reproducible pH measurement is essential if pH is to be used in conjunction with the constants to be discussed in Section 9.2.3.3 in order to estimate properties of the carbonate system since these properties are very pH sensitive.

All methods for pH determination rest on the comparison of two measurements, one on the sample and the other on a standard. Detailed descriptions of earlier methods are to be found in the standard texts (Dole, 1941; Kolthoff

TABLE 9.6

Composition of normal sea water, the sea water standard buffer and solutions C and D used in preparing the buffer. (Hansson, 1973b).

	Solution			
Species	"Normal" sea water	Standard buffer	Solution C	Solution D
Na^+	468·04	463	368	468
K^+	10·00	10	10	10
Mg^{2+}	53·27	54	54	54
Ca^{2+}	10·33	10	10	10
Sr^{2+}	0·10	—	—	—
Cl^-	545·88	550	548	550
SO_4^{2-}	28·20	28	28	28
Br^-	0·83	—	—	—
F^-	0·07	—	—	—
HCO_3^-	2·40	—	—	—
$B(OH)_3 + B(OH)_4^-$	0·43	—	—	—
H^+	—	—	98	—
B(tris)	—	5	—	—
BH^+	—	5	—	—

The buffer solution prepared by mixing 10·00 mmol of tris, 50·00 cm^3 of solution C and sufficient solution D to give a total volume of 1000 cm^3. The compositions of the solutions given (in mmol kg^{-1}) correspond to 35‰ salinity. For salinities 10, 20, 25, 30 and 40‰, the concentrations of ($Na^+ + H^+$), K^+, Mg^{2+}, Ca^{2+}, Cl^- and SO_4^{2-} were multiplied by the factors 2/7, 4/7, 5/7, 6/7 and 8/7 respectively. The concentration of H^+ in solution C was always 0·1 M independent of salinity. For details of standardization, see Hansson (1973b). Interpolated pH values for ranges of salinity and temperature given in Appendix Table A9.1.

and Laitinen, 1941; Britton, 1956; Gold, 1956; Bates, 1973) and the following account is concerned mainly with currently used methods employing glass electrodes. For details of earlier work of oceanographical interest employing indicators see e.g. Buch *et al.* (1932).

The glass electrode consists of a thin envelope of conductive glass containing a buffer electrolyte into which an electrode (e.g. Ag/AgCl) dips. The assembly constitutes a half cell which is immersed in the test solution. The cell is completed by means of a suitable reference electrode (e.g. calomel electrode). The high internal resistance of the glass electrode means that special measuring circuitry capable of tolerating a high input resistance is required. It is desirable to standardize the elctrode with at least two buffer solutions having pH values lying on each side of that of the unknown, since the pH response (slope) may not correspond to the theoretical E.M.F. versus pH slope. It should be recognized that the electrode may respond towards other cations (e.g. Na^+) if these are present at high concentrations, particularly at high pH values. Because of strain in the glass, these electrodes usually show an asymmetry potential; that is, when there is an identical electrolyte solution on each side of the membrane, a definite potential is shown. This may change with time and usage, and frequent standardization is necessary.

The reproducibility with which pH measurements can be made is an important factor in limiting the confidence with which estimates of the various components of the carbonate system can be made. Considerable care is necessary if work of the highest precision is being attempted, and recognition must be given to the response time for the electrode to equilibrate on transfer from the buffer to the test solution. This equilibration time varies from electrode to electrode and depends on the previous treatment to which it has been subjected. It is an ageing effect, probably connected with the sorption and desorption of ions on the glass surface, and it is influenced by the drawing of current by the electrode (Pytkowicz et al., 1966). Subjecting an electrode to rapid changes of pH impairs its subsequent response (Hansson, 1973d).

When careful attention is paid to eliminating and controlling as far as possible factors likely to impair performance, a reproducibility for a single electrode of better than ± 0.003 pH unit can be obtained; ± 0.006 pH unit is possible even when different selected electrodes are used (Pytkowicz et al., 1966; see also Takahashi et al., 1970). The conventional oceanographical practice has been to calibrate electrodes with dilute buffers before (and after) immersing them in the high ionic strength sea water test solution. The large, uncertain, liquid junction potential discussed above limits the fundamental meaning that can be extracted from the measured pH, and from a practical point of view the drastic change in ionic strength experienced by the electrode may result in it giving less than its best performance. Difficulties arising in this way are largely overcome by using sea water buffer systems as described above (Hansson, 1973b; Almgren et al., 1974a, b).

Typical glass electrodes are fragile, and without modification they are not suited to operation at high pressures. The demand for high-pressure glass electrode systems arises from the need to study events in situ at great depth and the complementary need for laboratory studies of the pressure dependence of equilibrium constants used in the interpretation of in situ measurements. An assembly designed for in situ work has been described (Distèche, 1959, 1962, 1964; Distèche and Dubuisson, 1960; Distèche and Distèche, 1965; see also Ben-Yaakov and Kaplan, 1968b). A simple apparatus for laboratory studies in which the problem of pressure equilibration across the membrane is overcome by using rubber stoppers was used by Culberson and

Pytkowicz (1968) (see also Pytkowicz, 1968). Electrodes suitable for field work at less extreme pressures have been described by Manheim (1961) and Grasshoff (1964). Descriptions of several patterns of electrodes suitable for specific field applications are given in Section 20.3.1.1.

In the determination of pH at high pressures using glass electrodes, the investigator is again confronted with the asymmetry potential, liquid junction potential and electrode slope problems, but there is now the additional need to know the pressure dependence of these factors. These aspects have been examined by Distèche and Distèche (1965) and Culberson *et al.* (1967), and have been discussed in a review by Pytkowicz (1968). The pressure dependence of the asymmetry potential can be determined by measuring the E.M.F. of the cell as a function of pressure when the same reference solution is used on each side of the membrane. Distèche (1959) demonstrated the constancy of the electrode slope with pressure. In the investigation made by Culberson *et al.* (1967), this constancy of slope with pressure was assumed, and the slope at 1 atmosphere pressure was established by a calibration procedure based on the use of buffer solutions. It was assumed that the asymmetry and liquid junction potentials were unchanged when sea water replaced the dilute buffer. The pH at pressure P was calculated using the expression

$$(\text{pH})_P = (\text{pH})_1 + \frac{1}{s}\left[E_p - e_A(P - 1) - E_1\right] \tag{9.33}$$

where E_p and E_1 are the measured E.M.F. values at pressures of P and 1 atm respectively, $(\text{pH})_P$ and $(\text{pH})_1$ are the corresponding pH values, e_A is the pressure coefficient of the asymmetry potential and s is the electrode slope (given theoretically by $2 \cdot 303\, RT/F$).

$(\text{pH})_P$ so determined is attended by the same uncertainties of interpretation in terms of hydrogen ion activity as are measurements at 1 atmosphere pressure. However, for the main purposes for which these determinations are intended, these uncertainties are of little consequence. Usually, these pH values are used in conjunction with apparent equilibrium constants (Section 9.2.3.4) which embody similar uncertainties arising from the way in which they are defined and determined. Provided that the measured pH values and the apparent equilibrium constants are determined in the same way (i.e. same pH scale used) the uncertainties largely cancel when they are combined to give information on the carbonate system.

The effect of temperature on the pH of sea water has been commented on by various workers (Buch and Nynas, 1939; Gieskes, 1969, 1970; Ben-Yaakov, 1970; Murray, 1970). According to Murray, the *in situ* pH of sea water at all chlorinities can be calculated by using the equation

$$\text{pH}\,(t_2\,(\textit{in situ})) = \text{pH}\,(t_1) + 0\cdot0113\,(t_1 - t_2)$$

where $pH(t_1)$ is the measured pH at the laboratory temperature $t_1\,°C$ and $t_2\,°C$ is the *in situ* sea water temperature.

The effect of various samplers on pH has been investigated on a Geosecs Intercalibration cruise (Takahashi *et al.*, 1970).

9.2.3.2. *Alkalinity, carbonate alkalinity, ΣCO_2, P_{CO_2} and p_{CO_2}*

The amphiprotic properties of water facilitate solvolytic reactions towards weak acid anions, e.g.,

$$HCO_3^- + H_2O = H_2CO_3 + OH^-$$

Similar reactions occur for the CO_3^{2-} and borate ions. This reaction is equivalent to the combination (9.7) and (9.9). As a result of these equilibria, the pH of sea water is on the alkaline side of neutrality. The alkalinity (total alkalinity, titration alkalinity) of a sea water sample is the amount of hydrogen ion (eq kg^{-1}) needed to convert all the anions of the weak acids to their unionized acids, and is given by an expression which recognizes the electroneutrality of the solution, viz.,

$$A = c_{HCO_3(T)} + 2c_{CO_3(T)} + c_{B(OH)_4} + (c_{OH} - c_H) + c_{SA} \qquad (9.34)$$

where the concentration terms refer to the total equilibrium concentrations and c_{SA}, the surplus alkalinity, refers to the sum concentrations of all weak acid anions other than those of carbonic and boric acids, Because the second and third ionization constants of boric acid are very low, contributions to the alkalinity made by other anions derived from boric acid can be ignored (Ingri, 1963), although the high precision now attainable for the determination of alkalinity and pH means that attention can be paid to contributions arising from changes in the concentrations of the nutrients nitrate and phosphate (Gunderson and Mountain, 1973; Dyrssen and Gunderson, 1974; Almgren *et al.*, 1974b). The term surplus alkalinity was used by Gripenberg (1960) who considered that in water from the Gulf of Bothnia there was a small but significant additional contribution to the alkalinity possibly attributable, in part, to complexes formed between organic compounds and boric acid. Such compounds were thought likely to have an enhanced acid strength in much the same manner as do complexes formed, for example, between polyhydroxy compounds, such as mannitol, and boric acid. However, Williams and Strack (1966) have convincingly demonstrated that the amount of boric acid which is complexed in this way in the sea is insiginificant ($<0.1\%$). In this account the term surplus alkalinity refers to contributions arising from all other minor anions.

For uncontaminated sea water of pH 8·3 at 10°C, $c_{HCO_3(T)}$, $c_{CO_3(T)}$ and $c_{B(OH)_4}$ are about 20×10^{-4}, 2.0×10^{-4} and 1.0×10^{-4} mol l^{-1} respectively, whereas $(c_{OH} - c_H)$ is only some $2 \times 10^{-6} \, mol \, l^{-1}$ and can be disregarded. For most ocean waters the surplus alkalinity is also small enough to be ignored except in very refined measurements, and to a good approximation, (9.35) reduces to

$$A = c_{HCO_3(T)} + 2c_{CO_3(T)} + c_{B(OH)_4} \qquad (9.35)$$

For water from anoxic regions c_{SA} can no longer be ignored and will contain contributions from sulphide, ammonia and phosphate, and alkalinities as high as $12.66 \, meq \, l^{-1}$ have been reported (Gaines and Pilson, 1972). Estuarine, industrially contaminated or atypical waters may also contain significant surplus alkalinity.

Specific alkalinity defined as

$$\text{Specific alkalinity} = \frac{\text{Alkalinity} \times 10^3}{Cl\%_0} \qquad (9.36)$$

is a term which is often used although it has the disadvantage of employing "mixed" units. Its value is about 0.126 for most ocean waters, although considerable deviations from this occur, and the range may be $0.119–0.130$ (Koczy, 1956; see also Section 9.6). Particularly large deviations from average ocean values occur in circumstances when solution of calcium carbonate occurs.

The carbonate alkalinity, CA, is the sum total of the contributions made to the alkalinity by HCO_3^- and CO_3^{2-}. That is,

$$CA = c_{HCO_3(T)} + 2c_{CO_3(T)} \qquad (9.37)$$

It is normally obtained from the alkalinity by subtracting the contribution arising from the boron content (see below).

The total carbon dioxide content (total inorganic carbon), ΣCO_2, is defined by

$$\Sigma CO_2 = c_{CO_2(aq)} + c_{H_2CO_3} + c_{HCO_3(T)} + c_{CO_3(T)} \qquad (9.38)$$

$$= c_{CO_2(T)} + c_{HCO_3(T)} + c_{CO_3(T)}$$

At the pH of most ocean waters, less than 1% of the inorganic carbon exists as $CO_2(aq) + H_2CO_3$ and the H_2CO_3 concentration is only ca. 0.2% of that of $CO_2(aq)$. Thus, it is reasonable for many approximate purposes to write

$$\Sigma CO_2 = c_{HCO_3(T)} + c_{CO_3(T)} \qquad (9.39)$$

It will be appreciated that although CA and ΣCO_2 are numerically similar, under normal circumstances $CA > \Sigma CO_2$.

Some comment should be made on the depth (pressure) dependence of alkalinity, carbonate alkalinity and total carbon (see Pytkowicz, 1968) since generally these quantities are determined at atmospheric pressure whereas for the purposes of the calculations outlined in subsequent sections it is the *in situ* values that are required. Two possible corrections need to be considered: (i) The volume of a given sample is reduced with depth because of the compressibility of sea water, and (ii) even when allowance has been made for compressibility, the pressure dependence of the acid ionization constants of carbonic and boric acid causes changes in the concentrations of bicarbonate, carbonate and borate. The first of these corrections can be catered for by assuming a decrease in volume of about 4% per 10000 m (Buch and Gripenberg, 1932), or by expressing the concentrations in terms of unit weight of sea water. When this is done it is clear that ΣCO_2 is, of course, independent of pressure. The definition of alkalinity by (9.34) and the condition of electroneutrality requires that the total alkalinity is also conservative and depth independent, although the relative proportions of bicarbonate, carbonate, borate, hydrogen and hydroxyl ions which contribute to the alkalinity are depth dependent. The ionization of boric acid increases with pressure and there is a small compensating reduction in CA (Pytkowicz, 1968). It might also be noted that as carbon dioxide is taken up by or lost from sea water the carbonate alkalinity changes, although over the range of such exchanges normally encountered, the effect is generally sufficiently small to be ignored. Total alkalinity is, of course, conserved during such exchange.

The partial pressure of CO_2 in a water sample is that pressure of CO_2 in the gas phase with which the sample would be in equilibrium, and in this account is denoted by P_{CO_2}. The actual partial pressure of CO_2 in the atmosphere is denoted by p_{CO_2}. Although these two quantities may be of similar magnitude (*ca.* 320×10^{-6} atm), they are rarely equal and quite large variations particularly of P_{CO_2} often occur (Section 9.6).

Determinations

(a) *Alkalinity.* This has been estimated by treating the sample with a measured excess of strong acid, driving off the liberated CO_2 and back titrating to a pH of 6·0 to 7·0 at which the boric acid is negligibly ionized (Gripenberg, 1937). Alternatively, in earlier work the sample was titrated in the presence of the liberated CO_2 to a pH of 4·5 to 5·0 (Greenberg *et al.*, 1932). A more rapid method based on a single pH determination was proposed by Thompson and Anderson (1940), and has subsequently been modified by later groups of workers (West and Robinson, 1941; Anderson and Robinson, 1946; Bruneau *et al.*, 1953; Culberson *et al.*, 1970). In the modification proposed by Culberson *et al.*, 30 cm^3 of 0·01 M HCl is added to a 100 cm^3 sample, the CO_2

purged with CO_2-free water-saturated air, and the final pH measured. The alkalinity is calculated from

$$A = \frac{1000}{V_s} \cdot V.M. - \frac{1000}{V}(V_s + V)\frac{a_H}{\gamma_{H(T)}} \qquad (9.40)$$

where V_s and V are the sample and HCl volumes respectively and M is the HCl molarity. The activity of the hydrogen ions in the final solution (a_H) is estimated from the pH (NBS scale) measured using a glass electrode and assuming the validity of equation (9.31). The quantity, $\gamma_{H(T)}$ is an activity coefficient defined by

$$c_{H(T)} = \frac{a_H}{\gamma_{H(T)}} = c_{H(f)} + c_{HSO_4} + c_{HF} \qquad (9.41)$$

That is, the term $a_H/\gamma_{H(T)}$ in (9.40) is effectively the total excess hydrogen ion concentration over that necessary to titrate anions of the weak acids and comprises the 'true' hydrogen ion concentration, $c_{H(f)}$ plus that involved in HSO_4^- and HF formation. Culberson et al. determined γ_H at 25°C to be 0·741. This value agrees well with those calculated from the activity coefficient of the hydrogen ion in sea water for ranges of temperature and salinity (Leyendekkers, 1973). This method for the alkalinity determination is claimed to give a precision of $\pm 0·008$ meq l^{-1} on the basis of combined standard deviations of replicates containing 9, 5, 3 and 3 samples. A modification of this procedure to make it suitable for application to highly alkaline anoxic waters has been described by Gaines and Pilson (1972).

Hansson (1972) considers potentiometric titration methods to offer higher precision and accuracy than do single point methods such as that just described. A high precision potentiometric titration method based on the procedure originally described by Gran (1952) (see Section 20.3.2.2) has been used by Edmond (1970) to give a precision in the alkalinity estimation of $\pm 0·17\%$. Dyrssen (1965); Dyrssen and Sillén, (1967) also advocate the use of Gran type potentiometric titrations. Hansson and Jagner (1973) have described an improved procedure based on Gran functions, and shipboard application of this procedure has been described by Almgren et al. (1974b). The high level of precision now attainable means that the technique is virtually capable of allowing comments on the contribution of phosphate (and other protolytes) to the alkalinity to be made. (See also Dyrrsen and Hansson, 1974).

It should be noted that care is necessary in the choice of materials for the constructure of samplers for the collection of water for alkalinity determinations as changes in the alkalinity and pH may accompany attack on brass by dissolved oxygen (Park, 1968a). Care in the choice of samplers and containers for handling samples for carbonate analysis is essential at the levels of precision currently attainable (Almgren *et al.*, 1974a). Freedom from Zn, Cd, Ba or organic acid additives in plastic coated materials is important, and smooth inner surfaces are desirable to minimize the risk of CO_2 loss through bubble formation. Such loss may appreciably influence the pH.

An intercomparison of methods for alkalinity determination has been made on a recent Geosecs cruise (Takahashi *et al.*, 1970). This investigation also considered the influence of sampler type and method of sample treatment to ensure stability for subsequent examination.

(b) *Carbonate alkalinity.* In order to derive this quantity from the measured alkalinity, it is necessary to obtain an estimate for the $B(OH)_4^-$ contribution to the alkalinity. The total boron content of sea water is given by (Culkin, 1965) ΣB (mg kg^{-1})/Cl‰ = 0·222 for most ocean waters, although local deviations from this may occur. For the ionization

$$H_3BO_3 + H_2O = B(OH)_4^- + H^+ \tag{9.42}$$

the apparent ionization constant, K_B', is

$$K_B' = \frac{a_H \cdot c_{B(OH)_4}}{c_{H_3BO_3}}. \tag{9.43}$$

Since

$$c_{\Sigma B} = c_{H_3BO_3} + c_{B(OH)_4} \tag{9.44}$$

then

$$c_{B(OH)_4} = \frac{K_B' \cdot c_{\Sigma B}}{(K_B' + a_H)} \tag{9.45}$$

Thus,

$$CA = c_{HCO_3} + 2c_{CO_3}$$

$$= A - \frac{K_B' \cdot c_{\Sigma B}}{(K_B' + a_H)} - (c_{OH} - c_H) \tag{9.46}$$

or, between pH 5·5 and 8·5,

$$CA = A - \frac{K_B' \cdot c_{\Sigma B}}{(K_B' + a_H)} \tag{9.47}$$

The apparent ionization constant of boric acid is discussed in Section (9.2.3.4), and Tables of K_B' have been given by Buch (1933a), Lyman (1956) and Hansson (1973a, c), (see Tables A9.5 and A9.8 of the Appendix to this Chapter). Care is necessary in their use in order to ensure that the pH scale used when equation (9.47) is applied is consistent with that on which the Tables are based.

The borate contribution to the alkalinity is usually 5–10% of the total alkalinity; this percentage falls as the pH of the sample is decreased since this reduces the degree of ionization of the boric acid.

(c) *Total carbon dioxide*, ΣCO_2. In early work ΣCO_2 was determined by measuring the total volume of CO_2 liberated when a known volume sample was treated with hydrochloric or phosphoric acid (Buch *et al.*, 1932) using a van Slyke type apparatus (van Slyke and Neill, 1924; van Slyke and Sendroy, 1927; Strickland and Parsons, 1960). A Conway microdiffusion method has been described (Saruhashi, 1953), but most current routine methods are based on gas chromatography (Swinnerton *et al.*, 1962; Park *et al.*, 1964; Menzel and Vaccaro, 1964; Curl and Davey, 1967; Reusmann, 1968; Weiss and Craig, 1973) or infra-red spectrophotometry (Smith, 1953; Kanwisher, 1960; Takahashi, 1961; Ibert and Hood, 1963; Waterman, 1965; Rulolf, 1971). The method described by Swinnerton *et al.* made use of an injection port comprising a small glass chamber containing a horizontally mounted sintered disc. The total dissolved gas was stripped from the acidified sample injected on to the upper side of the disc by carrier gas which passed through the disc from below. The columns used by these authors were designed for the separation and analysis of the total dissolved gas, but a version described by Park *et al.* used a single silica gel column which was adequate for the separation of the CO_2. This could be determined with a coefficient of variation of 0·7% from 5 replicate determinations. The method is rapid (5 min. per analysis) and is suited for shipboard use. A similar method was used by Takahashi *et al.* (1970) on the 1969 GEOSECS Intercalibration cruise. Curl and Davey have described a special sampling valve, and using it they were able to obtain ΣCO_2 measurements to better than 1% in routine use.

For infra-red-based determinations of ΣCO_2, Wong (1970) vacuum extracted the CO_2 from an acidified 30 cm³ sample, the liberated gas being quantitatively condensed in a cold trap at $-196°C$. Trap-to-trap distillation of the condensate removed water, and the CO_2 was determined from its infra-red absorption. The method has an accuracy of $\pm 0·15\%$. A potentiometric acid titration method (Edmond, 1970; Takahashi *et al.*, 1970; Hansson and Jagner, 1973) has an accuracy for ΣCO_2 determination of better than $\pm 0·6\%$. A method for extracting the total carbon dioxide from sea water samples for ^{13}C analysis has been described by Kroopnick (1974).

Unless precautions are taken the ΣCO_2 estimates of stored sea water samples may exceed those made at the time of sampling because of oxidation of dissolved organic matter. Stability during storage for up to three months can be ensured by the addition of 100 p.p.m. of $HgCl_2$ to the sample (see also Takahashi *et al.*, 1970).

(d) *The partial pressure of* CO_2, P_{CO_2}. The equilibrium partial pressure of CO_2 in sea water can be simply determined by analysis of the CO_2 content of a bubble of air shaken with the sample (Krogh, 1910; Buch, 1939a, b). More recent methods make use of infra-red analysis of air which has been brought into intimate contact with sea water in specially designed equilibrators and dried before examination (Smith, 1953; Kanwisher, 1960; Takahashi, 1961; Ibert and Hood, 1963; Waterman, 1965; Keeling and Waterman, 1968; Rulolf, 1971). Two different types of equilibrator were used in the GEOSECS Intercalibration Cruise (Takahashi *et al.*, 1970) depending on whether surface or deep water samples were to be examined. For surface water, P_{CO_2} values were determined continuously using a system developed by Broecker and Takahashi (1966). In this, N_2 carrier gas was recirculated in a closed system through which a shower of sea water droplets was passed continuously. Water vapour was removed by means of a cold trap, and the CO_2 content of the carrier was determined with an infra-red analyser. For examination of deep water, discrete samples of 20 l were used (Li, 1967; Li *et al.*, 1969). A stream of finely divided N_2 carrier gas bubbles was recirculated through the sample and when equilibrium was reached (10 min) the CO_2 content was determined. For both systems the measured CO_2 content of the carrier gas was adjusted to give the P_{CO_2} of the sample by applying corrections (Li, 1967) to allow for the removal of water vapour, changes in the total pressure of equilibration and the loss of CO_2 from the water to the carrier gas. Both methods gave an accuracy of better than 1%. Data obtained on the 1969 GEOSECS Intercalibration cruise indicated that P_{CO_2} could be determined to better than $\pm 2\%$. If the temperature at the measurement point differs from the *in situ* value, account should be taken of the temperature coefficient of P_{CO_2} (see Section 9.2.3.5).

Electrode systems for monitoring P_{CO_2} have been described (see Section 20.3.1.1), but generally these are less precise than are other currently available methods.

(e) *Atmospheric partial pressure of* CO_2, p_{CO_2}. This can be determined by conventional analysis of the air sample (Fonselius *et al.*, 1955). Usually, infra-red analysis of the dried sample is used (Fonselius *et al.*, 1956; Takahashi, 1961; Ibert and Hood, 1963; Keeling, 1961; Bolin and Keeling, 1963; Bischof, 1970); this method is particularly suitable for continuous automatic monitoring which is desirable for the detection of rapid or short-term variations. The need to be able to sample at various altitudes and over long tracks has led to the development of aircraft-borne sampling techniques (Bischof, 1970, 1971). Interest in possible long and short term changes in p_{CO_2} has resulted in attention being given to precision in monitoring programmes. For this purpose and to facilitate intercomparison of sets of analyses the use of a common standard is helpful (Keeling, 1960; Bolin and Keeling, 1963; Bolin and Bischof, 1970; Bischof, 1971). In suitably controlled sets of determinations the relative accuracy is *ca.* 0·2 p.p.m. (Bolin and Bischof, 1970), although the absolute accuracy is considerably less than this.

It is customary to express the CO_2 content of a gas sample in units of atmospheres or of p.p.m. by volume. If the pressure of sampling is one atmosphere, then the *in situ* CO_2 partial pressure expressed in atm $\times 10^6$ is numerically the same as the concentration expressed in p.p.m. by volume. When the sampling pressure deviates appreciably from one atmosphere (for example, in high altitude air sampling, in connection with determinations of the P_{CO_2} of high altitude lakes or as a result of variations in atmospheric pressure at sea level) the numerical values in the two systems of units differ unless the partial pressure is expressed in terms of the sample reduced to S.T.P.

It is important to indicate whether reported values for p_{CO_2} refer to the gas as sampled or to the dried gas. Often it is assumed that the gas sampled at the sea surface or air emergent from an air–sea equilibrator is saturated with respect to water vapour. Removal of 23 torr of water vapour from a sample of air having a CO_2 concentration of 320 p.p.m. by volume will increase the concentration by about 9 p.p.m.

(f) *Carbonate saturometry.* An area of progressively growing importance concerns the degree of calcium carbonate saturation at various depths in the ocean (Section 9.2.4.3). Information on this topic can be obtained on the basis of shipboard measurements of alkalinity from which, by means of the equations given in Section 9.2.3.5 and the pressure adjusted values of the appropriate equilibrium constants, the *in situ* value for $c_{CO_3(T)}$ can be calculated. When combined with $c_{Ca(T)}$ a value for $c_{Ca(T)} \cdot c_{CO_3(T)}$ is obtained. An alternative approach makes use of *in situ* carbonate saturometry (Weyl, 1961; Chave *et al.*, 1962; Schmalz and Swanson, 1969; Ben-Yaakov, 1970; Ben-Yaakov and Kaplan, 1968a, 1969, 1971; Ben-Yaakov *et al.*, 1974). The basis of this method is the monitoring of the pH change which occurs when sea water is equilibrated *in situ* with calcium carbonate. In a recent version (Ben-Yaakov *et al.*, 1974), the saturometer is provided with two pH electrodes; one measures the pH of the surrounding water and the other follows the pH change within the saturometer cell as sea water, which is periodically pumped through the cell, equilibrates with the solid phase. For the essential theory underlying the method the above references should be consulted.

The method is promising and attractive since it is direct, but as noted by Ben-Yaakov et al. (1974), a number of features merit careful scrutiny before results obtained using it can be interpreted with full confidence. There is the possibility that the history of the mineral phase may influence the results. Sufficient time must be allowed for equilibrium to become established or an extrapolation procedure to infinite time must be applied to the signal. The design of the instrument should be such that spurious signals arising from metal–solution potentials do not arise. There is the possibility that freshly deposited Mg-rich calcite coatings from super-saturated sea water may influence subsequent measurements.

At present there is still some disagreement between results found using shipboard data (via alkalinity etc.) and those obtained using the saturometer; controlled intercomparison of the two approaches is required.

9.2.3.3. Equilibrium constants in the sea water carbonate system

If ion pairing is disregarded, the essential steps in the sea water carbonate equilibria are

$$CO_2(g) = CO_2(aq) \tag{9.5}$$

$$H_2O + CO_2(aq) = H_2CO_3 \tag{9.6}$$

$$H_2CO_3 = H^+ + HCO_3^- \tag{9.7}$$

$$HCO_3^- = H^+ + CO_3^{2-} \tag{9.8}$$

Each step is characterized by an equilibrium constant. However, for certain equilibria, particularly the ionizations, there is more than one way in which the constants can be defined. This proliferation of constants is a potential source of confusion, but indicates the importance of this area and the need felt by investigators for reliable constants to facilitate the interpretation of routine measurements.

The dissolved inorganic carbon is distributed between $CO_2(aq)$, H_2CO_3, HCO_3^- and CO_3^{2-}; ion pairing of HCO_3^- and CO_3^{2-} with Ca^{2+}, Mg^{2+} and Na^+ will occur (Section 9.2.3.7), but for the moment this will be ignored beyond recognizing that the total HCO_3^- and CO_3^{2-} concentrations are the sums of the "free" concentrations of these ions plus the amounts of them bound as ion pairs.

The relative proportions of the four major forms of dissolved inorganic carbon depend on the temperature, pressure, pH and salinity (e.g. Fig. 9.2), and by making use of the first and second ionization constants of carbonic

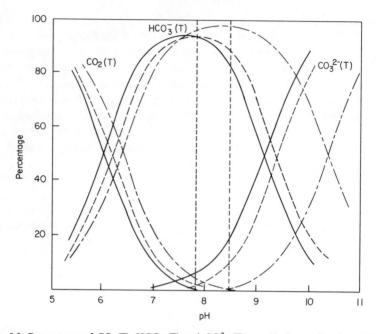

Fig. 9.2. Percentages of $CO_2(T)$, $HCO_3^-(T)$ and $CO_3^{2-}(T)$ vs. pH. Curves based on Lyman's apparent constants. The vertical broken lines indicate the approximate pH range of sea water. The extention beyond these limits is for purposes of illustration only, and assume the constants to be pH independent (see, e.g. Pytkowicz et al., 1974). ———, 25°C and 19‰ Cl; ─ ─ ─ ─, 0°C and 19‰ Cl; ─ ─ ─ ─ ─, 25°C and pure water. The change at 25°C on changing the medium from pure water to 19‰ should be noted. This change reflects the more marked effect of salinity on the second as compared with the first apparent constant. Use of other constants (e.g. those of Hansson, 1973a, d or of Mehrbach et al., 1973) leads to slightly different distributions, but does not greatly affect the general picture.

acid it can be shown that, at one atmosphere pressure and normal temperatures, for pH values below about 3 only negligible ionization occurs, the dissolved inorganic carbon existing almost entirely as $CO_2(aq) + H_2CO_3$. Even at pH 5·5 only some 15% of the total inorganic carbon is in the form of HCO_3^-, and the percentage which forms CO_3^{2-} is not significant below pH 7·5. Above this value carbonate (and borate) become significant, and predominate at pH >9.

The solubility of CO_2. Most determinations of CO_2 solubility are made using acidified solutions so as to ensure that the total CO_2 approximates to $c_{CO_2(aq)} + c_{H_2CO_3} = c_{CO_2(T)}$. The equilibrium constant for (9.6) is such that less than 0·2% of the dissolved CO_2 is hydrated to H_2CO_3 (see below), and consequently little error is introduced by identifying $c_{CO_2(T)}$ with $c_{CO_2(aq)}$.

Over the range of CO_2 pressures of interest to oceanographers, CO_2 dissolves in aqueous solutions in accordance with Henry's Law. The linear relationship between the concentration of dissolved CO_2 and the equilibrium of CO_2 is expressed by

$$c_{CO_2(T)} = \alpha \cdot P_{CO_2} \qquad (9.49)$$

where α, the solubility coefficient $(mol\,kg^{-1}\,atm^{-1})$, is a function of temperature hydrostatic pressure and salinity. Several sets of values have been reported (Bohr, 1899; Geffken, 1904; Markham and Kobe, 1941; Harned and Davis, 1943; Morrison and Billett, 1952; Cox and Head, 1962; Enns et al., 1965; Murray and Riley, 1971; Li and Tsui, 1971). Most of the earlier measurements relate to pure water or to simple salt (NaCl) solutions, and tables prepared by Buch et al. (1932) for the solubility of CO_2 in sea water are based on Bohr's determinations for sodium chloride solutions by making use of the assumption that the effect of a given weight of the salts in sea water on the solubility is the same as that of the same weight of NaCl. Li et al. (1969) have summarized Buch's estimates of α made in this way for the chlorinity range 0–35% and the temperature range 0–35°C by

$$\log(\alpha/\alpha_0) = (0 \cdot 0806 - 0 \cdot 00074t) \times Cl\%/20$$

where α_0, the solubility in distilled water at t°C, is given by

$$\alpha_0 = (770 - 29 \cdot 5t + 0 \cdot 685t^2 - 0 \cdot 0075t^3) \times 10^{-4}\,mol\,l^{-1}\,bar^{-1}$$

An experimental investigation of the solubility in sea water by Li and Tsui (1971) using an infra-red based method gave values in agreement with those tabulated by Buch (see also Krogh, 1904; Li, 1967), although the accuracy of these determinations has been questioned (see below).

Edmond and Gieskes (1970) have also examined the earlier results on CO_2 solubility and have attempted to adapt data obtained by Harned and Davis (1943) for NaCl solutions to the uptake by sea water by applying an ionic strength correction. Thus, the functional relationship suggested by them is

$$-\log \alpha = -2385 \cdot 73/T + 14 \cdot 0184 - 0 \cdot 0152642T$$

$$+ I(0 \cdot 28596 - 6 \cdot 167 \times 10^{-4}T)$$

where T is in K and I is the ionic strength of sea water as given by Lyman and Fleming (1940), viz,

$$I = 0 \cdot 00147 + 0 \cdot 03592(Cl\%) + 0 \cdot 000068(Cl\%)^2$$

Detailed agreement between the tabulations prepared by Buch and precise determinations for sea water would not be expected. It has long been known that, in principle, the assumptions made by Buch—that the salt in sea water has the same effect on the solubility of CO_2 as does the same weight of NaCl —is not strictly valid, and Lyman (1956) has suggested that the value recommended by Buch for 36‰ water is really more correctly applicable to water of salinity 35‰. Edmond and Gieskes have questioned whether uncertainties in the basic data allow this distinction to be made. However, recent determinations (using a gravimetric method), of the solubility in acidified sea water (Murray and Riley, 1971) confirm that, as suspected by Lyman, the values given by Buch *et al.* are somewhat too high (*ca.* 3%). Keeling (1973a) has proposed that the results obtained by Murray and Riley can be summarized by

$$\log_{10} \alpha = 2622 \cdot 38/T - 15 \cdot 5873 + 0 \cdot 017841T$$
$$- Cl‰(0 \cdot 0117950 + 2 \cdot 77676 \times 10^{-5}T)$$

Weiss (1974) has recently examined the solubility of carbon dioxide in sea water, and his results confirm the accuracy of Murray and Riley's determinations as opposed to those made by Li and Tsui. However, Weiss has pointed out that neither of the above two groups of workers, acidified the water used in their determinations of the solubility in pure water and that some correction is needed on this account. In addition, for accurate interpretation of the results obtained by Murray and Riley, correction is necessary to allow for non-ideal behaviour in the gas phase. According to Weiss, the fully corrected results obtained by Murray and Riley are summarized by the equation

$$\ln \alpha = -58 \cdot 0931 + 90 \cdot 5069(100/T) + 22 \cdot 2940 \ln (T/100)$$
$$+ [0 \cdot 027766 - 0 \cdot 025888(T/100) + 0 \cdot 0050578(T/100)^2] \, S$$

where T is the absolute temperature and S is the salinity (‰). Values of the solubility calculated by Weiss using this equation are given in Appendix Table A9.2a of this Chapter. It is becoming increasingly the practice to express concentrations on a weight (rather than volume) of solvent basis. Appendix Table A9.2b (Weiss, 1974) gives values for the CO_2 solubility expressed as mol (kg sea water)$^{-1}$ atm^{-1}. Values for this quantity (α') can be obtained via the equation

$$\ln \alpha' = -60 \cdot 2409 + 93 \cdot 4517(100/T) + 23 \cdot 3585 \ln(T/100)$$
$$+ [0 \cdot 023517 - 0 \cdot 023656(T/100) + 0 \cdot 0047036(T/100)^2] \, S$$

The solubility decreases with increase of both salinity and temperature. Comparison of the solubilities of the major atmospheric gases shows that, for a given pressure, the uptake of CO_2 is much greater than that of O_2 and N_2.

Thus, whereas the relative atmospheric proportions of $N_2:O_2:CO_2$ are *ca.* 2400:630:1, the proportions in water in equilibrium with the atmosphere are *ca.* 28:19:1, the precise value depending on the salinity and temperature. The equilibrium molar ratios of CO_2 in sea water (19‰ Cl) to that in the atmosphere are 1·5 and 0·70 at 0° and 25°C respectively.

The dependence of the solubility coefficient on hydrostatic pressure is not, at present, of great oceanographical interest since, for most purposes, the act of equilibration occurs at the surface. However, Enns *et al.* (1965) have shown that the solubility coefficient of CO_2 in water decreases by about 16% per 100 atmospheres of applied hydrostatic pressure.

The CO_2 hydration equilibrium. The thermodynamic equilibrium constant for (9.6), $K_{0\,(TD)}$, is

$$K_{0\,(TD)} = \frac{a_{H_2CO_3}}{a_{CO_2(aq)} \cdot a_{H_2O}} \qquad (9.50)$$

and has a value of about 2×10^{-3}, the exact value depending on the temperature (see Lyman, 1956). Thus, since the activity of water is close to unity for most solutions of oceanographical interest, the apparent constant K_0' defined by

$$K_0' = \frac{c_{H_2CO_3}}{c_{CO_2(aq)} \cdot a_{H_2O}} \qquad (9.51)$$

$$\approx \frac{c_{H_2CO_3}}{c_{CO_2(aq)}} \qquad (9.51a)$$

(effectively the ratio of the rate coefficients of reactions (9.17) and (9.18)) is also about 2×10^{-3}. Kern (1960) has reviewed this and other aspects of the CO_2/H_2O system.

The ionization of carbonic acid

Ionization constants of carbonic acid are important as they allow the concentrations of the individual components of the carbonate system to be evaluated on the basis of a few analytical observations. In principle, use could be made of either thermodynamic constants (defined entirely in terms of activities—see e.g. equation (9.65) below) or apparent constants (in which, for certain species, concentration rather than activity is used—see e.g. equation (9.66) below). Both types of constant are functions of temperature and pressure, but on a suitably defined activity scale thermodynamic constants are independent of the salinity. In general, apparent constants are not independent of the salinity and, consequently, it is necessary, even at fixed temperature and pressure, to use a different value of the constant at each salinity. Although, superficially, it might be thought that thermodynamic constants have greater fundamental significance than the corresponding apparent constants and also that they would offer greater convenience in use, neither of these views is strictly correct. In the first place, thermodynamic constants owe their constancy to the importation of the artifical concept of activity—a device intended to preserve the formal simplicity of the framework of thermodynamic relationships established for ideal systems. Furthermore, the primary oceanographical requirement of equilibrium constants is that they should facilitate the calculation and the inter-relating of concentrations on the basis of routine analytical measurements; knowledge of activities, activity products or activity ratios is rarely the main goal. If thermodynamic constants are to be used to determine concentrations it is essential that the appropriate activity-concentration and ion pair equilibrium relationships should be known. This information is usually lacking. Apparent constants meet practical requirements better than do their thermodynamic counterparts (Pytkowicz, 1973a, b), particularly as tables of their values for ranges of temperature and salinity are available. However, those values in general use in oceanographical work are, understandably, based on solutions having the relative ionic composition and pH range of sea water. It should be remembered that apparent constants are functions not simply of ionic strength, but also of the ionic composition of the solution. For this reason, published apparent constants cannot always be used confidently in connection with salt solutions atypical of sea water or for interpreting measurements made on hypersaline solutions having ionic strengths above the range embraced in the determination of the constants.

Discussion of proposed constants is conveniently approached by setting aside for the moment the distinction between activity and concentration in order to see the general form in which the definition of the ionization constants can be cast. These generalized forms can then be related to the particular forms used by different investigators when the appropriate terms are replaced by activities or concentrations.*

Most definitions of the first ionization constant are based on one of the forms

---

* The symbolism used in this account for the ionization constants has been selected in an attempt to maintain some sort of internal consistency and system but yet not to conflict too violently with symbolisms used by previous workers. In general, a prime indicates an apparent constant, subscripts 1 and 2 indicate respectively first and second ionizations and the lower case subscripts b, h and l refer respectively to constants based on the definitions used by Buch, Hansson and Lyman. Since the definitions used by both Buch and Lyman for the second ionization constant of carbonic acid are the same (except for the pH scales used) the symbol K_2' is used for both.

(i)
$$K_{1(i)} = \frac{\{H^+\}\{HCO_3^-\}}{\{H_2CO_3\}} \tag{9.52}$$

(ii)
$$K_{1(ii)} = \frac{\{H^+\}\{HCO_3^-\}}{\{H_2CO_3 + CO_2(aq)\}} \tag{9.53}$$

(iii)
$$K_{1(iii)} = \frac{\{H^+\}\{HCO_3^-\}}{\{H_2O\}\{CO_2(aq)\}} \tag{9.54}$$

derived from equations (9.6), (9.7) and (9.8) above. The second ionization constant is based on

$$K_2 = \frac{\{H^+\}\{CO_3^{2-}\}}{\{HCO_3^-\}} \tag{9.55}$$

If, in (9.52) activities are used throughout, the constant becomes one form of the first thermodynamic ionization constant, whereas if activity is used for H^+ and concentration for HCO_3^- and H_2CO_3, the definition becomes that of one form of the first apparent ionization constant, i.e.

$$K'_{1(i)} = \frac{a_H \cdot c_{HCO_3(T)}}{c_{H_2CO_3}} \tag{9.56}$$

At low ionic strengths and at ordinary temperatures $K'_{1(i)}$ is about 2×10^{-4} (Mills and Urey, 1940; Roughton, 1941; Berg and Patterson, 1953; Wissburn *et al.*, 1954). This shows carbonic acid to be much stronger than is generally appreciated. It is, in fact, about as strong as acetic acid, and the illusion of low acid strength arises from the more general usage made of constants based on (9.53) which, expressed as an apparent constant, becomes

$$K'_{1l} = \frac{a_H \cdot c_{HCO_3(T)}}{c_{CO_2(aq)} + c_{H_2CO_3}} \tag{9.57}$$

$$= \frac{a_H \cdot c_{HCO_3(T)}}{c_{CO_2(T)}} \tag{9.58}$$

This definition, which has been used by Lyman (1956) and by Mehrbach *et al.* (1973), reflects the fact that the investigator is usually more concerned with the total dissolved carbon dioxide than with the carbonic acid alone. Since $c_{CO_2(aq)}$ is so very much greater than $c_{H_2CO_3}$, K'_{1l} will be very much smaller than $K'_{1(i)}$. The relationship between $K'_{1(i)}$, K'_{1l} and K'_0 can be seen by inserting the expression obtained for $c_{CO_2(aq)}$ from (9.51) into (9.57) when K'_{1l} becomes

$$K'_{1l} = \frac{a_H \cdot c_{HCO_3(T)} K'_0 \cdot a_{H_2O}}{c_{H_2CO_3} \cdot (1 + K'_0 \cdot a_{H_2O})}. \tag{9.59}$$

Making use of (9.56) above,

$$K'_{1l} = \frac{K'_{1(i)}K'_0 \cdot a_{H_2O}}{1 + K'_0 a_{H_2O}}$$ (9.60)

$$\approx K'_{1(i)}K'_0 \cdot a_{H_2O}$$

the last approximation being possible because $K'_0 \cdot a_{H_2O} \ll 1$.

A definition of the first apparent constant based on (9.53) was also adopted by Dyrssen and Sillén (1967) and subsequently used by their school (see Hansson, 1973a, c, d). These workers expressed (9.53) in the form

$$K'_{1h} = \frac{c_{H(T)} \cdot c_{HCO_3(T)}}{c_{CO_2(T)}}$$ (9.61)

in which $c_{H(T)}$ is the total sea water hydrogen ion concentration comprising not only the various solvated species of H^+, but also HSO_4^-. As used by the Swedish school in conjunction with the scale of activities in sea water defined by them, K'_{1h} is also a thermodynamic constant *for a particular salinity* since, on this activity scale, the activity coefficients of H^+, $HCO_3^-(T)$ and $CO_2(T)$ are each virtually unity. It will also be noted that the ratio K'_{1l}/K'_{1h} is equal to the activity coefficient of the hydrogen ion on the infinitely dilute solution in pure water activity scale.

Definitions of the first ionization constant of carbonic acid of the form indicated by (9.54) are based on a combination of (9.6) and (9.7) to give

$$CO_2(aq) + H_2O \rightleftharpoons H^+ + HCO_3^-$$ (9.62)

and the corresponding thermodynamic ionization constant is

$$K_{1(TD)} = \frac{a_H \cdot a_{HCO_3}}{a_{H_2O} \cdot a_{CO_2(aq)}}$$ (9.63)

The great importance of this and the second thermodynamic ionization constant in respiratory and other physiological and biological processes has led to careful determinations of their values (Shedlovsky and McInnes, 1935; Harned and Scholes, 1941; Harned and Davis, 1943; Nasanen, 1946, 1947) some of which are given in Table 9.7.

The first apparent ionization constant corresponding to (9.54), viz,

$$K'_{1b} = \frac{a_H \cdot c_{HCO_3(T)}}{a_{H_2O} \cdot a_{CO_2(aq)}}$$ (9.64)

was used by Buch et al. (1932) in their classic investigation of the sea water carbonate system and is discussed in the next Section. When allowance is

TABLE 9.7

First and second thermodynamic dissociation constants of carbonic acid.

t (°C)	$K_1 \times 10^7$ (1)	(2)	(3)	$K_2 \times 10^{11}$ (4)
0	2·647	—	2·613	2·36
10	3·430	—	—	3·24
15	3·802	3·79	3·722	3·71
25	4·452	4·48	4·310	4·69
35	4·914	4·89	—	5·62

K_1 refers to the process $CO_2(aq) + H_2O \rightleftharpoons H^+(aq) + HCO_3^-(aq)$.
K_2 refers to the process $HCO_3^-(aq) \rightleftharpoons H^+(aq) + CO_3^{2-}(aq)$.
(1) Harned and Davis, 1943.
(2) Nasanen, 1947.
(3) Shedlovsky and McInnes, 1935.
(4) Harned and Scholes, 1941.

made for the different pH scales this is related to the Lyman constant by the reasonable approximation

$$K'_{1l} = K'_{1b}(p_s/p_0) \cdot (\alpha_0/\alpha) \qquad (9.64a)$$

where p_s and p_0 are the saturated vapour pressures of sea water and pure water respectively and α and α_0 are the carbon dioxide solubilities in sea water and pure water respectively.

All definitions of the second ionization of carbonic acid are based on (9.55) above. The second thermodynamic ionization constant is

$$K_{2(TD)} = \frac{a_H \cdot a_{CO_3}}{a_{HCO_3}} \qquad (9.65)$$

and values for this are given in Table 9.7. The corresponding definition of the second apparent constant used by Buch *et al.* (1932) and Lyman (1956) is

$$K'_2 = \frac{a_H \cdot c_{CO_3(T)}}{c_{HCO_3(T)}} \qquad (9.66)$$

concentrations being used for the carbonate and bicarbonate terms. Dyrssen and Sillén (1967) and, subsequently, Hansson (1973a, c, d) have made use of

$$K'_{2h} = \frac{c_H \cdot c_{CO(T)}}{c_{HCO_3(T)}} \qquad (9.67)$$

in which c_H has the same significance as in the definition of K'_{1h} given above in equation (9.61). K'_2 and K'_{2h} are related through the activity coefficient of the hydrogen ion on the infinitely dilute solution in pure water activity scale.

9.2.3.4. *Determination of the first and second apparent ionization constants of carbonic acid*

The main compilations of values of the first and second apparent ionization constants at atmospheric pressure are those due to Buch (Buch *et al.*, 1932; Buch, 1929, 1933a, b, c, 1938, 1939a, b, 1942, 1945, 1951), Lyman (1956), Hansson (1973a, d) and Mehrbach *et al.* (1973). In addition, values have been reported by other workers (Moberg *et al.*, 1934; Distèche and Distèche, 1967) and, recently, investigations have been extended to the determination of the constants at higher pressures (see Distèche and Distèche, 1967; Culberson and Pytkowicz, 1968).

Determinations by Buch et al. These workers based their definition of the first apparent constant (equation (9.64)) on the equilibrium (9.62). In the determination of K'_{1b} the hydrogen ion activity was assumed to be related to psH (the Sørensen pH) by psH $= -\log_{10} a_H$ and was determined by an indicator method. The activity of the water was based on a scale in which the standard state was that of pure water. That is,

$$a_{H_2O} = \frac{p_s}{p_0} \tag{9.68}$$

where p_s and p_0 are the vapour pressures of the sample and pure water respectively. This is related to the chlorinity by

$$a_{H_2O} = 1 - 9 \cdot 69 \times 10^{-4} \, Cl\%_0$$

(Harvey, 1955), although other relationships are available (Arons and Kientzler, 1954; Robinson, 1954). Buch *et al.* defined the activity of CO_2 by

$$a_{CO_2} = \alpha_0 \cdot p_{CO_2} \tag{9.69}$$

where α_0 is the solubility coefficient of CO_2 in pure water. On this basis the activity coefficient of the dissolved CO_2 is

$$\gamma_{CO_2} = \frac{\alpha_0}{\alpha} \tag{9.70}$$

where α is the solubility coefficient for the sea water sample. Some additional comment on this activity coefficient is desirable. The chemical potential of $CO_2(aq)$ in pure water is given by

$$\mu_{CO_2(aq)} = \mu^0 + RT \ln \gamma' \cdot c_{CO_2(aq)}$$

where γ' is the activity coefficient intended to cater for deviations from Henry's Law behaviour and μ^0 is the standard chemical potential. Since, for the range of CO_2 concentrations of interest Henry's Law is obeyed, it is legitimate to set $\gamma' = 1$ and thus,

$$\mu_{CO_2(aq)} = \mu^0 + RT \ln c_{CO_2(aq)}$$

In view of equation (9.51a) and the fact that $c_{CO_2(T)} = c_{CO_2(aq)} + c_{H_2CO_3}$,

$$\mu_{CO_2(aq)} = \mu^0 + RT \ln \frac{c_{CO_2(T)}}{(1 + K_0')}$$

$$= \mu^{0\prime} + RT \ln c_{CO_2(T)}$$

$$= \mu^{0\prime} + RT \ln \alpha_0 \cdot p_{CO_2}, \tag{9.71}$$

where

$$\mu^{0\prime} = \mu^0 - RT \ln (1 + K_0')$$

and p_{CO_2} is the applied CO_2 pressure. For the uptake of CO_2 by sea water under the same applied CO_2 pressure the chemical potential will also be $\mu_{CO_2(aq)}$ since the two solutions are in thermodynamic equilibrium, and an argument similar to that immediately above gives for the sea water system

$$\mu_{CO_2(aq)} = \mu^{*\prime} + RT \ln \alpha \cdot p_{CO_2} \tag{9.72}$$

$$= \mu^{*\prime} + RT \ln c_{CO_2(T)} \tag{9.72a}$$

where $\mu^{*\prime}$, the standard chemical potential of CO_2 in sea water, will differ from that in pure water ($\mu^{0\prime}$). By equating (9.71) and (9.72) $\mu^{*\prime}$ is obtained, viz,

$$\mu^{*\prime} = \mu^{0\prime} + RT \ln \frac{\alpha_0}{\alpha}, \tag{9.73}$$

and insertion of this expression in (9.72a) gives

$$\mu_{CO_2(aq)} = \mu^{0\prime} + RT \ln \frac{\alpha_0}{\alpha} \cdot c_{CO_2(T)}. \tag{9.74}$$

Thus, α_0/α is a sort of activity coefficient, although its function is not to accommodate departures of the CO_2/solvent system from ideal behaviour, but rather to cope with effects which stem from changes in the nature of the solvent (and therefore of the Henry's Law constant) as the salinity changes. The activity coefficient has also been discussed by Spencer (1965) and Hansson (1973a).

Equation (9.64) can also be written in the form

$$K_{1b}' = CA \frac{a_H}{\alpha_0 \cdot p_{CO_2} \left(1 + \dfrac{2K_2'}{a_H}\right) a_{H_2O}}$$

by making use of the above activity relationships and the expression for $c_{HCO_3(T)}$ given in a later Section. By confining measurements to solutions of pH less than about 5, even quite rough values for K'_2 are adequate to enable K'_{1b} to be obtained via this expression and measured values of CA, pH and P_{CO_2}. Buch et al. (1932) determined K'_{1b} at twelve salinities at 20°C and at four temperatures (10, 20, 25 and 35°C) at salinities close to 34‰.

The original determinations of K'_2 by Buch et al. (1932) were in error because it was not appreciated at that time that the borate contribution to the alkalinity was so important at the higher pH values which they employed in order to obtain appreciable carbonate ionization. Later investigations using borate-free water at pH > 8.5 made use of the expression

$$K'_2 = a_H\left[\left\{CA\left(1 + \frac{\alpha}{\alpha_0} \cdot \frac{a_H}{a_{H_2O}} \cdot \frac{CA}{K'_{1b}}\right) - \Sigma CO_2\right\}\middle/(2\Sigma CO_2 - CA)\right]$$

in conjunction with measurements of pH, CA and ΣCO_2 to obtain K'_2 values for four salinities at 20°C. Values for K'_B, the first ionization constant of boric acid, were obtained by making measurements in boric acid-enriched natural sea water.

The final tabulations of K'_{1b} and K'_2 (Buch, 1945, 1951) are interpolations based on the measurements described above, but also rely on measurements of the temperature coefficients of the first and second ionization constants made by Shedlovsky and McInnes (1935) and Harned and Scholes (1941) respectively.* These tables have been widely reproduced (Buch, 1951; Harvey, 1960; Saruhashi, 1955; Barnes, 1959), but when using them it should be remembered that the constants are based on the Sørensen pH scale, and it is necessary to apply the appropriate correction (see Section 9.2.3.1 and also Bates, 1973) if the measured pH is on the NBS scale.

Determinations by Lyman (1956). Lyman criticized the determinations of the first apparent ionization constant made by Buch et al. on the grounds that the procedure employed required three separate determinations (pH, alkalinity

---

* In principle, the temperature coefficient for the thermodynamic constant will differ from that for the apparent constant. Thus, since $K'_{1b} = K_{1(TD)}[1/\gamma_{HCO_3(T)}]$, where $\gamma_{HCO_3(T)}$ is the activity coefficient of HCO_3^-, then

$$\frac{d \ln K'_{1b}}{dT} = \frac{\Delta H}{RT^2} + \frac{d \ln \gamma_{HCO_3(T)}}{dT},$$

where ΔH is the enthalpy change accompanying reaction when all reactants are in their standard states. Only if the second term on the R.H.S. of this expression is zero is it strictly correct to identify the two temperature coefficients. Since $\gamma_{HCO_3(T)}$ is a stoichiometric activity coefficient, its value is influenced not only by ion–ion interactions, but also by the degree of ion pairing in which HCO_3^- is involved. It would, therefore, be expected to be temperature dependent. In practice, the temperature coefficients of $K_{1(TD)}$ and K'_{1b} do not appear to be very different.

and P_{CO_2}) each on a separate sample. In addition, the pH determinations and, therefore, the constants themselves were based on the Sørensen pH scale. In Lyman's approach use was made of the fact that during the titration of a weak dibasic acid, the pH is a function of the two ionization constants; these can be obtained by suitable interpretation of the pH-titration curve (van Slyke, 1922; Britton, 1956). His experiments were designed to allow correction to be made for the boric acid in sea water. Three titrations were performed on each water sample, the pH being followed by means of a glass electrode calibrated at pH 4·0:

(i) titration of the sample with HCl to an endpoint slightly beyond that corresponding to the stoichiometric equivalent of a solution of carbon dioxide in the presence of a mixture of salts (pH *ca.* 4),

(ii) the solution resulting from (i) was freed from CO_2 and was titrated with CO_2-free NaOH to a pH of about 10,

(iii) the solution resulting from titration (ii) was treated with sufficient HCl to neutralize the excess NaOH (equivalent approximately to the amount of NaOH used in (ii)), mannitol was added, and titration (ii) was then repeated.

From titrations (ii) and (iii) the total boron content and the apparent ionization constant of boric acid could be determined. In addition, the results of these titration allowed the evaluation of the corrections which needed to be applied to titration (i) before the first and second ionization constants of carbonic acid could be calculated. The results were interpreted in terms of the definition of the first apparent constant given by (9.58), viz,

$$K'_{1l} = \frac{a_H \cdot c_{HCO_3(T)}}{c_{CO_2(T)}}$$

and of the second apparent constant of the same form as that used by Buch *et al.*, viz,

$$K'_2 = \frac{a_H \cdot c_{CO_3(T)}}{c_{HCO_3(T)}}$$

The first apparent ionization constant of boric acid was defined by

$$K'_B = \frac{a_H \cdot c_{B(OH)_4}}{c_{B(OH)_3}}$$

Throughout the investigation the NBS pH scale was used. Tabulations of the first and second apparent ionization constants of carbonic acid and the first apparent ionization constant of boric acid given by Lyman are shown in Tables A9.3 to A9.5 of the Appendix to this chapter. Lyman also gave values for the second apparent ionization constant of boric acid (K'_{2B}).

C

However, Ingri (1963) was unable to confirm these estimates, and it has been suggested that here Lyman's work may have been in error because of a large sodium error in his glass electrode (Hansson, 1972).

There are a number of complaints in the literature about poor agreement between the Buch and the Lyman constants (Takahashi *et al.*, 1970; Hansson, 1972). Takahashi *et al.* have noted that the second apparent constants determined by Buch are *ca.* 20% greater than those determined by Lyman and that observations made on the first GEOSECS intercalibration cruise can be made internally consistent by assuming second apparent constants about 30% greater than those found by Lyman. Berner and Wilde (1972) have questioned the values given by Lyman for the second apparent constant.

Edmond and Gieskes (1970) have made a detailed comparison of the basic data for the apparent ionization constants of carbonic and boric acid obtained by Lyman with those obtained by Buch *et al.*, the results of the latter workers being corrected for the different definitions of the first apparent constant and for the different pH scale used. Additionally, isolated measurements for the apparent constants (Distèche and Distèche, 1967), values for the thermodynamic constants (Harned and Bonner, 1945; Harned and Scholes, 1941; Owen and King, 1943) and values for the first apparent constant in NaCl solution (Harned and Bonner, 1945) were incorporated in their analysis which is summarized in Figs. 9.3 to 9.5 based on those given in their paper. At 20°C there is general agreement between the K'_1 values obtained after adjustment of the K'_{1b} measurements made by Buch *et al.* and the K'_1 values found by Lyman, although the results of the latter show more scatter.

The temperature dependence of Buch's results is closely similar to that observed by Harned and Bonner (1945) for $K_{1(TD)}$ but somewhat dissimilar from that found by Lyman. Adopting the temperature coefficient given by Harned and Bonner's work, Edmond and Gieskes proposed that the functional relationship for K'_{1l} is

$$pK'_1 = 3404.71/T + 0.032786 \times T - 14.7122 - 0.19178 \times (Cl‰)^{\frac{1}{3}} \quad (9.75)$$

where T is the temperature in K, and that the corresponding relationship for K'_2 is

$$pK'_2 = 2902.39/T + 0.02379 \times T - 6.4710 - 0.4693 \times (Cl‰)^{\frac{1}{3}} \quad (9.76)$$

which is based on the temperature dependence for $K_{2(TD)}$ found by Harned and Scholes (1941). Equation (9.76) is stated to agree with Lyman's pK'_2 values to within 0.04 pK'_2 units. The general agreement between the pH corrected pK'_B values given by Buch and those found by Lyman (Fig. 9.5) and between the temperature dependence of pK'_B and that for $pK_{B(TD)}$ (Owen

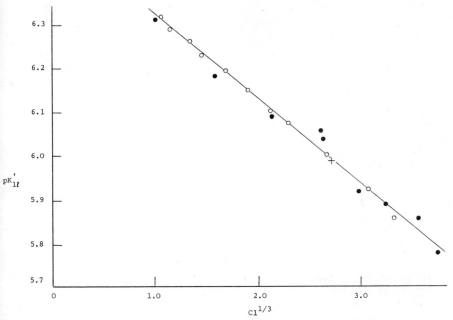

FIG. 9.3. pK'_{1l} as a function of $(Cl\%_0)^{1/3}$ at 20°C. ●, Lyman (1956); ○, Buch *et al.*, (1932); +, Distèche and Distèche (1967). The data based on the work of Buch *et al.* have been adjusted to make them compatible with those of Lyman (after Edmond and Gieskes, 1970).

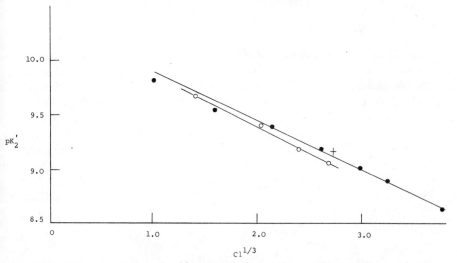

FIG. 9.4. pK'_2 as a function of $(Cl\%_0)^{1/3}$ at 20°C. ●, Lyman (1956); ○, Buch (1938); +, Distèche and Diestèche (1967). Buch's data adjusted (after Edmond and Gieskes, 1970).

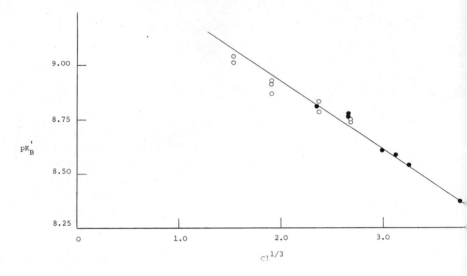

FIG. 9.5. pK'_B as a function of $(Cl‰)^{1/3}$ at 20°C. ●, Lyman (1956); ○, Buch (1938) (after Edmond and Gieskes, 1970).

and King, 1943) (Fig. 9.5) led Edmond and Gieskes to propose for the functional relationship for pK'_B

$$pK'_B = 2291 \cdot 90/T + 0 \cdot 01756 \times T - 3 \cdot 3850 - 0 \cdot 32051 \times (Cl‰)^{\frac{1}{3}}. \qquad (9.77)$$

Determinations by Mehrbach et al. (1973). Mehrbach et al. (1973) have recently made careful redeterminations of K'_{1l} and K'_2 for the temperature range 2–35°C and the salinity range 19–34‰. In their method, filtered sea water (adjusted by dilution or evaporation to obtain the desired salinity) was made alkalinity free by the addition of HCl followed by a purging with CO_2-free air, and the pH was adjusted by the addition of NaOH to bring it close to the value at which the buffer capacity of the carbonate system is a minimum. This solution was titrated with sodium bicarbonate solution, the pH being monitored by means of a calibrated glass electrode system. The pH quickly converged to a value pH^0, and from this value the product $K'_{1l} \cdot K'_2$ could be obtained. Thus, from equation (9.85) of Section 9.2.3.5,

$$R = \frac{\Sigma CO_2}{CA} = \frac{a_H^2 + K'_{1l} \cdot a_H + K'_{1l} \cdot K'_2}{K'_{1l} \cdot a_H + 2K'_2 \cdot K'_{1l}}$$

which, on re-arrangement, gives

$$K'_{1l} \cdot K'_2 = \frac{a_H^2 + K'_{1l} \cdot a_H(1 - R)}{2R - 1}$$

$$= \frac{10^{-2pH^0} + K'_{1l} \cdot 10^{-pH^0} \cdot (1 - R)}{2R - 1}. \tag{9.78}$$

Since the solution was adjusted almost to the point of pH convergence before the addition of the bicarbonate solution, the value of R was virtually unity (that of pure bicarbonate).

K'_{1l} values were obtained by a method based on a modified Gran titration of sea water (Gran, 1952; Dyrssen and Sillén, 1967; Edmond, 1970), and from these and the product $K'_{1l} \cdot K'_2$ values for K'_2 could be calculated. Tabulations of K'_{1l} and K'_2 were prepared by means of an interpolation procedure which made use of a non-linear least squares equation which embodied terms of several functions of temperature and salinity. The recommended values are given in Tables A9.9 and A9.10 of the Appendix to this Chapter. Values for K'_{1l} and K'_2 can be obtained from

$$pK'_{1l} = -13 \cdot 7201 + 0 \cdot 031334 \times T + 3235 \cdot 76/T + 1 \cdot 300 \times 10^{-5}$$
$$\times S \times T - 0 \cdot 1032 \times S^{\frac{1}{2}} \tag{9.79}$$

$$pK'_2 = 5371 \cdot 9645 + 1 \cdot 671221 \times T + 0 \cdot 22913 \times S + 18 \cdot 3802$$
$$\times \log_{10} S - 128375 \cdot 28/T - 2194 \cdot 3055 \times \log_{10} T - 8 \cdot 0944$$
$$\times 10^{-4} \times S \times T - 5617 \cdot 11 \times \log_{10} S/T + 2 \cdot 136 \times S/T \tag{9.80}$$

where S is the salinity (‰) and T is the temperature in degrees Kelvin.

Determinations by Hansson (1973a, c, d). The determinations described by Hansson (1973) were probably more comprehensive than those previously attempted, and included not only refined measurements of the ionization constants of carbonic and boric acids, but also the establishment of a pH scale and a set of reference buffers more suited for sea water work than are the normal low ionic strength buffer systems. In this approach attempts were made to eliminate, or at least very much reduce, uncertainties arising from liquid junction potentials during both the determination of the constants and in their application to sea water examination. In order to do this, some changes from previous procedures were made. The more commonly used dilute solution activity scale was abandoned in favour of a scale determined in such a way that for a minor constituent X ($= H^+$, OH^-, HCO_3^-, CO_3^{2-}, CO_2) the activity coefficient, defined by $\gamma_X = a_X/c_{X(T)}$, tends to unity as $c_{X(T)}$ tends to zero in sea water of a given salinity. On this convention, for the likely concentration ranges of these minor constituents in sea water, the high ionic strength of the medium ensures that the numerical value of the activity, a_X, is virtually the same as that of the concentration, c_X. It will be appreciated that $c_{X(T)}$ is, of course, the total concentration of the species X and includes not only the solvated forms of it, but also the amounts bound as ion pairs. Thus, for example, $c_{H(T)} = c_{H(f)} + c_{HSO_4}$ and $c_{CO_3(T)} = c_{CO_3(f)} + c_{NaCO_3} + c_{MgCO_3} + c_{CaCO_3}$, where the subscript (f) indicates the "true" concentration of the species. The first and second constants were defined by (9.61) and (9.67) respectively, viz,

$$K'_{1h} = \frac{c_{H(T)} c_{HCO_3(T)}}{c_{CO_2(T)}}$$

and

$$K'_{2h} = \frac{c_{H(T)} c_{CO_3(T)}}{c_{HCO_3(T)}}$$

and the experimental method for their determinations was essentially the potentiometric titration of solutions containing carbonate and bicarbonate dissolved in synthetic sea water. Numerical analysis of the EMF-titration results enabled values to be obtained for K'_{1h} and K'_{2h} which were consistent with the primary data. Synthetic sea water was used, the most significant differences from natural sea water being the replacement of K^+ by Na^+ and the omission of the minor constituents Sr, Br, F and B.

Increments of HCl were added from a syringe microburette to the solution of volume *ca.* 250 cm^3 contained in a closed vessel so designed that the solution occupied some 99·5% of the total space. This low ratio of gas/liquid volumes ensured minimal errors arising from the escape of CO_2 generated during the titration. The total volume of HCl added during the titration increased the solution volume by only *ca.* 0·2%. The titration was monitored by means of the cell

$$Ag \,|\, AgCl \,|\, \text{sea water saturated} \,|\, \text{sea water} \,|\, \text{Glass electrode with AgCl} \,|\, Ag$$
$$\text{with AgCl}$$

the glass electrodes being carefully selected for their reliability of performance. The salinity of the sea water used in the Ag | AgCl half of the cell was the same as that in the solution thereby reducing junction potential uncertainties.

Reagent solutions were checked for protolytic impurities by carrying out the following successive potentiometric titrations on the carbonate-free solutions,

(a) addition of 0·2 M NaOH until $c_{OH} = 2 \times 10^{-4}$ M
(b) back titration with 0·2 M HCl until $c_H = 2 \times 10^{-4}$ M.

Examination using computer analysis indicated the presence of a trace protolytic impurity. Neither its identity nor its origin were established, although quantitatively it was equivalent to a weak acid, HX, of concentration $2 \times 10^{-5}\,mol\,kg^{-1}$ and of K_{HX} of *ca.* 10^{-5}. These two titrations also enabled values to be obtained for K_W. The importance of the determination of these quantities (c_{HX}, K_{HX} and K_W) arose from the use made of them in the computer analysis of the results of the main carbonate titration. The observed EMF during the titration is a function of many factors which include not only the two ionization constants, K'_{1h} and K'_{2h}, but also c_{HX}, K_{HX} and K_W. The programme LETAGROP (Sillén, 1963) sought suitable values for K'_{1h} and K'_{2h} by adjusting parameters relevant to the titration so as to obtain the best fit between the experimental and calculated curves. The programme operation was facilitated if reliable values could be assigned to some of the parameters at the time of the input.

Values for K'_{1h} and K'_{2h} were determined for the salinities 20, 25, 30 and 35 and 40‰ and at temperatures of 5, 10, 15, 20, 25 and 30°C. An interpolation procedure made use of the linear pK'_{1h} and pK'_{2h} versus $1/T$ K plots and the assumption that for each ionization, the standard enthalpy change was salinity independent. These interpolated values are given in Tables A9.6 and A9.7 of the Appendix to this Chapter.

It should be emphasized that when these constants are used to estimate the components of the carbonate system as outlined in Section 9.2.3.5 below, it is imperative that the pH measurements used in the determinations are based on a scale consistent with that used for the definition and determination of the constants. As with the Buch and Lyman constants, their application should be restricted to solutions having relative ionic compositions similar to those of the media in which they were determined.

The ionization of boric acid was also investigated by Hansson by potentiometric titration between a sodium hydroxide solution and a solution of boric acid in synthetic sea water. The reaction was followed using a cell again designed to minimize junction potential uncertainties, and the EMF titration results were processed using the LETAGROP (Sillén, 1963; Brauner *et al.*, 1969) programme so as to obtain ionization constants consistent with the experimental data. Measurements were made at several temperatures and salinities and used to obtain the interpolated values given in Table A9.8 of the Appendix to this Chapter. It should be noted that although there is evidence (Ingri, 1963) that in boric acid containing solutions the species $B(OH)_3$, $B(OH)_4^-$, $B_2O_3(OH)_4^-$, $B_2O_3(OH)_5^{2-}$, $B_4O_5(OH)_4^{2-}$ and $B_5O_6(OH)_4^-$ can exist. Hansson's work suggests that, in sea water, only negligible polynuclear formation occurs, the only boron species of significance being $B(OH)_3$ and $B(OH)_4^-$.

Hansson (1973a, c) has also examined the ionization of carbonic and boric acids in sodium chloride solution in order to give a comparison with the sea water values and thereby enable ionic medium effects such as ion pairing to be quantified (Section 9.2.3.7).

Direct comparison of these constants with the Buch–Lyman–Mehrbach values is not straightforward because of the different ways in which they are defined, and the relative merits of the four sets is probably best judged from their performance in practice. Neither the Mehrbach nor the Hansson constants have been available for sufficient time for them to receive the test of routine usage. Nevertheless, general considerations—the known divergence between estimates made on the basis of the Buch and the Lyman constants (see e.g. Takahashi *et al.*, 1970) and the attention paid during the establishment of the Hansson constants and the associated pH scale to the minimization of junction potential uncertainties—are strong arguments in favour of adopting them. Comparative field tests of the four sets of constants is desirable. A comparison made by Gieskes (1974) of the $c_{CO_2(T)}$ and the $c_{CO_3(T)}$ estimates given by the Hansson and earlier constants showed the agreement to be sometimes unsatisfactory.

Mehrbach *et al.* (1973) have attempted to compare values of the first and second apparent ionization constants of carbonic acid obtained by different workers by adjusting them so as to take account of the different pH scales used.* When this is done (Table 9.8), it is seen that the Mehrbach values for the first apparent constant lie below those based on the data given by Buch. Relative to the Lyman values, those of Mehrbach *et al.* are lower at low temperatures and higher at high temperatures. The K_2' values reported by Mehrbach *et al.* are somewhat lower than those given by other groups of workers.

An indirect comparison of the various reported constants can be made via the ratio K_1'/K_2' $(= K_r')$, that is, the stoichiometric equilibrium constant for the process

$$CO_2(T) + H_2O + CO_3^{2-}(T) \rightleftharpoons 2HCO_3^-(T)$$

This is an important reaction because of its role in controlling the carbonate ion concentration (and, consequently, the degree of carbonate saturation) as carbon dioxide is taken up from the atmosphere by the sea (Section 9.2.4.5), and reliable values of K_r' as a function of both temperature and salinity are desirable. However, in the present context the importance of K_r' is that it allows some comparison to be made between different workers' sets of apparent ionization constants in a way which is not dependent on the pH scale used in the determinations of the individual constants. Fig. 9.6 has been

---

* Note the small difference between the pH scale used by Lyman and that used by Mehrbach *et al.* (see Mehrbach *et al.*, (1973).The Hansson constants can be adjusted to make them comparable with those based on the NBS pH scale by multiplying them by the activity coefficient of the hydrogen ion (on the dilute solution activity scale).

TABLE 9.8

Comparison of reported values for the first and second ionization constants of carbonic acid (Mehrbach et al., 1973)

K'_1 values

Source	Salinity (‰)	Temp (°C)				
		5	10	20	25	35
Lyman (1956)	35·0	0·776 (0·760)*	0·832 (0·842)*		1·00 (1·085)*	1·07 (1·158)*
Buch (1932)†	21·81 33·96 34·49		0·832 (0·788)	0·839 (0·786) 0·995 (0·936)	1·064 (0·988)	1·15 (1·069)
Hansson (1973a, d)‡	35·0	0·693 (0·716)	0·760 (0·797)	0·897 (0·941)	0·962 (0·999)	

K'_2 values

Source	Salinity (‰)	Temp (°C)				
		5	10	20	25	35
Lyman (1956)	35·0	4·68 (4·48)*	5·37 (5·16)*		8·13 (8·12)*	11·48 (10·12)*
Buch (1938)†	25·13 34·96			6·63 (4·99) 9·06 (6·58)		
Hansson (1973a, d)‡	35·0	4·18 (4·04)	4·97 (4·72)	6·79 (6·59)	7·82 (7·68)	

* Mehrbach et al. values corrected to pH scale used by Lyman (1956).
† Values recalculated by Edmond and Gieskes (1970).
‡ Values adjusted to NBS pH scale.

constructed from K_1' and K_2' values quoted in the tables given by Lyman (1956), Hansson (1973a, d) and Mehrbach *et al.* (1973), but a number of cautionary remarks are necessary before an attempt is made to interpret it. The tabulated values (see e.g. the Appendix to this Chapter) are based on empirical interpolation equations. This exerts a smoothing effect on the data and allows no account to be taken of the errors associated with the use of a particular interpolation equation. Lyman's pK' values were cited to only two decimal places; the values given by Hansson were cited to three. For this reason the plotted values based on Lyman's tabulations show more scatter than those based on either those of Hansson or of Mehrbach *et al.*

Although a fuller statistical analysis would make use of the original basic data of the various groups of workers or would take into account the errors associated with the interpolation equations used in the preparation of their tabulations, the tentative summary given by Fig. 9.6 indicates a less than satisfactory agreement between the various K_r' estimates

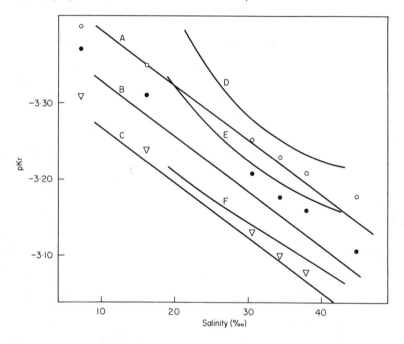

Fig. 9.6. pK_r' as a function of salinity. Based on the recommended tabulations for the first and second apparent ionization constants of carbonic acid given by Lyman (1956), Hansson (1973a) and Mehrbach *et al.* (1973). The linear plots correspond to Hansson's data (A, 5°C; B, 15°C; C, 25°C), the curved lines to the data of Mehrbach *et al.* (D, 5°C; E, 15°C; F, 25°C) and the symbols to the data of Lyman (○, 5°C; ●, 15°C; ▽, 25°C).

Effect of ionic composition on apparent constants

Because the ionic effect of the medium influences the magnitude of apparent ionization constants not only through the ionic strength (salinity), but also through the ionic composition, circumstances which give rise to marked changes in the concentration ratios of pair-forming ions (see Section 9.2.3.7 and also Garrels *et al.*, 1961; Pytkowicz, 1969b) may materially alter the values to be assigned to the apparent constants. The first and second apparent constants of carbonic acid can be written (see e.g. Ben-Yaakov and Gold-haber, 1973)

$$K'_{1l} = K_{1(TD)} \cdot \frac{c_{HCO_3(T)} \cdot c_{CO_2(f)} \cdot \gamma_{CO_2(f)}}{c_{CO_2(T)} \cdot c_{HCO_3(f)} \cdot \gamma_{HCO_3(f)}}$$

and

$$K'_2 = K_{2(TD)} \frac{c_{CO_3(T)} \cdot c_{HCO_3(f)} \cdot \gamma_{HCO_3(f)}}{c_{HCO_3(T)} \cdot c_{CO_3(f)} \cdot \gamma_{CO_3(f)}}$$

where $c_{i(f)}$ indicates the free (unassociated) concentration of i and $c_{i(T)}$ is the total (stoichiometric) concentration. Free ionic activity coefficients (as opposed to stoichiometric activity coefficients) are denoted by $\gamma_{i(f)}$. Present evidence shows that the values to be assigned to the appropriate $\gamma_{i(f)}$ terms, although depending on the identity of i, change only slightly with change of ionic strength at ionic strengths close to that of sea water. However, changes in $c_{HCO_3(f)}$ and $c_{CO_3(f)}$ (which will accompany alteration in the amount of ion pair formation involving bicarbonate and carbonate consequent on an alteration of the relative ionic proportions) will cause changes in K'_{1l} and K'_2. Calculations have been made by Ben-Yaakov and Goldhaber (1973) of the sensitivity of K'_{1l} and K'_2 to small changes in various ionic concentrations. Their approach made use of a sea water model and calculation procedure similar to that described by Garrels and Thompson (1962) and Berner (1971), and was applied to water of 19‰ Cl and 25°C. Their conclusions were expressed in terms of a sensitivity parameter, viz,

$$s_{K'} = \frac{\Delta K'/K'}{\Delta c_i/c_i}$$

where K' is the appropriate apparent constant and c_i is the concentration of the species whose concentration effect is being investigated. Sensitivity parameters for the carbonic acid ionization constants and for the apparent solubility product of $CaCO_3$ (Section 9.2.4) are given in Table 9.9.

As expected from its known tendency to form ion pairs in sea water, magnesium has the largest sensitivity parameter, whereas potassium, which

TABLE 9.9

Sensitivity of K'_{1l}, K'_2, K'_{calc} to a change in sea water composition

$$s_{K'} = \frac{\Delta K'/K'}{\Delta c_i/c_i}$$

Ion (i)	$s_{K'} \times 10^3$		
	K'_{calc}	K'_{1l}	K'_2
K^+	$-0 \cdot 05$	$0 \cdot 11$	$0 \cdot 32$
Na^+	$180 \cdot 67$	$88 \cdot 42$	$111 \cdot 06$
Mg^{2+}	$640 \cdot 54$	$155 \cdot 05$	$442 \cdot 24$
Ca^{2+}	$70 \cdot 90$	$33 \cdot 73$	$38 \cdot 85$
SO_4^{2-}	$-9 \cdot 15$	$-18 \cdot 93$	$-53 \cdot 98$
HCO_3^-	$2 \cdot 08$	$-0 \cdot 97$	$-2 \cdot 75$
CO_3^{2-}	$-0 \cdot 34$	$-0 \cdot 56$	$-0 \cdot 17$
Cl^-	0	0	0

shows a much smaller tendency to pair with other ions has only a small $s_{K'}$ value. Those for anions are also small.

The ionic composition of most open ocean waters is sufficiently constant to justify concluding that during their examination no correction need be applied to the conventionally determined apparent constants. However, there are a number of systems to which it would not, strictly, be legitimate to apply directly the apparent constants. These include waters which have suffered enrichment or impoverishment of particular ions due to intimate contact with dissolving, precipitating or adsorbing solid phases (e.g. pore solutions) and also hypersaline solutions. In addition, a number of laboratory systems in which, for reasons of experimental convenience, the concentration of a particular ion is increased far beyond that found in natural sea water should not be examined in terms of the unadjusted constants. However, over the range of pH values of ocean water the apparent constants are independent of the hydrogen ion concentration (Pytkowicz *et al.*, 1974).

9.2.3.5. *Evaluation of the components of the carbonate system*

The primary reason for requiring reliable compilations of the carbonic and boric acid ionization constants is that they enable the evaluation of $c_{HCO_3(T)}$, $c_{CO_3(T)}$ and $c_{CO_2(T)}$ (and also, by summation, of ΣCO_2) to be made on the basis of only a few shipboard measurements. Suitable combinations of equations (9.37) and (9.38) with the first and second ionization constants gives expressions for $c_{HCO_3(T)}$, $c_{CO_3(T)}$ and $c_{CO_2(T)}$ as functions of selected observables. The four parameters accessible to measurement are the carbonate alkalinity (via the alkalinity and a borate correction—see Section 9.2.3.2), the pH, ΣCO_2 and P_{CO_2}. In order to characterize a sample at least two of these must be determined. Although there are six ways in which pairs can be selected from the four parameters, the early investigators usually made use only of the pH-alkalinity combination. However, the later development of reliable and convenient instrumental methods for ΣCO_2 and P_{CO_2} analyses has broadened the experimental basis for carbonate investigation, and the use of these observables as parameters is now commonplace.

In the event of three observable parameters being available, the calculation of $c_{HCO_3(T)}$, $c_{CO_3(T)}$ and $c_{CO_2(T)}$ can be made without need for a knowledge of both ionization constants; if the selected parameters are the alkalinity, ΣCO_2 and P_{CO_2}, neither ionization constant need be known. Park (1969) has discussed this topic in some detail and has shown that by combination of two or more of the four observables alkalinity, pH, ΣCO_2 and P_{CO_2} it is possible to derive six equations for $c_{CO_2(T)}$, ten for $c_{HCO_3(T)}$ and eleven for $c_{CO_3(T)}$. The following account is based on that by Park. To simplify the presentation, the Lyman constants are used throughout. The corresponding expressions using the Buch constants can be obtained by using expression (9.64a).

The basic equations may be summarized

$$K'_{11} = \frac{a_H \cdot c_{HCO_3(T)}}{c_{CO_2(T)}} \tag{9.58}$$

$$K'_2 = \frac{a_H \cdot c_{CO_3(T)}}{c_{HCO_3(T)}} \tag{9.66}$$

$$CA = c_{HCO_3(T)} + 2c_{CO_3(T)} \tag{9.37}$$

$$c_{CO_2(T)} = P_{CO_2} \cdot \alpha \tag{9.49}$$

and

$$\Sigma CO_2 = c_{HCO_3(T)} + c_{CO_3(T)} + c_{CO_2(T)} \tag{9.38}$$

(a) *Alkalinity and* pH *as observables*
From (9.66) and (9.37)

$$c_{HCO_3(T)} = CA \frac{1}{1 + 2K'_2/a_H} \tag{9.81}$$

From (9.81) and (9.37)

$$c_{CO_3(T)} = CA \frac{K'_2/a_H}{1 + 2K'_2/a_H} \tag{9.82}$$

From (9.81) and (9.58)

$$c_{CO_2(T)} = CA \frac{a_H}{K'_{11}[1 + 2K'_2/a_H]} \tag{9.83}$$

From (9.83) and (9.49)

$$P_{CO_2} = CA \frac{a_H}{K'_{11} \cdot \alpha[1 + 2K'_2/a_H]} \tag{9.84}$$

Combination of (9.81) to (9.84) gives

$$\Sigma CO_2 = CA \frac{1 + K'_2/a_H + a_H/K'_{11}}{1 + 2K'_2/a_H} \tag{9.85}$$

(b) pH *and* ΣCO_2 *as observables*
From (9.58), (9.66) and (9.38),

$$c_{HCO_3(T)} = \Sigma CO_2 \frac{K'_{11} a_H}{a_H^2 + K'_{11} \cdot a_H + K'_{11} \cdot K'_2} \tag{9.36}$$

$$c_{CO_3(T)} = \Sigma CO_2 \frac{K'_{11} \cdot K'_2}{a_H^2 + K'_{11} \cdot a_H + K'_{11} \cdot K'_2} \tag{9.87}$$

$$c_{CO_2(T)} = \Sigma CO_2 \frac{a_H^2}{a_H^2 + K'_{11} \cdot a_H + K'_{11} \cdot K'_2} \tag{9.88}$$

and

$$P_{CO_2} = \frac{\Sigma CO_2}{\alpha} \frac{a_H^2}{a_H^2 + K'_{11} \cdot a_H + K'_{11} \cdot K'_2} \tag{9.89}$$

(c) pH *and* P_{CO_2} *as observables*
From the basic equations it is easily shown that

$$c_{HCO_3(T)} = \frac{P_{CO_2} K'_{11} \cdot \alpha}{a_H} \tag{9.90}$$

$$c_{CO_3(T)} = \frac{P_{CO_2} \cdot K'_{11} \cdot K'_2 \cdot \alpha}{a_H^2} \tag{9.91}$$

$$c_{CO_2(T)} = P_{CO_2} \cdot \alpha \tag{9.49}$$

$$\Sigma CO_2 = P_{CO_2} \cdot \alpha \left[1 + \frac{K'_{11}}{a_H}(1 + K'_2/a_H) \right] \tag{9.92}$$

(d) ΣCO_2 and alkalinity as observables

By combination of the basic equations

$$c_{HCO_3(T)} = \frac{\Sigma CO_2 . K_r' - Z}{K_r' - 4} \tag{9.93}$$

$$c_{CO_3(T)} = \frac{CA . K_r - \Sigma CO_2 . K_r' - 4CA + Z}{2(K_r' - 4)} \tag{9.94}$$

and

$$c_{CO_2(T)} = \Sigma CO_2 - CA + \frac{CA . K_r' - CO_2 . K_r' - 4CA + Z}{2(K_r' - 4)} \tag{9.95}$$

where

$$Z = [(4CA + \Sigma CO_2 . K_r - CA . K_r')^2 + 4(K_r' - 4) CA^2]^{\frac{1}{2}} \tag{9.96}$$

where K_r is the ratio K_{11}'/K_2'.

(e) Alkalinity and P_{CO_2} as observables

From equations (9.58), (9.66), (9.37) and (9.49),

$$c_{HCO_3(T)} = \frac{-K_r \alpha . P_{CO_2} + [(8CA + K_r' . \alpha . P_{CO_2}) K_r' . \alpha . P_{CO_2}]^{\frac{1}{2}}}{4} \tag{9.97}$$

$$c_{CO_3(T)} = \frac{4CA + K_r . \alpha . P_{CO_2} - [(8CA + K_r' . \alpha . P_{CO_2}) K_r' . \alpha . P_{CO_2}]^{\frac{1}{2}}}{8} \tag{9.98}$$

and

$$c_{CO_2(T)} = P_{CO_2} . \alpha \tag{9.49}$$

where K_r' is the ratio K_{11}'/K_2'.

(f) ΣCO_2 and P_{CO_2} as observables

From the basic equations

$$c_{HCO_3(T)} = \frac{-K_r' . \alpha . P_{CO_2} + Y}{2} \tag{9.99}$$

$$c_{CO_3(T)} = \Sigma CO_2 - P_{CO_2} . \alpha + \frac{K_r' . \alpha . P_{CO_2} - Y}{2} \tag{9.100}$$

and

$$c_{CO_2(T)} = P_{CO_2} . \alpha \tag{9.49}$$

where

$$Y = [K'_r.\alpha.P_{CO_2}(K'_r.\alpha.P_{CO_2} - 4\alpha.P_{CO_2} + 4CO_2)]^{\frac{1}{2}}$$

and K'_r is the ratio K'_{1l}/K'_2.

When three of the observable parameters are available, expressions under (g) to (j) below may be used.

(g) pH, *alkalinity and* ΣCO_2 *as observables*

Manipulation of the basic equations gives

$$c_{HCO_3(T)} = \frac{(2\Sigma CO_2 - CA)\,K'_{1l}}{2a_H + K'_{1l}} = \frac{CA.a_H}{a_H + 2K'_2}, \tag{9.101}$$

$$c_{CO_3(T)} = \frac{CA.a_H + CA.K'_{1l} - \Sigma CO_2.K'_{1l}}{2a_H + K'_{1l}} = \frac{CA.K'_2}{a_H + 2K'_2} \tag{9.102}$$

and

$$c_{CO_2(T)} = \frac{(2\Sigma CO_2 - CA)}{2a_H + K'_{1l}}a_H = \Sigma CO_2 - \frac{CA\,(a_H + K'_2)}{a_H + 2K'_2} \tag{9.103}$$

each being alternatively expressed in terms of K'_{1l} or K'_2.

(h) pH, *alkalinity and* P_{CO_2} *as observables*

The equations which can be derived are

$$c_{HCO_3(T)} = \frac{K'_{1l}.\alpha.P_{CO_2}}{a_H}, \tag{9.104}$$

$$c_{CO_3(T)} = \frac{CA.a_H - K'_{1l}.\alpha.P_{CO_2}}{2a_H} \tag{9.105}$$

and

$$c_{CO_2(T)} = \alpha.P_{CO_2}. \tag{9.49}$$

(i) pH, ΣCO_2 *and* P_{CO_2} *as observables*

As under (g) above, there are alternative expressions according to whether K'_{1l} or K'_2 is to be used.

$$c_{HCO_3(T)} = \frac{K'_{1l}.\alpha.P_{CO_2}}{a_H} = \frac{(\Sigma CO_2 - \alpha.P_{CO_2})\,a_H}{a_H + K'_2}, \tag{9.106}$$

$$c_{CO_3(T)} = \frac{\Sigma CO_2.a_H - (a_H + K'_{1l}).\alpha.P_{CO_2}}{a^H} = \frac{(\Sigma CO_2 - \alpha.P_{CO_2})K'_2}{a_H + K'_2} \tag{9.107}$$

and

$$c_{CO_2(T)} = \alpha . P_{CO_2}. \tag{9.49}$$

(j) *Alkalinity, ΣCO_2 and P_{CO_2} as observables*
From (9.37), (9.38) and (9.49), equations which do not embody constants K'_{1l} and K'_2 can be derived, viz,

$$c_{HCO_3(T)} = 2\Sigma CO_2 - CA - 2\alpha . P_{CO_2} \tag{9.108}$$

$$c_{CO_3(T)} = CA - \Sigma CO_2 + \alpha . P_{CO_2} \tag{9.109}$$

and

$$c_{CO_2(T)} = \alpha . P_{CO_2}. \tag{9.49}$$

Not all of the equations listed under (a) to (j) above have been exploited. Of these groups, those under (a) enjoy a sort of traditional distinction, and those under (b) were originally advocated by Buch (1951) for use with waters likely to be contaminated with sufficient amounts of humus, organic matter or protolytic impurity to make the usual method of carbonate alkalinity estimation (subtraction of the borate contribution from the observed alkalinity) unreliable. However, the development of convenient and improved instrumental methods for ΣCO_2 estimation have made it increasingly attractive to use this approach (Saruhashi, 1955; Park, 1965). It might be noted that Saruhashi has given tables from which the proportions of HCO_3^- (T), CO_3^{2-} (T) and CO_2(T) can be read off for given values of temperature, salinity and ΣCO_2. In his discussion of the selection of suitable expressions to use, Park (1969) has drawn attention to the attractiveness of group (c) because of the relatively large range of P_{CO_2} found in sea water. Groups (d) and (e), in common with (a), are useful only when reliable values for the carbonate alkalinity can be obtained. Although (a) has been much used, (d) and (e) have received little attention. Infrared CO_2 analysis of water samples lends itself well to the application of group (f), and these equations have been used by Li (1967). Groups (g) to (j), although demanding more invested time since the determination of three parameters is necessary, offer the advantage of independence of one or more of the carbonate ionization constants.

The recent papers by Almgren *et al.* (1974a, b, c) should be consulted for details of their precise and sophisticated instrumental procedures for processing primary data (obtained from alkalinity and total carbonate determinations by hydrochloric acid titration). It might be noted that the work of this group is an extension of that of Dyrssen and Sillén (1967), Hansson (1973a, b, c, d) and Hansson and Jagner (1973) and is based throughout on the $mol\,kg^{-1}$

of sea water scale which this group advocate for the expression of concentration.

Simpson and Broecker (1973) have described a method for the determination of the carbonate concentration in sea water which is based on knowledge of ΣCO_2 and three P_{CO_2} determinations. The first of these measures the P_{CO_2} of gas in equilibrium with the sample solution; the second two are made after the addition of a known amount of strong base and then of boric acid. Using this information it is possible to calculate $c_{CO_3(T)}$ without the need either to make EMF (pH) measurements or to know the ionization constants of carbonic acid. Electrode methods for the estimation of c_{CO_2} have been described (see Chapter 20 and Herman and Rechnitz, 1974).

It should be noted that although it is often the practice to record the pH of sea water, this quantity is rarely of particular interest in itself, the main purpose of its determination usually being to facilitate the calculation of quantities of more direct chemical, environmental and ecological interest through the relationships given above. Considerable care is necessary when pH is used as an observable parameter since small errors in its determination result in large errors in the corresponding estimate of, say, P_{CO_2}. There are in the literature several comments on poor agreement observed between estimates of P_{CO_2} obtained directly and those obtained via the pH and alkalinity method, this possibly reflecting not only difficulties associated with the making of meaningful pH measurements but also uncertainties associated with the available carbonate system ionization constants. For this reason, direct determinations of P_{CO_2} are desirable. However, if the temperature at the measurement point differs from that of the water *in situ*, correction of the indicated P_{CO_2} is necessary. The need to know the temperature coefficient of P_{CO_2} is stimulated not only by the necessity of making this correction, but also by interest in the way in which the P_{CO_2} of sea water changes as it warms or cools under natural circumstances (Harvey, 1955; Revelle and Suess, 1957; Kanwisher, 1960; Takahashi, 1961; Ibert and Hood, 1963; Kelley, 1970; Gordon and Jones, 1973). The carbon dioxide partial pressure increases with increase of temperature, but the magnitude of the increase depends on the constraints imposed on the system. In general, for short term temperature changes (such as those caused by diurnal solar heating or those which might arise in a sampling system) one is interested in the temperature coefficient at constant salinity, alkalinity and ΣCO_2. A change of temperature under these constraints will result in a change of pH and, therefore, it is not correct to make use of the temperature coefficient at constant pH (stated to be about 3·2 p.p.m. $(^{\circ}C)^{-1}$—Harvey, 1955).

An experimental determination of the temperature coefficient made by Kanwisher (1960) using a system in which constancy of salinity, alkalinity and ΣCO_2 were maintained indicated a value of 4.5% $(°C)^{-1}$. This is in excellent agreement with the expression

$$\frac{d(P_{CO_2})}{dT} = 4.4 \times 10^{-2}(P_{CO_2}) - 4.6 \times 10^{-6}(P_{CO_2})^2$$

which has been proposed by Gordon and Jones (1973) on the basis of a numerical analysis using Lyman's constants and the solubility coefficients used by Buch. The expression is inappreciably changed if the newer solubility data given by Murray and Riley (1971) and Li and Tsui (1971) (see also Weiss, 1974) are used. For P_{CO_2} values close to 320 p.p.m., the correction factor is about 13 p.p.m. $(°C)^{-1}$.

9.2.3.6. *Effect of pressure on the carbonate equilibria*

Some additional comments are necessary in order to account for changes in the physical chemistry of the carbonate system with increasing depth. Compression of the water at depth (*ca.* 4% in 10000 m) results in a small increase in ionic concentrations, although this has only a small influence on the overall chemistry compared with changes consequent on the pressure dependencies of the equilibrium constants. Because of these changes, the distribution in deep water of the various carbon containing species (dissolved CO_2, HCO_3^-, CO_3^{2-} and ion pairs) differs from that in an otherwise similar mass of water at surface pressure.

Although in principle the effect of pressure on an equilibrium constant can be predicted from partial molar volume data (see below), the absence until recently of detailed information on many aspects of the carbonate system (particularly ion pairing effects) and the lack of sufficiently reliable partial molar volume data has meant that reliance has been placed on direct or semi-direct determinations of pressure dependencies (Buch and Gripenberg, 1932; Distèche, 1959, 1972, 1974; Distèche and Distèche, 1965, 1967; Culberson *et al.*, 1967; Culberson and Pytkowicz, 1968; Pytkowicz, 1968).

The need for information on the pressure dependencies of the carbonic acid ionization constants was appreciated by Buch and Gripenberg (1932). They outlined a procedure whereby *in situ* values of the components of the carbonate system could be calculated from shipboard measurements made on a sample which had been brought up from depth (pressure P) and decompressed to atmospheric pressure. The decompression was assumed to be without effect on the alkalinity, the carbonate alkalinity and ΣCO_2 except insofar as the volume changed (see comments on pp. 26–27 and also Pytkowicz, 1968). The ratio $\Sigma CO_2/CA$ (denoted below by Q) was determined at atmospheric pressure and was inserted in an expression equivalent to

$$(a_H)_P^2 - (Q - 1)(K'_{1l})_P(a_H)_P - (2Q - 1)(K'_{1l})_P(K'_2)_P = 0 \qquad (9.110)$$

from which pH_P, the pH at depth, could be calculated provided the pressure dependencies of K'_{1l} and K'_2 were known. On the basis of Brander's (1932) determination of the pressure dependence of the first ionization constant of carbonic acid in distilled water, Buch and Gripenberg wrote

$$\Delta pK'_{1l} = -0.48 \times 10^{-4}\Delta z$$

where Δz is the depth in metres. Because experimental determinations of the pressure dependency of the second ionization constant were not available, the dependency used by them was based on that determined by Brander for acetic acid, the assumption being that weak acids would behave similarly. Thus, Buch and Gripenberg wrote

$$\Delta pK'_2 = -0.18 \times 10^{-4}\Delta z$$

In this way they established values for $\Delta pH = pH_P - pH_1$ as a function of depth for different surface pH values from which *in situ* estimates of the pH at depth could be obtained.

The alkalinity and carbonate alkalinity determinations made on the sample at atmospheric pressure were adjusted to the *in situ* values by multiplying by the ratio of the *in situ* to surface water densities. These adjusted values were then combined with the estimate of the pH at depth and the corresponding pressure-corrected values of the two apparent ionization constants in order to obtain the *in situ* values for the carbonate components (see Sections 9.2.3.4 and 9.2.3.5).

Pytkowicz (1968) has drawn attention to a number of unsatisfactory features associated with the Buch and Gripenberg procedure. The corrections of the two ionization constants for depth were based on Brander's measurements for solutions in distilled water at 20°C, and there was no direct knowledge of the pressure coefficients built in to the method. The method implicitly assumed that the change in partial molar volumes accompanying the first ionization in distilled water was the same as that which occurs in sea water, an assumption now known to be invalid because of the dependence of partial molar volumes on ionic strength. The equating of the second apparent ionization constant pressure dependency in sea water with that of acetic acid in distilled water is an assumption which, however necessary at the time, is clearly unsatisfactory.

Since the original work of Buch and Gripenberg, other groups of investigators have attempted to obtain more direct information on the pressure dependencies of the carbonate system parameters (Distèche, 1959; Distèche and Distèche, 1965, 1967; Culberson *et al*, 1967; Culberson and Pytkowicz,

1968; Pytkowicz, 1968). In particular, the school associated with Pytkowicz have made a comprehensive investigation of the pressure dependence of ionization constants of carbonic and boric acid in sea water. Their investigation made use of the high pressure cell referred to in Section 9.2.3.1 represented by

Ag, AgCl | Reference solution ‖ glass ‖ sea water | AgCl, Ag

as a means of the laboratory determination of the pH of the sea water at high pressures. The reference solution contained 0·01 M HCl plus sufficient NaCl to make the ionic strength the same as that of the artificial sea water being examined. The composition of the sea water solution was based on that described by Kester et al. (1967), but was modified by replacing the KBr with an equivalent amount of KCl since it is known that Br^- ions influence the performance of Ag, AgCl electrodes. For the determination of the pressure coefficients of the carbonic acid ionization constants, boric acid was omitted from the sea water solution; for the determination of the pressure dependence of the boric acid ionization the concentration of boric acid was increased to $3 \, mmol \, l^{-1}$.

The potential (E) of the above cell is given by

$$E = s[\log (a_H a_{Cl})_i - \log (a_H a_{Cl})_o] + E_{asym}$$

where the subscripts i and o refer respectively to the inner (reference) and outer (test) solutions. E_{asym} is the asymmetry potential, that is, the potential shown when the reference and test solutions are identical in composition (Section 9.2.3.1) and s is the electrode slope ($= 2\cdot303 \, RT/F$). By making the two assumptions (i) the effect of pressure on the activity coefficients of chloride ion is the same in the inner and outer compartments of the cell, and (ii) the activity coefficient of H^+ in the reference compartment is pressure-independent, the change in pH on increasing the pressure from atmospheric pressure to P atmospheres is

$$pH_1 - pH_P = \frac{1}{s}[E_1 - E_P - \Delta E_{asym}]$$

and provided that the pressure dependence of the asymmetry potential is known (Section 9.2.3.1) the pH at pressure P could be calculated.

The apparent ionization constants at pressure P (($K'_{1)})_P$ and $(K'_2)_P$) were determined by measuring the pH in pairs of sea water solutions which had been treated so as to have different carbonate alkalinities and, therefore, different a_H and Q values. The values of Q were determined at atmospheric pressure from the measured pH using Lyman's values for the apparent constants. The pair of simultaneous equations generated when these values were fed into equation (9.110) were solved to give $(K'_{1})_P$ and $(K'_2)_P$.

Values for the apparent ionization constant of boric acid at high pressures were obtained by inserting into the definition of $Q\ (= \Sigma\ CO_2/CA)$ the expression for CA (Section 9.2.3.2) viz.

$$CA = A - \frac{\Sigma B \cdot (K'_B)_P}{(a_H)_P + (K'_B)_P} \tag{9.111}$$

where A is the alkalinity. When divided by the total boron content, ΣB, this gives

$$\frac{\Sigma CO_2}{\Sigma B} = \left[\frac{A}{\Sigma B} - \frac{(K'_B)_P}{(a_H)_P + (K'_B)_P} \right] \times Q \tag{9.112}$$

which can be re-arranged to give

$$(K'_B)_P = \frac{(a_H)_P \cdot F}{1 - F} \tag{9.113}$$

where F is defined by

$$F = \left[\frac{A}{\Sigma B} - \frac{\Sigma CO_2}{Q \cdot \Sigma B} \right] \tag{9.114}$$

The ratios $\Sigma CO_2/\Sigma B$ and $A/\Sigma B$ are temperature and pressure independent. The ratio $\Sigma CO_2/\Sigma B$ was calculated from equation (9.112) using Lyman's apparent constants for the boric acid ionization at atmospheric pressure and the measured value of pH_1. The appropriate value of F was calculated for each pressure from (9.114) by inserting the values for $(K'_{1l})_P$ and $(K'_2)_P$ determined as above and the ratios $A/\Sigma B$ and $\Sigma CO_2/\Sigma B$. Insertion of F into (9.113) allowed calculation of $(K'_B)_P$.

Culberson and Pytkowicz summarized the pressure dependencies of K'_{1l}, K'_2 and K'_B found by them at two salinities by

34·8‰

$$\Delta pK'_{1l} = 0.013 + 1.319 \times 10^{-3}\,P - 3.061 \times 10^{-6}\,PT - 0.161 \times 10^{-6}\,T^2$$
$$- 0.020 \times 10^{-6}\,P^2$$

$$\Delta pK'_2 = -0.015 + 0.839 \times 10^{-3}\,P - 1.908 \times 10^{-6}\,PT + 0.182 \times 10^{-6}\,T^2$$

$$\Delta pK'_B = 1.809 \times 10^{-3}\,P - 4.515 \times 10^{-6}\,PT - 0.169 \times 10^{-6}\,P^2$$
$$+ 1.759 \times 10^{-12}\,P^2\,T^2$$

38·5‰

$$\Delta pK'_{1l} = 0.467 \times 10^{-3}\,P - 4.4 \times 10^{-8}\,P^2$$

$$\Delta pK_2' = 0\cdot280 \times 10^{-3} P$$

$$\Delta pK_B' = 0\cdot492 \times 10^{-3} P - 1\cdot4 \times 10^{-8} P^2$$

Within this salinity range each constant changes only slightly with salinity (ca. $0\cdot004$ pK' unit for a 1% salinity change) and their recommended values for the fractional changes in the constants as a function of pressure (Table 9.10) are applicable.

Since, in most investigations, the pH of the sample is measured after decompression to atmospheric pressure, values for the pressure coeffficient of pH are needed to facilitate the calculation of the carbonate system component concentrations at depth. These are given in Table 9.11.

The effect of applying pressure corrections when estimating the pH, $HCO_{3(T)}^-$ and $CO_{3(T)}^{2-}$ concentrations and ΣCO_2 at depth is illustrated by the plots of Fig. 9.7 (Culberson and Pytkowicz, 1968) in which field data from a North Pacific station have been used to obtain these quantities both with and without applying the corrections. The pressure correction is seen to lead to a considerable pH shift compared to that measured at surface pressure. For water of about 6000 m depth, application of the pressure corrections leads to a decrease in the estimated CO_3^{2-}(T) concentration of about 15%; the HCO_3^-(T) concentration increases by about 0·6%.

Other summaries of the available data on ionization constant pressure dependencies have been given. These often make use of the equation

$$\left(\frac{\partial \ln K'}{\partial P}\right)_{T,C} = -\frac{\Delta \overline{V}}{RT} \qquad (9.115)$$

in which K' is the apparent constant and $\Delta \overline{V}$ is effectively a partial molar volume associated with the reaction. The thermodynamical justification for this is given below, but in the immediate context $\Delta \overline{V}$ can be regarded as an empirical quantity which satisfies the available data. If $\Delta \overline{V}$ is pressure independent, integration of (9.115) gives

$$\ln \frac{K_{(P)}'}{K_{(1)}'} = -\frac{\Delta \overline{V}}{RT}(P - 1) \qquad (9.116)$$

where $K_{(P)}'$ and $K_{(1)}'$ correspond to the apparent constant values at P and 1 atmosphere pressure respectively. For this particular system it seems reasonable to assume that $\Delta \overline{V}$ is pressure independent (see e.g. Distèche and Distèche, 1967) although more complex expressions which allow for the possibility of a dependence of the partial molar volume change on the applied pressure are available (see Chapter 2 and also Owen and Brinkley, 1941).

Edmond and Gieskes (1970) have proposed that the best summary in

TABLE 9.10

Values of the pressure coefficients of the apparent dissociation constants. Interpolated values of the pressure coefficients should be accurate to $\pm 2\%$ (Culberson and Pytkowicz, 1968)

Pressure (atm)	Temp (C)		
	0	5	10
	$(K'_{1t})_P/(K'_{1t})_1$		
100	1·12	1·11	1·11
200	1·25	1·24	1·23
300	1·39	1·38	1·36
400	1·55	1·53	1·50
500	1·73	1·69	1·66
600	1·92	1·88	1·84
700	2·13	2·08	2·03
800	2·37	2·30	2·23
900	2·62	2·54	2·46
1000	2·91	2·80	2·70
	$(K'_2)_P/(K'_2)_1$		
100	1·07	1·07	1·07
200	1·15	1·15	1·15
300	1·24	1·23	1·23
400	1·34	1·33	1·32
500	1·44	1·42	1·41
600	1·55	1·53	1·51
700	1·66	1·64	1·62
800	1·79	1·76	1·73
900	1·93	1·89	1·86
1000	2·07	2·03	1·99
	$(K'_B)_P/(K'_B)_1$		
100	1·14	1·13	1·13
200	1·30	1·28	1·27
300	1·47	1·45	1·43
400	1·67	1·64	1·61
500	1·90	1·85	1·81
600	2·14	2·09	2·03
700	2·42	2·35	2·28
800	2·73	2·64	2·55
900	3·07	2·96	2·85
1000	3·45	3·31	3·18

terms of equation (9.116) is given when

$$\Delta \overline{V}(K'_1) = -(24\cdot2 - 0\cdot085t)\,\text{cm}^3\,\text{mol}^{-1}$$

TABLE 9.11

Calculated values of $(pH_1 - pH_p)$ at 34·8‰ salinity (Culberson and Pytkowicz, 1968)

Temp (C)	Pressure (atm)	pH at atmospheric pressure				
		7·6	7·8	8·0	8·2	8·4
0	250	0·112	0·107	0·103	0·100	0·098
	500	0·222	0·213	0·205	0·200	0·196
	750	0·330	0·318	0·308	0·300	0·294
	1000	0·437	0·422	0·409	0·399	0·391
5	250	0·107	0·102	0·098	0·096	0·094
	500	0·212	0·203	0·197	0·192	0·189
	750	0·316	0·304	0·294	0·288	0·283
	1000	0·417	0·402	0·391	0·383	0·376
10	250	0·102	0·098	0·094	0·092	0·091
	500	0·203	0·195	0·189	0·185	0·182
	750	0·302	0·291	0·283	0·277	0·272
	1000	0·401	0·387	0·376	0·369	0·362

$$\Delta \overline{V}(K'_2) = -(16\cdot4 - 0\cdot040t)\,cm^3\,mol^{-1}$$

$$\Delta \overline{V}(K'_B) = -(27\cdot5 - 0\cdot095t)\,cm^3\,mol^{-1}$$

where t is the temperature in °C. A similar summary (but based on the work by Distèche and Distèche (1967)) has been given by Li et al. (1969). There is reasonable agreement between the $\Delta pK'$ $(= \log_{10}[(K')_{100}/(K')_1])$ estimates made by Culberson and Pytkowicz (1968) and by Distèche and Distèche (1967) for both the ionization of boric acid and the first ionization of carbonic acid. The agreement between these two groups for $\Delta pK'_2$ is not quite so good, that obtained by Culberson and Pytkowicz being some 10% higher than the value calculated from results given by Distèche and Distèche.

Thermodynamic basis of the pressure effect. Expression (9.115) can be justified from considerations of the effect of pressure on the partial molar volumes of the species involved in the reaction. To be pedantic, the term "species involved" needs qualification since, in sea water, carbonate and bicarbonate are each involved in more than one type of species (see Chapters 2 and 3 and also Section 9.2.3.7). Consideration of this point will be deferred until the partial molar volume concept itself has been examined.

The partial molar volume, $\overline{V}_i$, of a component i of a solution is defined (Denbigh, 1971) by

$$\overline{V}_i = (\partial V/\partial n_i)_{T, P, n_j} \qquad (9.117)$$

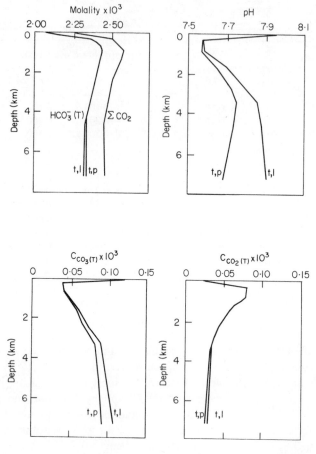

FIG. 9.7. Vertical distribution of ΣCO_2, $HCO_3^-(T)$, $CO_3^{2-}(T)$, pH and $CO_2(T)$. The curves labelled t, p are corrected for both temperature and pressure. Those labelled t, 1 are corrected for temperature only (after Culberson and Pytkowicz, 1968).

where V is the total volume of the solution and the subscripts T, P and n_j imply respectively constancy of temperature, pressure and the amounts of all components other than i during the addition of the increment of i. The partial molar volume of i is usually a function of c_i, but its value under certain limiting conditions is of special interest. For investigations of dilute solutions in which pure water is the solvent, $\overline{V}_i$ at infinite dilution is denoted by $\overline{V}_i^0$. For oceanographical work it is useful to consider sea water to be the solvent; on this basis the partial molar volume at a particular concentration of i is denoted by $\overline{V}_i^*$ and the value at infinite dilution in sea water by $\overline{V}_i^{*0}$. It is not to be expected that $\overline{V}_i^0$ and $\overline{V}_i^{*0}$ will have the same numerical value, and it is important to remember that $\overline{V}_i^{*0}$ will depend on the salinity. For the ranges of concentration in ocean water of most of the ions relevant to the carbonate system, $\overline{V}_i^*$ will be virtually the same as $\overline{V}_i^{*0}$.

The partial molar volume concept has a number of properties which may be summarized as follows. Thus, often $\overline{V}_i$ (and also $\overline{V}_i^*$) is less than the corresponding molar volume. This is particularly so for electrolytes, and the net contraction of volume on the addition of the solute to the solution which this implies, when seen against the background of the low compressibility of liquids, is indicative of the strong forces which exist between ions and the solvent molecules. When the solvent medium is such that complex or ion pair formation between ions is unimportant, the partial molar volumes of electrolytes are additive. The need to maintain electroneutrality means that, strictly, the operation implied by (9.117) cannot be carried out for individual ions. Nevertheless, it is useful to be able to consider the partial molar volume of an electrolyte as the sum of those of its individual anionic and cationic components. By assuming a $\overline{V}_i^0$ value for a particular ion and making use of the additivity principle, a scale can be established for the partial molar volumes of the other ions. Often it has been the practice to set $\overline{V}_H^0$ equal to zero (see, e.g. Owen and Brinkley, 1941), although it has been proposed that by taking not zero, but $-4 \cdot 5 \text{ cm}^3 \text{ mol}^{-1}$ for $\overline{V}_H^0$ (Mukerjee, 1961, 1966), a scale can be established on which both anions and cations show the same functional relationship between partial molar volume and some ionic property such as crystal radius. Millero (1969, 1971a, b) has extended this approach to solutions in sea water and has suggested that $\overline{V}_H^{*0}$ should be taken as $-3 \cdot 7 \text{ cm}^3 \text{ mol}^{-1}$ since, when this is done, a common smooth plot of $\overline{V}_i^{*0}$ against the cube of the crystal radius is obtained for both anions and cations.

In a model which has proved useful in discussions of ionic interaction in solution (Millero and Drost–Hansen, 1968; Millero, 1969, 1971a, b), the partial molar volumes of ions are regarded as arising from two principal contributions, viz,

$$\overline{V}_i^{*0} \text{ (ion)} = \overline{V}_i^{*0} \text{ (int)} + \overline{V}_i^{*0} \text{ (elec)}$$

where $\overline{V}_i^{*0}$ (int) is the intrinsic partial molar volume (the crystal molar volume plus void space contribution) and $\overline{V}_i^{*0}$ (elec) is the electrostriction partial molar volume (a decrease in volume arising from the strong forces of interaction between an ion and the solvent molecules). The partial molar volume of an ion can be related to its crystal radius (Mukerjee, 1961; Spiro et al., 1968; Millero, 1969, 1971a, b) by

$$\overline{V}_i^{*0} \text{ (ion)} = Ar^3 - B(Z^2/r)$$

where Z is the charge on the ion, r is the crystal radius and A and B are constants. When an ion is transferred from pure water to sea water, the resultant change in partial molar volume $(\bar{V}_i^{*0} - \bar{V}_i^0)$ is denoted by $\Delta \bar{V}_i^*$ (trans). The relationship

$$\Delta \bar{V}_i^* \text{ (trans)} = 0.37 (Z^2/r) + 0.83$$

applies to the transfer of most ions relevant to sea water, important exceptions being SO_4^{2-}, CO_3^{2-} and HCO_3^-. These show much larger $\Delta \bar{V}_i^*$ (trans) values than are predicted by this equation (Millero, 1969); the most likely explanation is that in sea water these ions undergo pairing (with Ca^{2+}, Mg^{2+} and Na^+) and release some of their electrostricted water as they do so. It might be noted that the partial molar volumes of ions in sea water are greater than the corresponding values in pure water. That is, in sea water, which is less compressible than pure water, they have a smaller electrostrictive effect (Millero, 1969).

Insofar as it is legitimate to attach a meaning to $\bar{V}_i^*$ for a single ion, the value derived by the hypothetical operation (9.117) embodies contributions from all the various physical states and chemical associations with which the added ion becomes involved, each making a suitably weighted contribution to the observed $\bar{V}_i^*$. For instance, carbonate in sea water exists partly as "free" (i.e., hydrated) carbonate, but also as $MgCO_3^0$, $CaCO_3^0$ and $NaCO_3^-$ ion pairs. The added dn_i moles of i become distributed between these various species, and the observed volume change comprises that consequent on the changes in the amounts of these individual species (plus, of course, that consequent on the changes of the amounts of free cation species). Thus, the detailed interpretation of $\bar{V}_i^*$ for certain ions may be quite complex, and the value obtained by the operation (9.117) is not necessarily that for the free ion (for further comments see e.g., Duedall, 1972).

Below, simple expressions are given relating the pressure dependence of equilibrium constants to the partial molar volume change accompanying reaction. If use is to be made of these expressions, estimates are required for the individual partial molar volumes of the reactants and products. Not all the necessary values have been experimentally determined, and the purpose of the above remarks is to indicate that models and procedures exist whereby theoretical estimates can sometimes be made of those partial molar volumes for which determinations are not available.

Because a particular equilibrium may be described in different ways (i.e., by either thermodynamic or apparent constants, and also by using different standard states in the definition of a constant—compare, for example, the Lyman and Hansson constants), strictly, it is necessary to ensure that in a proposed expression for the pressure dependence of an equilibrium constant, there is proper matching of the constant and the partial molar quantities. For the usual thermodynamic constants (equations (9.63) and (9.65)), and also for both the Buch and Lyman apparent constants, the standard state is based on a reference state which is an infinitely dilute solution of i in pure water. Thus, the chemical potential, μ_i, of component i is given by*

$$\mu_i = \mu_i^0 + RT \ln \gamma_i c_i \tag{9.118}$$

where μ_i^0, the standard chemical potential, is a function of temperature and pressure and is the chemical potential of a hypothetical ideal solution of i of unit concentration in pure water. For the Hansson constants the standard state is a hypothetical ideal solution of i of unit concentration in sea water, and the reference state is an infinitely dilute solution of i in sea water. The chemical potential itself is independent of the choice of standard state, but expressed in terms of the sea water standard state it becomes

$$\mu_i = \mu_i^{*0} + RT \ln \gamma_i^* c_i \tag{9.119}$$

where μ_i^{*0} the chemical potential at the standard state on this convention is also a function of temperature and pressure. It will be seen that in order to ensure that μ_i given by (9.118) and (9.119) is the same despite the shift in standard states, the change in the standard chemical potential is accommodated by a corresponding change in the activity coefficient. That is,

$$\ln \gamma_i^* = \ln \gamma_i + \Delta\mu_s/RT$$

where $\Delta\mu_s = \mu_i^0 - \mu_i^{*0}$.

Differentiation of (9.118) with respect to pressure gives

$$\frac{d\mu_i}{dP} = \frac{d\mu_i^0}{dP} + RT \frac{d \ln \gamma_i}{dP} \tag{9.120}$$

and making use of the relationships

$$\frac{d\mu_i}{dP} = \overline{V}_i^* \tag{9.121}$$

and

$$\frac{d\mu_i^0}{dP} = \overline{V}_i^0 \tag{9.122}$$

it follows that

$$\frac{d \ln \gamma_i}{dP} = \frac{\overline{V}_i^* - \overline{V}_i^0}{RT} \tag{9.123}$$

---

* Strictly (9.118) should be written $\mu_i = \mu_i^0 (P, T) + RT \ln \gamma_i c_i$ in order to emphasize the dependence of μ_i^0 on both pressure and temperature.

Similar treatment of (9.119) leads to the additional equations

$$\frac{d\mu_{i.}^{*0}}{dP} = \bar{V}_i^{*0} \tag{9.124}$$

and

$$\frac{d \ln \gamma_i^*}{dP} = \frac{\bar{V}_i^* - \bar{V}_i^{*0}}{RT} \tag{9.125}$$

Expressions (9.121) to (9.125) can now be applied to obtain relationships for the pressure dependencies of the various ionization constants.

Remembering that, for a particular process at equilibrium

$$\Sigma v_p \, \mu_{\text{(products)}} - \Sigma v_r \, \mu_{\text{(reactants)}} = 0$$

where v_p and v_r are the stoichiometric coefficients of the products and reactants respectively, we have, for the first ionization of carbonic acid

$$\mu_{H_2O}^0 + \mu_{CO_2(aq)}^0 - \mu_{HCO_3}^0 - \mu_H^0 = RT \ln \frac{\gamma_H c_H \cdot \gamma_{HCO_3} c_{HCO_3}}{a_{H_2O} \cdot \gamma_{CO_2(aq)} c_{CO_2(aq)}}$$

$$= RT \ln K_{1(TD)} \tag{9.126}$$

Differentiation with respect to P and use of (9.122) leads to

$$\left(\frac{\partial \ln K_{1(TD)}}{\partial P} \right)_{T, C} = -\frac{\Delta \bar{V}}{RT} \tag{9.127}$$

where

$$\Delta \bar{V} = \bar{V}_H^0 + \bar{V}_{HCO_3}^0 - \bar{V}_{H_2O}^0 - \bar{V}_{CO_2(aq)}^0$$

Similar expressions for the pressure dependencies of the Buch and Lyman first apparent constants can be obtained by combining the definitions of these constants (equations (9.64) and (9.57) respectively with (9.127) and (9.123)). By starting from equation (9.119) and using arguments similar to those outlined above an expression for the pressure dependency of the Hansson constant can be obtained. These expressions and those which can also be derived for the second ionization constant of carbonic acid and for the ionization of boric acid are all of the form (9.115) above where the meaning to be attached to $\Delta \bar{V}$ is indicated in Table 9.12.

In order for equations (9.115) and (9.116) to be used for the prediction of the pressure dependencies of the acid ionization constants rather than simply as devices for summarizing observed behaviour of the constants, values are

TABLE 9.12

Interpretation of the partial molar volume changes associated with equation (9.115)

First ionization of carbonic acid

$$K_{1(TD)}, \quad \Delta\bar{V} = \bar{V}^0_H + \bar{V}^0_{HCO_3} - \bar{V}^0_{H_2O} - \bar{V}^0_{CO_2}$$

$$K'_{1b}, \quad \Delta\bar{V} = \bar{V}^0_H + \bar{V}^*_{HCO_3} - \bar{V}^0_{H_2O} - \bar{V}^0_{CO_2}$$

$$K'_{1l}, \quad \Delta\bar{V} = \bar{V}^0_H + \bar{V}^*_{HCO_3} - \bar{V}^*_{H_2O} - \bar{V}^*_{CO_2}$$

$$K'_{1h}, \quad \Delta\bar{V} = \bar{V}^{*0}_H + \bar{V}^*_{HCO_3} - \bar{V}^*_{H_2O} - \bar{V}^*_{CO_2}$$

Second ionization of carbonic acid

$$K_{2(TD)}, \quad \Delta\bar{V} = \bar{V}^0_H + \bar{V}^0_{CO_3} - \bar{V}^0_{HCO_3}$$

$$K'_{2}, \quad \Delta\bar{V} = \bar{V}^0_H + \bar{V}^*_{CO_3} - \bar{V}^*_{HCO_3}$$

$$K'_{2h}, \quad \Delta V = \bar{V}^{*0}_H + \bar{V}^*_{CO_3} - \bar{V}^*_{HCO_3}$$

First ionization of boric acid

$$K_{B(TD)}, \quad \Delta\bar{V} = \bar{V}^0_H + \bar{V}^0_B - \bar{V}^0_{HB}$$

$$K'_{B}, \quad \Delta\bar{V} = \bar{V}^0_H + \bar{V}^*_B - \bar{V}^*_{HB}$$

$$K'_{Bh}, \quad \Delta\bar{V} = \bar{V}^{*0}_H + \bar{V}^*_B - \bar{V}^*_{HB}$$

needed for the appropriate partial molar volumes. Shortage of information has made reliable and comprehensive prediction for a range of temperatures impossible at present, and the best that can be done is to compare observed pressure coefficients ($\Delta\bar{V}$) (obtained for example from the work of Distèche and Distèche (1967) and of Culberson and Pytkowicz (1968)) with those which would be expected on the basis of what reliable information is available. Millero and Berner (1972) have attempted such a comparison. They point out that on the basis of the equation proposed by Edmond and Gieskes $\Delta\bar{V}(K'_{1l})$ at 25°C was $-22\cdot1$ cm^3 mol^{-1}. The corresponding value derived from measurements made by Disteche and Disteche by applying a temperature correction was *ca.* $-18\cdot8$ cm^3 mol^{-1}. An estimate based on partial molar volume data was $-22\cdot4$ cm^3 mol^{-1}. This estimate made use of a value for $\bar{V}^*_{CO_2(aq)}$ of $31\cdot3$ cm^3 mol^{-1} chosen from a consideration of values given by Kritchevsy and Iliiskaya (1945) and Enns *et al.* (1965) for a solution of CO_2 in pure water by making the assumption that $\bar{V}^*_{CO_2(aq)} = \bar{V}^0_{CO_2(aq)}$. Recently, a direct determination of the partial molar volume of CO_2 in sea water of 35‰ salinity at 23°C (Bradshaw, 1973) has indicated a value for $\bar{V}^*_{CO_2(aq)}$ of $33\cdot4$ cm^3 mol^{-1}. Use of this value would increase the discrepancy between the theoretical and observed estimates.

Comparison of the observed pressure coefficients for K'_2 with estimates based on partial molar volume data is difficult because of the large divergence between the observed $\Delta\overline{V}(K_2')$ values ($\Delta\overline{V}(K_2')$ (Distèche and Distèche) $=$ $-10\cdot6$ cm^3 mol^{-1}; $\Delta\overline{V}(K_2')$ (Culberson and Pytkowicz) $= -15\cdot4$ cm^3 mol^{-1}) and uncertainty about the best value to take for $\overline{V}^*_{CO_3}$ (see Millero and Berner, 1972). However, from estimates made by Duedall (1972) for the partial molar volumes in sea water of Na_2CO_3 and K_2CO_3, Millero and Berner estimated $\overline{V}^*_{CO_3}$ to be $19\cdot2$ cm^3 mol^{-1}. Since the values for this calculated from the K_2' pressure coefficients derived from the work of Culberson and Pytkowicz and of Distèche and Distèche respectively are $19\cdot7$ and $24\cdot5$ cm^3 mol^{-1}, they concluded that the work of Culberson and Pytkowicz gives the more reliable coefficient.

9.2.3.7. *Ion pair formation* (see also Chapters 2 and 3)

HCO_3^- and CO_3^{2-} concentrations obtained *via* the equations given in Section 9.2.3.5 are total concentrations (i.e., 'free' plus ion pair bound forms). That is

$$c_{HCO_3(T)} = c_{HCO_3(f)} + \Sigma c_{MHCO_3}$$

and

$$c_{CO_3(T)} = c_{CO_3(f)} + \Sigma c_{MCO_3}$$

where $c_{HCO_3(f)}$ and $c_{CO_3(f)}$ are the concentrations of "free" (hydrated) bicarbonate and carbonate respectively, and the summation terms imply the sums of the concentrations of the appropriate ion pairs. Pair formation is known to be marked for many ions, and occurs particularly between non-polarizable ("hard") cations having an inert gas electronic structure such as Na^+, Ca^{2+} and Mg^{2+} and hard anions such as CO_3^{2-} and SO_4^{2-} (see Chapter 3). It is an accepted phenomenon in natural aqueous systems (Garrels *et al.*, 1961; Garrels and Thompson, 1962; Garrels and Christ, 1965; Thompson, 1966; Thompson and Ross, 1966; Pytkowicz and Kester, 1969; Kester and Pytkowicz, 1968, 1969, 1970; Berner, 1965, 1971; Butler and Huston, 1970; Hawley, 1973; Pytkowicz and Hawley, 1974; Dyrssen and Hansson, 1973), and its occurrence is the principal reason for the dependence of apparent constants not only on the ionic strength of a solution, but also on the ionic composition.

The activity of CO_3^{2-} can be written

$$a_{CO_3} = c_{CO_3(T)}\gamma_{CO_3(T)} = c_{CO_3(f)} \cdot \gamma_{CO_3(f)}$$

and thus,

$$\gamma_{CO_3(T)} = \frac{c_{CO_3(f)} \cdot \gamma_{CO_3(f)}}{c_{CO_3(T)}}$$

where the symbol (f) indicates the free ion. Similar expressions apply to the bicarbonate and sulphate ions. It follows that if, as seems likely, the activity coefficient of the free ion depends mainly on the ionic strength and not on the composition, (see above references), the total activity coefficient may be quite small, particularly if $c_{CO_3(f)} \ll c_{CO_3(T)}$. Furthermore, since the $c_{CO_3(f)}/c_{CO_3(T)}$ ratio depends on solution composition, particularly on the concentrations of pair forming cations, both $\gamma_{CO_3(T)}$ and any apparent constant which is related to the corresponding thermodynamic constant through $\gamma_{CO_3(T)}$ will be compositionally dependent (see Section 9.2.3.3).

Pair formation also affects various physical properties such as ultrasonic absorption (Fisher, 1965, 1967, 1972) and the Raman spectra of solutions (Daly *et al.*, 1972), and it bears more or less directly on other aspects of the chemistry of natural waters such as mineral solubility. Quantitative understanding of the speciation in natural solutions is essential if values for the various apparent constants to use in connection with unusual solutions (pore waters and highly saline solutions) are to be predicted. This understanding is still far from complete, although the last couple of decades has seen considerable progress and a number of attempts have been made to establish models for sea water in terms of the speciation likely to be present (see Chapters 2 and 3 and also Garrels and Thompson, 1962; Kester and Pytkowicz, 1969; Berner, 1971; van Breeman, 1972; Hawley, 1973; Lafon, 1970; Ben-Yaakov and Goldhaber, 1973), and to cater for the effects of temperature and pressure.

The main pairs of concern in this Section are those formed between the cations Na^+, Ca^{2+} and Mg^{2+} and the anions HCO_3^- and CO_3^{2-}. Additionally, for the sake of completeness the sulphate ion pairs $NaSO_4^-$, $MgSO_4^0$ and $CaSO_4^0$ are also considered.

The ion pair association equilibrium

$$M^{x+} + A^{y-} = MA^{(x-y)+}$$

can be described using either thermodynamic or apparent constants ($K_{MA(TD)}$ and K'_{MA} respectively), viz.

$$K_{MA(TD)} = \frac{a_{MA}}{a_M \cdot a_A} = K'_{MA} \frac{\gamma_{MA}}{\gamma_{M(f)} \cdot \gamma_{A(f)}}$$

where

$$K'_{MA} = \frac{c_{MA}}{c_{M(f)} \cdot c_{A(f)}},$$

the qualifying symbol (f) again indicating that the quantity refers to the "free" ion. As with the ionization constants of carbonic acid, apparent rather than

D

thermodynamic constants are the more convenient to use, despite their dependence on ionic strength, since they are directly related to concentrations.

An examination of the dependence of the total activity coefficients $\gamma_{HCO_3(T)}$ and $\gamma_{CO_3(T)}$ on solution composition was made by Garrels et al. (1961). They expressed the activity coefficient $\gamma_{HCO_3(T)}$ by

$$\gamma_{HCO_3(T)} = \frac{K_{1(TD)} \cdot a_{CO_2} \cdot a_{H_2O}}{a_H \cdot c_{HCO_3(T)}}$$

where $K_{1(TD)}$ is the first thermodynamic ionization constant of carbonic acid. The experimental conditions chosen allowed control of a_{CO_2}, a_{H_2O} and $c_{HCO_3(T)}$, and by assuming that the pH was related to a_H by equation (9.31), $\gamma_{HCO_3(T)}$ values were obtained. Insertion of these and the values for $c_{CO_3(T)}$, $c_{HCO_3(T)}$ and $K_{2(TD)}$ (the second thermodynamic ionization constant of carbonic acid) into

$$\frac{\gamma_{CO_3(T)}}{\gamma_{HCO_3(T)}} = \frac{K_{2(TD)} \cdot c_{HCO_3(T)}}{a_H \cdot c_{CO_3(T)}}$$

enabled values for $\gamma_{CO_3(T)}$ to be obtained. Plots were obtained of these activity coefficients as functions of the ionic strength for electrolyte solutions containing NaCl, MgCl$_2$ and NaCl + MgCl$_2$ in 10:1 molar ratio (that of sea water) as well as earlier data for NaHCO$_3$, KHCO$_3$, Na$_2$CO$_3$ and K$_2$CO$_3$ solutions (from Walker et al., 1927) and for sea water (Sverdrup et al., 1942). A Debye–Hückel plot was also given, and the agreement between this and the $\gamma_{HCO_3(T)}$ and the $\gamma_{CO_3(T)}$ estimates for solutions containing KHCO$_3$ and K$_2$CO$_3$ when major cations other than K$^+$ were absent indicated that neither HCO$_3^-$ nor CO$_3^{2-}$ formed ion pairs with K$^+$. The lower values for these activity coefficients in the presence of Na$^+$ and, particularly, Mg^{2+} was consistent with the occurrence of pairing to give NaCO$_3^-$ and MgCO$_3^0$.

The results of this and other investigations (Greenwald, 1941; Davies, 1962) were used by Garrels and Thompson (1962) to establish a set of thermodynamic ion pair dissociation constants for the major sulphate, bicarbonate and carbonate pairs in sea water. By combining these with estimates of the individual ion activity coefficients (mean salt method) and assuming that the activity coefficients of uncharged pairs were closely similar to those of neutral molecules (see however, Kester, 1969) and that those of singly charged pairs were equal to that of bicarbonate they set up a model for the speciation in sea water in terms of the proportions of the ions Na$^+$, K$^+$, Ca^{2+}, Mg^{2+}, SO$_4^{2-}$, HCO$_3^-$ and CO$_3^{2-}$ involved in pairing (see values, recalculated by Whitfield, in Chapter 2, Table 2.3.8). Their conclusions relate to surface water of 19‰ Cl at 25°C. Weaknesses of their approach are the methods used for estimating the activity coefficients of the single ion species, the charged pairs and the uncharged pairs (see above and Pytkowicz and Hawley, 1974). In the light of later work the conclusions to be derived from this model should be regarded as rather good first approximations. They suggest that Mg^{2+} is strongly paired whereas little of the K^+ participates. Most of the carbonate, about one-half of the sulphate and one-third of the bicarbonate is bound in this way.

Subsequent investigators agree with the general features of this picture, although at the present time there are differences between the proposed values for some of the association constants, and inevitably this must be reflected in differences in proposed species distributions. Some of this later work has extended understanding by examining the influence of temperature and pressure on certain of the equilibria. This extension is to be welcomed, not only because it allows thermodynamic parameters such as the appropriate enthalpy and partial molar volume changes to be estimated, but also because it is likely to make association constant measurements more directly applicable to field situations.

Sulphate pair formation. Pytkowicz and his co-workers (Kester and Pytkowicz, 1968, 1969, 1970; Pytkowicz and Kester, 1969) have examined the effect of temperature on the apparent formation constants of $NaSO_4^-$ and $MgSO_4^0$ in artificial sea water of effective ionic strength (i.e., the ionic strength corrected for pairing) of 0·608. K'_{NaSO_4} (25°C) (Kester and Pytkowicz, 1969) and K'_{NaSO_4} (2·5°C) (Kester and Pytkowicz, 1970) were determined by a potentiometric method in which the Na^+ concentration of a test solution was compared with that of a standard in which no pairing was thought to occur. Addition of a concentrated NaCl solution to the standard was made until the test and standard gave the same reading with a sodium sensitive electrode. Under these circumstances it was assumed that

$$c_{Na(f)\,(test)} = \frac{\gamma_{Na(std)}}{\gamma_{Na(f)\,(test)}} \cdot c_{Na(std)}$$

The ratio of the activity coefficients was estimated by the mean salt method, and from the estimates of $c_{Na(f)\,(test)}$ and a knowledge of the total (i.e. paired plus unpaired) Na^+ in the test solution, K'_{NaSO_4} could be determined. Measurements of K'_{NaSO_4} at 1·5°C and at 500 and 1000 atmospheres were made using a pressure cell similar to that used in the examination of the pressure dependence of pH (Section 9.2.3.1) but in which the membrane was Na^+ sensitive glass. K'_{MgSO_4} was determined in a similar way at 25°C and one atmosphere (Kester and Pytkowicz, 1968) and at 1·5°C and one atmosphere (Kester and

Pytkowicz, 1970). Instability of the Mg^{2+} sensitive electrode at higher pressures precluded experimental determination of the pressure coefficient of K'_{MgSO_4}. This was estimated (Kester and Pytkowicz, 1970) from theoretical values for the change in the partial molar volume associated with the pairing process although the validity of this estimation has been questioned by Millero (1971a). K'_{CaSO_4} (25°C) was also determined using a potentiometric method and a Ca^{2+} sensitive electrode (Kester and Pytkowicz, 1969). Instability of this electrode at 2°C prevented direct determination of the temperature coefficient; this was assumed to be the same as that for the $MgSO_4$ pair. The pressure coefficient was also assumed to be the same as that for K'_{MgSO_4}.

The main conclusions obtained by Pytkowicz and Kester and the modification suggested by Millero (1971a) on the basis of his theoretical estimates of the pressure effect are summarized in Table 9.13 (see also Pytkowicz, 1972a).

TABLE 9.13

The effect of temperature and pressure on sulphate speciation in sea water* (Millero, 1971a)

T (°C)	P (atm)	%Free SO_4^{2-}	%$NaSO_4^-$	%$MgSO_4^0$	%$CaSO_4^0$
25	1	39·0	38·0	19·0	4·0
2	1	28·0	47·0	21·0	4·0
2	1000	39·0	32·0	24·0	5·0
		42·0†	35·0†	19·0†	4·0†

  * Given as % of the total sulphate as given species; results at 25° and 2° (1 atm) are taken from Kester and Pytkowicz (1970).
  † Calculated by Millero (1971a).

Sulphate pair formation at salinities close to 34‰ is exothermic and consequently the net amount of pairing increases with decrease in temperature (cf. the predictions made by Lafon, 1970). The formation constants of $NaSO_4^-$, $MgSO_4^0$ and $CaSO_4^0$ decrease with increase of pressure (positive change of partial molar volume accompanying pair formation). The conclusions with regard to the effect of pressure on $MgSO_4^0$ and $CaSO_4^0$ pair formation are based on theoretical assumptions, and experimental determinations are desirable (Millero, 1971a; Pytkowicz, 1972a).

Bicarbonate and carbonate pair formation. A recent investigation of the ion pairing properties of HCO_3^- and CO_3^{2-} in solutions having the ionic strength of sea water has been made by Hawley (1973), who based his method on the compositional dependence of the apparent ionization constants of carbonic acid (see Pytkowicz and Hawley, 1974). He showed that the observed value of K_1' ($\equiv K_{1l}'$)* is related to the various ion association constants by

$$\frac{K_1'}{a_{H_2O} \cdot \gamma_{CO_2(T)}} = \frac{K_{1(TD)} \cdot k}{\gamma_{HCO_3(f)}} [1 + K_{NaHCO_3}' \cdot c_{Na(f)} + K_{MgHCO_3}' \cdot c_{Mg(f)}$$

$$+ K_{CaHCO_3}' \cdot c_{Ca(f)}] \quad (9.128)$$

where $\gamma_{HCO_3(f)}$ is the activity coefficient of the 'free' HCO_3^- ion, and k relates the operational pH and the hydrogen ion activity, a_H, by

$$X = k \cdot a_H$$

where
$$pH = -\log X$$

(Pytkowicz, 1968; Hawley, 1973). $K_{1(TD)}$ is independent of the ionic strength and k and $\gamma_{HCO_3(f)}$ are assumed to be so. In equation (9.128), K_{NaHCO_3}' etc. are the appropriate stoichiometric ion association constants and $c_{Na(f)}$ etc. are the free (unassociated) concentrations of the cations. Determination of K_1' by a potentiometric titration procedure for a number of solutions of different ionic composition but each of ionic strength equivalent to that of sea water enabled a series of simultaneous equations of the form of (9.128) to be generated. Solution of this set enabled the individual ion association constants to be obtained.

An extension of this approach to the determination of carbonate ion association constants made use of the equation

$$\frac{K_1' \cdot K_2'}{a_{H_2O} \cdot \gamma_{CO_2(T)}} = \frac{k^2 \cdot K_{1(TD)} \cdot K_{2(TD)}}{\gamma_{CO_3(f)}} [1 + K_{NaCO_3}' \cdot c_{Na(f)} + K_{MgCO_3}' \cdot c_{Mg(f)}$$

$$+ K_{CaCO_3}' \cdot c_{Ca(f)}] \quad (9.129)$$

where $\gamma_{CO_3(f)}$ is the activity coefficient of the free CO_3^{2-} ion and is virtually independent of the solution composition at fixed ionic strength. K_1' and K_2' are functions of the solution composition, and a set of simultaneous equations of the form of (9.129) was generated by determining $K_1' \cdot K_2'$ for a number of solutions of the same effective ionic strength (that of sea water) but of differing ionic composition. Solution of the set of equations enabled values for the individual carbonate ion association constants to be determined. A satisfactory solution could be obtained only by assuming that, in addition to the pairs indicated in (9.129), the triple associates $Mg_2CO_3^{2+}$, $MgCaCO_3^{2+}$ and possibly $Ca_2CO_3^{2+}$ were formed.

Association constants at 25°C found by Hawley are summarized in

* The symbol K_{1l}' has been reserved for the Lyman-type apparent constants determined in water having the relative ionic composition of sea water.

Table 9.14 together with values obtained by other workers (Garrels and Thompson, 1962; Kester and Pytkowicz, 1969; Butler and Huston, 1970; Hansson, 1972). Pytkowicz and Hawley (1974) used their bicarbonate and carbonate constants in conjunction with (i) the sulphate association constants

TABLE 9.14

Stoichiometric association constants for the major chemical species in sea water of 34‰ salinity at 25°C (Pytkowicz and Hawley, 1974)

	(1)	(2)	(3)		(4)
Ionic strength	0·72	0·66	0·5	1·0	
$NaHCO_3^0$	0·280	0·26	0·39	0·21	—
$MgHCO_3^+$	1·62	5·22	—	—	1·04
$CaHCO_3^+$	1·96	5·10	—	—	1·04
$NaCO_3^-$	4·25	4·16	1·38	1·86	—
$MgCO_3^0$	112	160	—	—	32·5
$Mg_2CO_3^{2+}$	387	—	—	—	—
$CaCO_3^0$	162	78	—	—	32·5
$MgCaCO_3^{2+}$	1040	—	—	—	—

(1) Pytkowicz and Hawley (1974).
(2) Garrels and Thompson (1962)
(3) Butler and Huston (1970)
(4) Dyrssen and Hansson (1973).

for Na^+, Mg^{2+} and Ca^{2+} found by Pytkowicz and Kester (1969) and Kester and Pytkowicz (1968, 1969), (ii) a potassium sulphate association constant based on Garrel and Thompson's estimates of the ion activity coefficients and the thermodynamic association constant and (iii) magnesium and calcium fluoride association constants quoted by Elgquist (1970); in this way they obtained the sea water distribution of species given in Table 9.15 (see also Hawley, 1973).

The distribution of cations differs only slightly from a speciation suggested by Kester and Pytkowicz (1969); the distribution of anions differs more significantly from this earlier tabulation, partly because Kester and Pytkowicz took no account of the formation of triple associates. The distribution indicated in Table 9.15 differs even more from that suggested by Garrels and Thompson and reflects the fact that Hawley's determination of the bicarbonate and carbonate association constants show calcium association to be more important relative to that of magnesium than it was thought to be by Garrels and Thompson.

TABLE 9.15

Species in sea water of 34‰ salinity at 25° C (Pytkowicz and Hawley, 1974)

	Na^+	Mg^{2+}	Ca^{2+}	K^+
Total molality	0·4822	0·05489	0·01063	0·01062
% Free Metal	97·70	89·11	88·35	100·00
% MSO_4	2·25	10·35	10·87	—
% $MHCO_3$	0·05	0·24	0·29	—
% MCO_3	0·01	0·17	0·41	—
% Mg_2CO_3	—	0·03	—	—
% $MgCaCO_3$	—	0·01	0·07	—
% MF	—	0·07	0·02	—

	SO_4^{2-}	HCO_3^-	CO_3^{2-}	F^-
Total molality	0·02906	0·00213	0·000171	0·000080
% Free anion	39·19	81·33	7·99	51·04
% NaX	37·29	10·73	15·99	—
% MgX	19·55	6·44	43·86	46·94
% CaX	3·97	1·50	20·96	2·02
% Mg_2CO_3	—	—	7·39	—
% $MgCaCO_3$	—	—	3·82	—

As might be expected, a good deal of the magnesium and calcium of sea water is involved in ion association, but little of the potassium is involved. Of the anions, more than 90% of the carbonate participates in pair or triple ion formation and about 60% of the sulphate associates. Only about 20% of the bicarbonate forms ion pairs.

Although most of the ion paired carbonate and bicarbonate in sea water is associated with the major ions Na^+, Ca^{2+} and Mg^{2+}, it should be appreciated that other ions present at much lower concentrations also participate in carbonate pair formation (e.g., $PbCO_3^0$, $Pb(CO_3)_2^{2-}$, $CuCO_3^0$, $ZnCO_3$— see e.g. Chapter 3 and also Stumm and Bilinksi, 1972; Zirino and Healy, 1970; Zirino and Yamamato, 1972).

Borate pair formation. Borate is known to participate in ion pair formation (Sillén and Martell, 1964; Kester and Byrne, 1972; Dyrssen and Hansson, 1973), and recently Byrne and Kester (1974) have determined the formation constants for the species $NaB(OH)_4^0$, $CaB(OH)_4^+$ and $MgB(OH)_4^+$ (0·57, 13·0 and 8·0 respectively at 25°C and ionic strength 0·68). Approximately 44% of the borate ion in sea water is complexed in this way.

9.2.4. CALCIUM CARBONATE PRECIPITATION AND DISSOLUTION

The upper layers of much of the world's oceans are saturated or super-saturated with respect to $CaCO_3$ whereas often the lower layers (below ca. 300 m in much of the Atlantic and Central Pacific and 200–1000 m in the North Pacific—see e.g. Pytkowicz, 1965b, 1970) are undersaturated. Frequently, sediments found above a certain depth—the compensation depth —are relatively rich in calcareous material; below this horizon there is a dramatic decrease in their $CaCO_3$ content. Clearly, these features pose problems of $CaCO_3$ solution and transport. Thus, the conditions governing $CaCO_3$ precipitation and solution are important because of their bearing on carbonate sedimentation and solution, the mobilization of Ca^{2+} during post-depositional diagenesis, the origin of the compensation depth and the transfer of Ca in the oceans.

Removal of the dissolved $CaCO_3$ occurs in response to the biological demand of calcium carbonate utilizing organisms, and ultimately leads to the deposition of calcareous debris. Additionally, there is the possibility of chemical precipitation when certain conditions (thermodynamic and kinetic) are satisfied. These are that the water should be at least saturated with respect to $CaCO_3$ and that the rate of generation and growth of nuclei for carbonate deposition are adequate. It is possible that such calcium carbonate formation does occur in a number of restricted areas (e.g. the Bahama Banks), but although here the shallowness of the banks and the accessibility of pre-existing nuclei and particles make heterogeneous precipitation via oolite formation possible, it is not certain that aragonite needle deposition results from purely inorganic precipitation since the rate of homogeneous nucleation is likely to be too low (see Berner, 1971).

Uptake of $CaCO_3$ by sea water results from the terrestrial weathering of Ca-containing rocks and from the attack of sea water on sediments (Gieskes, 1974; Edmond, 1974) and possibly also on calcereous tests and $CaCO_3$ particles as they fall through the water column (Pytkowicz, 1973b). Here again, kinetic factors may be important (see also Edmond, 1974).

Since the sea water/calcium carbonate system is intimately linked with the more general carbonate system, the addition or removal of $CaCO_3$ is accompanied by changes in the alkalinity and ΣCO_2. The change in carbonate alkalinity is related to the moles of $CaCO_3$ dissolving, Δc_{Ca}, by $\Delta CA = 2 \times \Delta c_{Ca}$, and the corresponding change in ΣCO_2 is $\Delta(\Sigma CO_2) = \Delta c_{Ca}$. The carbonate which enters solution will, of course, become part of the general carbonate–bicarbonate system, but it is evident that since in sea water the stoichiometric concentrations of Ca^{2+}, HCO_3^- and CO_3^{2-} are about 10^{-2}, 20×10^{-4} and $2 \times 10^{-4} \ mol \ l^{-1}$ respectively, small uptakes of $CaCO_3$ may cause changes of several hundred percent in $c_{CO_3(T)}$ but of only a few per cent in $c_{Ca(T)}$ and $c_{HCO_3(T)}$. In fact the $c_{Ca(T)}$/chlorinity ratio of sea water does not vary by more than about 1%.

9.2.4.1. Calcium carbonate solution–solid equilibria

The thermodynamic requirement for solution equilibrium is

$$\mu_{CaCO_3(s)} = \mu_{Ca} + \mu_{CO_3},$$

where $\mu_{CaCO_3(s)}$ is the chemical potential of the solid carbonate and the μ_{Ca} and μ_{CO_3} are those of the calcium and carbonate ions in solution respectively. The chemical potentials of the dissolved Ca^{2+} and CO_3^{2-} are given by

$$\mu_{Ca} = \mu_{Ca}^0(P, T) + RT \ln a_{Ca}$$

$$\mu_{CO_3} = \mu_{CO_3}^0(P, T) + RT \ln a_{CO_3}$$

respectively, $\mu_{Ca}^0(P, T)$ and $\mu_{CO_3}^0(P, T)$ being the standard chemical potentials. Combination of these three expressions leads to

$$\exp\frac{\mu_{CaCO_3(s)} - (\mu_{Ca}^0 + \mu_{CO_3}^0)}{RT} = a_{Ca} \cdot a_{CO_3}$$
$$= K_{CaCO_3}$$

where K_{CaCO_3} is the activity solubility product. The value for $\mu_{CaCO_3(S)}$ (and therefore, of K_{CaCO_3}) depends on the crystalline modification of $CaCO_3$. Natural calcium carbonate shows three main modifications—vaterite, aragonite and calcite. Of these, vaterite is not sufficiently widespread to be of general interest. Of the other forms, aragonite is metastable with respect to calcite at normal pressures and temperatures and, consequently, has a higher chemical potential and, therefore, a higher solubility product. The thermodynamic (activity) solubility products of calcite and aragonite (K_{calc} and K_{arag}) at 25°C are about 4.0×10^{-9} and 6.3×10^{-9} respectively (Berner, 1971). Both solubility products decrease with increase of temperature.

Because interest is generally focused on concentrations rather than activities, use is commonly made of apparent (stoichiometric) solubility products, K'_{calc} and K'_{arag} (Pytkowicz, 1969a, b). These are related to the corresponding thermodynamic quantities by

$$K'_{calc} = [c_{Ca(T)} \cdot c_{CO_3(T)}]_{calc} = \frac{K_{calc}}{\gamma_{Ca(T)} \cdot \gamma_{CO_3(T)}}$$

and

$$K'_{arag} = [c_{Ca(T)} \cdot c_{CO_3(T)}]_{arag} = \frac{K_{arag}}{\gamma_{Ca(T)} \cdot \gamma_{CO_3(T)}}$$

where $c_{Ca(T)}$ and $c_{CO_3(T)}$ are the total concentrations of the ions in equilibrium with the appropriate solid forms, and $\gamma_{Ca(T)}$ and $\gamma_{CO_3(T)}$ are the stoichiometric activity coefficients.

Reliable values for K'_{calc} and K'_{arag} are vital for many aspects of the investigation of carbonate solubility, and many attempts have been made to determine them as functions of salinity, temperature and pressure (Wattenberg, 1933, 1936; Wattenberg and Timmerman, 1936; Smith, 1940; Hindman, 1943; Kramer, 1958; MacIntyre, 1965; Broecker and Takahashi, 1966). Much of the earlier work has been critically examined by Edmond and Gieskes (1970) (see also Li et al., 1969; Ben-Yaakov and Goldhaber, 1973; Ingle et al., 1973).

Solubility product measurements made in natural sea water are subject to the possible risk of uncertainties arising from changes on the surface of the solid as a result of calcium ion exchange with other cations, particularly Mg^{2+}, of sea water. Several investigators have drawn attention to this possibility (see, e.g., Berner, 1966, 1971; Pytkowicz and Fowler, 1967; Pytkowicz et al., 1967; Pytkowicz, 1968, 1969a), and it has been suggested that this may cause solubility behaviour which is not well-defined in the thermodynamic sense. Solubility hysteresis may be present, the solubility product determined on approach from undersaturation being different from that determined on approach from oversaturation. However, hysteresis was not detected by MacIntyre (1965). Although the results obtained by MacIntyre are probably the most reliable ones available up to 1973 (see e.g. Edmond and Gieskes, 1970), their strict applicability to the conditions which obtain in sea water has been questioned (Ben-Yaakov and Goldhaber, 1973—see below).

Over an ionic strength range up to that of ocean water K'_{calc} and K'_{arag} increase with increasing salinity (partly because of non-specific Debye–Hückel type interactions, but mainly because of ion pair formation), and at the salinity of ocean water they are some 100-fold greater than the corresponding values for solutions in distilled water. Plots of the available data given by Edmond and Gieskes (1970) for K'_{calc} and K'_{arag} indicated that there was reasonable agreement between the measurements made by Hindman (1943) and those made by MacIntyre but, relative to these, the earlier determinations made by Wattenberg and Timmerman (1936) showed an apparent systematic error possibly attributable to the quinhydrone electrode used in their work (Edmond and Gieskes, 1970).

Edmond and Gieskes concluded that K'_{calc} and K'_{arag} are given by

$$K'_{calc} = (0.1614 + 0.02892 \times Cl\% - 0.0063t) \times 10^{-6}\,mol^2\,l^{-2}$$

and

$$K'_{arag} = (0.5115 + 0.02892 \times Cl\% - 0.0063t) \times 10^{-6}\,mol^2\,l^{-2},$$

where t is the temperature (°C) for the chlorinity range 15–25‰ and the temperature range 0 to 40°C. Similar relationships have also been proposed by Li *et al.* (1969). However, more recent determinations (Ingle *et al.*, 1973) suggest that both the value for K'_{calc} and the temperature coefficient are less than the above expression implies (see below).

The strict validity of applying certain laboratory determinations of K'_{calc} and K'_{arag} to natural water systems has been questioned by Ben-Yaakov and Goldhaber (1973). The basis of their criticism was that apparent constants are functions of both ionic strength and solution composition. Consequently, experimentally determined apparent constants should really only be applied to media having both a composition and an ionic strength similar to that used in the determination of the constants. Now, for reasons of experimental convenience MacIntyre made part of his investigation using partial pressures of carbon dioxide of 1 atmosphere with the result that the calcium concentration and, particularly, the carbonate alkalinity were enhanced compared with those of natural sea water. Ben-Yaakov and Goldhaber consider that these compositional changes so affect K'_{calc}, K'_{1l} and K'_2 (the latter being used in the calculation of the solubility product) that adjustment is needed before MacIntyre's results are applicable to sea water. A correction procedure based on the Garrels and Thompson (1962) model indicated a value for K'_{calc} (applicable to sea water of 19 Cl‰ at 25°C) of $5 \cdot 2 \times 10^{-7} \, mol^2 \, l^{-2}$—about $6 \cdot 3\%$ below that given by MacIntyre. However, Ingle *et al.* have pointed out that in making this correction, Ben-Yaakov and Goldhaber did not take account of the fact that some of MacIntyres work was done at 18 Cl‰.

The determinations by Ingle *et al.* (1973) were made in the light of previous investigations, and particular care was taken to eliminate sources of uncertainty arising from the condition of the solid (analytical grade calcium carbonate and Iceland spar) by a washing procedure and ultra-sonic treatment intended to remove strain. Equilibrium was approached from both under and supersaturation. Because accurate knowledge of the second ionization constant of carbonic acid is crucial if reliable values for K'_{calc} were to be obtained, these workers based their measurements on the determination of K'_2 made by Mehrbach *et al.* (1973). According to Ingle *et al.*, the value of K'_{calc} at 25°C is $(4 \cdot 59 \pm 0 \cdot 05) \times 10^{-7} \, mol^2 \, (kg \, sea \, water)^{-2}$ with a temperature coefficient of $-0 \cdot 0108 \times 10^{-7} \, (°C)^{-1}$ between 2 and 25°C. The average values reported by Ingle *et al.* are given in Table 9.16 and according to them, K'_{calc} at one atmosphere pressure is given by

$$K'_{calc} = (1 \cdot 390 - 3 \cdot 096 \times 10^{-3} \, t) \times S \times 10^{-8} \, mol^2 \, kg^{-2}$$

when t is the temperature (°C) and S is the salinity (‰).

TABLE 9.16

The apparent solubility product of calcite in sea water of 35‰ salinity. (From Ingle et al., 1973)

Temperature (°C)	$K'_{calc} \times 10^7$ (mol^2 (kg sea water)$^{-2}$)
2	4·83
13	4·75
25	4·59

The dependence of the apparent solubility products of calcite and aragonite on pressure has been examined by Pytkowicz and Connors (1964) (aragonite oolites, 6 and 21°C), Pytkowicz *et al.* (1967) (aragonite oolites, 2°C) Pytkowicz and Fowler (1967) (calcitic foraminifera, 22°C) and Hawley and Pytkowicz (1969) (aragonite oolites, 2°C). The original investigation by Pytkowicz and Connors was based on a decompression technique in which aragonite oolites were equilibrated with the solution at high pressures, the system decompressed and the pH measured. It was considered that the rate of nucleation of the supersaturated decompressed solution was low enough not to invalidate the results. However, Edmond and Gieskes (1970) consider that the slightly lower values for K'_{arag} at high pressures obtained by Pytkowicz and Connors compared with later determinations made by the Pytkowicz school implies that there may have been some reprecipitation. Later determinations of K'_{arag} and K'_{calc} at high pressures $((K'_{arag})_P$ and $(K'_{calc})_P)$ made by Pytkowicz and his co-workers were based on *in situ* measurements of the pH in a high pressure cell. For aragonite, the pressure coefficient (defined here as the ratio of the apparent solubility product at pressure P to that at one atmosphere—see also Pytkowicz (1968)—is given by

$$\frac{(K'_{arag})_P}{(K'_{arag})_1} = \frac{(c_{Ca(T)})_P \cdot (c_{CO_3(T)})_P}{(c_{Ca(T)})_1 \cdot (c_{CO_3(T)})_1} \tag{9.130}$$

Since the calcium and bicarbonate concentrations change only slightly during the equilibration whereas that of the carbonate changes several fold, expression (9.130), when combined with that for the second apparent ionization constant for carbonic acid $(K'_2 = a_H \cdot c_{CO_3(T)}/c_{HCO_3(T)})$ gives, to a good approximation,

$$\frac{(K'_{arag})_P}{(K'_{arag})_1} = \frac{(K'_2)_P \cdot (a_H)_1}{(K'_2)_1 \cdot (a_H)_P}$$

A similar expression applies for calcite. Somewhat more complex expressions were needed for the pressure coefficient when borate was present in the water (Hawley and Pytkowicz, 1969). The appropriate values for $(K'_2)_P$ were taken from the paper by Culberson et al. (1967).

Pressure coefficients for the calcite and aragonite apparent solubility products are given in Table 9.17 and also in Fig. 9.8 (Edmond and Gieskes, 1970) which shows the log of the ratio to be a linear function of applied pressure up to 1000 atm. Some hysteresis was noted on the relaxation of

TABLE 9.17

Pressure coefficients of the apparent solubility products of calcite and aragonite

	Temp (°C)	Cl (‰)	P (atm)		$(K'_i)_P/(K'_i)_1$
Aragonite oolites					
			500		1·79
(a)	22	19·16	1000		3·18
			500		1·22
					2·05
					2·17
(b)	2	19·16	500		2·17
					2·04
				av.	2·11
					3·94
					4·48
(b)	2	19·16	1000		4·34
					4·14
				av.	4·23
					1·80
(b)	22	19·16	500		1·81
				av.	1·80
					3·17
(b)	22	19·16	1000		3·14
				av.	3·16
Calcite					
			100		1·10
			200		1·25
			500		1·88
(a)	22	19·16	750		2·62
			1000		3·56
			500		2·12
			1		1·18

(a) Pytkowicz and Fowler (1967).
(b) Hawley and Pytkowicz (1969).
Note the hysteresis shown by the calcite data. Pressure coefficients found with increasing applied pressure are more relevant to the conditions experienced by descending carbonate particles.

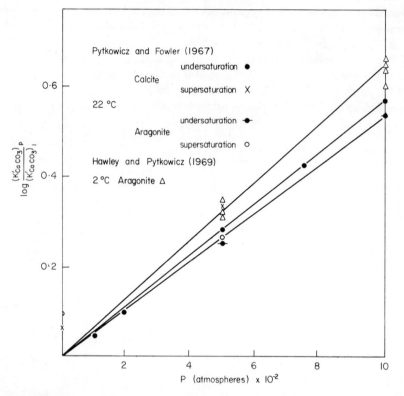

FIG. 9.8. Log $[(K'_{CaCO_3})_p/(K'_{CaCO_3})_1]$ as a function of pressure for calcite and aragonite. (After Edmond and Gieskes, 1970).

pressure, possibly because of some change of the particle surface. However, in practice, descending tests experience only a progressive pressure increase. This linear relationship implies that $\Delta \overline{V}$, the partial molar volume change accompanying solution, is pressure independent and allows the simple integrated form of

$$\frac{\mathrm{d} \ln (K')_P}{\mathrm{d}P} = -\frac{\Delta \overline{V}}{RT}$$

viz.

$$\log [(K')_P/(K')_1] = -\Delta \overline{V}(P - 1)/2 \cdot 303RT$$

to be used for the evaluation of the pressure coefficient at any particular pressure. Values for $\Delta \bar{V}$ for calcite and aragonite considered by Edmond and Geiskes to best describe the available data are*

$$\Delta \bar{V}_{calc} = -(35 \cdot 4 - 0 \cdot 23t)\, cm^3\, mol^{-1}$$

and

$$\Delta \bar{V}_{arag} = -(32 \cdot 8 - 0 \cdot 23t)\, cm^3\, mol^{-1}$$

where t is the temperature (°C). The temperature dependence estimate is based on the experiments of the Pytkowicz school for aragonite at 2 and 22°C, and was assumed to apply also to calcite.

In their analysis of the available data, Li et al. (1969) gave $\Delta \bar{V}$ to be $-29 \cdot 7$ $cm^3\, mol^{-1}$ and $-27 \cdot 0\, cm^3\, mol^{-1}$ for calcite and aragonite respectively, and considered the temperature coefficients of $\Delta \bar{V}$ to be negligible.

In the above relationships, $\Delta \bar{V}_{calc}$ would be expected to correspond to $\bar{V}^*_{CaCO_3} - V_{calc}$ where V_{calc} is the molar volume of calcite ($36 \cdot 9\, cm^3\, mol^{-1}$ —Weast (1965)) and $\bar{V}^*_{CaCO_3}$ is the partial molar volume of $CaCO_3$ in sea water. A determination of $\bar{V}^*_{CaCO_3}$ by Duedall (1972) gave a value of $-6 \cdot 5 \pm 1 \cdot 7$ $cm^3\, mol^{-1}$ which, when combined with the molar volume of calcite, leads to a predicted value for $\Delta \bar{V}_{calc}$ of $-43 \cdot 4\, cm^3\, mol^{-1}$ at 20°C. Millero and Berner (1972) have calculated this quantity on the basis of an ion pairing model to be $-39 \cdot 4\, cm^3\, mol^{-1}$ for calcite and $-36 \cdot 7\, cm^3\, mol^{-1}$ for aragonite (both at 25°C). On the basis of the Edmond and Gieskes equation cited above, the values for $\Delta \bar{V}_{calc}$ derived from the measurements of the Pytkowicz group are $-30 \cdot 8\, cm^3\, mol^{-1}$ at 20°C and $-29 \cdot 6\, cm^3\, mol^{-1}$ at 25°C. A satisfactory explanation of the difference of about $10\, cm^3\, mol^{-1}$ between the predicted and observed values is not immediately obvious. The fairly close agreement between the Millero and Berner calculations and the Duedall dilatometric based estimates implies that the assumptions underlying the calculations made by Millero and Berner are sound. Millero and Berner have suggested that the observed value of $\Delta \bar{V}_{calc}$ may be in error, and Duedall has suggested the possibility that true chemical equilibrium was not reached in the experiments of the Pytkowicz group. In his view this would lead to a numerically smaller volume change than the equilibrium $\Delta \bar{V}_{calc}$. Estimates made by Bradshaw (1973) for the volume change associated with the solution by sea water of calcite and aragonite agree with Duedall's measurements. In Bradshaw's view a possible explanation of the difference between the Millero and Berner (1972), Duedall (1972) and Bradshaw (1973) estimates of the partial molar volume change and that derived by Edmond and Gieskes (1970) from the measurements made by the Pytkowicz group is that the "theoretical" estimates apply to pure phases whereas the measurement by Pytkowicz

---

* Subsequently (Gieskes, 1974) modified to $\Delta \bar{V}_{calc} = -(47 \cdot 5 - 0 \cdot 23t)\, cm^3\, mol^{-1}$ and $\Delta \bar{V}_{arag} = -(45 \cdot 0 - 0 \cdot 23\, t)\, cm^3\, mol^{-1}$.

and Fowler, for example, was based on natural samples into the surfaces of which Mg^{2+} ions might have become incorporated.

Whatever the explanation of the discrepancy between the observed and predicted values, the implications in the interpretation of $CaCO_3$-depth profiles (see later) are important, and further investigation is needed.

In the main, experimental investigations of the solubility product have been concerned with pure calcite or aragonite and it must be remembered that often the solid phase in contact with the water may have incorporated in it considerable magnesium. The dependence of the solubility product on the magnesium content of calcites is not yet fully understood, but it seems likely that high magnesium calcites have higher solubilities than does aragonite (Chave *et al.*, 1962).

9.2.4.2. *Kinetic factors associated with* $CaCO_3$ *solution and precipitation*

Most of the carbonate particles which fall through the water column are skeletal tests, and only in a few restricted areas, such as the Bahama Banks (Smith, 1940; Cloud, 1962a, b), where abundant nuclei and seed crystals exist does chemical precipitation occur. Whether or not chemical precipitation occurs from a supersaturated solution at an appreciable rate, and even the form which first appears (calcite or aragonite), depends on kinetic rather than thermodynamic factors. These are determined by environmental conditions, such as the degree of supersaturation, the concentrations of other ions (particularly Mg^{2+}) (Pytkowicz, 1965a, 1973a), the nature and amount of the organic matter in the water and the temperature. Generally, the rate of nucleation is extremely small, and Pytkowicz (1965a) has estimated that at the degree of supersaturation which exists in surface ocean water some 10^5 years would be necessary for precipitation to occur, although Berner (1971) considers that these laboratory determinations are not likely to reflect conditions in the natural environment. When precipitation does occur, the metastable aragonite is, of course, the form to appear first.

Magnesium markedly retards the rate of calcite crystallization from dilute solutions, the rate of aragonite growth being much less affected (Smallwood, 1972). A similar influence by Mg^{2+} ions is noted in artificial sea water solutions (Kitano and Hood, 1962; Chave and Suess, 1970). Other inorganic materials including phosphorus compounds (Simkiss, 1964; Pytkowicz, 1973a) also influence the relative rates of crystallization of calcite and aragonite.

It is known that organic materials may influence polymorph formation (Kitano and Hood, 1965) and that many substances (e.g. stearic acid and albumin) are adsorbed by both calcite and aragonite (Suess, 1970). It is likely that such sorption on growth sites will interfere with the rate of crystallization. Smallwood (1972) has shown that in controlled experiments using solutions of low ionic strength, albumin retarded the growth of aragonite crystals but was without appreciable influence on the growth of calcite. Chondroitin sulphate, sodium alginate and agar addition retarded the growth of both polymorphs, and agar additions inhibited the formation of seed crystals. Kitano *et al.* (1970) have described the effect of a wide range of additives on the crystal form of the carbonate crystallizing from $CaCO_3$-rich solutions, and Pytkowicz (1973a) has shown that stearic acid is without effect on the rate of crystallization. Citric acid, when present at high concentration, had an inhibiting effect similar to that of magnesium. This effect was not attributed to complexing of the calcium ions.

The possibility that organic additives retard crystallite formation by sorption at growth sites is given support by results obtained by Chave and Suess (1970) who showed that as precipitation occurred, organic matter is removed from solution together with the precipitate, and that rapid $CaCO_3$ precipitation does not occur until much of the organic matter has been removed. This is also consistent with the observation that the rate of adsorption of organic matter onto carbonate surfaces is faster than the deposition of $CaCO_3$ on the same surface (Chave and Suess, 1967). In the absence of organic additives the rate of nucleation in seed-free (filtered) artifical sea water is influenced by the degree of supersaturation.

The dissolution of calcium carbonate is also a kinetically controlled process, although the mechanisms which apply in the oceans are not yet fully understood. The rate is likely to be a function not only of the kinetics of attack of ions (such as H_3O^+) immediately adjacent to the surface, and possibly also of the rate of diffusion of reactants and products down the chemical potential gradients which must be established in the boundary layer region (although see Section 9.2.4.4). In the event of organic coatings being present, diffusion through these must also be considered. During the course of attack, if diffusion is the rate controlling process, the environmental conditions (pH and ionic concentrations) immediately adjacent to the surface of the dissolving particle and within the boundary layer will differ from those in the bulk solution. The thickness of the boundary layer and, therefore, the rate of attack will depend on particle shape and size and on the amount of turbulence in the water, this being greater for descending than for static particles. Inhibition of attack by specifically adsorbed species and charge separation phenomena may further complicate interpretation of events at the interface (see also Section 9.2.4.4).

Similar, but if anything more difficult, problems of calcium carbonate solution and deposition exist for reactions taking place in the pore solutions of sediments. The rates of these reactions are almost certainly transport controlled, although little is known concerning their detailed mechanisms, the sort of concentration gradients which exist and the relevant diffusion coefficients. The overall processes which occur in carbonate-rich sediments in contact with sea water are complex, as illustrated by observations made by Pytkowicz (1971b) on events which occur during the interaction of sea water with Bahama beach deposits. These deposits contained aragonite and both high and low magnesium calcites. The surface near-shore water was super-saturated with respect to both aragonite and pure calcite, but despite this there was no evidence of calcium carbonate deposition or cementation (beachrock formation), at least to depths of ca. 1 m in the intertidal zone and ca. 0·3 m in the subtidal zones of several beaches. The absence of $CaCO_3$ deposition was attributed to protective organic coatings possessed by the particles. The pore water of the upper few inches of most sands of the inter-tidal zones had similar pH, alkalinity and carbonate saturation values to those of adjacent near-shore water, but below this level the low pH and higher alkalinity implied that bio-oxidation of organic material had led to the generation of CO_2. Under these high P_{CO_2}, low pH, conditions the carbonate concentration was depressed sufficiently to make the water undersaturated with respect to the high magnesium calcite fraction of the deposits thereby leading to dissolution of calcium carbonate and hence an increased alkalinity.

9.2.4.3. Calcium carbonate-depth profiles

A convenient measure of the degree of saturation of sea water with respect to calcium carbonate is the ratio of the observed stoichiometric ionic product $(c_{Ca(T)} \cdot c_{CO_3(T)})$ to the corresponding apparent solubility product. Thus, if for calcite the degree of saturation is denoted by Ω_{calc} and the observed ionic product by IP, then

$$\Omega_{calc} = \frac{c_{Ca(T)} \cdot c_{CO_3(T)}}{[c_{Ca(T)} \cdot c_{CO_3(T)}]_{calc}} = \frac{IP}{K'_{calc}}$$

A similar expression applies for aragonite. In using this expression, $c_{CO_3(T)}$ is the carbonate ion concentration estimated for example by the apparent constant methods outlined in previous sections, and K'_{calc} is the value of the concentration (stoichiometric) solubility product appropriate to the par-ticular depth (pressure). The carbonate system can also be handled within a framework comprising the thermodynamic constants for the ionization of carbonic acid and the activity product for calcite by making use of individual and stoichiometric activity coefficients ($\gamma_{i(f)}$ and $\gamma_{i(T)}$) obtained as described by Berner (1971). In this way the ionic activity product $a_{Ca} \cdot a_{CO_3}$ (=IAP) can be computed, and the ratio of this to the activity solubility product of calcite (K_{calc}) is entirely equivalent to Ω_{calc}.

Other methods of quantifying the degree of saturation with respect to

$CaCO_3$ are sometimes used. The quantity ΔS, the number of moles of calcium carbonate which must dissolve or precipitate in order to achieve saturation, has certain merits (see Pytkowicz, 1968 for a discussion of this and other aspects of calcium carbonate/sea water problems).

A value for Ω of unity indicates saturation; values less than or greater than unity indicate undersaturation or supersaturation respectively. Under equivalent conditions of temperature and pressure, $K'_{arag} > K'_{calc}$ and, therefore, the degree of saturation with respect to calcite will be greater than that with respect to aragonite. For this reason certain waters may be saturated with respect to calcite but not with respect to aragonite.

There is general agreement that most oceanic surface waters are super-saturated with respect to $CaCO_3$ (Pytkowicz, 1965a, 1968, 1970, 1972b; Berner, 1965, 1971; Li et al., 1969; Edmond and Gieskes, 1970; Pytkowicz and Fowler, 1967; Ben-Yaakov and Kaplan, 1971; Ben-Yaakov et al., 1974; Ingle et al., 1973), precipitation being retarded because of the factors noted above. However, intercomparison of Ω_{calc} estimates from different stations is often difficult because of the dependence of the calculated profile on the values chosen for the surface pressure solubility product of calcite and the pressure coefficient of the solubility product, and satisfactory agreement between saturometer-based and shipboard analysis based saturation profiles has not yet been attained (Ben-Yaakov et al., 1974). A number of published profiles may need revision as more refined values for the constants become available.

Recently, Ingle et al. (1973) have recalculated depth-saturation profiles based on results obtained on the YALOC-69 (Wyatt et al., 1970) and Circe (Edmond, 1970) cruises using the recent determinations of K'_{calc} and K'_2 values determined by Mehrbach et al. (1973). The results of their calculations are summarized in Fig. 9.9 together with a profile for the Eastern Pacific off southern California (Ben-Yaakov and Kaplan, 1971) obtained by means of a saturometer method. Although the profile obtained by Ben-Yaakov and Kaplan should be comparable with those obtained from the other Pacific profiles (e.g. YALOC-69), it shows a more pronounced maximum at about 2000 m and a somewhat lower saturation in the surface waters. It is apparent from Fig. 9.9 that, although the general forms of the profiles are similar for both the Atlantic and Pacific stations, there is a greater degree of saturation in the Atlantic, and only below 4200 m does the water become under-saturated. In both oceans the fall in saturation below the first few hundred metres coincides with the increase in ΣCO_2 and the fall of pH caused by the oxidation of organic matter; the fall-off of Ω with depth at greater depths is a consequence of the effect of pressure and temperature on the equilibrium constants governing the system. (See also discussion by Edmond (1974)).

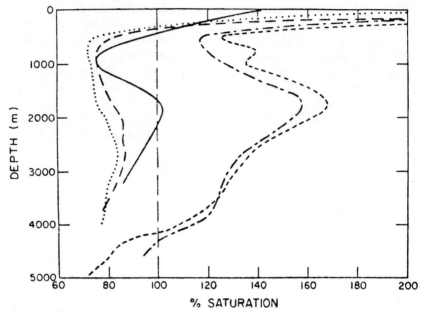

FIG. 9.9. Percentage saturation with respect to calcite versus depth. ———, YALOC-69 stn. 121;
········, YALOC-69 stn. 127; ———, Ben-Yaakov and Kaplan (1971); —·—·—, "Circe" stn. 223;
-----, "Circe" stn. 246 (Ingle *et al.*, 1973).

9.2.4.4. *Carbonate saturation and compensation depths*

The depth at which the water is just saturated with respect to $CaCO_3$ (supersaturation and undersaturation existing respectively above and below it) is the saturation depth. The compensation (or vanishing) depth corresponds to the horizon at which an abrupt decrease in the abundance of $CaCO_3$ in the sediment occurs. Both of these depths depend on location and on the form of $CaCO_3$ being considered, but the compensation depth often lies 2000 m or so below the saturation depth (Fig. 9.10). Thus, sedimentary calcareous material exists at depths at which the bulk ocean water is undersaturated. It has long been realized that calcareous sediments persist to great depths (Murray and Hjort, 1912; see also Arrhenius, 1963; Turekian, 1965). The coexistence of undersaturated water and calcareous material implies that kinetic rather than thermodynamic factors determine the compensation depth, although other possibilities have been considered (see the review by Pytkowicz, 1968, and also Li *et al.*, 1969 and Edmond, 1974). A low dissolution rate would be expected to result from slow diffusional transport of material through the boundary layer which surrounds descending particles, protective organic coatings which impede both transport and solution attack, the inhibiting effect of

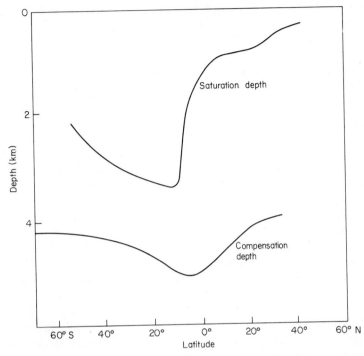

FIG. 9.10. The carbonate compensation depth (Lisitsin and Petelin, 1967) and the saturation depth (Hawley and Pytkowicz, 1969) in the Pacific (Pytkowicz, 1970).

specifically adsorbed species such as PO_4^{3-} (Morse and Berner, 1972), the slow replacement of the calcium carbonate saturated layers formed at the surface of the sediments, or the partial protection of calcareous material by its burial with inert material.

Somewhat above the compensation depth is the lysocline (at about 4000 m depth in the tropics on the western side of the East Pacific Rise). This is an important zone from a kinetic point of view since it marks a depth region at which there is a sharp increase in the rate of dissolution of calcium carbonate (Berger, 1967, 1968, 1970, 1971). Since the position of the lysocline is not always associated with the boundary of major water masses, the increase in the dissolution rate is unlikely to be connected with a change of mass type. The position of the compensation depth is almost certainly connected with that of the lysocline and, in kinetic terms, corresponds to the depth at which the rate of sedimentation equals that of dissolution.

Various attempts have been made to recognize the factors which influence the rate at which calcium carbonate particles enter solution (Heath and

Culberson, 1970; Pond *et al.*, 1971; Berner, 1971; Morse and Berner, 1972). Knowledge of this rate would be useful in the interpretation of alkalinity profiles and would help to decide whether observed increases of alkalinity with depth reflected the properties of the source waters of the deeper masses or whether they were consequences of the *in situ* uptake of calcium carbonate (Pytkowicz, 1968). Information on the particle size distribution as a function of depth would be useful in this respect, but in the absence of such information Pond *et al.* (1971) attempted to calculate the rate of dissolution of spherical particles as they settle. They concluded that significant dissolution, particularly of small, thin-shelled foraminifera was likely to occur and that only the large, thick-shelled, species would reach depths below *ca.* 3500 m. However, Berger's (1970) investigations of the dissolution rates of foraminifera as a function of depth indicated little solution attack on thin-shelled specimens suspended at depths shallower than 2500 m for periods as long as 4 months, and Morse and Berner (1972) consider that diffusion is not the rate limiting factor governing attack, at least in the upper part of the water column. In view of the observations made by Berger (1967, 1968, 1970, 1971), those by Peterson (1966) using calcite spheres suspended at various depths and those by Bramlette (1961) on Pacific core tops, it seems clear that some dissolution occurs at all but the shallowest depths, but that the rate increases rapidly in the lysocline zone above the compensation depth. Evidently some critical change in the solution–solid relationship occurs in this zone and makes more rapid attack possible.

A general kinetic view of events below the lysocline can be obtained by focusing attention not on the rate of accumulation of calcite in the sediments, but rather on the percentage of the original detrital particles which dissolve (Heath and Culberson, 1970). By adopting a simple model in which the rate of dissolution was assumed to be zero above the saturation depth, very low between the saturation depth and the lysocline and to increase linearly with depth below the lysocline, Heath and Culberson were able to show that a sharp fall-off in the percentage of calcite remaining undissolved would occur at a certain depth (corresponding to the compensation depth). This general conclusion would not be expected to be very sensitive to the way in which the solution rate depended on depth below the lysocline.

The nature and mechanism of the increase rate below the lysocline was not specified, but recent experiments by Morse and Berner (1972) have given information which bears on this aspect of the problem. Laboratory studies of the rates of dissolution in sea water of both reagent grade calcite and deep sea carbonate sediments were made under conditions of constant P_{CO_2} and pH. For each experiment the parameter ΔpH defined by

$$\Delta pH = \overline{pH} - pH$$

could be calculated where $\overline{pH}$ was the pH of a saturated calcite solution having the same P_{CO_2} and $c_{Ca(T)}$ values as those used in the experiment, and pH was the constantly maintained value. Morse and Berner were able to show that there was a pronounced increase in the rate of dissolution when ΔpH was 0·14 to 0·16. By making use of ΔpH values calculated by Berner and Wilde (1972) for a vertical section at the station used by Peterson (1966) and Berger (1967), Morse and Berner showed that when their laboratory results were expressed in terms of equivalent depth, the depth-rate of dissolution relationship was closely similar to those given by the field experiments. In their view some change in the mechanism of dissolution involving the particle surface occurs when ΔpH reaches a critical value. Possibly this is associated with the removal of some surface active inhibitor. Morse and Berner consider that PO_4^{3-} is in some way involved. (See also Morse, 1974a, b; Morse and Berner, 1974).

Recently, Edmond (1974) has discussed evidence bearing on the origin of the compensation depth. He considers that the non-coincidence of the saturation and compensation depths discount the thermodynamic (instantaneous equilibrium) model (Li *et al.* 1969) and in place of the kinetic model developed by Berner, he favours a steady state model in which much of the dissolution occurs at, or near, the sediment surface through the agency of bottom water motion. (See also Takahashi (1974)).

Problems associated with the oceanic distribution of calcium and carbonate are linked to the dissolution of descending particles and have been discussed in some detail by Pytkowicz (1967, 1968, 1973b); only the main features are outlined below.

Essentially, calcium and carbonate are distributed in accordance with two interlocking natural cycles, the main elements of which are summarized first in much simplified form in Fig. 9.11. Calcium derived from terrestrial reservoir A by weathering, mainly of carbonate, but also to a lesser extent of igneous rocks, enters (via route 1) the upper layers of the sea (reservoir B) where it joins calcium delivered from deeper water by advection and upwelling and becomes available for bio-incorporation in carbonate secreting organisms. Ultimately, following the death of these organisms, calcium carbonate particles precipitate via route 2; of this precipitated material, part becomes more or less permanently incorporated (via route 3) in the sediments (reservoir C), the remainder dissolving (route 4) and joining admixed water from B in reservoir D. As a result of oceanic turn-over, water of high calcium content is returned to the surface (route 5) where, because of the reduced hydrostatic pressure, higher temperature and higher $c_{CO_3(T)}$ supersaturation develops (Pytkowicz, 1963). In the long term, sea floor spreading will result in some of the calcium and carbonate being re-introduced (via metamorphic and tectonic activity) to the weathering environment.

An idealized $CO_2/HCO_3^-/CO_3^{2-}$ cycle is shown in Fig. 9.11(b). This indicates that CO_2 enters the sea surface either via terrestrial weathering as bicarbonate or directly from the atmosphere. Within each zone of the sea interchanges between the various sub-reservoirs of carbon (i.e., organic carbon, dissolved CO_2, HCO_3^- and CO_3^{2-}) occur in accordance with physical and biological processes. Thus, removal of $CaCO_3$ from solution by shell-building fauna results in the generation of CO_2 by

$$Ca^{2+} + 2HCO_3^- = CaCO_3 + H_2O + CO_2,$$

and the liberated CO_2 may be used in photosynthesis, escape to the atmosphere or remain in solution. Descending $CaCO_3$ tests carry CO_3^{2-} to the intermediate and deep reservoirs, but since below the photic zone, and particularly in the intermediate zone, organic debris, which also falls through the water column, is oxidized to CO_2 in accordance with the approximate representation

$$CH_2O + O_2 \rightarrow CO_2 + H_2O,$$

conditions favourable to the solution of calcium carbonate thus occur. Solution of $CaCO_3$ is most intense in the deep water, about 50% of the oxidation generated CO_2 of the deep reservoir being used in this way (Pytkowicz, 1967); the rest joins that formed in the intermediate layer and is returned to the surface by advection and in regions of upwelling.

These two cycles interlock in the sense that certain routes are common to both (e.g., the transport of $CaCO_3$ by descending calcareous tests corresponds to the transport of both calcium and carbonate) and more comprehensive versions of them can be combined to give the overall representation used by Pytkowicz (1973b) shown in Fig. 9.12. In this, approximate values are given for the estimates of the various fluxes in units of 10^{14} g C year^{-1}; the reservoir sizes are in units of 10^{20} g C. It is evident that this combined cycle also interlocks with other natural cycles such as those for oxygen and those nutrients which are utilized in the photic zone and liberated in the lower zones where oxidation and decomposition occurs.

Clearly, both carbonate and calcium enter into a complex interlocking system. Parts of the overall system can be adequately treated on a thermodynamic equilibrium basis (e.g., the basic CO_2—HCO_3^-—CO_3^{2-} ionizations), but other parts are demonstrably not in a state of thermodynamic equilibrium, particularly when viewed within the larger context of the overall system (note, for example, the supersaturation of surface waters with respect to $CaCO_3$, the variable distribution of the degree of calcium carbonate saturation in the oceans and the persistence of calcium carbonates in sediments when the overlying water is undersaturated). According to Pytkowicz (1973b)

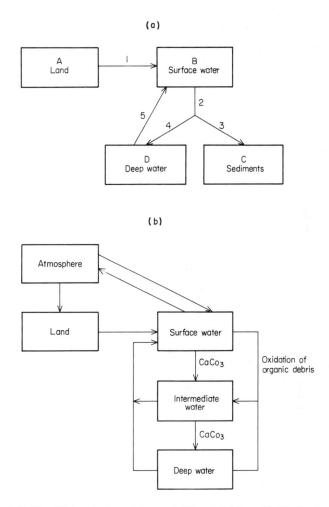

FIG. 9.11. Simplified cycles for calcium and CO_2. (a) Calcium; (b) CO_2 (see text).

the complex cycle visualized by him and summarized in Fig. 9.12 was approximately in a steady state before becoming perturbed by human activity, and the long-term pH stability of the oceans is a consequence of the steady state properties of the weathering, biogenic and chemical reaction components of the cycle rather than of simple mineral-sea water thermodynamic equilibrium interactions. (cf. Sillén, 1961, 1967).

Many facets of the proposed cycle are conjectural or uncertain (including the magnitudes of some of the proposed fluxes), and the cycle as a whole is not

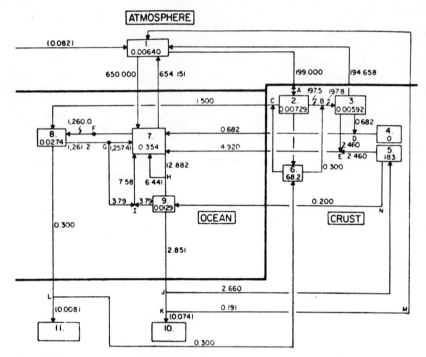

Fig. 9.12. Oceanic calcium and CO_2 cycles. The figures indicate the sizes of the carbon reservoirs (10^{20} g C) and the values assumed for the carbon fluxes (10^{14} g C year^{-1}) (Pytkowicz, 1973b).

The symbols have the following meaning:

A	Photosynthesis on land	1	Atmospheric CO_2
B	Oxidation on land	2	Organic carbon on land
C	Physical weathering of organic carbon	3	Dissolved inorganic carbon on land
D	Weathering of aluminium silicates	4	CO_2 content of aluminium silicates
E	Weathering of carbonates	5	Sedimentary carbonates
F	Photosynthesis in the oceans	6	Lithified organic carbon
G	Oxidation in the oceans	7	Dissolved inorganic carbon in the oceans
H	Formation of calcareous tests	8	Organic carbon in the oceans
I	Dissolution of tests	9	Calcareous tests plus detrital carbonates
J	Junction	10	Carbonate sink
K	Junction	11	Organic carbon sink
L	Junction		
M	Metamorphism		
N	Physical weathering of $CaCO_3$		

yet of proven validity. Nevertheless, it is a useful concept since it draws together many otherwise isolated aspects of carbonate geochemistry and it provides a broad basis from which further examinations of likely circulation paths and inter-relationships between the various biogenic, chemical and mineralogical reactions can be made.

The properties of cyclic systems are amenable to algebraic analysis. For an examination of features of cyclic and equilibrium systems and consideration of particular aspects of Fig. 9.12 (such as reverse weathering) the original paper by Pytkowicz (1973b) should be consulted (see also Pytkowicz, 1971a).

9.2.4.5. *Effect of carbon dioxide uptake on the dissolution of calcium carbonate*

Recently, the possibility has been considered that the progressive uptake of carbon dioxide by the surface waters of the sea as a result of its growing industrial output (Section 9.3) may ultimately lead to the removal of super-saturation with respect to calcium carbonate through the process

$$CO_3^{2-} + CO_2 + H_2O \rightleftharpoons 2HCO_3^-$$

(Broecker *et al.*, 1971; Fairhall, 1973a, b; Zimen and Altenhein, 1973a, b). According to Broecker *et al.*, when $c_{CO_3(T)}$ has been reduced to below about 6×10^{-5} mol l^{-1}, the dissolution of aragonite is potentially capable of occurring; attack on calcite will commence when $c_{CO_3(T)}$ falls below about 3.5×10^{-5} mol l^{-1}. Although it has been suggested that if the present trends continue, the supersaturation may be removed in the near future (Fairhall, 1973a, b; Zimen and Altenhein, 1973a, b), this conclusion would follow only if the above reaction proceeded entirely to the right, that is, if *all* the added carbon dioxide reacted to remove carbonate ion. In fact, the above reaction is an equilibrium (Whitfield, 1974b, c; Skirrow and Whitfield, 1974), the stoichiometric equilibrium constant, K'_r, being given by K'_{1l}/K'_2, and the problem can be analysed by examining the dependence of $c_{CO_3(T)}$ on P_{CO_2}. This can be done by making the assumption that the total alkalinity remains constant throughout the carbon dioxide addition although conclusions drawn in this way differ only negligibly from those which follow if it is assumed that the carbonate alkalinity remains constant and P_{CO_2} and $c_{CO_3(T)}$ are related through equation (9.97) (Skirrow and Whitfield, 1974). Plots of P_{CO_2} versus $c_{CO_3(T)}$ made using this relationship (Fig. 9.13) show that at 25°C (i.e. for tropical waters, which are of the greatest interest in this respect), for carbonate alkalinities likely to be encountered, the carbonate ion concentration will not fall below the critical value until the partial pressure of carbon dioxide in the water is some 3000×10^{-6} atm or more. Presumably the atmospheric partial pressure of carbon dioxide must be at least of a similar magnitude. Whitfield (1974b, c) has arrived at a similar conclusion, and present evidence suggests that a biological catastrophe in tropical waters on this account is not immediately imminent. There is the possibility that, because of the temperature dependencies of the solubility coefficient of carbon dioxide and of K'_r, the saturation level for aragonite in more temperate waters may be approached early in the next century (Whitfield, 1974b, c). This

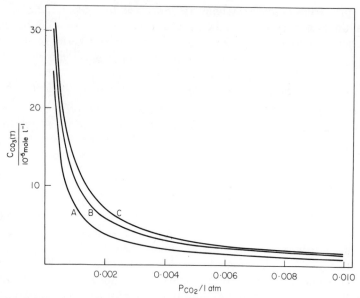

FIG. 9.13. Equilibrium concentration of CO_3^{2-} (T) as a function of P_{CO_2}. A: T = 10°C, CA = 2·4 meq^{-1}, K_r' = 1531; B: T = 25°C, CA = 2·2 meq^{-1}, K_r' = 1230; C: T = 25°C, CA = 2·4 meq^{-1}, K_r' = 1230. (Skirrow and Whitfield, 1974).

topic is controversional and discussion in the literature continues (see e.g., Fairhall, 1973b; Whitfield 1974d; Fairhall and Erickson, 1974).

Whitfield (1974b, c) has made an estimate of the pH change likely in the surface waters of the oceans by the year 2010 in the event of the realization of several pessimistic assumptions concerning (i) the total output of fossil fuel carbon dioxide and (ii) the fraction of this output which is assimilated into, and retained by, the mixed layer. According to him a pH change of less than 0·1 unit is to be expected. Calculations by Dyrssen (1974) agree with this estimate.

9.3. ATMOSPHERIC CARBON DIOXIDE

Atmospheric carbon dioxide occurs to the extent of about 320 p.p.m. (total amount *ca.* 0·0068 × 10^{20} g C), and relative to the Chicago standard (Craig, 1957) it shows a $\delta^{13}C$ enrichment of about −7‰. These are approximate values since carbon dioxide exchanges between natural carbon reservoirs and is produced in industrial areas. Consequently, its atmospheric concentration and isotopic composition show temporal, local and regional fluctuations. Although the total size of the atmosphere reservoir is very much less than some of the other natural carbon reservoirs (see Table 9.1), the atmosphere occupies a crucial position in the carbon cycle. Much of the total world carbon must at some time have passed through it. The size of the atmospheric reservoir and the magnitude of its response time towards interaction with adjacent reservoirs means that changes taking place in it can be detected over relatively short time scales, particularly as monitoring of the carbon dioxide content of it has now reached a high degree of precision. The intimate connection between the carbon dioxide of the atmosphere and that of the sea demands that some attention should be given to its atmospheric concentration and its short- and long-term changes, to variations in its isotopic composition and to the rate of exchange between the atmosphere and the sea.

The time scales for atmospheric mixing are short compared with those for the turnover of the sea. Nevertheless, the fact that the atmospheric CO_2 content depends on both geographical location and altitude shows that something equivalent to mixing barriers exist in the atmosphere. Formal, albeit somewhat arbitrary, recognition is given to these barriers when box models are used for discussions of atmospheric mixing. Some of these models (e.g. Fig. 9.14) are relatively complex (particularly if account is also taken of exchange with the sea) and attempt to take account of the finite transfer times not only between the hemispheres, but also between the stratosphere and the troposphere and between the troposphere, the biosphere and the sea. Exchange times which have been proposed (Walton et al., 1970) are shown in Table 9.18.

TABLE 9.18

Atmospheric carbon dioxide exchange times (Walton et al., 1970)

Stratospheric exchange times (years)	
upper stratosphere–lower stratosphere	2·3
lower stratosphere–troposphere	2·1
stratosphere–troposphere	4·0
Tropospheric exchange times (years)	
troposphere N–S	4·4
troposphere–ocean	9·3
troposphere N–ocean N	10·7
troposphere S–ocean S	8·0
troposphere–biosphere N	5·8
troposphere–biosphere (overall)	8·9

Some of these values are approximate, and there is evidence (Baxter and Walton, 1971) that the stratosphere–troposphere exchange time may not be constant (see also Nydal, 1963, 1967, 1968).

9.3.1. SHORT TERM VARIATIONS

Diurnal variations are an established phenomenon over most land areas, particularly when the sampling station is close to a major CO_2 source or sink. The range observed depends on the locality of the station, the season, prevailing weather and wind conditions, the height of sampling above ground level and the type of vegetation cover. Typical illustrations of the daily range at ground level have been given by Keeling (1958,1961) for woodlands, grasslands, mountain and coastal areas (e.g., Figs. 9.15 and 9.16).

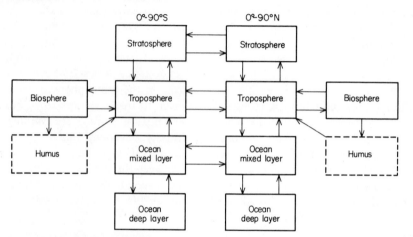

FIG. 9.14. Block model for the exchange of CO_2 within the atmosphere and between the atmosphere and other reservoirs (after Nydal, 1968).

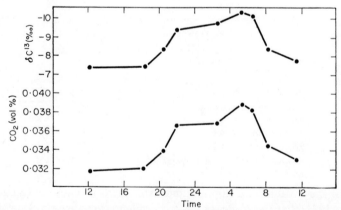

FIG. 9.15. Diurnal variation of the concentration and isotopic ratios of atmospheric CO_2 in a coastal redwood forest of California, 18–19 May 1955, Big Spur St. Pk. (Keeling, 1958).

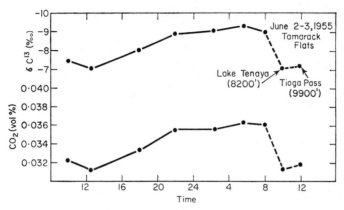

FIG. 9.16. Diurnal variation of the concentration and isotopic ratios of atmospheric CO_2 in a forest of the Sierra Nevada, 2–3 June 1955, Tamarack Flats (Keeling, 1958).

The daily range, which may be as great as 100 p.p.m., is strongly influenced by plant cover and is reduced when the air is derived directly from over the sea or is sampled over relatively barren ground. The $\delta^{13}C$ data (also included in Figs. 9.15 and 9.16) vary in sympathy with the CO_2 concentrations, and show that near-ground samples may be strongly influenced by the local environment (see also Leith, 1963).

In regions of fairly dense vegetation cover, the atmospheric carbon dioxide content increased from about 320 p.p.m. at mid-day to almost 400 p.p.m. during the night. This increase is, of course, related to plant and soil respiratory processes. Variations followed a similar pattern for different types of plant cover but were more intense in rain forest areas than over grassland and were small over barren land, mountains and the sea. The similarity between plots of the carbon dioxide content and those for $\delta^{13}C$ suggests that the two effects have a common origin in the addition to the atmosphere of carbon dioxide having a $\delta^{13}C$ value in the range 21–25‰—that generally found for terrestrial plants. Turbulence during the day causes mixing with the free atmosphere (see also Huber, 1952; Leith, 1963) and was responsible for the more or less general minimum value of 309–317 p.p.m. and a $\delta^{13}C$ value of -7‰ found at most stations including one receiving air directly from the Pacific. The pattern of the diurnal change depends on the season, and a peak value may be observed during the dark hours and a minimum during the day in autumn, whereas little diurnal variation may occur during early spring because of the absence of photosynthesis. Where diurnal variations resulting from plant respiratory processes do occur, it is to be expected that the range will become progressively smaller with height of sampling, although the

local effect has been detected at 20 m (Huber, 1952) and at 150 m by Chapman
et al. (1954).

Bischof and others (Bischof, 1960, 1962, 1965, 1970; Bolin and Bischof,
1970; Bolin and Keeling, 1963; Fonselius *et al.*, 1956) have commented on
seasonal variations in the atmospheric carbon dioxide content, and often the
measurements have included observations made at altitudes of up to several
km. Over Scandinavia considerable changes in the atmospheric content were
detected at heights of 3 km, but at 1000 m the range was 308–320 p.p.m. in
the 1950s (Bischof, 1960, 1962). The range and average increased towards
the ground, and a seasonal variation with a minimum in summer and a
maximum in the winter was noted. Even at the South Pole a small seasonal
variation of about 2 p.p.m. is detectable (Brown and Keeling, 1965).

Analysis of upper troposphere and lower stratosphere air samples collected
on commercial air routes in the northern hemisphere (Bolin and Bischof, 1970)
shows that distinct seasonal variations exist, these decreasing with increasing
altitude (Fig. 9.17). In addition, a phase shift of 25–30 days between the variations at 2 km height and the upper troposphere was noted. The seasonal
variation in the stratosphere is very much less than that in the upper troposphere. The tropospheric changes are similar to ones previously reported by
Bolin and Keeling (1963) on the basis of measurements made by continuous
infra-red gas analysers sited at Barrow (Alaska), Mauna Loa (Hawaii),
Little America (Antarctica) and the South Pole. This earlier survey which
covered the years 1957 to 1962, was also supported by examination of samples
taken on numerous aircraft flights at various altitudes and of samples taken at
surface level during the Monsoon, Downwind and other cruises as part of an
investigation of seasonal and latitudinal variations.

Superimposed on the seasonal changes is a gradual long-term increase of
about 0·7 p.p.m. yr^{-1} (see below) which needs to be taken into account when
seasonal and latitudinal data for a period of several years are being analysed.
Latitudinal gradients of the atmospheric carbon dioxide content are to be
expected if global circulation occurs in response to regional CO_2 sources
and sinks. Bolin and Keeling summarized their examination by giving
smoothed curves showing the carbon dioxide content for each month as a
function of latitude (Fig. 19.18). They explained their observations in terms of
a model in which there is a source of carbon dioxide in the tropics, an industrial
source in the mid-latitudes of the northern hemisphere and sinks in the polar
regions. They concluded that some $1·5 \times 10^{10}$ tons of carbon dioxide are
assimilated by the vegetation of the northern hemisphere north of 45°N
during the summer months. The time averaged sources and sinks proposed by
Bolin and Keeling are shown in Fig. 9.19.

Postma (1964) has made estimates of the global regions in which carbon-dioxide is transferred from the sea to the atmosphere. His estimates were based on consideration of the relationship between oxygen consumption and phosphate and carbon dioxide liberation in the sea. The relationships were established by reference to Pacific deep water, and deviations in the surface layers were attributed to carbon dioxide exchange with the atmosphere. The main escape routes were thought to lie in a broad band in the equatorial

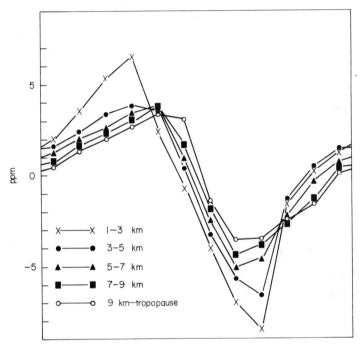

FIG. 9.17. Seasonal variations of the CO_2 content of the atmosphere at different levels in the troposphere (Bolin and Bischof, 1970).

region and in somewhat narrower bands at higher latitudes. The progressive increase in the total atmospheric carbon dioxide content (see below) will reduce these escape regions, and according to Postma, an increase of 25% in the total atmospheric CO_2 would eliminate the high latitude escape regions. An increase of more than 60% was thought to be necessary before the equatorial escape region would disappear.

9.3.2. LONG TERM VARIATIONS

In addition to the diurnal and seasonal changes in the atmospheric carbon dioxide content there exists a long-term increase. This and its possible con-

E

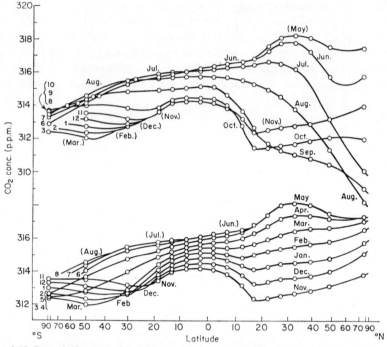

FIG. 9.18. Smoothed curves of the concentration of atmospheric CO_2 as a function of latitude for each calendar month (Bolin and Keeling, 1963).

nection with the increasing use of fossil fuels has been considered by several of the early investigators (Callendar, 1938, 1940, 1949, 1957, 1958; Bray, 1959; Slocum, 1955). Callendar considered that between 1860 and the 1930's an increase in atmospheric carbon dioxide of about 10% had occurred, and that this was mainly attributable to industrial output. Fonselius *et al.* (1956) also concluded that most of the fossil fuel produced carbon dioxide had remained in the atmosphere. The results of subsequent investigations agree that although industrial output has increased the atmospheric content, much of the output has been assimilated by other reservoirs. Detailed consideration of the trend over the last century is hampered by the lack of generally accepted atmospheric measurements for the early part of this period, particularly as many of the older measurements were made with no particular regard to local conditions, time or sampling method. However, careful observations made over the last two decades have established that a progressive increase in the concentration of atmospheric carbon dioxide exists. Thus, Pales and Keeling (1965) have shown that over the period 1959 to 1963 the average carbon dioxide content of air sampled at Mauna Loa increased

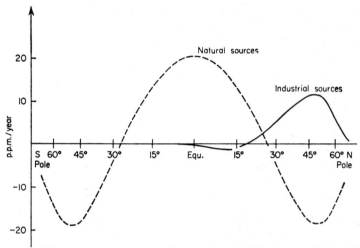

FIG. 9.19. Computed time average sources and sinks for CO_2 as a function of latitude. The solid horizontal line is the latitudinal average of the natural source function. The abscissa is in units of p.p.m. yr^{-1} averaged over the total air column (Bolin and Bischof, 1963).

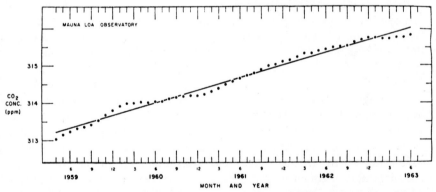

FIG. 9.20. Twelve month running mean of the concentration of CO_2 at Mauna Loa Observatory. Means plotted versus the sixth month of the appropriate 12-month interval. The straight line indicates a rate of increase of 0·68 p.p.m. yr^{-1} (Pales and Keeling, 1965).

at a rate of about 0·68 p.p.m. yr^{-1} (Fig. 9.20), and more recent measurements by Bolin and Bischof (1970) covering the period 1963 to 1968 indicate an average annual increase of 0·7 ± 0·1 p.p.m. (See also Bischof, 1973).

There are several possible causes of a long-term increase. The most important of these is the increased usage of fossil fuels. Changes in the biosphere and soil resulting from deforestation or changes in agricultural practice (Hutchinson, 1954), and a gradual adjustment of the average temperature of the sea may play some part, although it is not easy to quantify their contributions. Alteration of the mean rate at which juvenile carbon dioxide is being produced is not thought to be significant (Revelle and Suess, 1957); the rate of input of juvenile CO_2 (0.082×10^{14} g C yr^{-1}—Pytkowicz (1973b)) is very much less than the current rate of discharge of industrial CO_2 (*ca.* 40×10^{14} g C yr^{-1}).

Brannon *et al.* (1957) estimated that the production of carbon dioxide resulting from the burning of fossil fuels was exponential from about 1860 to 1910 after which the production rate deviated slightly from the exponential form. The cumulative total given by these estimates for 1957 was about 14% of the total carbon dioxide of the atmosphere. Figures used by Revelle and Suess (1957) and based on a United Nations estimate were in reasonable agreement with the conclusions of Brannon *et al.*, and the forecast for the years 2000 to 2009 suggested that over this decade the carbon dioxide produced from artificial sources may be as much as 20% of the present atmospheric total. According to Arnold and Anderson (1957), industrial carbon dioxide production has increased exponentially with a half-time of seventeen years up to 1910 and of forty since that date. Fergusson (1958) combined these periods and assumed a half-time of 28 years for the period up to 1954, and represented the annual rate of industrial carbon dioxide production by $C \exp(\mu t)$ where μ is 0.025 yr^{-1} and t is the time in years since 1850. With this zero, C has a value of 0.024% yr^{-1}. In a similar expression used by Bolin and Eriksson (1959) and derived by them from the United Nations data given by Revelle and Suess, μ is 0.029 yr^{-1}. According to Keeling (1973a), the function proposed by Bolin and Eriksson, although agreeing reasonably with the actual production rate between 1860 and 1920, gives a considerable over-estimate of industrial output after this period. In part this arises because a simple smooth exponential curve takes no account of the rather slow rise in the production rate between 1930 and 1950. Broecker *et al.* (1971) have commented on the trends of fossil fuel utilization, and for the purpose of assessing the consequences of a progressively increasing consumption they assumed that a 4.5% per year increase estimated for the period 1958–1966 will continue into the future. They estimated that the total CO_2 production by 2016 will be about 210×10^{16} g, some 7-fold that produced up to 1960. The rate of production in 2016 would be about 4 times that in the 1960's.

The rate of release of inactive carbon dioxide to the atmosphere as a result of industrial activity has important consequences for the atmospheric carbon-14 concentration (Suess, industrial or dilution effect), and in connection with their examination of this effect, Baxter and Walton (1970) have re-examined the rate of industrial carbon dioxide production. They took account of the estimated world production of coal, lignite, petroleum, natural gasoline, natural gas and of the decomposition of limestone during cement manufacture and made assumptions concerning the carbon contents of the various fuels and the proportions of these contents which were converted into carbon dioxide during combustion. Their conclusions, which agree to within a few per cent with those arrived at by Revelle and Suess, are summarized in Fig. 9.21. According to these estimates the production rate of industrial CO_2 in 1969 was twice that in 1954, and the total production of fossil fuel carbon dioxide between 1860 (the date chosen by many investigators as a reference time before which industrial production was negligible) and 1969 amounts to about 22% of the normal atmospheric content. Baxter and Walton consider that the annual excess carbon dioxide production is approximately equal to $0.05 \exp (0.03\ t)$ where t is the time in years which has elapsed since 1880.

Keeling (1973b) has critically reviewed information bearing on the likely rate of input of carbon dioxide to the atmosphere from various sources. Apparently, many of the earlier estimates have given values which were too high because of incorrect assumptions made concerning the percentage of the raw material which is converted into CO_2. Keeling considers that when proper account is paid to relevant factors (such as the carbon content of the fuel, the non-CO_2 productive use of some of them and handling losses), then the estimate for the cumulative addition up to 1970 to the natural carbon system as a consequence of human activity is $1.12 \pm 0.14 \times 10^{17}\ \mathrm{g\,C}$ (*ca.* 18% of the pre-industrial atmospheric content). Rotty (1973) has given an account of various refinements which should be considered when assessing the amount of carbon dioxide which is released to the atmosphere via the combustion of fossil fuels.

If it is accepted that the atmospheric carbon dioxide concentration before industrial output became appreciable (*ca.* 1860) was 290 p.p.m., then the percentage increase to the level of the 1970's (*ca.* 320 p.p.m.) is about 10%. The total output of industrial carbon dioxide to this time has been about 20% of the pre-industrial atmospheric total, and it is evident that much must have entered other reservoirs—mainly either the biosphere or the sea. Bolin and Bischof (1970) have drawn attention to the fact that the rate of increase of atmospheric CO_2 has remained close to 0.7 p.p.m. yr^{-1} for the two decades up to about 1970 even though the rate of industrial output has increased by 50% over this period, and have suggested that possibly a smaller percentage of the atmospheric total now stays in the atmosphere than was the case in the 1950's. They considered that of an accumulated industrial output to the atmosphere up to 1969 of 22% of the pre-industrial total, at least 8% has remained in the atmosphere. Although it is possible that there may be an increased vegetative assimilation as a result of the higher atmospheric total, quantitative evidence is not available. It is known that the rate of growth of

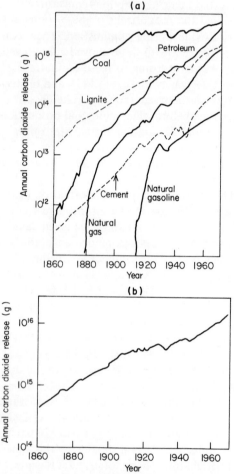

FIG. 9.21. The Suess effect. (a) Annual release of fossil carbon dioxide from various sources. (b) Total annual release of fossil carbon dioxide (Baxter and Walton, 1970).

many plants is a function of the ambient carbon dioxide pressure, but it is hard to accept that the net rate of growth of the biosphere is simply and directly related to the atmospheric carbon dioxide partial pressure, particularly as one would expect the growth rate to be very much influenced by the transport and availability of water and other nutrients. According to Bolin and Bischof, at most one-quarter of the assimilated carbon dioxide enters the biosphere and at least one-half of the industrial output enters the sea. It is possible that the absence of parallel increases in the atmospheric partial pressure and the industrial output may be the result of relatively small temperature changes in the surface layers of the sea. Calculations made by Broecker *et al.* (1971) using a layered model for the sea (see Fig. 9.22e) indicated that, of the fossil fuel carbon dioxide generated up to 1960, about $40 \pm 7\%$ had passed into the sea. Keeling *et al.* (1973) consider that about one-half of the carbon dioxide produced between 1959 and 1969 has remained in the atmosphere, and that the cumulative airborne fraction for the period 1700 to 1954 was 45%.

Possible long-term consequences of an input of carbon dioxide to the atmosphere in keeping with the present roughly exponential trend has excited the attention of several commentators (see e.g., Revelle, 1965; Broecker *et al.*, 1971; Pytkowicz, 1972c; Fairhall, 1973a, b). This topic provides a fruitful area for speculation and conjecture although, cluttered as it is with many imponderables, it is difficult to be quantitative. No attempt is made here to give an independent analysis, and the following comments merely draw attention to features which have been considered in past discussions. It might be noted that, almost invariably, forecasts of the ultimate effects on the human condition are gloomily pessimistic.

If reserves of fossil fuels prove adequate for the present trends to continue, by the year 2000 the atmospheric carbon dioxide content will have increased some 16% above the present level. According to one view, if the level rises very much beyond this there may be unpleasant repercussions on the climate because of the moderating effect of carbon dioxide on the earth's albedo. The reduced back radiation of energy into space consequent on the infra-red absorption of carbon dioxide will lead to an enhanced mean surface temperature. This may lead to a reduction of the polar ice-caps, a rise in mean sea level and severely altered oceanic circulation patterns as a result of which stratification will develop and a reduced nutrient circulation would occur. Possibly the higher temperature may lead to a higher evaporation rate and increased cloudiness and precipitation; increased cloudiness may, by reflection of incoming radiation, to some extent offset the temperature increase. The higher carbon dioxide content and the increased precipitation would be expected to increase the rate of atmospheric weathering. Part of the increased carbon dioxide content will be accommodated by uptake by the oceans, and the possibility has been discussed of adverse consequences arising because of the changed H^+, HCO_3^- and CO_3^{2-} concentrations which would lead to enhanced sub-aqueous weathering.

These and other consequences are undeniably undesirable, and there is an obvious need to keep an eye on things, particularly in view of the exponential character of the input. It is comforting to note that up to the 1960's at any rate, Broecker *et al.* (1971) were unable to report any major climatic effect which could definitely be laid at the door of the industrial effect. The prediction that uptake of CO_2 by the sea may lead to loss of calcium carbonate supersaturation in the very near future (Fairhall, 1973a, b; Zimen and Altenhein, 1973a, b) is not generally accepted, at least for tropical waters (Whitfield, 1974b, c, d; Skirrow and Whitfield, 1974) and has been discussed in Section 9.2.4.5.

The limited nature of the earth's resources means that exponential exploitation of them cannot proceed indefinitely and doubtless a combination of scarcity, economic factors and social pressures will lead to a fall-off from the present trend of the rate of usage of fossil fuels. According to Zimen and Altenhein (1973b), when the finite supply of reserves is taken into account the rate of exploitation is expected to pass through a maximum about A.D. 2400.

9.4. EXCHANGE OF CARBON DIOXIDE BETWEEN NATURAL RESERVOIRS

This problem has been discussed in some detail in recent reviews (Broecker *et al.*, 1971; Keeling, 1973a), and will be examined in outline only here. The principal features of the exchanges may be stated as follows. To a first approximation it is assumed that a steady state with respect to carbon distribution between exchange reservoirs (the atmosphere, the biosphere plus humus and the sea) existed before the onset of appreciable industrial carbon dioxide output. As a consequence of this output the steady state has become perturbed. The net rate of passage of carbon dioxide into the sea is increased, thereby increasing the oceanic ΣCO_2, but the properties of the carbonate equilibria are such that even if complete atmosphere–sea equilibrium were to be restored, the fractional increase in the partial pressure of carbon dioxide would be much greater than the fraction increase in ΣCO_2 (see below). Furthermore, the predicted rate of approach of equilibrium with the sea is a function of the structure assumed for the sea. For example, if a two-layer model for the ocean is assumed, the mixing rate between the upper layer and the deep sea will exert a throttling effect on the rate of attainment of overall equilibrium.

Carbon dioxide also exchanges between the land biosphere and the atmosphere and, in fact, the flux of CO_2 through the biosphere is large compared with the present industrial output (Keeling, 1973a). However, it is not yet known just how the biosphere responds to an atmospheric CO_2 increase, and calculational difficulties are accentuated by lack of precise knowledge of its size and also by the fact that within it there is a whole spectrum of carbon ages ranging from that for leaves and annual plants to those for trees, long-lived plants and humus.

The output of inactive (fossil) carbon dioxide has repercussions on the ^{14}C activities of the various reservoirs. Immediately obvious is the reduction in those of the atmosphere and contemporaneous wood; less obvious is the effect on the ^{14}C activity of the surface layers of the sea. Entry of inactive carbon dioxide to this reservoir and its participation in the carbonate equilibria results in disequilibria between the carbon-14 of the sea and that of the atmosphere such that ^{14}C passes from the sea to the atmosphere.

The problem confronting the investigator is to set up a suitable model for the interacting system. The model should be reasonably realistic and yet mathematically tractible, and should explain the observable features in terms of likely exchange coefficients. These include (see, e.g. Keeling, 1973a) the fraction of industrial carbon dioxide which remains in the atmosphere, the magnitude of the Suess effect (the percentage reduction of the ^{14}C activity consequent on industrial carbon dioxide production), the relative $^{14}C/^{12}C$ ratios of the atmosphere and the sea, and the distribution of carbon specific activity within the sea.

Typical models which have been used are shown in Fig. 9.22 (see also Revelle and Suess, 1957; Arnold and Anderson, 1957; Craig, 1957, 1958, 1963; Brannon et al., 1957; Broecker et al., 1960; Broecker, 1963; Broecker et al., 1971; Bolin and Eriksson, 1959; Nydal, 1968; Young and Fairhall, 1968; Rafter and O'Brien, 1970; Walton et al., 1970; Baxter and Walton 1970; Broecker and Li, 1970; Machta, 1972; Fairhall, 1973a; Keeling, 1973a). It is usual to assume that, apart from the Suess and nuclear ^{14}C effects, a stationary state with respect to ^{14}C distribution has prevailed. This implies that there have not been periods of rapid oceanic mixing or overturn alternating with periods of relaxation towards equilibrium, and that variations in the natural ^{14}C production rate have not been too dramatic.

The models shown in Fig. 9.22 differ considerably in complexity. The simple model (a) in which no recognition is given to the infra-structure of the sea (Revelle and Suess, 1957) is inadequate for most purposes and leads to predictions of the atmospheric CO_2 accumulation which are too small. However, some compromise with reality is obtained by postulating the existence of a mixing barrier (as in (b)) corresponding to the thermocline so as to give an upper (mixed) layer ca. 75 m deep containing a total amount of inorganic carbon roughly equal to that in the atmosphere and a lower deep layer containing about 50 times the carbon of the atmosphere. In the event of the upper layer being assumed to be too small or the exchange with the lower layer too slow, the predicted rate of accumulation of CO_2 in the atmosphere will be too large. Further complexity can be introduced by making allowance for the biosphere and humus (c), by postulating the occurrence of outcrops of deep water so as to allow for some direct (high latitude) atmosphere-deep carbon dioxide exchange (d), by interposing an intermediate layer of appreciable thickness (Broecker et al., 1971) which is not vertically well-mixed (e) or by adopting more comprehensive world-wide models (f) and (g) which attempt to be more specific with regard to geographical areas and water masses, and which direct attention towards sub-surface events.

Generally, inter-reservoir transfer rates are assumed to be first order with respect to the immediately transferable carbon in the reservoir from which transfer is occurring. Thus, the rate of direct transfer from reservoir i to reservoir j is $k_{ij}N_i$ where the rate coefficient k_{ij} is the reciprocal of the mean residence time (τ_{ij}) with respect to this transfer mode and N_i is a measure of the concentration of immediately transferable carbon in reservoir i (in units consistent with those chosen for the flux). Uptake by the biosphere is not a simple physical process and it is unlikely that transport to and from it are governed by simple first order laws. For transport from the sea to the atmosphere, the rate is proportional not to the total inorganic carbon of the sea

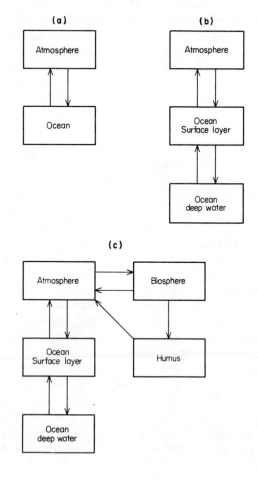

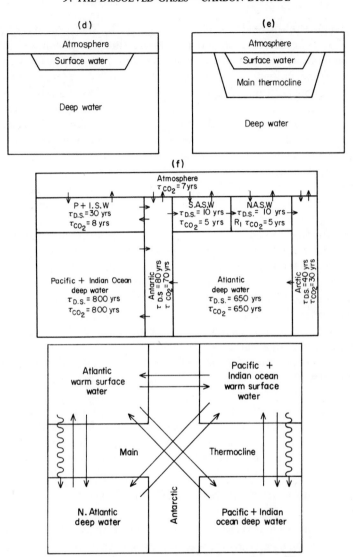

FIG. 9.22. Models for the transfer of carbon dioxide between natural reservoirs. (a) Simple atmosphere–uniform sea model (see, e.g. Revelle and Suess, 1957). (b) Atmosphere-two layer sea model. The upper layer of the sea (*ca.* 75–100 m depth) is assumed to be well mixed but to exchange only slowly with the deep sea. (c) Inclusion of the biosphere and humus reservoirs (see, e.g. Craig, 1957; Bolin and Eriksson, 1959; Keeling, 1973a). (d) Provision for direct entry from the atmosphere to the deep sea (see. e.g. Craig, 1958). (e) Inclusion of the main thermocline (which is not well mixed) through which CO_2 passes only slowly (see, e.g. Broecker *et al.*, 1971). (f) World wide model (see, e.g. Broecker *et al.*, 1960). (g) Interchange between major water masses through the main thermocline, account being taken of particulate flux (wavy arrows) (see, e.g. Broecker and Li, 1970).

but to the amount present as $CO_2(aq)$ plus H_2CO_3. When there is only one mode of loss from a reservoir, the residence time has a simple meaning. When there are two or more exit routes, the total rate of transfer is the sum of those of the individual routes (i.e. $k_{ij}N_i + k_{il}N_i + \ldots$, etc.), and τ_t (the total mean residence time with respect to all modes of loss) is related to those with respect to the individual routes by $1/\tau_t = 1/\tau_{ij} + 1/\tau_{il} + \ldots$, etc. Thus, the concept of residence time in the sea as judged from radio–carbon estimates is dependent on the choice of model used to interpret the data, and since carbon may be transported by routes not available for dissolved solids (e.g., via the atmosphere), the residence time for carbon may differ from the normally accepted water residence (flushing) time. Clearly, even in terms of simple models the radio–carbon age of a water mass is not necessarily the same as the oceanographical age.

For a particular model the non-steady state distribution of inactive carbon is described in terms of the appropriate set of differential equations, allowance being made for the rate of injection to the atmosphere of industrial carbon dioxide; a similar set describes the distribution of radio–carbon, allowance now being made for the natural decay of ^{14}C and for the fractionation of carbon isotopes which accompanies the crossing of phase boundaries (Keeling, 1973a). The transfer between adjacent water masses is a bulk process, and the possibility of fractionation does not arise.

A complete description of the CO_2 circulatory system demands models more complex than those shown in Fig. 9.22 in order to recognize the possibility of slow mixing between the stratosphere and the troposphere and also within each of these zones (inter-hemisphere mixing). Account should also be paid to the gravitational settling of organic matter from the surface to the deep water and its oxidation to carbon dioxide (Keeling and Bolin, 1968) and to the advective and diffusive fluxes of carbon dioxide in the sea (Craig, 1969).

9.4.1. RESPONSE OF THE SEA TO CHANGES IN THE ATMOSPHERIC CARBON DIOXIDE CONTENT

It is important to understand the extent to which the sea has the potential capacity to buffer, by readjustment of equilibrium, an increase in atmospheric carbon dioxide. The buffering effect is described by the equation

$$\frac{\delta P_{CO_2}}{P_{CO_2}} = \varepsilon \frac{\delta \Sigma CO_2}{\Sigma CO_2}$$

in which δP_{CO_2} is the shift in the equilibrium value of P_{CO_2} which follows the imposition on the atmosphere of a small incremental increase in the partial pressure of carbon dioxide, $\delta \Sigma CO_2$ is the consequential increase of ΣCO_2

and ε is the buffer factor (Bolin and Eriksson, 1959; Broecker *et al.*, 1971; Keeling, 1973a). By assuming the total alkalinity to remain constant (but ignoring the borate contribution), Bolin and Eriksson calculated ε to be about 12·5. Since the CO_2 which enters the sea participates not only in the equilibrium

$$CO_2 + H_2O + CO_3^{2-} \rightleftharpoons 2HCO_3^-$$

but also in

$$CO_2 + B(OH)_4^- \rightleftharpoons HCO_3^- + B(OH)_3$$

account should also be taken of the borate contribution to the alkalinity. When this is done (Broecker *et al.*, 1971; Keeling, 1973a), ε is calculated to be somewhat smaller (*ca.* 9–10). Keeling (1973a) has pointed out that ε depends on ΣCO_2 and that it will show a progressive increase as the amount of CO_2 taken up by the sea increases. Although it also depends somewhat on the water mass, the magnitude of ε indicates that a relatively large fractional increase in the partial pressure of CO_2 is balanced by a much smaller fractional increase in the total inorganic carbon of the water. If ΔP_{CO_2} is the magnitude of an instantaneous injection of carbon dioxide to the atmosphere and $\Delta P_{CO_{2(e)}}$ is the amount of this which must be transferred to regain equilibrium, then

$$\frac{\Delta P_{CO_{2(e)}}}{\Delta P_{CO_2}} = \frac{1}{1 + \varepsilon r}$$

where r is the ratio of the total carbon content of the atmosphere to that of the aqueous reservoir. If, tentatively, the upper layer is assumed to have a carbon content roughly equal to that of the atmosphere and transfer to the upper layer only is considered, $r \sim 1$, and for an 11% initial atmospheric increase, balance will be restored when almost one-tenth has entered the sea. If equilibrium were to be established between the atmosphere and the entire sea (which contains about 50 times as much carbon as does the atmosphere), then about four-fifths of the increment would enter the sea and the resultant increase in P_{CO_2} would amount to only about 2%. Thus, although the potential ability of the sea to buffer an increase of atmospheric carbon dioxide is appreciable, if the rate of establishment of equilibrium between the surface layer and the deep ocean is slow (time scale of centuries), then even quite rapid equilibration between the atmosphere and the surface layer will not appreciably reduce the atmospheric increase. Few laboratory scale verifications of the buffer relationship appear to have been made, although experiments made by Kanwisher (1960) (see also Kanwisher, 1963a, b) were reasonably consistent with predictions based on the above relationship.

Over very long periods of time it would not be expected that the alkalinity would remain constant throughout the progressive addition of carbon dioxide since, in the event of supersaturation with respect to calcium carbonate being removed, interaction with sedimentary carbonates may become important (see Bolin and Eriksson, 1959; Broecker et al., 1971; Fairhall, 1973a, b; Whitfield, 1973b, c, d; Skirrow and Whitfield, 1974).

A further point in connection with the buffer factor which is worth examining is the influence which ε has on the rate of attainment of equilibrium between a gas phase and a single well-mixed aqueous phase. Anticipating Section 9.5, the rate of change of p_{CO_2} as carbon dioxide is taken up by an aqueous phase (in which the partial pressure of carbon dioxide is P_{CO_2}) is given by an expression equivalent to 9.142, viz,

$$-\frac{dp_{CO_2}}{dt} = k[(p_{CO_2(i)} - \delta p_{CO_2}) - (P_{CO_2(i)} + \delta P_{CO_2})]$$

Using the subscript i to indicate the initial partial pressures ($t = 0$), the above equation can be written

$$-\frac{dp_{CO_2}}{dt} = k[(p_{CO_2(i)} - \delta p_{CO_2}) - (P_{CO_2(i)} + \delta P_{CO_2})]$$

where δp_{CO_2} is the decrease in p_{CO_2} at time t and δP_{CO_2} is the corresponding increase in P_{CO_2}. If the total number of moles of CO_2 in the gas phase is initially n and if the volume of the liquid phase is V, then approximately

$$\delta \Sigma CO_2 . V = \frac{\delta p_{CO_2}}{p_{CO_2(i)}} . n .$$

Making use of the buffer relationship one obtains from this

$$\delta P_{CO_2} = \varepsilon \frac{P_{CO_2(i)} . n . \delta p_{CO_2}}{p_{CO_2(i)} . \Sigma CO_2 . V},$$

and provided that the difference in the magnitudes of p_{CO_2} and P_{CO_2} is small compared with the absolute magnitude of either, then

$$\delta P_{CO_2} \approx \varepsilon . r . \delta p_{CO_2}$$

where r is the ratio of the total amounts of inorganic carbon in the gaseous and aqueous phase.

Insertion of this expression for δP_{CO_2} into the rate expression followed by substitution of the relationship $\delta p_{CO_2} = p_{CO_2(i)} - p_{CO_2}$ gives, on integration and a little further manipulation,

$$\frac{\delta p_{CO_2}}{\Delta_i} = \frac{1}{(1 + \varepsilon . t)} \{1 - \exp[-k(1 + \varepsilon . r)t]\}$$

where Δ_i is the initial difference between the partial carbon dioxide pressures of the gaseous and liquid phases. It can be seen that at $t = \infty$,

$$\frac{\delta p_{CO_2}}{\Delta_i} = \frac{1}{(1 + \varepsilon . r)}$$

as expected (see above), and that at time t the fraction of the equilibrium uptake which has been achieved is given by

$$1 - \exp\left[-k(1 + \varepsilon . r)t\right].$$

It is evident that for a given value of k the rate of approach to equilibrium increases as ε increases but, of course, a high value for ε means that only a relatively small transfer of CO_2 is necessary for the attainment of equilibrium because of the high response of P_{CO_2} towards uptake. (see also Broecker et al., 1971).

9.4.2. CARBON DIOXIDE RESIDENCE TIMES

It should be appreciated that estimated reservoir residence times are sensitive to the model used to interpret the data. Calculations by Craig (1957) made use of the steady state balance between the rate of introduction of radio-carbon to the atmosphere by natural production plus evolution from natural reservoirs (the sea, humus and the biosphere) and its loss from the atmosphere by physical removal and decay. A value for τ_{as} (the air–sea exchange time) of about six years was obtained. Use of a somewhat different model in which some high latitude direct atmosphere–deep water contact was proposed (Craig, 1963) indicated τ_{as} to be about 15 years.

Other methods of estimating the residence time have made use of the Suess effect (Brannon et al., 1957; Revelle and Suess, 1957; Fergusson, 1958; Bolin and Eriksson, 1959; Keeling, 1973a) and the dispersal of radio-carbon produced by nuclear explosions (Rafter and Fergusson, 1958a, b; Nydal, 1967, 1968; Munnich and Roether, 1967; Walton et al., 1970). Keeling's (1973a) account is a more exact development of the approach originally used by Bolin and Eriksson. In it an attempt is made, using a two-layer ocean model, to inter-relate the Suess effect and the increase in inactive atmospheric carbon dioxide in terms of likely air–sea (mixed layer) and deep sea–mixed layer residence times, and appropriate values for the buffer factor, the effective volume of the mixed layer and a land biosphere growth factor. Keeling's account, which is essential reading for those interested in CO_2 exchange, should be consulted for details, but the following might be noted. Transfer times for the air–mixed layer and deep sea–mixed layer exchanges of about 7 and 1500 years respectively and a mixed layer–atmosphere carbon content ratio of about 2·0 (cf. 1·2 in the simple two layer ocean model as as used by Craig (1957) and Bolin and Eriksson (1959)) are not too inconsistent with the available data. Decrease of ε had little effect on the predicted Suess effect, but led to a reduction in the predicted percentage increase of CO_2 in the atmosphere. The land biosphere probably plays some part in moderating the increase in atmospheric carbon dioxide.

For certain purposes, the residence time of CO_2 in the atmosphere with respect to all modes of physical loss (i.e. to both the biosphere and to the surface layer of the sea) is more important than that with respect to loss to the sea alone. In a theoretical investigation of the Suess effect, Baxter and Walton (1970) used a model in which the biosphere and the surface layer were united in a composite reservoir, and made use of previous measurements on the dispersal of bomb-produced carbon-14 (Nydal, 1967, 1968; Munnich and Roether, 1967; Walton et al., 1970) in order to calculate the rate coefficient for the troposphere-(surface-layer + biosphere) exchange. This combination of the biosphere and the surface layer was suited to their investigation since it avoided difficulties arising from uncertainties about the relative importances of the individual biosphere and surface layer uptakes. The best value for the combined coefficient was taken to be $0·23 \text{ yr}^{-1}$. When multiplied by the ratio of the carbon content of the troposphere to that of the total atmosphere (0·84), this gave an atmosphere-(surface-layer + biosphere) exchange coefficient of $0·19 \text{ yr}^{-1}$, corresponding to an atmospheric residence time of 5·3 years. The residence time with respect to uptake by the sea will, of course, be greater than this.

Radio-carbon estimates of the mean atmospheric residence time with respect to transfer to the sea range from about three to twenty years, and a figure somewhat less than a decade is used as a working value for many purposes, and is not inconsistent with experimental determinations of the invasion coefficient for CO_2 into a stirred solution (see Section 9.5). However, a radio-carbon estimate is an average value for entry to the sea, and since the real rate of entry of carbon dioxide into the sea is a function of temperature, turbulence and the partial pressure of carbon dioxide in the atmosphere it will vary from region to region. The thickness of the boundary layer through which diffusion must occur (Section 9.5) and the rate of hydration of dissolved CO_2 are possible rate determining factors (Bolin, 1960).

From an examination of the available data (Craig, 1958; Walton et al., 1970), Baxter and Walton (1970) considered that in terms of the two-layer model, the residence time of CO_2 in the upper layer with respect to transfer to the deep water is about 17 years. At the present time this residence time is not known with certainty, and further measurements as bomb-produced ^{14}C

is assimilated are needed. There are indications that this transfer time may be as low as 5 years. The residence time in the deep water with respect to transfer to the upper layer is generally considered to be of the order of 1000 years. (Craig, 1957, 1958; Bolin and Eriksson, 1959; Broecker et al., 1960, Broecker, 1963; Keeling, 1973a). However, this estimate is an oceanic average and it is likely that the residence time in the deep water depends on the water mass, and the interpretation of radio-carbon data depends on the model used.

By making use of specific activity measurements by a number of different groups of workers (Rafter and Fergusson, 1958a, b; Fonselius and Ostlund, 1959; Burling and Garner, 1959; Bien et al., 1960; Broecker et al., 1960), Broecker et al. were able to set up a world-wide circulation model which attempted to give attention to particular water masses. Thus, within the oceans there is a flux of water from the Antarctic via the South and North Atlantic surfaces and a return at depth via the Atlantic Deep Water. The Pacific and Indian Oceans are combined into one system in which the surface and sub-surface waters are separated by the thermocline across which little mixing is supposed to occur. Both the surface and deep waters are in exchange with the vertically mixed Antarctic reservoir through which the Pacific and Atlantic waters communicate. Together with the atmosphere, the entire system comprises eight reservoirs, and the stationary state condition is described by eight conservation equations—one for each reservoir—in which the rate at which carbon-14 enters a reservoir is balanced by its rate of exit plus its rate of radio-decay in the reservoir. Corrections were made to allow for the introduction to the surface layers of industrial CO_2 and of active carbon arising from nuclear explosions. The residence times of carbon and of dissolved solids in each of the water masses indicated in Fig. 9.22(f) differ for certain of the masses because carbon is subject to transport through the atmosphere as well as by direct mixing. A number of simplifications and assumptions are necessarily built into this model. In particular, it assumes the existence of a steady state, that the activity of the source region of each mass is characteristic of the water added to the mass, that mixing between adjacent masses does not affect their respective activities and that the solution of sedimentary carbonates or the oxidation of organic detritus does not result in significant changes of activity.

9.4.3. CHANGES IN THE ATMOSPHERIC ^{14}C CONTENT (see also Chapter 18)

Reductions in the ^{14}C activity of modern (but pre-bomb) organic matter is an established phenomenon (Suess, 1955; Hayes et al., 1955; Brannon et al., 1957; Fergusson, 1958; Baxter and Walton, 1970) and the effect has also been noted for the shells of organisms grown in shallow restricted bays and estuaries (see, e.g. Brannon *et al.*, 1957). Depletions in wood amounting to as much as 4% have been reported. The effect has also been reported to be to some extent geographical, tending to be smaller in the southern than in the northern hemisphere. Possibly this is associated with the heavier concentration of industry in the northern hemisphere and the finite inter-hemisphere transfer times, and results obtained from rings sampled from trees growing near industrial areas often show a higher effect (Arnold and Anderson, 1957, see also Young and Fairhall, 1968; Walton *et al.*, 1970). Fergusson showed that rings from trees grown in New Zealand, Peru and North America were increasingly impoverished in ^{14}C over the last century and had an average Suess effect of about 2%. The depletion in activity roughly paralleled the usage of fossil fuels.

Several groups of workers have sought to explain the magnitude of the effect in terms of suitable mixing models (see above and also Bolin and Eriksson, 1969; Baxter and Walton, 1970; Keeling, 1973a). According to Baxter and Walton's calculations, it is possible to recognize three distinct periods of exponential growth of the Suess effect—(i) 1880 to 1915 during which time there was a doubling each 15 years and the effect at time t is given by $S_t = -0.30 \exp(0.046\,t)\%$ where t is in years since 1880, (ii) 1915 to 1950 when the doubling time was 33 years and $S_t = -1.55 \exp(0.021\,t)\%$ where t is in years since 1915, and (111) 1950 to 1969 when the doubling time was 21 years and $S_t = -3.10 \exp(0.034\,t)\%$ where t is in years since 1950.

The agreement between the magnitude of the predicted Suess effect from Baxter and Walton's model and that observed is good, particularly when account is taken of other, natural, perturbing effects on the atmospheric ^{14}C level (see below). Predictions of the future value of the Suess effect are, of course, highly speculative and rest on assumptions concerning the future rate of usage of fossil fuels. However, on the basis of United Nations estimates (1955) of future fuel consumption, Baxter and Walton calculated the effect to be -8.6%, -16.0% and -31.1% in 1980, 2000 and 2025 respectively. These are probably underestimates since the rate of increase of usage of fossil fuels appears to be greater than originally thought (Baxter and Walton, 1970).

The Suess effect takes account only of the dilution of natural $^{14}CO_2$ by inactive fossil fuel carbon dioxide. Superimposed on this effect is the bomb effect, and combination of these two results in the atmospheric ^{14}C-time plot shown in Fig. 9.23. The distribution and assimilation of nuclear device produced $^{14}CO_2$ has been often investigated (Broecker and Olsen, 1960; Munnich and Vogel, 1959; Walton *et al.*, 1970; Nydal, 1963, 1967, 1968; Nydal and Lövseth, 1970; Young, 1967; Munnich and Roether, 1967) and advantage has been taken of the phenomenon to evaluate CO_2 transfer times within the atmospheric reservoirs and between the atmosphere and the sea (or sea plus biosphere). Some of these investigations have underlined the often highly variable nature of some of the transfer times within the atmosphere. Figure 9.23 shows that in recent times the Suess effect has been completely swamped by the bomb effect, but that in the absence of future injections of artificial ^{14}C to the atmosphere, a relaxation to the natural (plus Suess effect) level will occur as the atmospheric CO_2 redistributes itself between the various reservoirs (Nydal, 1968; Nydal and Lövseth, 1970; Munnich and Roether, 1967).

Over past millenia marked changes in the atmospheric carbon-14 concentration have occurred. This has been shown from analysis of tree ring data (see, e.g. Dyck, 1965), and Baxter and Walton (1971) have made a detailed examination of short term (Suess effect corrected) variations of atmospheric $^{14}CO_2$ over the last century by analysing wines, spirits and seeds. Over this period fluctuations in the natural concentration of about 3% occurred, and these were negatively correlated with sunspot activity. They considered that since the natural changes in production rate were insufficient to account for the atmospheric changes, the rates of stratospheric–tropospheric mixing were solar sensitive and were primarily responsible for short-term tropospheric carbon-14 changes. Approximately, the maximum input of carbon-14 to the troposphere occurs during the two year period following a sunspot maximum; minimum input from the stratosphere occurs about three years before maximum solar activity.

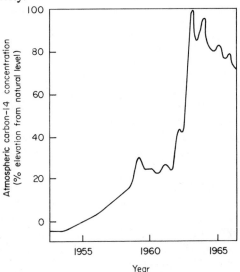

FIG. 9.23. Carbon-14 concentration in the northern troposphere (Baxter and Walton, 1971).

Other factors which may have imposed changes on atmospheric $^{14}CO_2$ activity might be briefly noted. Thus, it is possible that the above noted ^{14}C variations are modulations imposed on a much longer time-scale (millenia) and larger amplitude variation which results from changes in the production rate of natural ^{14}C. The change in production rate may be related to changes in the earth's geomagnetism (see, e.g. Suess, 1973). In addition, catastrophic overturning of the seas as a result of prolonged periods of cold (see, e.g. Worthington, 1954; Broecker et al., 1958; de Vries, 1958) would cause deep water of relatively low carbon-14 activity to be brought to the surface and the subsequent readjustment of the atmosphere–sea surface equilibrium would cause a lowering of the atmospheric specific activity. An attempt has been made by de Vries to correlate past atmospheric activity variations with periods of glacial advance and retreat.

9.4.4. CARBON-14 ACTIVITY IN THE SEA (see also Chapter 18).

Nuclear device generated $^{14}CO_2$ has markedly affected the activity of surface waters. Even before the early 1960's there was some scatter in surface $\Delta^{14}C$ values (Bien et al., 1960, 1963a, b; 1965), possibly because of the uptake of nuclear bomb produced $^{14}CO_2$. At this time surface values tended to centre around $-50\permil$ except in low latitudes where low values of about $-170\permil$ were observed (see also Rafter, 1955; Rafter and Fergusson, 1958a, b; Brodie and Burling, 1958; Garner, 1958; Burling and Garner, 1959). Low values for surface water samples taken from the Antarctic south of the Atlantic have also been reported by Broecker et al. (1960), and are probably more characteristic of outcrops of the deep sea. More recent collections of $\Delta^{14}C$ data for surface water given by Nydal and Lövseth (1970) have included measurements made by them and other workers (Munnich and Roether, 1967; Bien and Suess, 1967) over a period of time during which nuclear device generated ^{14}C was being taken up from the atmosphere by the sea (see also Young and Fairhall, 1968; Fairhall et al., 1970, 1971; Rafter and O'Brien, 1970). As with the earlier data, considerable scatter was evident, but despite this, an upward trend in the $\Delta^{14}C$ values for the water was apparent and a peak $\Delta^{14}C$ value was probably attained in the late 1960's. A radio-carbon profile determined in connection with the GEOSECS Intercalibration project off the Pacific coast of California in September 1969 (Ostlund and Niskin, 1970) also showed that there had been marked uptake of bomb-produced ^{14}C since surface water values for $\Delta^{14}C$ were about $+270\permil$. At about 400 m the value was $-100\permil$ and below 1000 m it was about $-225\permil$.

According to Fairhall (1973a, see also Fairhall et al., 1972; Nydal and Lövseth, 1970) loss of ^{14}C from the surface layers is occurring more rapidly than has been predicted by box models for oceanic carbon-14 distribution, most of which indicate a residence time in the mixed (surface) layer of 10 years or more. Much of the bomb-produced carbon-14 has entered the main thermocline where its concentration decreases exponentially with depth, and the residence time in the mixed layer may be less than 5 years, rather less than had previously been thought. The rate of dispersal of bomb-produced $^{14}CO_2$ through the thermocline will depend on the stability of this layer towards vertical mixing. The stability will be higher in warmer equatorial regions than in waters of higher latitudes.

Pacific deep water samples examined by Bien et al. (1963a, b) were in the range -190 to $-250‰$ and showed a progressive increase in apparent age from south to north which, at the time of their work, was assumed to correspond to a water movement of about $0·06$ cm s^{-1} in a northerly direction. The apparent radio-carbon age was probably affected by the oxidation of organic debris falling from above since the northerly movement was accompanied by a decrease in the dissolved oxygen, but Craig (1969) has drawn attention to more serious limitations associated with closed system approaches to the estimation of the rate and direction of movement and the "age" of water masses by ^{14}C specific activity measurements. In Craig's view, the sampling stations selected by Bien et al. were so positioned relative to the deep water circulation pattern that even if all other assumptions underlying the model are valid, it is not correct to identify the differences in the ^{14}C "ages" as the time taken for water to move directly between the two stations. In addition, Craig examined both vertical and horizontal transport in the deep water in terms of a diffusion–advection model. He has shown that despite the large range of $\Delta^{14}C$ values (ca. $60‰$) found by Bien et al. for the Pacific, the absolute ^{14}C concentration is, within the limits of measurement precision, almost independent of both depth (below $1·5$ km) and latitude (between 60°S and 50°N). This constancy arises because of the approximate equality for ^{14}C of its production rate in deep water via particulate transport and its radioactive decay rate, and the absence of a ^{14}C gradient means that there is no vertical diffusive flux of radio-carbon. The horizontal changes in the $\Delta^{14}C$ of bottom water were considered by Craig to reflect changes in ΣCO_2. Stable carbon diffuses downwards from the ΣCO_2 maximum (Section 9.6) and dilutes the specific activity of carbon in the bottom water. For this reason it is invalid to assume that specific activity changes of bottom water can be interpreted in terms of a closed system flow time.

According to Bien et al., Indian Ocean deep water samples were slightly more active than those of Pacific deep water of similar latitudes, and water of low activity (high apparent age) detected at 10°S at a depth of 400 m was probably representative of an influx of Northwest Intermediate Indian Ocean water which has a relatively low activity.

A detailed study of the radio-carbon distribution in the Atlantic before 1960 was made by Broecker *et al.* (1960) who examined 135 samples comprising deep and surface waters taken from the major water masses. The lowest surface activity at that time ($\Delta^{14}C$, $-120\%_0$) was found in a sample taken south of the Antarctic Convergence where water of low activity rising from below probably sinks again before a steady state with respect to transfer of atmospheric carbon can be attained. Many of the activities of the surface water samples at that time were interpreted in terms of known water movements. Thus, the Falkland current brought water of relatively low carbon-14 content ($\Delta^{14}C$, $-78\%_0$) northwards from the Drake Passage, and where it meets the Brazil Current which had a $\Delta^{14}C$ value of $-39\%_0$, water of intermediate activity was formed. The average $\Delta^{14}C$ values for the South Atlantic surface samples was about $-57\%_0$ whereas that for those from the North Atlantic was about $-49\%_0$; the Caribbean, which draws water from both the South and North Atlantic had an intermediate value. Higher northern latitudes investigated by Fonselius and Ostlund (1959) at about this time showed $\Delta^{14}C$ values of about $-35\%_0$.

The core of each of the deep water masses of the Atlantic examined by Broecker *et al.* (1960) showed a standard deviation in $\Delta^{14}C$ values only slightly greater than the measurement error, and since clear differences were observed between the various masses they concluded that circulation within each mass was probably rapid relative to the residence time within each mass. Deep water originating in the high latitudes of the southern hemisphere had a lower carbon-14 content than did that originating at high latitudes in the northern hemisphere, and bottom water of the Eastern Atlantic Basin showed a $\Delta^{14}C$ value about $27\%_0$ lower than that of the Western Basin. The lowest activity found in the Atlantic system ($\Delta^{14}C$, $-144\%_0$) was that of Antarctic bottom water taken from the South Western Atlantic. Some abnormally low values (*ca.* $-350\%_0$) for surface and near-surface water from under Antarctic pack-ice indicate only slow vertical and horizontal exchange of this water.

The carbon-14 contents of the Black Sea (Ostlund, 1969), the Mediterranean Sea (Ostlund, 1969; Broecker and Gerard, 1969) and the Caribbean Sea (Szabo *et al.*, 1967) have been examined, and Broecker and Gerard have interpreted their results for the Mediterranean in terms of a model in which there is carbon dioxide exchange between the atmosphere and surface water and between the surface and intermediate waters. The intermediate water is in exchange contact with both the Eastern and Western Basins which are not themselves in direct contact. The Western Basin supplies water to the Tyrrhenian Basin which is in contact with the intermediate water. These measurements were made in the late 1950's before appreciable bomb-produced carbon-14 was evident in the water, and the carbon-14 activity of the surface Mediterranean water was close to that of adjacent North Atlantic surface water.

The carbon-14 activities of marine organisms will reflect those of the environments in which they grew (Broecker, 1963). For samples taken before recent increases of artificially produced ^{14}C, those originating from warm surface waters would be expected to have $\Delta^{14}C$ values of about -55%, whereas those from colder regions where deep water of low ^{14}C content outcrops will show high apparent 'ages'. Samples taken from brackish coastal waters or near the discharge of large rivers may show anomalous carbon-14 activities in relation to their true ages if the amount of inactive carbonate discharged from the land is high. For samples taken during the increase of atmospheric carbon-14 activity these generalizations will need modification.

9.5. TRANSFER OF CO_2 ACROSS THE AIR–SEA INTERFACE

9.5.1. GENERAL COMMENTS

The passage of CO_2 (and other atmospheric gases) across the air–sea interface is a dynamic, two-way, process even when there is equilibrium between the atmospheric and dissolved gas. Nowhere on the sea surface is there long term equilibrium between the dissolved CO_2 and its partial pressure in the atmosphere. Dis-equilibria arise from factors such as the upwelling of CO_2-rich water, consumption of carbon dioxide during photosynthesis and changes of water temperature. The system responds by giving rise to a net flux of CO_2 across the interface in one direction or the other, although this flux is small compared with the absolute fluxes in each direction. Knowledge of these fluxes is important because the rate of entry of CO_2 into the sea helps to determine its mean atmospheric life-time and its rate of accumulation in the atmosphere. Other materials also participate in air–sea transfers, and the problem is of general interest and has marked environmental overtones. Presumably the sea is a major sink for many substances (SO_2, NO_2, industrial solvent vapours, pesticides—see Chapter 17; Liss, 1972, 1973; Liss and Slater, 1974) which would otherwise accumulate more rapidly in the atmosphere. The problem is important in gaseous transfer in many industrial processes and also in biological systems.

Because of incomplete understanding it is still not possible to calculate the rate of transfer in either direction from first principles, but it is clear that the simple picture of a gas–liquid interface as a single plane surface separating two phases, each well mixed, is inadequate, and that the regions in the immediate vicinity of the interface differ in important respects from the bulk phases. The current, most commonly used, models (see, e.g. Liss and Slater, 1974; Hoover, 1966; Hoover and Berkshire, 1969; Quinn and Otto, 1971; Broecker and Peng, 1974) are variants of one proposed by Lewis and Whitman (1924). Essentially, this considers that the boundary layers exist on both sides of the interface (Fig. 9.24). Above and below these layers each phrase is rapidly mixed by eddy diffusion and turbulence; transport within the layers is assumed to be by the much slower process of molecular diffusion, and it is this aspect of the migration which is rate determining. Although the two-layer picture provides a working model on which to base theoretical treatments, the real physical situation almost certainly lacks the crispness which the model implies. It is unlikely, for example, that the liquid boundary layer has a uniform, well-defined thickness, and it has been suggested (Quinn and Otto, 1971) that solvent evaporation at the surface may complicate the simple picture by generating temperature and density gradients which promote convective mixing. Other models have been proposed (Higbie, 1935; Danckwerts, 1970; Kishinevski, 1955; Dobbins, 1964) but will not be considered here.

For the sake of completeness, two other possible barriers to transport should be indicated. The first can be regarded as a resistance to transport across the phase boundary itself. This process is probably activated and this implies that only a fraction of those molecules striking the surface succeed in penetrating. The second arises if surface films of contaminant interfere with transport. The first of these effects is not likely to be controlling for gas transfer under natural conditions. The second is thought to be important only in restricted areas when relatively thick films (e.g. oil slicks) develop (see Section 10.5.7). Passage across the actual phase boundary itself is likely to be rate determining for the evaporation of water.

The steady-state transport of an unreactive species through two idealized boundary layers can be discussed by reference to Fig. 9.24. This shows that when concentration is used as a parameter there is, in general, a concentration discontinuity at the interface; if partial pressure is used as a parameter, the partial pressures of the gaseous and dissolved forms of the migrating substances will be the same at the interface.

At the steady-state the one-dimensional flux, j, is the same in both layers. If D_g and D_l are the diffusion coefficients in the gaseous and liquid layers respectively, the flux in the gaseous layer is given from Fick's Law as

$$j = -D_g \cdot \frac{dc_g}{dz} \qquad (9.131)$$

or, if the concentration profile is linear, by

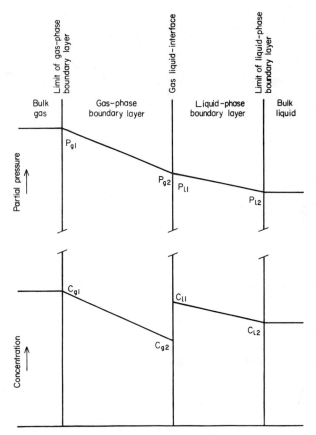

FIG. 9.24. The boundary layers at the gas–liquid interface (diagrammatic). Transport occurs via diffusion down concentration gradients in both the gaseous and liquid boundary layers. Generally, a discontinuity exists in the concentration profile at the interface. No discontinuity exists in the partial-pressure profile. The figure refers to an ideal system in which there is no chemical reaction between the gas and the solvent to give ionic species. When such a reaction does occur, account should be taken of the gradients (not necessarily constant) of the various ionic species (see text).

$$j = \frac{D_g}{\delta_g} \cdot [c_{g1} - c_{g2}], \qquad (9.132)$$

where c_{g1} and c_{g2} are the concentrations of the migrating species at the upper and lower limits of the gaseous boundary layer respectively, and δ_g is the layer thickness. For the liquid phase boundary layer the corresponding equations are

$$j = -D_l \cdot \frac{dc_l}{dz} \qquad (9.133)$$

and

$$j = (D_l/\delta_l) \cdot [c_{l1} - c_{l2}] \tag{9.134}$$

where c_{l1} and c_{l2} are the concentrations at the upper and lower limits of the layer and δ_l is the thickness of the layer. Provided that equilibrium between the gaseous and dissolved forms of the species is maintained at the interface,

$$c_{l1} = sc_{g2} \tag{9.135}$$

where s is the appropriate solubility coefficient. By combining (9.132), (9.134) and (9.135), the flux is seen to be given by

$$j = \frac{(D_g/\delta_g)(D_l/\delta_l)}{(D_g/\delta_g) + (D_l s/\delta_l)} [sc_{g1} - c_{l2}] \tag{9.136}$$

$$= Q[sc_{g1} - c_{l2}] \tag{9.137}$$

where Q is a composite rate coefficient for the two-layer system. More conveniently, the reciprocal of Q can be regarded as the effective resistance to transfer through the system and

$$\frac{1}{Q} = \frac{\delta_l}{D_l} + s\frac{\delta_g}{D_g} \tag{9.138}$$

where the quantities $s\dfrac{\delta_g}{D_g}$ and $\dfrac{\delta_l}{D_l}$ are measures of the effective resistances to transport offered by the two individual boundary layers. It is seen that, as is intuitively obvious from the nature of the problem, except in the fortuitous circumstance that these two quantities are of similar magnitude, one or other of them will exert the rate determining influence. Diffusion coefficients for the migration of small molecules within liquid systems are of the order of 10^{-5} cm^2 s^{-1} whereas those for the corresponding gaseous transport are $ca.$ 10^{-2} cm^2 s^{-1}. Evidently, only if the ratio δ_g/δ_l has an improbably large value or if s is extremely large is the gaseous layer likely to contribute significantly to the total resistance to transport. On the whole, it seems likely that, for the transfer of N_2, O_2, and CO_2, diffusion through the liquid layer is the important factor (Liss, 1973), and when this is so, $Q \approx Q' (= D_l/\delta_l)$ and (9.136) reduces to

$$j = Q'[sc_{g1} - c_{l2}] \tag{9.139}$$

$$= Q'\Delta c_l.$$

where Δc_l is the concentration change across the thickness of the liquid boundary layer. However, there is evidence (Liss, 1972; Brindlecombe and Spedding, 1972) that the uptake of SO_2 is particularly rapid and that gaseous diffusion at least partially controls the rate. This is probably a consequence of its high solubility and of its high reactivity with water—see below. For the uptake of water vapour by the sea, the major impediment to transfer is almost certainly gaseous diffusion (Liss, 1973) since the liquid boundary layer has no part to play in the process.

An expression equivalent to (9.139) is readily derived from simple kinetic assumptions which give more direct recognition to the dynamic, two-way nature of the phenomenon. If it is assumed that exit from each phase is a first order process and that the rate of each transfer is proportional to the bulk concentration in that phase, the net flux in the downward (gas–liquid) direction is

$$j = k_d \cdot p_g - k_u \cdot c_l \tag{9.140}$$

where k_d and k_u are the appropriate rate coefficients and p_g and c_l are respectively the pressure of the migrating species in the gas phase and its concentration in the bulk of the solution. Since $c_l = \alpha \cdot P_l$ where P_l is the partial pressure of the gas in solution and α is the appropriate solubility coefficient then

$$j = k_d p_g - \alpha \cdot k_u \cdot P_l \tag{9.141}$$

At equilibrium, $j = 0$ and $P_g = P_l$. Consequently, $k_d = \alpha k_u$ and

$$j = k_d \cdot \Delta P \tag{9.142}$$

where $\Delta P = p_g - P_l$. Equation (9.142) is a form of the Bohr (1899) expression for the invasion of liquids by gases. When due regard is paid to the units (9.142) is seen to be equivalent to (9.136) or (9.139) and, if it is accepted that diffusion through the liquid boundary layer is the rate determining process, it follows that $k_d = (D_l/\delta_l)\alpha$. Expressions (9.139) and (9.142) suggest that the transfer coefficients for invasion and evasion should be the same, at least for the sorption and desorption of simple, inert, gases which behave ideally and do not react with the solvent other than by simple solution. Moreover, because of the similarity of the expected values of the diffusion coefficients of the major atmospheric constituents N_2 and O_2 in water, the transfer coefficients expressed in terms of Q (but not in terms of k_d) of these substances might be expected to differ only slightly, and to be close to that of CO_2, provided that the latter behaves as an unreactive gas (see below).

The concept of a surface film of a well defined thickness in which turbulence is entirely absent and in which transport is solely by molecular diffusion is somewhat artificial. The difficulty of assessing independently of transport experiments a thickness, δ_l, to be assigned to the layer means that prediction of invasion and evasion rates from first principles is not possible. Thus, most investigations of the subject have been restricted to the determination of invasion and evasion coefficients as a function of conditions.

Expression of results in terms of Q'—the piston concept—(or k_a) avoids the need to make a separation into D and δ_l. However, despite uncertainties associated with the static layer model Broecker and Peng (1974) have found it to be a useful device in the interpretation of gas-liquid transfer processes. Laboratory experiments (see e.g. Kanwisher, 1963a, b; Broecker and Peng, 1974) indicates that film thicknesses are some 50–300 μm and decrease with increased stirring; mean thicknesses estimated for the ocean surface layer on the basis of (i) the steady state distribution of $^{14}CO_2$, (ii) the dispersal of nuclear device generated $^{14}CO_2$ and (iii) the magnitude of radon loss through the sea surface (Broecker and Peng, 1974) indicate values of about 40 μm. The residence time of a gas in the atmosphere with respect to transfer to the mixed layer is given by

$$\tau_{as} = \frac{\text{amount in atmosphere}}{\text{invasion rate}}$$

$$= \frac{k.p_g}{(D_l/\delta_l).p_g.\alpha}$$

$$= \frac{k.\delta_l}{D_l.\alpha}$$

where k relates the sea level partial pressure of a gas to the total amount of it in the atmosphere. The low values of α for e.g. O_2, N_2, H_2, Ar etc. mean that they have atmospheric life times of ca. 300 yr. More soluble gases such as CO_2 and SO_2 have much shorter mean atmospheric life times (ca. 1 decade for CO_2 and even less for SO_2) even if the possibility of hydration augmented uptake (see below) is ignored. The mean residence time of a constituent in the mixed layer with respect to transfer to the atmosphere (τ_{sa}) is given by

$$\tau_{sa} \text{ (gas exchange)} = \frac{h\delta_l}{D_l}$$

where h is the thickness assumed for the mixed layer (ca. 80 m, Broecker and Peng, 1974). For "unreactive" gases τ_{sa} is about one month, but for CO_2 this value needs further qualification. Thus, the one month refers to the *gas exchange* residence time of the dissolved CO_2 as such and pays no regard to exchange with dissolved carbonate and bicarbonate in the sea. When exchange of the total carbon is considered the *isotropic equilibration* residence time is calculated to be about 12 yr. Broecker and Peng have also shown that the *chemical equilibration time*—the time required for CO_2 contents of the mixed layer of the sea to equilibrate with an increase in atmospheric p_{CO_2} through the equilibrium

$$CO_2 + H_2O + CO_3^{2-} \rightarrow 2HCO_3^-$$

is given approximately by

$$\tau_{sa\,(chemical\,equilib)} = \tau_{sa\,(gas\,exchange)} \frac{c_{CO_3\,(T)}}{c_{CO_2\,(T)}}$$

and is about 1·5 yr.

9.5.2. INFLUENCE OF GAS–SOLVENT REACTIVITY ON INVASION RATES

For the uptake or loss of reactive gases such as CO_2 or SO_2 by aqueous solutions, additional factors need to be taken into account. Carbon dioxide reacts via hydration to give, on ionization, HCO_3^- and CO_3^{2-}, and two extreme circumstances can be visualized. The first would arise if the rate of hydration (Section 9.2.2.2) was so slow that the CO_2 passed through the liquid phase boundary layer before appreciable reaction took place. If this occurred the CO_2 would behave towards the uptake process as an inert gas. The other extreme would be encountered if the hydration rate was rapid enough for carbonate equilibrium to be attained at all points during passage through the surface film. This would result in the establishment of concentration gradients not only of the dissolved CO_2, but also of the other components of the equilibria. Since the net rate of transport is the sum of the individual transport rates of each carbon containing species down its own concentration gradient, the uptake rate would be enhanced relative to that expected if the hydration reactions of CO_2 were slow. An intermediate, more general, situation would arise if the hydration reaction is fast enough for some reaction of the dissolved gas to occur during its passage, but yet not sufficiently fast for equilibrium to be attained within the film.

The importance of hydration rate in augmenting the total CO_2 flux depends not only on the magnitude of the rate constant, but also on the thickness of the boundary layer and the pH of the solution. For a sufficiently thin surface film (such as would correspond with high turbulence), little reaction will occur during transfer even if the hydration and dehydration rates are high. At very low pH values the hydration and dehydration rates will have little influence on the invasion rate because of the low equilibrium concentrations of CO_3^{2-} and HCO_3^-; however, for the pH range of ocean waters, less than 1% of ΣCO_2 exists as dissolved gas, and provided that the surface film is not too thin, the possibility that the rate of hydration might considerably augment the invasion and evasion rates is a real one.

Theoretical examinations of this problem in terms of the boundary layer model have been made by Bolin (1960), Hoover (1966), Hoover and Berkshire (1969) and Quinn and Otto (1971). Of these treatments, only that of Quinn and Otto takes into account the need to preserve electroneutrality at all points in the film. The effect of this restraint is to require that the pH is not constant across the film, and this model leads to predictions which differ somewhat from those which follow when the assumption of constant pH is made.

Sorption and desorption rates increase as the rates of hydration and dehydration increase, and Quinn and Otto estimated an upper limit for the additional augmentation of the CO_2 flux consequent on chemical equilibrium being attained at all points in the film (infinite reaction rate). Their analysis made use of two dimensionless parameters. The first, θ, defined by

$$\theta = \frac{c_{CO_2(0)} - c_{CO_2(b)}}{c_{CO_2(b)}} \qquad (9.143)$$

(in which the subscripts 0 and b refer respectively to the surface and bulk concentrations of CO_2) can be regarded as the driving force behind the transport. The second parameter was the augmentation of the flux which occurs (because of the attainment of equilibrium) relative to the maximum desorption rate. This maximum rate corresponds to that which would occur if CO_2-free air were to be blown across the surface. Quite different predictions arise from the constant pH and electroneutrality models. For the constant pH model, the augmentation is a linear function of θ and consequently the same augmentation should occur for sorption and desorption for a given value of the driving force. However, when the electroneutrality restraint is introduced, the predicted augmentation is no longer a linear function of θ and, at a pH value of 8·0, for a given value of the driving force augmentation is greater for desorption than for sorption.

Quinn and Otto were not able to give a simple analytical solution embracing the intermediate (more general) situation when hydration, although appreciable, is only partially complete during the passage of CO_2 through the film. However, the numerical examination given by them showed that, provided the surface layer was thin ($< 400 \, \mu m$) the predictions of the constant pH and the electroneutrality models agreed since the amount of reaction in the film would then be insufficient for the differences in the two models to be apparent. Moreover, for very thin films the exchange rate should be uninfluenced by the hydration and dehydration rates. For thicker films the electroneutrality model predicts a greater augmentation for desorption than for sorption.

9.5.3. LABORATORY AND FIELD STUDIES

The aims of particular investigations in this area have been diverse, but in the main have been concerned with the determination of the Bohr transfer coefficient as a function of conditions (e.g. wind speed), comparison of this coefficient for CO_2 with those obtained for the transfer of other gases (N_2, O_2, SO_2, Rn, H_2O), the search for a possible augmentation of CO_2 flux under conditions when the rates of hydration and dehydration are likely to be critical, and comparison of uptake rates with those expected from the known mean atmospheric life-time of CO_2 as estimated from $^{14}CO_2$ data (Section 9.4).

Intercomparison of the results of different groups of workers is not easy because of the difficulty of controlling conditions or even knowing the value to assign to certain variables. In particular, there is no means of directly measuring the film thickness and, as pointed out by Quinn and Otto (1971), this thickness is simply a correlating parameter which may have little relevance to the actual physical situation.

The exchange appears to be a first order process. This has been verified by Sugiura et al. (1963) who examined in laboratory experiments the rate of loss of $^{14}CO_2$ from an atmosphere in chemical equilibrium with a stirred sea water solution. As might be expected, the uptake rate was strongly dependent on the stirrer speed, but by back extrapolation of plots of their results against stirrer speed to zero speed they were able to estimate an uptake rate and exchange coefficient into sea water at 20°C of 0.316 mg cm^{-2} atm^{-1} min^{-1}, in reasonable agreement with a value determined earlier by Bohr (1899) (see Table 9.19). The exchange constant was uninfluenced by chlorinity in the range 12 to 18‰, but increased with increasing temperature. Field estimates of invasion and evasion rates were obtained from experiments in which air, either slightly depleted or enriched in carbon dioxide relative to its partial pressure in sea water, was circulated above the water using a canopy floating on the surface, the CO_2 content of the air being monitored by an infrared gas analyser. Values of the exchange coefficient obtained in this way were somewhat higher than those obtained from the laboratory studies for zero stirrer speed, but were comparable with values obtained by a similar method by Park and Hood (1963) in which the uptake of $^{14}CO_2$ from gas circulating through a floating canopy was followed (Table 9.19).

All workers agree that increase in shear on the surface induced either by increase in air speed (in wind tunnel experiments) or by increase in turbulence in the water facilitates exchange (Hanya and Ishiwatari, 1961; Sugiura et al. 1963; Kanwisher, 1963a, b; Hoover and Berkshire, 1969; Liss, 1973), and there is an approximate square law dependence on the wind speed. This is often

TABLE 9.19

Rate constants for the transfer of carbon dioxide through the air–sea interface

	mmol cm^{-2} atm^{-1} min^{-1}	
Field experiment 1 (invasion)	0·044	a
Field experiment 2 (invasion)	0·021	a
Field experiment 3 (evasion)	0·0315	a
Bohr (1899)	0·0060	b, c
Becker (1924)	0·00075	b, c
Miyake and Hamanda	0·00045	b, c
Guyer and Tobler (1934)	0·15	b, c
Sugiura et al. (1963)	0·0073	c
Hoover and Berkshire (1969)	0·0032	c
Liss (1973)	0·034	c
7 year mean atmospheric residence time	0·0125	

a see Sugiura et al. (1963).
b Quoted by Sugiura et al. (1963).
c Laboratory determinations.
Figures quoted in this table for particular workers should be regarded as approximate since, generally, ranges of values were observed according to the conditions (pH, temperature, air flow and water turbulence).

interpreted in terms of a reduced film thickness, and it is significant that it is only at lower air speeds (corresponding to larger film thicknesses) that there is evidence for an augmented rate of CO_2 exchange attributable to its reactivity with the water. Using $^{14}CO_2$ as a tracer, Hoover and Berkshire (1969) showed that for solutions of pH < 4 (when the CO_2 behaves effectively as an inert gas), the transfer rate was lower than that found when the solution pH was between 6·3 and 6·8 only for the lower air speeds; at higher air velocities the rate of transfer was independent of the pH (Fig. 9.25). Similar observations have been made by Liss (1973). His investigation was essentially a simultaneous comparison of the exchange coefficients of O_2, CO_2 and H_2O. Under circumstances such that the reactivity-augmented flux of CO_2 would not be expected to be evident (low pH solutions or high air velocities), the ratio of the exchange coefficients for CO_2 and O_2 was approximately unity, but for relatively quiescent water conditions and pH values >5 there was a noticeable enhancement of the rate of exchange of CO_2 relative to that of O_2. In passing, it might be noted that in all Liss's experiments the exchange behaviour of water vapour was radically different from that of the permanent gases. The exchange coefficient for water vapour was always several hundred fold greater than that for CO_2 and O_2. In addition, breaking of the surface of the water in the tank used in Liss's work produced a much smaller enhancement of the rate of H_2O exchange than of that of oxygen. These observations support the view that the exchanges of O_2 and CO_2 are controlled by events

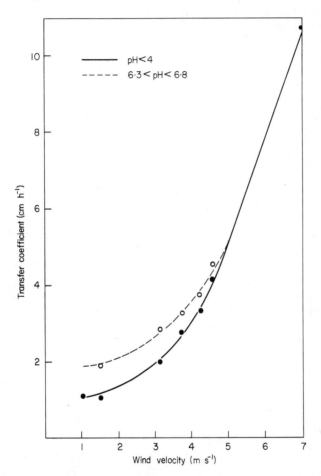

FIG. 9.25. Transfer coefficient of CO_2 at high and low pH values as a function of wind speed (after Hoover, 1966).

in the surface layer of the water whereas the exchange of water vapour is controlled at least partly by events in the gas phase. In many respects the behaviour of SO_2 more closely resembles that of water vapour than that of

F

the major atmospheric gases, presumably because of its high solubility, this effect being reinforced by a reactivity-enhanced uptake similar to that shown by CO_2 under high pH conditions. Evidence has been given that the reactivity enhanced uptake of SO_2 is suppressed somewhat when the solution pH is very low (Liss, 1972; Brindlecomb and Spedding, 1972; Spedding, 1972).

The mean atmospheric life-time of CO_2 molecules is about 7 years. If, for the purpose of rough comparison with the laboratory and field estimates of the CO_2 exchange coefficient, it is assumed that the sole route by which CO_2 is lost from the atmosphere is by solution in the sea, then this life time corresponds to an uptake rate averaged over the ocean surface of about 1.77×10^{-4} mg cm^{-2} min^{-1} (Sugiura et al., 1963). Taking P_{CO_2} at the ocean surface to be about 300 p.p.m., this uptake rate corresponds to an exchange coefficient of about 0.013 mol cm^{-2} atm^{-1} min^{-1}. This value may be compared with the experimentally observed values given in Table 9.19. Clearly, a mean atmospheric life-time of the order of several years is not incompatible with the observed exchange coefficients and the view that the sea is the principal sink for atmospheric CO_2. However, the rate of entry to the sea surface is likely to vary considerably with locality and conditions. It seems unlikely that under most natural conditions the surface boundary layer on the sea is sufficiently thick for the reactivity augmented uptake of CO_2 to be apparent unless the reactivity of the dissolved CO_2 is increased by the catalytic activity of other dissolved materials. The possibility that the exchange rate of CO_2 may be further enhanced by catalytically active materials was suggested by Kanwisher (1963a, b) who investigated the effect of the addition of carbonic anhydrase on the rate of uptake of CO_2 by a stirred solution. Although he was unable to detect any appreciable effect on the rate unless excessive amounts were added, it is possible that his solutions were too efficiently stirred for the effect to be apparent since more recent work by Berger and Libby (1969) has shown that the uptake rate can be increased as much as 20-fold on the addition of carbonic anhydrase.

Other effects which might influence the transfer rates might be noted. According to Kanwisher (1963a) small droplets and bubbles formed under turbulent conditions may assist transport. Hoover (1966) found that the growth of algal films which occurred during laboratory experiments retarded the transfer process. Such films were not thought likely to be important under natural conditions except possibly for sheltered coastal waters. Hoover also noted that when the water temperature in the tank he used was below the atmospheric dew point, there was a reduction in the rate of carbon dioxide transfer. Presumably condensation of water takes place at the interface at higher humidities, but it is not clear how this influences the carbon dioxide exchange.

Oil films on the sea surface, unless unusually thick, do not seem to influence the transfer of atmospheric gases.

9.6. FIELD OBSERVATIONS

The buffer capacity of sea water is insufficient to prevent some pH variation, and for open ocean surface water values range from about 7·8 to 8·4. Variations also occur in the specific alkalinity (*ca.* 0·119–0·130 meq l^{-1} (Cl‰)$^{-1}$, ΣCO_2 (*ca.* 1·95–2·20 mmol l^{-1}) and P_{CO_2} (less than 250 to more than 450 × 10^{-6} atm), this last quantity being particularly sensitive to those physicochemical and biological processes which affect the carbonate system. These ranges should be regarded as approximate only since if account is also taken of intermediate and deep water (and particularly of anoxic and pore waters), much larger variations are found.

Because of the temperature and pressure dependencies of the carbonate equilibria, some adjustment of the relative values for $c_{HCO_3(T)}$, $c_{CO_3(T)}$, pH and P_{CO_2} accompanies change of temperature and depth, even when there is no material exchange. When material exchange does occur, usually it is carbon dioxide or calcium carbonate loss or gain which needs to be considered. Carbon dioxide loss, either by evasion to the atmosphere or by photosynthetic uptake, leads to a pH increase and to a decrease of P_{CO_2} and ΣCO_2; the generation of carbon dioxide (by respiration or the oxidative decay of organic debris) has the converse effect. Uptake or loss of calcium carbonate affects not only ΣCO_2, but also the carbonate alkalinity and the relative proportions of HCO_3^-, CO_3^{2-} and H^+. Biological withdrawal of calcium carbonate from solution is common in many surface waters, and occasionally inorganic carbonate deposition occurs although more usually its deposition involves a biological mechanism, (e.g., test formation). The dissolution of calcium carbonate takes place as tests fall through the water column; ultimately oceanic circulation and advection returns the calcium carbonate (and CO_2) enriched water to the surface.

The mixing of water masses of different carbonate alkalinities, ΣCO_2 contents and temperatures will generate a mass having intermediate values for these quantities, but the complex interdependence of $c_{HCO_3(T)}$, $c_{CO_3(T)}$, ΣCO_2 and temperature means that the concentrations of the individual components in the newly generated water mass are not related in a simple fashion to those of the parent masses (see e.g. Mook and Koene, 1974).

On a localized scale, variations originate from the presence of pH influencing materials. Brackish water, estuarine, tidal and harbour waters, shallow marine ponds, pore waters, water subject to localized industrial contamination and waters of anoxic basins may have values for the components of the carbonate system which differ considerably from those for the open ocean.

9.6.1. SURFACE AND SHALLOW WATERS

The range of pH values for open ocean surface water is seen from Fig. 9.26a which is taken from a paper by Postma (1964). Although the bulk of the measurements fall within the expected range, there is a good deal of scatter, some of which may result from measurement error, and this underlines the difficulties associated with the use of pH as an observable parameter when values of the carbonate system components are to be calculated and emphasizes the need to exploit the levels of measurement precision which are now available. However, some particular sets of determinations show systematic trends (see the upper curve of Fig. 9.26b, also taken from the paper by Postma and compiled by him from data obtained on *Ob* and *Vityaz* cruises). Other ocean surface water pH determinations differ little from those shown in Fig. 9.26. Thus, in the seas between the Phillipines and Japan, the value is 8·1–8·3 (Akiyama *et al.*, 1966, 1968) and the surface pH (adjusted to 25°C) noted on the GEOSECS Pacific Station was *ca.* 8·15 (Takahashi *et al.*, 1970). Surface water above the New Hebrides Trench has a pH of about 8·3 (Rotschi, 1965). Typical surface water pH values quoted by Keeling (1973a) are close to 8·2 (see also Almgren *et al.*, 1974a).

Diurnal variations of the pH, dissolved CO_2 content, ΣCO_2 and, often, dissolved oxygen are noted in certain circumstances. During daylight hours photosynthesis leads to a reduction in the concentration of dissolved CO_2 to a late afternoon minimum, and this is accompanied by a rise in pH (Schmalz and Swanson, 1969). During the dark hours, respiratory processes predominate, the dissolved carbon dioxide content of the water increases and the pH falls. According to Schmalz and Swanson, the diurnal pH range may be 0·15 units, and the high night-time carbon dioxide concentrations may lead to calcium carbonate dissolution; during daylight hours supersaturation may develop. Diurnal effects have previously been reported by Park *et al.* (1958) for certain shallow water Texas bays. Marked variations of the pH, dissolved oxygen and temperature were noted, the peak values occurring in the afternoon (Fig. 9.27). The effect is seen to be more pronounced in summer than in winter.

Over the Bahama Banks some diurnal changes of P_{CO_2} associated with tidal movement have been reported by Broecker and Takahashi (1966) although the range was usually less than 30 p.p.m. In this area some CO_2 liberation might be expected as calcium carbonate deposition occurs. In fact, P_{CO_2} measurements made by Broecker and Takahashi for the water

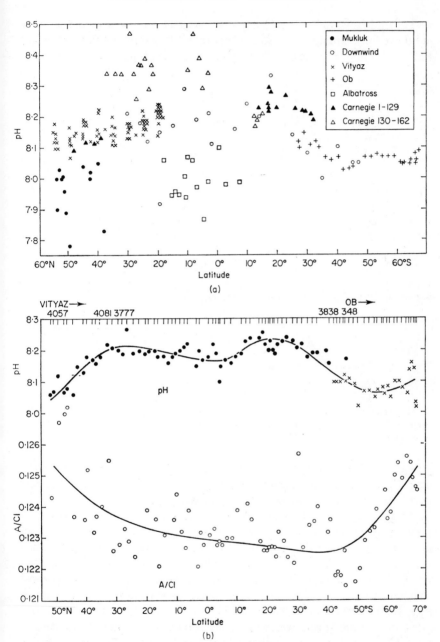

FIG. 9.26. (a) pH (*in situ*) of ocean surface water in the western Pacific Ocean between 110° and 150° W. (b) pH *in situ* and specific alkalinity of surface water in the Pacific. (Postma, 1964).

west of Andros Island indicated that evasion to the atmosphere (and possibly photosynthetic uptake) were sufficiently rapid to maintain a fairly constant P_{CO_2}, and values in excess of 400 p.p.m. were exceptional (see below).

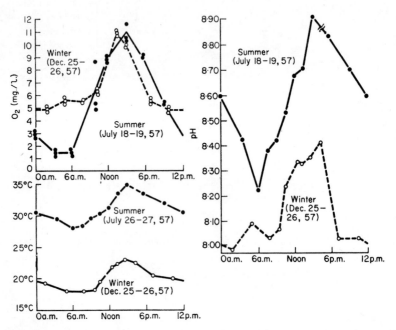

FIG. 9.27. Diurnal oxygen, pH and temperature variations at Redfish Bay, Texas. Solid lines for summer data for July 18–19, 1957. Dotted lines for winter data for December 25–26, 1957 (Park *et al.*, 1958).

Seasonal effects are common. Thus, in a small body of water, stratification may develop during the summer growing season, and as a result there will be a retarded circulation and a noticeable pH gradient. For a marine pond investigated by Orr (1947), the pH at the surface in June was increased by photosynthetic uptake to *ca.* 9·9 whereas in the oxygen deficient bottom layers (*ca.* 3 m) the pH was about 7·4. The P_{CO_2} of the water of Izembek Lagoon on the Alaskan Peninsula is also subject to seasonal influences, and Gordon *et al.* (1971) have shown the existence of pronounced seasonal changes in the P_{CO_2} of surface water in the North Pacific. An increase in P_{CO_2} (of about 18×10^{-6} atm) during the period March to April was attributed to the spring warming of the surface water, and an increase of 60×10^{-6} atm between October (1968) and March (1969) was probably a combination of the effects of the cessation of biosynthetic activity and the onset of storm generated vertical mixing of surface water with the carbon dioxide-rich lower layers.

Not all diurnal changes in the carbonate system can be attributed to photosynthesis and respiration, and short-term variations of the P_{CO_2} of Atlantic water noted by Takahashi (1961) were only rarely of the type expected for photosynthetic activity (Fig. 9.28). The more usual observation (Fig. 9.29) was a daylight maximum (more pronounced in surface water than in water from 5 or 10 m), the implication being that physicochemical effects associated with temperature adjustment sometimes outweigh biological factors in determining the diurnal range.

Shallow water masses tend to have their own peculiarities which depend on local climatic and circulation conditions. The Bahama Bank is a special example, and is one of the few marine localities in which calcium carbonate precipitation (possibly inorganic) is occurring at the present time; another such area is in the Persian Gulf (see Wells and Illing, 1964). Over the Bahama Bank west of Andros Island the average water depth is *ca.* 4·5 m and the annual temperature range of the water is 20–30°C. Open ocean surface water flows onto the Bank, and during its period of residence—estimated from measurements based on the uptake of bomb-produced ^{14}C to be about 250 days immediately to the west of Andros Island, but to be only about 10 days or less as Bimini Island is approached (Broecker and Takahashi, 1966)— the salinity was found to increase to about 42‰ as a result of evaporation

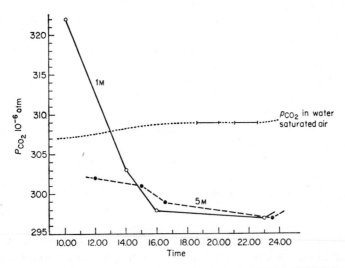

FIG. 9.28. Diurnal variation of the partial pressure of CO_2 in the sea 15°30′ N, 40°33′ W (November 26, 1957) (Takahashi, 1961).

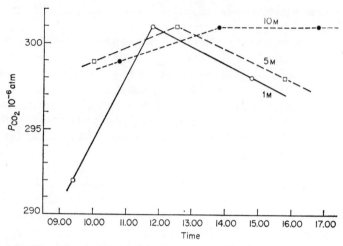

Fig. 9.29. Diurnal variations of the partial pressure of carbon dioxide in the sea 14°26′ N, 38°52′ W (November 27, 1957) (Takahashi, 1961). The partial pressure of CO_2 in air above the water $= 309 \times 10^{-6}$ atm (water saturated air).

loss. This loss corresponded to about 120 cm H_2O y^{-1}, although deviations from the almost linear salinity-residence time relationship were noted after periods of heavy rainfall. Clearly, the physico-chemical conditions of the water over the Banks differ considerably from those of the oceanic source water, and this difference was reflected in the P_{CO_2}, ΣCO_2 and alkalinity values of water from over the Banks, and the fact that calcium carbonate (aragonite) was deposited. Thus, whereas P_{CO_2} and ΣCO_2 for the sea off the perimeter of the Bank were 330×10^{-6} atm and 2.00 mmol l^{-1} respectively, as the Bank was crossed and the water salinity increased, P_{CO_2} increased to a directly measured value of slightly more than 400×10^{-6} atm and ΣCO_2 (normalized to a salinity of $36.4\%_0$) fell to about 1.20 mmol l^{-1}. There was also a decrease in the (normalized) alkalinity (2.38–1.60 meq l^{-1}). Somewhat higher P_{CO_2} values were reported by Cloud (1962a, b), but the estimates which he gave were obtained indirectly, and in view of the proven reliability of the direct method used by Broecker and Takahashi and the fact that their P_{CO_2} measurements were in reasonable agreement with those made by earlier investigators (Smith, 1941; Seibold, 1962), it seems likely that the estimates given by Broecker and Takahashi are correct.

Although the P_{CO_2} of water from over the Bank is higher than is that of adjacent ocean surface water, it is not abnormally high, and, of course, it rises only to a level such that the rate of evasion from the surface is able to keep pace with the rate of carbonate deposition. However, there was some difficulty in accounting for the CO_2 budget. As water crosses the Bank, inorganic carbon is lost by carbonate precipitation and by CO_2 loss. That is for the salinity-normalized system, the relationship

$$\Delta(\Sigma CO_2) = \Delta CaCO_3 + \Delta CO_2$$

applies. According to Broecker and Takahashi, each mole of carbonate which precipitates is accompanied by the loss from the system of about 0·6 mol of CO_2. This carbon dioxide loss must be accounted for either by evasion to the atmosphere or by biosynthetic uptake. There was no convincing evidence for the latter, but only about one-half of the CO_2 loss could be accounted for in terms of evasion through the surface.

There is, as yet, no unanimity of opinion on the mechanism of carbonate precipitation. An inorganic mechanism has an intuitive appeal since it requires the minimum of hypothesis. The solution is already abnormally supersaturated, in shallow water there should exist abundant nuclei, and the relatively high summer temperatures should encourage those processes which have an appreciable activation energy. Additional, albeit partial, support for an inorganic mechanism is given by the observation that the kinetics of the precipitation followed a second order law (Broecker and Takahashi, 1966). The rate of deposition was proportional to the degree of super-saturation, and precipitation reduced the activity product to about $1·0 \times 10^{-8}$ in the more sheltered areas in the lee of Andros Asland. This value corresponds to a supersaturation of about 15%.

The precipitation of $CaCO_3$ has been attributed to biogenic processes by Lowenstam and Epstein (1957). They considered that the precipitation temperature of the carbonate muds as estimated from $^{18}O/^{16}O$ measurements was too high for them to have been generated in isotopic equilibrium with the water, and suggested that organic processes involving algae must have been responsible. However, if, as Cloud (1962a) has suggested, chemical precipitation occurs mainly during the summer when the water temperature approaches 28–30°C, the isotopic composition of the aragonite muds could be accounted for in terms of a purely inorganic mechanism. If the process is purely inorganic, it is unlikely that it is homogeneous since, as Berner (1971) has argued, the rate of nucleation expected for water of the composition of that over the Banks is far too low.

In both the Bahama Bank and Persian Gulf carbonate sedimentation areas the phenomenon of "whitings"—a turbidity in the water caused by suspended aragonite needles—is encountered. Possible explanations are (i) the resuspension of sediments and (ii) chemical relaxation of the thermodynamically unstable condition of supersaturation by sudden, large-scale, precipitation (see Cloud, 1962a, b; Wells and Illing, 1964), the onset of the process being

catalysed by some photosynthetic process. Evidence in favour of the resuspension hypothesis is now convincing (Broecker and Takahashi, 1966). A
sudden, relatively large-scale precipitation should be accompanied by a CO_2
release into the water. Since the transfer of this to the atmosphere is a slow
process, there ought to be a noticeable increase in the P_{CO_2} of the water (of
about 70×10^{-6} atm), and a pH decrease (of about 0·7 pH units) as an area
affected by a whiting is traversed. Such marked changes do not appear
to have been reported. Furthermore, Broecker and Takahashi have shown
that the organic content of the suspended material is less than might have been
expected if biogenic processes were involved and that the ^{14}C activity of the
suspension was more characteristic of the bottom sediments than of the
bomb-^{14}C-enriched water.

Oceanic P_{CO_2} surveys. Many oceanic surveys made during the last decade or
so have taken advantage of direct methods for P_{CO_2} determination (see
Takahashi, 1961; Ibert, 1963; Hood et al., 1963; Keeling et al., 1965; Keeling
and Waterman, 1968; Kelley, 1970; Kelley and Hood, 1970a, b, c; Kelley
et al., 1971), and from the accumulated knowledge a picture is emerging of
the overall world-wide distribution of surface water P_{CO_2} values (Keeling,
1968). Temperature, P_{CO_2} and p_{CO_2} traces obtained by Takahashi (1961)
on an Atlantic cruise are shown in Fig. 9.30, the appropriate vessel track
being shown in Fig. 9.34. Some sections and cruise tracks for the Lusiad
expedition (on which over 14 000 atmospheric and 12 000 oceanic measurements were made) are shown in Figs 9.31 to 9.33 (Keeling and Waterman,
1968).

A feature common to these sets of observations and also to those of other
workers (Akiyama, 1966, 1968, 1969a, b; Miyake and Sugimura, 1968;
Gordon et al., 1971; Lyakhin, 1971) is the general inequality between the
partial pressure of carbon dioxide in the atmosphere and that in the sea.
Furthermore, it is clear that whereas horizontal gradients in p_{CO_2} are
generally small, both the range and the gradients of P_{CO_2} are often pronounced,
and distinct sources and sinks for carbon dioxide can be recognized. The P_{CO_2}
range for the Atlantic track shown in Fig. 9.30 is *ca.* 100 p.p.m., and that for
the Pacific traverse of the Lusiad expedition is *ca.* 68×10^{-6} atm with a
minimum P_{CO_2} of 252×10^{-6} atm at 175°E.

It is now possible to understand many aspects of the oceanic P_{CO_2} distribution in terms of known water circulation patterns (see e.g., Keeling and
Waterman, 1968; Keeling, 1968). For example, in the Pacific traverse (Fig.
9.31), the fall in P_{CO_2} which occurs in an easterly direction up to 175°E
is explained, at least in part, by a progressive cooling of the water which
moves northwards and eastwards under the influence of the Kuroshio

current. The P_{CO_2} decrease was somewhat smaller than that estimated from the temperature dependence of P_{CO_2} (as indicated by the tables given in Harvey, 1955), the difference being partially accounted for by uptake of carbon dioxide from the atmosphere. The amount of transfer needed to account for the difference between the observed and predicted P_{CO_2} decrease is not inconsistent with the likely rate of invasion of CO_2 from the atmosphere and the rate of water movement; however biological activity may play an, as yet, unquantified part.

Park *et al.* (1967) also observed low P_{CO_2} values in the central North Pacific during a traverse between Hawaii and the Aleutians. Since high chlorophyll pigment and low plant nutrient contents were observed where P_{CO_2} was low,

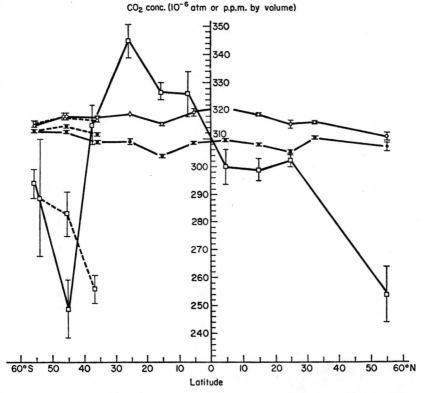

FIG. 9.30. Partial pressures of CO_2 in the surface water and the CO_2 concentration in the air over the North and South Atlantic Oceans (Takahashi, 1961). ———, measurements in the Western Hemisphere; – – – –, measurements in the Eastern Hemisphere; —O—O—, carbon dioxide concentration in the atmosphere (p.p.m. by volume); —●—●—, partial pressure of carbon dioxide in atmosphere saturated with water at surface temperature (10^{-6} atm); —□—□—, partial pressure of CO_2 in sea water (10^{-6} atm).

they concluded that biological activity contributes significantly to the regulation of P_{CO_2} and suggested that the effect may be seasonal.

In the equatorial Indian Ocean the distribution of the partial pressure of carbon dioxide in the surface water is dependent on the season, a band of high P_{CO_2} being associated with the northeast monsoon, when an eastward flowing undercurrent was known to be present; during the southwest monsoon the high pressure region of P_{CO_2} and the undercurrent were absent. The high P_{CO_2} during the northeast monsoon was thought to be linked with the onset of vertical mixing between the CO_2-rich undercurrent and the surface water. The fact that the transition from the high to low P_{CO_2} conditions is too rapid to be accounted for by carbon dioxide exchange with the atmosphere is supporting evidence for the vertical mixing hypothesis. Similar vertical mixing from an undercurrent probably accounts for the high partial pressure of carbon dioxide observed in the equatorial Atlantic, and Keeling

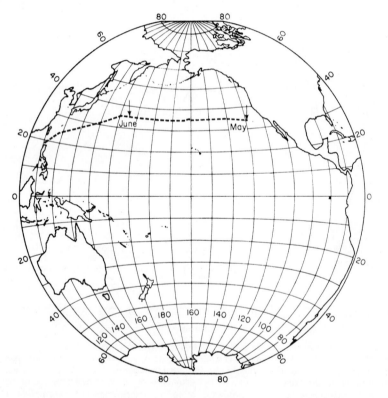

FIG. 9.31a.

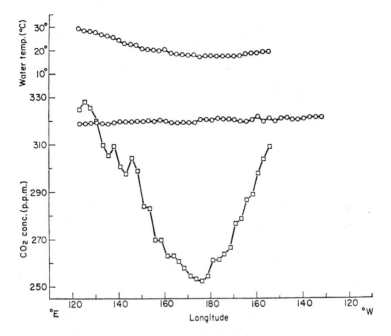

Fig. 9.31. (a) Track of R.V. *Argo*, Lusiad Expedition, May 18 to June 9, 1962 (Keeling and Waterman, 1968). (b) Average partial pressure of CO_2 in sea water (□) and in the atmosphere near the sea surface (O) versus longitude along the track shown in (a) (Keeling and Waterman, 1968).

and Waterman (1968) consider that a band of high P_{CO_2} in equatorial surface water is evidence of the presence of an undercurrent and *vice versa*.

Data obtained on a series of cruises from the Scripps Institution of Oceanography have been combined by Keeling (1968) with published results from other surveys (Krogh, 1904; Buch, 1939a; Takahashi, 1961; Hood *et al.*, 1963; Ibert, 1963; Keeling *et al.*, 1965; Keeling and Waterman, 1968) together with the results from the Eastpac and Nova Expeditions. The combined results were given in the form of a chart (see Figs. 9.34 and 9.35) showing the summer P_{CO_2} distributions for the surface water of the major oceans. Specifically excluded from this survey are shallow water masses (< 100 m depth) and polar seas. Although the chart is tentative (since it is based solely on the cruise tracks shown in Fig. 9.34 and refers to summer conditions) it is an important stage in the development of the study of the oceanic carbonate system since it succeeds in correlating and systematizing, without serious conflict, many hitherto isolated sets of measurements, and it provides a world-wide picture which, on

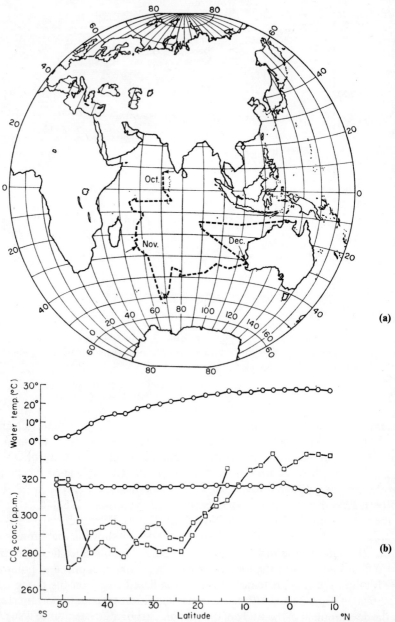

FIG. 9.32. (a) Track of R.V. *Argo*, Lusiad Expedition, October 6 to December 21, 1962 (Keeling and Waterman, 1968). (b) Average partial pressure of CO_2 in the sea (□) and in the atmosphere near the sea surface (O) vs. latitude along the track shown in (a) (Keeling and Waterman, 1968).

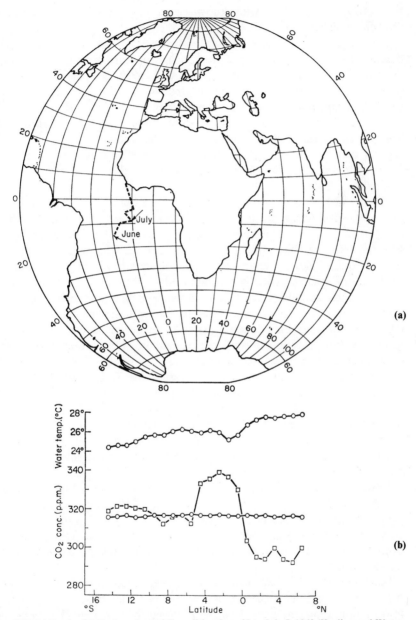

FIG. 9.33. (a) Track of R.V. *Argo*, Lusiad Expedition June 28 to July 7, 1963 (Keeling and Waterman, 1968). (b) Average partial pressure of CO_2 in the sea (□) and in the atmosphere near the surface (○) versus latitude along the track shown in (a) (Keeling and Waterman, 1968).

the whole, is consistent with known patterns of water movement and phosphate distribution.

It is seen that belts of high surface water P_{CO_2} exist in the equatorial Atlantic and Pacific Oceans, but that during the northern summer no such equatorial P_{CO_2} high is apparent in the Indian Ocean, the partial pressure of carbon dioxide of the water falling progressively in the southward direction to 30°S. Except where upwelling in coastal regions brings CO_2-rich water to the surface, low P_{CO_2} ($< p_{CO_2}$) regions exist in the sub-tropics of all oceans. The partial pressure of carbon dioxide in the water of certain upwelling regions may reach extremely high values. Over 600×10^{-6} atm has been reported for water off the Chilean coast (Kelley and Hood, 1971c).

The P_{CO_2} distribution shows a good correlation with that of PO_4^{3-}–P as described by Reid (1962) (see Keeling, 1968). Some relationship between the two distributions would be expected if both carbon dioxide and phosphate are generated during the decay of organic matter (Redfield et al., 1963; Postma, 1964). The biosynthetic uptake of carbon dioxide is accompanied by the removal of phosphate and of other nutrients from the water, and the oxidation of organic debris is accompanied by the liberation of phosphate and the generation of carbon dioxide. Because carbon dioxide is continually being withdrawn from surface waters by biosynthesis and transferred (via the oxidation of sinking organic material) to deeper layers, these deeper layers show a prevailing overall enrichment in both carbon dioxide and phosphate relative to surface water. In regions of upwelling and vertical mixing the surface water shows high values for ΣCO_2 and PO_4^{3-}–P. If biological processes are absent, the phosphate is more or less conserved, but the fate of the excess carbon dioxide depends on the circumstances. P_{CO_2} will be high because of the CO_2 generation during the sub-surface oxidation, but it will be further enhanced if the upwelling is followed by rapid warming of the water. In this way the region may become an effective source of carbon dioxide to the atmosphere (Bolin and Keeling, 1963; Keeling, 1968).

Low P_{CO_2} values (as found in sub-tropical gyres) may be the result of either a high rate of photosynthesis or rapid cooling of the water. Useful guiding principles given by Keeling are that where the surface water is moving equatorwards, low P_{CO_2} values are probably the consequence of photosynthesis; where the movement is polewards, cooling is probably the more important factor.

Not included on Keeling's chart were P_{CO_2} values for waters of northern and Arctic Polar regions. Recently, a number of expeditions have provided information about these areas (Ku et al., 1969; Kelley, 1970; Kinney et al., 1971; Kelley and Hood, 1971a, b, c; Kelley et al., 1971; Park et al., 1967; Gordon et al., 1973). In the Barents and Kara Seas, a cruise track which

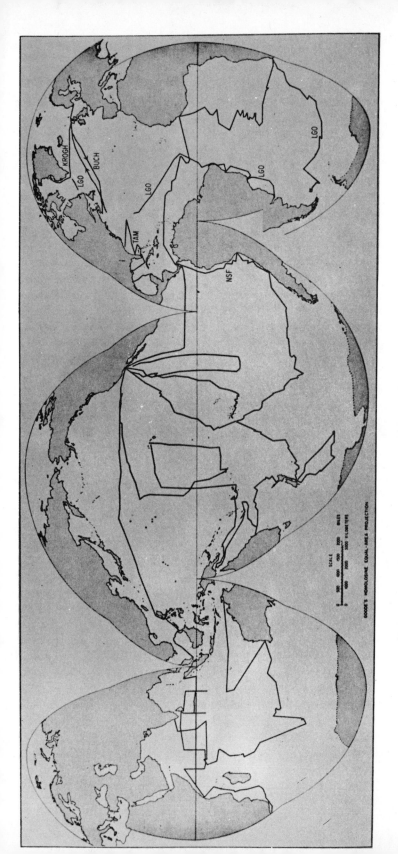

FIG. 9.34. Cruise tracks of vessels from which the data summarized in Fig. 9.35 was obtained. The labelled tracks correspond to those studied by KROGH, Krogh (1904); BUCH, Buch (1939a); LGO, Takahashi (1961); TAM, Ibert (1963); NSF, Hood et al. (1963). Other tracks are those of vessels from the Scripps Institution of Oceanography (Keeling 1968).

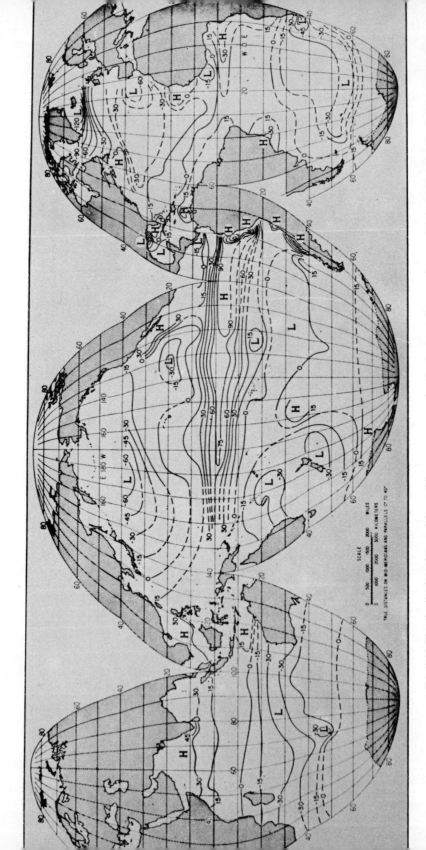

FIG. 9.35. The distribution of P_{CO_2} of the world's oceans expressed as the departure in p.p.m. from equilibrium with atmospheric CO_2. H indicates high; L, low (Keeling, 1968).

included the northern extension of the Gulf Stream, the cold East Spitzbergen Current and the outflow regions of the Ob and Yenesi rivers (Kelley, 1970) gave data showing that during the late summer and early autumn, those surface waters which were not directly influenced by river outflow generally had a low partial pressure of carbon dioxide relative to that of the atmosphere; this low value (*ca.* 230–310 × 10^{-6} atm) persisted to the limit of the pack ice (Kelley and Hood, 1970a). However, near the mouths of the Ob and Yenesi rivers, P_{CO_2} was generally high relative to p_{CO_2} (Kelley, 1970). Similar supersaturations near the outflows of rivers and in estuaries have been found in other instances (Park *et al.*, 1969; Kelley, 1970; Kelley and Hood, 1971a), and where the influx of river water is of sufficient importance the specific alkalinity may deviate from that of the open ocean (Park, 1968b; Park *et al.*, 1969) and the high carbon dioxide content of the source water may cause the development of high P_{CO_2} values such as the 900 × 10^{-6} atm reported for the Straits of Jaun de Fuca where the influence of water from the Frazer river is probably felt (Kelley and Hood, 1971c).

The partial pressure of carbon dioxide in the surface water of leads and polynia of the Bering Sea during winter (Kelley and Hood, 1971b) showed the water to be supersaturated with respect to the atmosphere, and observations from ice island T3 (*ca.* 78°N, 175°W) by Kelley (1968) indicated the water under the ice to be supersaturated, the P_{CO_2} value showing large fluctuations (from 330–400 × 10^{-6} atm) during the course of two weeks in January.

In and adjacent to the Bering Sea a number of sources and sinks for carbon dioxide have been recognized (Kelley and Hood, 1971a; Kelley *et al.*, 1971), and a series of measurements made in the late spring and the early autumn near the Unimak, Amukta and Samala passes of the eastern Aleutian Islands indicated wide variations of the P_{CO_2} of surface water. Through the Amukta pass the partial pressure of carbon dioxide in the water rose from under 200 × 10^{-6} atm to slightly more than 500 × 10^{-6} atm. This high partial pressure, and that observed near the Samala pass (where partial pressures of 450 × 10^{-6} atm were detected) implies that there is upwelling of deep water which had been fed into the Bering Sea from the Pacific and subsequently returned to the Pacific as surface water. The water was characterized by a low temperature and a low dissolved oxygen concentration and by a high salinity and NO_3^-–N content. Kelley and Hood (1971c) have drawn attention to similar high P_{CO_2} areas associated with upwelling off the coasts of Chile and Peru. The shallow Unimak pass receives northwards flowing water from the Pacific, and since this water had been somewhat impoverished in carbon dioxide by the growth of autotrophic plants, its P_{CO_2} was less than that of the atmosphere.

Seasonal variations, presumably attributable to biological activity were also noted, particularly where vertical mixing was not well emphasized and where productivity was high (e.g., Izembek Lagoon on the Alaskan Peninsula, near the Unimak pass and also along the south coast of the Alaskan Peninsula where carbon dioxide undersaturation in June was more pronounced than it was in September).

These recent investigations of northern waters provide a useful extension of the world-wide distribution of P_{CO_2} as proposed in the chart prepared by Keeling (1968). (See also Gordon *et al.*, 1973).

9.6.2. VERTICAL PROFILES

Vertical profiles of the carbonate system components (and of oxygen) reflect the quasi-stationary state which exists at a particular locality. The forms of these profiles depend on the flux of organic matter derived from the upper layers of the water column, its rate of oxidation, the rate of dissolution of calcium carbonate and the characteristics of, and circulation patterns in, the various water masses through which the profile passes. Typically, below the surface layers, both ΣCO_2 and P_{CO_2} first increase with depth, and P_{CO_2} sometimes shows a maximum at a depth roughly coincident with that at which the dissolved oxygen concentration shows a minimum. Below this depth both ΣCO_2 and P_{CO_2} may either fall before becoming more or less constant, or fall progressively with depth, the detailed behaviour depending on the characteristics of the individual water masses which the profile transects. In the south Pacific the ΣCO_2 maximum and the carbonate alkalinity maximum are roughly coincident with the oxygen minimum; in the north Pacific, the oxygen minimum lies somewhat above the ΣCO_2 and carbonate alkalinity maxima (see Craig, 1969).

Data obtained by Li *et al.*, 1969) showed that for the north Atlantic and the Caribbean, ΣCO_2 reached *ca.* $2·28$ mmol l^{-1} (normalized to a salinity of $35\permil$) at a depth of 100 m and then changed only slightly with further increase of depth. For the Pacific ocean, ΣCO_2 at 1000 m was *ca.* $2·45$ mmol l^{-1} and for a northern North Pacific station the maximum ΣCO_2 value was $2·52$ mmol l^{-1}. The peak P_{CO_2} value in the Atlantic was 450×10^{-6} bar whereas that in the Pacific was much higher (1044×10^{-6} bar).[*] Plots given by Takahashi *et al.* (1970) for data obtained at the 1969 GEOSECS Inter-calibration Station ($28° 20'$ N, $121° 41'$ W) for alkalinity, ΣCO_2 and P_{CO_2} are shown in Figs. 9.36a, b and c. A principal aim of the GEOSECS study was to intercompare different methods for sampling and for the estimation of various

---

[*] According to Takahashi *et al.* (1970), faulty sample storage makes some of the ΣCO_2 measurements made by Li *et al.* (1969) too high by several percent. This applies mainly to the Atlantic and Antarctic samples.

parameters, and Figs. 9.36a to c indicate the ranges observed. Fig. 9.36d shows *in situ* and calculated pH-depth data for the north-eastern Pacific (Ben-Yaakov *et al.*, 1974). Samples taken at the GEOSECS station and analysed for oxygen (Craig and Weiss, 1970) indicated an oxygen concentration maximum (*ca.* 5·8 cm^3 (STP) kg^{-1}) at about 7 m depth; below this point the concentration fell to a minimum of 0·22 cm^3 (STP) kg^{-1} at a depth of 700 m and then rose progressively with depth to *ca.* 3·2 cm^3 (STP) kg^{-1} at 400 m.

Plots similar to those given in Figs. 9.36a to d have been obtained or described by other investigators (see e.g. Rotschi, 1965; Park, 1965; Edmond and Gieskes, 1970; Weiss and Craig, 1968; Craig, 1969; Culberson and Pytkowicz, 1970), and many authors have commented on the vertical gradients of specific alkalinity which are often found (Koczy, 1956; Hood *et al.*, 1963; Pytkowicz, 1968). Pytkowicz (1968) has discussed in some detail the changes of specific alkalinity which occur with increasing depth in terms of the dissolution of calcium carbonate, the properties of the source waters and the movements of the water masses. Indications of the uptake of calcium carbonate are given not only by the vertical increase of specific alkalinity, but also by horizontal gradients since these latter are often consistent with a specific alkalinity increase in the direction of movement and also with the age of the water mass as judged from its low oxygen content and high PO_4^{3-}–P content (see Pytkowicz, 1967, 1968). An examination of alkalinity profiles in the Pacific along sections at 43° S and 28° S (Turekian *et al.*, 1971) showed there to be an alkalinity maximum at depths of 2000–3500 m and higher alkalinities in the eastern part of the Pacific basin than in the western part. These observations were thought to reflect a flow pattern in which low alkalinity water enters the abyss via a western boundary current. The uptake of calcium carbonate by this water as it rises results in an alkalinity increase and a maximum at intermediate depths.

The quantitative interpretation of carbonate system profiles is that photosynthesis in the upper layers leads to the consumption of dissolved carbon dioxide and to the generation of organic matter; decaying organic material falls through the water column and leads to the consumption of dissolved oxygen, to an increase in ΣCO_2 and to the liberation of phosphate and other nutrients. Dissolution of calcite particles also occurs and is encouraged by the high CO_2 content. This increases the alkalinity and also contributes to ΣCO_2. The partial pressure of carbon dioxide in the water and the *in situ* pH are determined by the values of the other carbonate system components, the temperature and the appropriate equilibrium constants. These constants are, of course, temperature, pressure and salinity dependent. Individual profiles are modified by the physical properties and source characteristics of the various water masses through which they pass.

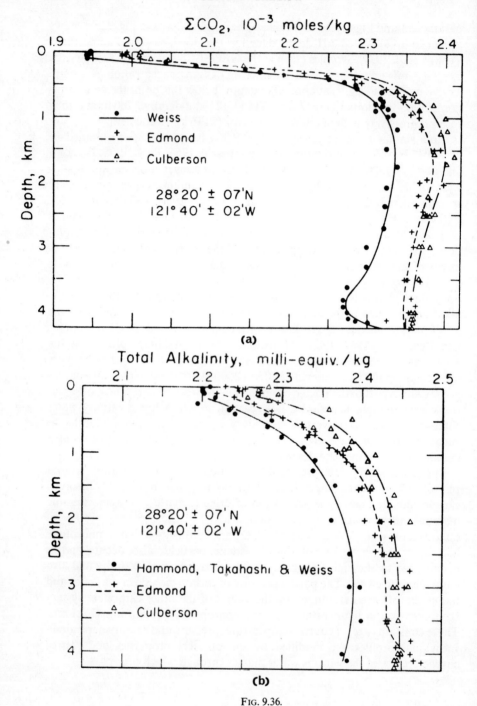

Fig. 9.36.

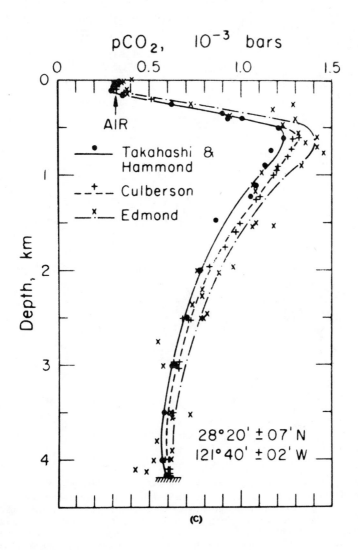

Fig. 9.36.

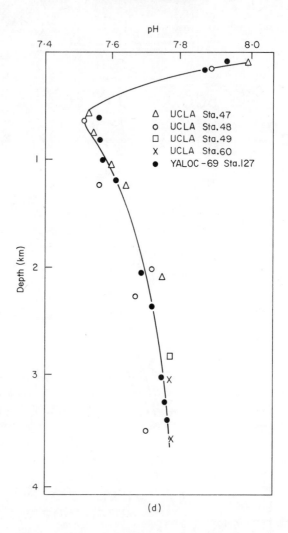

FIG. 9.36. Vertical profiles for ΣCO_2, total alkalinity and P_{CO_2} at the GEOSECS Pacific Station 28°20′ N, 121°40′ W. The changes of these quantities with depth and inter-comparison of different methods for their estimation are given. (a) ΣCO_2-depth profiles. ●, Gas chromatographic method (Weiss); +, potentiometric titration method (Edmond); △, calculations based on pH and alkalinity data (Culberson) using Lyman's apparent constants. (b) Total alkalinity-depth profiles. +, potentiometric titration method (Edmond); △, pH method (Culberson); ●, calculated from the P_{CO_2} and ΣCO_2 values (Hammond, Takahashi and Weiss). (c) P_{CO_2}-depth profiles. +, from pH and alkalinity data (Culberson); ×, from alkalinity and ΣCO_2 data (Edmond); ●, measurements based on the equilibrator-infrared method (Takahashi *et al.*, 1970). (d) *In situ* pH in northeastern Pacific as measured by *in situ* probe (UCLA) and as calculated by the method of Culberson (1972) from the Yaloc-69 data (see Culberson and Pytkowicz, 1970).

Additional field data which are useful in the interpretation of sub-surface events are provided by ^{13}C analysis of the inorganic and organic carbon. Isotope fractionation accompanies the hydration reaction of carbon dioxide (Deuser and Degens, 1967) and also the bioutilization of dissolved inorganic carbon (Williams and Gordon, 1970; Weber and Woodhead, 1971). As a result distinct $\delta^{13}C$ differences occur between carbons of different origins, variations in the $\delta^{13}C$ value of the dissolved inorganic carbon occur (e.g. Sackett and Moore, 1966), and diurnal variations in the $\delta^{13}C$ of surface waters are sometimes detectable (Weber and Woodhead, 1971). The depletion of ^{13}C observed when fractionation occurs is temperature dependent, but $\delta^{13}C$ is in the range -20 to -30% for planktonic material, the organisms of cooler polar waters showing the greatest depletion (see Sackett et al., 1965). The $\delta^{13}C$ values for carbonate secreting organisms are generally much less negative. The value depends on the organism (see Kroopnick (1974) for a short discussion), but an average value of about $2·0\%$ is sometimes assumed (Kroopnick, 1974). It will be appreciated that as contributions to ΣCO_2 are given by both the oxidation of organic matter and the dissolution of calcareous material, the observed $\delta^{13}C$ profiles can be used to give information on the relative contributions of these two sources (see also Deuser and Hunt, 1969; Craig, 1954).

A quantitative discussion of observed profiles has been given in terms of a vertical advection–diffusion model (Craig, 1969, 1970, 1971; Kroopnick et al., 1970; Kroopnick, 1974). For a reasonably well-defined water mass (i.e. one having a linear potential temperature-salinity relationship) the stationary state profile can be considered to be determined by upper and lower boundary concentrations. Between these boundaries, for a stationary state system,

$$\frac{\partial c_i}{\partial t} = 0 = K\frac{\partial^2 c_i}{\partial z^2} - w\frac{\partial c_i}{\partial z} + J - \lambda c_i$$

where K is the eddy diffusion coefficient, w is the vertical advection velocity and J is the production rate of component i (e.g. the rate of oxidative generation of CO_2) at the position z. The quantity λ is the radioactive decay constant (zero for non-radioactive species). The general solution of the above equation reduces to a simplifed special case for stable conservative tracers (e.g. salinity and temperature), and use of the available temperature and/or salinity data with the reduced form of the equation enabled Craig to obtain an estimate of the mixing parameter K/w. For stable non-conservative tracers for which $\lambda = 0$, (e.g. total CO_2 and dissolved oxygen), the general equation again reduces to a simplified form from which, by suitable combination with oxygen or ΣCO_2 profile data, estimates for J/w were obtained. From profiles of the absolute concentration of radio-carbon and the estimate of J/w, a value for the upwelling velocity w could be derived. Values for the various ratios and the absolute values for K, w and J were reasonably consistent with estimates obtained from geochemical considerations ($K \sim 1.6 \, cm^2 \, s^{-1}$; $w \sim 5 \, m \, yr^{-1}$). Craig extended his analysis to an examination of the carbon dioxide oxygen relationship in abyssal water and he concluded that the relationship proposed by Redfield et al. (1963)—that the changes in the $\Sigma CO_2/O_2$ ratio arising from the oxidation of organic matter should be 106/138—is not generally applicable. However, several groups of workers have obtained results which, at least to a good approximation, support this ratio (Sugiura, 1964, 1965a, b; Sugiura and Yashimura, 1964; Park, 1965; Li et al., 1969; Culberson and Pytkowicz, 1970). In order to test the Redfield relationship using profile data it is necessary to correct the observed ΣCO_2 changes so as to allow for contributions which arise from non-oxidative processes. When allowance was made for variations caused by mixing, for alkalinity changes caused by the dissolution of calcium carbonate and for the pre-formed ΣCO_2 and alkalinity (the values relevant to the surface sources of the waters), results obtained at two stations in the Pacific (see Culberson and Pytkowicz, 1970) were consistent with the model proposed by Redfield et al.

Dyrssen and Gunderson (1974) have examined the changes in alkalinity and ΣCO_2 which accompany the simultaneous oxidation of organic matter and the dissolution of calcium carbonate for stations near the Hawaiian Islands. As a slight modification of the Richards (1965) model they considered that one unit of organic matter (C:N:P = 106:16:1) comprised effectively 89 carbohydrate groups, 16 peptide or amino acid groups and one phosphate ester group according to

$$(CH_2O)_{89}(NHCO)_{16}C(H_2PO_4)^-.$$

Alkalinity, ΣCO_2, O_2, NO_3^- and PO_4^{3-} changes accompanying the formation of organic matter were summarized by

$$89CO_2 + 89H_2O \rightarrow 89CH_2O + 89O_2; \Delta A = 0; \Delta\Sigma CO_2 = -89u;$$

$$\Delta O_2 = +89u.$$

$$16CO_2 + 16NO_3^- + 16H^+ \rightarrow 16NHCO + 32O_2$$

$$\Delta A = +16u; \Delta\Sigma CO_2 = -16u; \Delta O_2 = +32u; \Delta NO_3^- = -16u.$$
$$CO_2 + MHPO_4 + H^+ \rightarrow CH_2PO_4^- + O_2 + M^{2+}$$

$$\Delta A = 0; \Delta\Sigma CO_2 = -u; \Delta O_2 = +u; \Delta PO_4^{3-} = -u.$$

giving overall changes for u units of organic matter production of

$$\Delta A = +16u; \Delta \Sigma CO_2 = -106u; \Delta O_2 = +122u; \Delta NO_3^- = -16u;$$
$$\Delta PO_4^{3-} = -u.$$

The reverse of these changes was considered to accompany the oxidation of organic matter in the water column. For the bio-uptake of n moles of calcium carbonate the relevant changes are

$$A = -2n; \Delta \Sigma CO_2 = -n; \Delta c_{Ca(T)} = -n.$$

Dyrssen and Gunderson were able to show that down to depths of 1200 m there was considerable covariation of the changes of alkalinity, ΣCO_2, NO_3^- and phosphate in sympathy with their proposed oxidation scheme, but that the changes of alkalinity and ΣCO_2 implied appreciable solution of $CaCO_3$. Other workers have previously commented on the relative contributions of the oxidation of organic matter and the dissolution of calcium carbonate to the increased ΣCO_2 of deep waters compared with that for surface water. By assuming that the major contribution to the alkalinity change was calcium carbonate dissolution but that both carbonate uptake and organic oxidation processes contributed to ΣCO_2, Li et al. (1969) estimated from data which they obtained that about 80% of the excess deep water CO_2 is derived from the oxidation of organic matter.

By making use of the observations by Almgren et al. (1974c) that, in the region of the Pacific North Trade Wind, the $\Delta A/\Delta c_{Ca(T)}$ value is 1.62 ± 0.38 together with the relationship

$$\Delta A/\Delta c_{Ca(T)} = (16u - 2n)/ - n$$

(see above) Dyrssen and Gunderson estimated that about 40% of the change in ΣCO_2 is derived from the dissolution of calcium carbonate. Calculations based on the observed change of nitrate indicated the calcium carbonate derived contribution to the total CO_2 change to be about 20%.

Kroopnick et al. (1970) have used carbon-13 measurements made at the North Pacific GEOSECS Intercalibration station to estimate the relative amounts of calcium carbonate-derived and organic carbon derived CO_2 in deep waters. Their results showed $\delta^{13}C$ to fall from its surface value of about $+2.26\%$ to a minimum of about -0.45% at 900 m and then to increase to about zero at 3800 m. By taking $\delta^{13}C$ for particulate organic matter to be -23% (Williams and Gordon, 1970) and that for foraminifera to be $+2.0\%$, and by making use of a vertical diffusion–advection analysis which took account of the downward flux of particulate matter from the surface, they calculated that about 70% of the additional organic carbon of the deep water at this station was derived from the dissolution of calcium carbonate. An estimate obtained by Craig (1970) for the South Pacific indicated about 30% to be derived from carbonate uptake. Alkalinity and ΣCO_2 profiles obtained by Culberson and Pytkowicz (1970) for water near the East Pacific rise have been used by Kroopnick (1974) in a vertical diffusion–advection analysis to obtain alkalinity and CO_2 production rates and hence the percentage of CO_2 which is derived from the *in situ* uptake of calcium carbonate. The value which he found (10–26%) was consistent with that obtained (20–40%) by comparing the *in situ* $\delta^{13}C$ of the added carbon (-14 to $-17 \cdot 5\permil$) with that of the source materials ($-23\permil$ for organic carbon and $+2 \cdot 0\permil$ for $CaCO_3$).

Kroopnick's investigation also gave information on the relationship between the rate of oxygen consumption with the depth. Below about 1 km (i.e. in the linear region of the salinity-potential temperature relationship) for most of the sections which he described an exponential decrease in the rate of oxygen consumption with depth gave the best fit to the observed data, but at all stations oxygen was non-conservative. Calculated rates of oxygen consumption fell in the range 0·029 to 0·14 μmol kg^{-1} yr^{-1}. The rate of carbon dioxide production (by oxidation plus carbonate dissolution) was between 0·03 and 0·12 μmol kg^{-1} yr^{-1}. When allowance was made for the fraction of CO_2 which was derived from carbonate uptake, the molar ratio of CO_2 generated by oxidation to oxygen consumed was calculated to lie between 0·5 and 0·9 (cf. 0·77—Redfield *et al.*, 1963).

Profiles at particular stations indicate the stationary state distribution of material at that geographical location only. However, it must be remembered that, in reality, the distribution is a three dimensional phenomenon since horizontal circulation in deep water should also be taken into account if a world-wide picture is to be obtained. The "excess" CO_2, alkalinity and nutrients generated at depth will be transported in accordance with deep water circulation patterns. Superimposed on this transport is the general upwards advection indicated above, but there will be a tendency for oxidation-generated and dissolution-generated material to increase progressively in concentration on the down-stream side of the deep water circulation pattern. The high ΣCO_2 of Pacific deep water as compared with that of the Atlantic may arise in this way. Thus, according to Li *et al.* (1969), whereas the ΣCO_2 values for both Atlantic and Pacific surface waters is close to $2 \cdot 05 \times 10^{-3}$ mol l^{-1}, the deep water values are $2 \cdot 28 \times 10^{-3}$ mol l^{-1} (Atlantic) and $2 \cdot 45 \times 10^{-3}$ mol l^{-1} (Pacific). These differing ΣCO_2 profiles for the two oceans, when taken in conjunction with the corresponding alkalinity profiles, have important consequences for the vertical distributions of P_{CO_2} and carbonate ion concentration. The calcium carbonate saturation profiles are, of course, related to the profiles for the carbonate ion concentration.

9.6.3. ANOXIC AND PORE WATERS (see also Chapters 15 and 16).

For poorly ventilated waters in which the circulation patterns are such that the rate of supply of dissolved oxygen is insufficient to cope with the rate at which organic debris is delivered, anaerobic, reducing, conditions develop (e.g., the Black Sea, Lake Nitinat and the Cariaco Trench—see Chapters 15 and 16). Similar anoxic conditions are also found in many recent marine sediments, where there is the additional complication of possible reaction between the various solid phases which are in intimate contact with the pore solutions. Under anaerobic conditions the bacterial conversion of sulphate to sulphide occurs, and the amounts of H_2S, NH_3 and CO_2 which are generated may be far greater than those found in normal sea water (see e.g., Table 9.2 and also Richards, 1965; Siever et al., 1965; Presley and Kaplan, 1968; Thorstenson, 1970; Berner et al., 1970; Nissenbaum et al., 1972; Ben-Yaakov, 1973; Almgren et al., 1974d), and the alkalinity may be very different from that of the overlying water. For these systems it is unlikely that the carbon dioxide system is solely responsible for pH control.

A number of precautions are necessary during the handling of pore water samples. The high carbon dioxide content means that often they are strongly supersaturated with respect to the atmosphere. For example, pH and alkalinity based determinations of the P_{CO_2} of the interstitial waters of certain Bermuda beaches indicated P_{CO_2} values of almost 2000×10^{-6} atm (Pytkowicz, 1971b), and the high ΣCO_2 values of many other reported analyses show that pore water P_{CO_2} values considerably in excess of p_{CO_2} are general. For this reason, delay in analysis or the use of unsatisfactory techniques may mean that the solution which is examined is not really representative of the *in situ* solution. Some of the handling problems have been considered by Berner et al. (1970). They noted that during the squeezing of sediments derived from silicate muds of the Connecticut coast there was a carbon dioxide loss and a pH rise (from ca. 7·5 to 8·2). The resultant solution was supersaturated with respect to calcium carbonate, but precipitation was slow and developed only over a period of days or weeks. Possibly if similar CO_2 loss occurred from pore waters derived from carbonate-rich sediments the delay in precipitation might be less, although the observations reported by Pytkowicz (1971b) suggest that, in Bermuda beach deposits deposition from solution is still slow even when the solution is supersaturated in the presence of calcium carbonate. This may be the result of the presence of protective organic coatings on the grains.

The changes in alkalinity suffered by pore waters have been discussed by Berner *et al.* (1970) in terms of the reactions

$$CO_2 + H_2O + CaCO_3 \rightleftharpoons Ca^{2+} + 2HCO_3^-$$
$$CaCO_3 \rightleftharpoons Ca^{2+} + CO_3^{2-}$$
$$\left.\right\} \quad \text{(i)}$$

$$2CH_2O + SO_4^{2-} \rightarrow S^{2-} + 2CO_3^{2-} + 2H_2O$$
$$S^{2-} + 2CO_2 + 2H_2O \rightarrow H_2S + 2HCO_3^-$$
$$\left.\right\} \quad \text{(ii)}$$

$$CH_2NH_2COOH + 2(H) \rightarrow NH_3 + CH_4 + CO_2$$
$$NH_3 + CO_2 + H_2O \rightarrow NH_4^+ + HCO_3^-$$
$$\left.\right\} \quad \text{(iii)}$$

$$3Al_2Si_2O_5(OH)_4 + 2K^+ (Mg^{2+}, Na^+) + 2HCO_3^-$$
$$\rightarrow 2KAl_3Si_3O_{10}(OH)_2 + 2CO_2 + 5H_2O \quad \text{(iv)}$$

$$HA + HCO_3^- \rightarrow A^- + CO_2 + H_2O \quad \text{(v)}$$

The pair of equilibria (i) needs no further comment. The pair of reactions under (ii) corresponds to the bacterial reduction of sulphate during the oxidation of carbohydrate (CH_2O) leading to the generation of sulphide and H_2S. The bacterial generation of ammonia and carbon via amino acids (CH_2NH_2COOH) is summarized by (iii) and authigenic silicate formation by (iv). In reaction (v) bacterially generated organic acids reduce the carbonate alkalinity. Of these processes, those involving the reactions of CO_2 with bases ($CaCO_3$, NH_3, S^{2-}) lead to an increase in the carbonate alkalinity whereas the loss of HCO_3^- and CO_3^{2-} by reaction with materials which, in the widest sense, are acids reduces the carbonate alkalinity.

The alkalinity may rise to 30-fold or more of that in the overlying water, and Berner *et al.* were able to show that the increase in carbonate alkalinity of organic-rich silicate muds from the central Connecticut coast and a Maine fjord were consistent with the changes in SO_4^{2-}, Mg^{2+}, Ca^{2+} and NH_4^+ interpreted in terms of the above scheme, although there was no evidence for authigenic silicate formation.

Gaines and Pilson (1972) have explained the high alkalinity (12·66 meq l^{-1}) and high total H_2S concentration (4·5 mmol l^{-1}) observed for water from anoxic pools in Pettaquamscutt River (Rhode Island) in terms of a model (Knull and Richards, 1969) for the bacterially assisted reduction of sulphate by organic matter, account being taken of the positive contribution to the alkalinity made by the ammonia which was generated. Almgren *et al.* (1974d) have explained their observations on the total alkalinity, ΣCO_2, pH, oxygen, sulphide, nitrate, ammonium and phosphate profiles in Byfjorden fjord (Sweden) in terms of an anoxic mechanism which is a modification of the scheme for oxic waters proposed by Dyrssen and Gundersen (1974) (see Section 9.6.2). The principal overall steps which were visualized are

$$89\, CH_2O + 44{\cdot}5\, SO_4^{2-} \rightarrow 89\, HCO_3^- + 44{\cdot}5\, H_2S$$

$$16\, NHCO + 32\, H_2O \rightarrow 16\, HCO_3^- + 16\, NH_4^+$$

$$C(H_2PO_4)^- + 0{\cdot}5\, SO_4^{2-} + H_2O \rightarrow HCO_3^- + H_2PO_4^- + 0{\cdot}5\, H_2S$$

For the reaction of u moles of organic material, the appropriate changes are

$$\Delta A = +106u;\, \Delta\Sigma CO_2 = +106u;\, \Delta H_2S = +45u;\, \Delta SO_4^{2-} = -45u;$$

$$\Delta NH_4^+ = +16u;\, \Delta PO_4 = +u.$$

Material which finds its way into a reducing environment will react in a direction which tends towards thermodynamic equilibrium, although the rates of equilibration may be slow. If sufficient thermodynamic data are available, it is possible in principle to calculate the equilibrium composition of a system as a function of pH and redox potential. An approach along these lines has been explored by Thorstenson (1970) in an attempt to explain the prevalent species encountered as the solutions becomes oxygen-impoverished. He concluded that in the pH-redox potential range of most natural waters, high molecular weight material, both dissolved and particulate, is thermodynamically unstable with respect to low molecular weight fragments which include CH_4, H_2CO_3, HCO_3^-, CO_3^{2-}, NH_3, NH_4^+, NO_3^-, H_2S and HS^-. The transition from predominantly oxidizing to predominantly reducing species occurs over a relatively small redox range, and tests of this approach using data derived from the Black Sea, the Cariaco Trench, Lake Nitinat and pore waters from Devils Hole (Bermuda) and from marine sediments from southern California indicated that, generally, equilibrium is closely approached.

The examination made by Thorstenson was primarily concerned with the equilibria which occurred as a function of redox potential and pH, and no account was taken of possible reactions between solution components and solid mineral phases. A somewhat different approach has been used by Ben-Yaakov (1973). He noted that despite the relatively low buffer capacity of sea water towards CO_2 addition (see Section 9.2.1), the pH range of the pore waters of many recent marine sediments is often restricted (ca. 7·0–8·0) even though ΣCO_2 may reach $60 \, mmol \, kg^{-1}$—some 20–30 times that of normal sea water—see Table 9.2. Ben-Yaakov assumed that the bacterially assisted reaction of sulphate with organic matter could be approximately summarized by

$$\tfrac{1}{53}(CH_2O)_{106}(NH_3)_{16}H_3PO_4 + SO_4^{2-}$$
$$\rightarrow CO_2 + HCO_3^- + HS^- + H_2O + \tfrac{16}{53}NH_3 + \tfrac{1}{55}H_3PO_4$$

The weak acids and bases so produced enter the solution in constant proportion, and since the initial concentrations of weak acids and bases are small compared with the amounts produced in the decomposition, the pH is well buffered. By making use of the relationship between the amount of sulphate lost and those of the other materials produced together with values for the appropriate ionization constants, Ben-Yaakov estimated that the pH should stabilize at 6·9 provided that the system could be regarded as closed in the sense that sulphide and carbonate loss from solution do not occur. If, as seems likely under most natural circumstances, some sulphide loss does occur (by, for example, reaction with the iron oxide of solid phases—see e.g. Berner et al., 1970), the pH will rise towards 8.3, the equilibrium value depending on the proportion of the sulphide which is removed. However, calcium carbonate precipitation may also assist the buffering and prevent excessive pH rise, and for sediments of low iron content (e.g., many carbonates), pH values not much higher than 6·9 are to be expected. Other processes which may also participate are reactions with silicates (Garrels, 1965) e.g.

$$2NaAlSi_3O_8 + H_2O + 2H^+ \rightarrow 2Na^+ + Al_2Si_2O_5(OH)_4 + 4SiO_2$$

and ionic diffusion between the interstitial and overlying waters. This last process would tend to restore the pH to that of the overlying water. From an examination of results based on pore waters of Saanich Inlet cores (Nissenbaum et al., 1972), Ben-Yaakov concluded that, in this instance, reactions of silicates make little contribution to the buffering. Neither Ben-Yaakov nor Berner et al. considered that organic acids play much part in determining either alklinity or the pH.

Often, sediment pore waters are highly supersaturated with respect to calcium carbonate (Berner et al., 1970), this condition persisting possibly because of the absence of suitable nuclei and the inhibiting effect on nucleation of dissolved organic matter.

9.7. ACKNOWLEDGEMENT

The author would particularly like to thank Andrew Dickson for his assistance with the proofs.

APPENDIX

TABLE A9.1

Interpolated pH *values for the sea water standard buffer* (*Hansson, 1973b*)

S (‰)	$t(°C)$					
	5	10	15	20	25	30
10	8·675	8·510	8·350	8·196	8·047	7·903
20	8·694	8·525	8·363	8·206	8·055	7·908
25	8·705	8·535	8·372	8·213	8·061	7·913
30	8·717	8·546	8·381	8·221	8·067	7·918
35	8·731	8·559	8·392	8·231	8·076	7·925
40	8·745	8·571	8·404	8·241	8·085	7·933

The pH scales are consistent with the activity scale chosen so that the activity coefficient approaches unity when $c_{H(T)}$ approaches zero in synthetic sea water of a certain salinity.

TABLE A9.2a

The solubility of carbon dioxide in water and (acidified) sea water (10^{-2} mol l^{-1} atm^{-1}). (Weiss, 1974; values based on determinations made by Murray and Riley, 1971)

$T(°C)$	0	10	20	30	34	35	36	38	40
					Salinity (‰)				
−1	—	—	7·273	6·903	6·760	6·724	6·689	6·620	6·551
0	7·758	7·364	6·990	6·635	6·498	6·465	6·431	6·364	6·298
1	7·458	7·081	6·723	6·382	6·251	6·219	6·187	6·123	6·060
2	7·174	6·813	6·469	6·143	6·017	5·986	5·955	5·894	5·833
3	6·905	6·558	6·229	5·916	5·795	5·766	5·736	5·677	5·619
4	6·650	6·317	6·001	5·701	5·585	5·557	5·528	5·472	5·416
5	6·408	6·088	5·785	5·497	5·386	5·358	5·331	5·277	5·223
6	6·178	5·871	5·580	5·303	5·196	5·170	5·144	5·092	5·040
8	5·751	5·469	5·200	4·945	4·846	4·822	4·797	4·749	4·702
10	5·366	5·105	4·857	4·621	4·529	4·507	4·485	4·440	4·396
12	5·017	4·776	4·546	4·327	4·243	4·222	4·201	4·160	4·119
14	4·700	4·477	4·264	4·062	3·983	3·964	3·945	3·906	3·869
16	4·412	4·205	4·008	3·820	3·747	3·729	3·712	3·676	3·641
18	4·149	3·958	3·775	3·600	3·533	3·516	3·499	3·466	3·434
20	3·910	3·732	3·562	3·400	3·337	3·322	3·306	3·275	3·245
22	3·691	3·526	3·368	3·217	3·158	3·144	3·130	3·101	3·073
24	3·491	3·337	3·190	3·050	2·995	2·982	2·968	2·942	2·915
26	3·307	3·164	3·027	2·897	2·846	2·833	2·821	2·796	2·771
28	3·138	3·005	2·878	2·756	2·709	2·697	2·685	2·662	2·639
30	2·983	2·859	2·741	2·627	2·583	2·572	2·561	2·540	2·518
32	2·840	2·725	2·615	2·509	2·468	2·457	2·447	2·427	2·407
34	2·708	2·601	2·498	2·400	2·361	2·352	2·342	2·323	2·305
36	2·587	2·487	2·391	2·299	2·263	2·254	2·246	2·228	2·211
38	2·474	2·382	2·292	2·207	2·173	2·165	2·157	2·140	2·124
40	2·370	2·284	2·201	2·121	2·090	2·082	2·074	2·059	2·044

TABLE A9.2b

The solubility of carbon dioxide in water and (acidified) sea water (10^{-2} mol kg^{-1} atm^{-1}) (Weiss, 1974; values based on determinations made by Murray and Riley, 1971)

T(°C)	0	10	20	30	34	35	36	38	40
0	7·758	—	7·158	6·739	6·579	6·539	6·500	6·422	6·345
1	7·458	7·305	6·880	6·479	6·325	6·287	6·249	6·175	6·101
2	7·174	7·024	6·616	6·232	6·085	6·048	6·012	5·941	5·870
3	6·904	6·758	6·367	5·999	5·857	5·822	5·788	5·719	5·651
4	6·649	6·506	6·131	5·777	5·642	5·608	5·575	5·509	5·444
5	6·407	6·267	5·907	5·568	5·438	5·405	5·374	5·310	5·248
6	6·177	6·040	5·695	5·369	5·244	5·213	5·182	5·122	5·062
8	5·752	5·825	5·493	5·180	5·060	5·031	5·001	4·943	4·885
10	5·367	5·427	5·120	4·831	4·720	4·693	4·666	4·612	4·558
12	5·019	5·067	4·784	4·516	4·413	4·388	4·363	4·313	4·263
14	4·703	4·741	4·479	4·231	4·136	4·112	4·089	4·042	3·997
16	4·416	4·446	4·202	3·972	3·884	3·862	3·840	3·797	3·755
18	4·155	4·177	3·951	3·738	3·655	3·635	3·615	3·575	3·536
20	3·916	3·933	3·723	3·524	3·448	3·429	3·410	3·373	3·336
22	3·699	3·715	3·515	3·330	3·258	3·241	3·223	3·189	3·154
24	3·499	3·507	3·325	3·152	3·086	3·069	3·053	3·021	2·989
26	3·317	3·321	3·151	2·990	2·928	2·912	2·897	2·867	2·837
28	3·149	3·150	2·992	2·841	2·783	2·769	2·755	2·727	2·699
30	2·995	2·994	2·846	2·705	2·651	2·638	2·624	2·598	2·572
32	2·854	2·850	2·712	2·580	2·530	2·517	2·505	2·480	2·455
34	2·723	2·718	2·589	2·466	2·418	2·406	2·395	2·372	2·349
36	2·603	2·596	2·476	2·360	2·316	2·305	2·294	2·272	2·250
38	2·492	2·484	2·371	2·263	2·221	2·211	2·201	2·180	2·160
40	2·389	2·381	2·275	2·174	2·134	2·125	2·115	2·096	2·077

Salinity (‰)

TABLE A9.3

First apparent dissociation constant of carbonic acid* in sea water (expressed as pK'_{11})
(Lyman, 1956)

Cl(‰)	$t(°C)$							
	0°	5°	10°	15°	20°	25°	30°	35°
0	6·58	6·52	6·47	6·42	6·38	6·35	6·33	6·31
1	6·47	6·42	6·37	6·33	6·29	6·26	6·24	6·23
4	6·36	6·32	6·28	6·24	6·21	6·18	6·16	6·15
9	6·27	6·23	6·19	6·15	6·13	6·10	6·08	6·07
16	6·18	6·14	6·11	6·07	6·05	6·03	6·01	5·99
17	6·17	6·13	6·10	6·06	6·04	6·02	6·00	5·98
18	6·16	6·12	6·09	6·06	6·03	6·01	5·99	5·97
19	6·15	6·11	6·08	6·05	6·02	6·00	5·98	5·97
20	6·14	6·10	6·07	6·04	6·01	5·99	5·97	5·96
21	6·13	6·09	6·06	6·03	6·00	5·98	5·96	5·95
25	6·09	6·05	6·02	6·00	5·97	5·95	5·93	5·92
36	6·00	5·97	5·94	5·92	5·89	5·87	5·86	5·84
49	5·92	5·88	5·86	5·84	5·82	5·80	5·78	5·77
64	5·84	5·80	5·78	5·76	5·74	5·72	5·71	5·70

* Values based on the N.B.S. pH scale.

TABLE A9.4

Second apparent dissociation constant of carbonic acid* in sea water (expressed as pK'_2) *(Lyman, 1956)*

Cl(‰)	$t(°C)$							
	0°	5°	10°	15°	20°	25°	30°	35°
0	10·62	10·55	10·49	10·43	10·38	10·33	10·29	10·25
1	10·06	9·99	9·93	9·87	9·81	9·76	9·71	9·66
4	9·78	9·72	9·67	9·61	9·54	9·49	9·43	9·38
9	9·64	9·58	9·52	9·46	9·40	9·34	9·27	9·21
16	9·46	9·40	9·35	9·29	9·23	9·17	9·10	9·02
17	9·44	9·38	9·32	9·27	9·21	9·15	9·08	9·00
18	9·42	9·36	9·30	9·25	9·19	9·12	9·06	8·98
19	9·40	9·34	9·28	9·23	9·17	9·10	9·02	8·95
20	9·38	9·32	9·26	9·21	9·15	9·08	9·01	8·92
21	9·36	9·30	9·25	9·19	9·13	9·06	8·98	8·89
25	9·29	9·23	9·17	9·11	9·05	8·98	8·91	8·82
36	9·12	9·06	8·99	8·93	8·86	8·79	8·72	8·63
49	8·95	8·89	8·82	8·75	8·68	8·61	8·53	8·43
64	8·77	8·71	8·64	8·57	8·50	8·42	8·34	8·23

* Values based on the N.B.S. pH scale.

TABLE A9.5

First apparent dissociation constant of boric acid* in sea water (expressed as pK'_B)
(Lyman, 1956)

Cl(‰)	$t°C$							
	0°	5°	10°	15°	20°	25°	30°	35°
0	9·50	9·44	9·38	9·33	9·28	9·24	9·20	9·16
1	9·40	9·34	9·28	9·23	9·18	9·14	9·10	9·06
4	9·28	9·22	9·16	9·11	9·06	9·02	8·98	8·94
9	9·14	9·08	9·03	8·98	8·93	8·88	8·85	8·82
16	9·00	8·95	8·89	8·84	8·80	8·76	8·72	8·69
17	8·98	8·93	8·88	8·83	8·78	8·74	8·70	8·67
18	8·96	8·91	8·86	8·81	8·76	8·72	8·69	8·66
19	8·95	8·90	8·85	8·80	8·75	8·71	8·67	8·64
20	8·94	8·88	8·83	8·78	8·74	8·69	8·65	8·63
21	8·92	8·87	8·82	8·77	8·72	8·68	8·64	8·61
25	8·85	8·80	8·75	8·70	8·66	8·62	8·59	8·56
36	8·71	8·66	8·61	8·57	8·53	8·49	8·46	8·43
49	8·56	8·52	8·47	8·43	$8·39^5$	8·36	8.33	8.30
64	8·41	8·37	8·33	8·30	$8·26^5$	8·23	8·20	8·17

* Values based on the N.B.S. pH scale.

TABLE A9.6

First ionization constant of carbonic acid (Hansson, 1973a, see p. 49)

$S‰$ \ $t°C$	pK'_{1h}					
	5	10	15	20	25	30
20	6·133	6·080	6·028	5·979	5·930	5·884
25	6·106	6·053	6·001	5·951	5·903	5·857
30	6·082	6·029	5·977	5·927	5·879	5·833
35	6·059	6·006	5·955	5·905	5·857	5·810
40	6·051	5·998	5·947	5·897	5·849	5·802

Values based on the Hansson sea water pH scale.

TABLE A9.7

Second ionization constant of carbonic acid (Hansson, 1973a, see p. 49)

pK'_{2h}

$t°C$ $S‰$	5	10	15	20	25	30
20	9·457	9·370	9·286	9·205	9·126	9·050
25	9·394	9·307	9·222	9·141	9·063	8·987
30	9·331	9·244	9·160	9·078	9·000	8·924
35	9·279	9·191	9·107	9·026	8·947	8·872
40	9·232	9·145	9·061	8·980	8·901	8·825

Values based on the Hansson sea water pH scale.
Minor refinements to these constants (Almgren, 1974b) give

$$pK'_{1h} = 841/T + 3·272 - 0·0101 \times S + 0·0001 \times S^2$$
$$pK'_{2h} = 1373/T + 4·854 - 0·01935 \times S + 0·000135 \times S^2$$

when T is in °K and S is the salinity (‰).

TABLE A9.8

First apparent ionization constants of boric acid (Hansson, 1973a, see p. 51)
pK'_{Bh}

$t°C$ $S‰$	5	10	15	20	25	30
20	8·96	8·90	8·84	8·77	8·72	8·66
25	8·92	8·86	8·79	8·73	8·67	8·62
30	8·89	8·82	8·76	8·70	8·64	8·58
35	8·86	8·79	8·73	8·67	8·61	8·55
40	8·85	8·78	8·72	8·66	8·60	8·54

Values based on Hansson sea water pH scale.
A minor refinement to these constants (Almgren, 1974b) gives

$$pK'_{Bh} = 1026/T + 5·527 - 0·0158 \times S + 0·00016 \times S^2$$

when T is in °K and S is the salinity (‰).

Temp. (°C)	Salinity (‰) 19·00	25·00	27·00	29·00	31·00	32·00	33·00	34·00	35·00	36·00	37·00	38·00	39·00	40·00	43·00
0	0·498	0·552	0·569	0·586	0·602	0·610	0·617	0·625	0·633	0·640	0·648	0·655	0·662	0·669	0·696
1	0·512	0·567	0·585	0·601	0·618	0·626	0·634	0·642	0·650	0·657	0·665	0·672	0·680	0·687	0·708
2	0·525	0·582	0·600	0·617	0·634	0·642	0·650	0·658	0·666	0·674	0·682	0·690	0·697	0·705	0·727
3	0·539	0·597	0·615	0·633	0·650	0·658	0·667	0·675	0·683	0·691	0·699	0·707	0·715	0·722	0·745
4	0·552	0·612	0·630	0·648	0·666	0·675	0·683	0·692	0·700	0·708	0·716	0·724	0·732	0·740	0·763
5	0·565	0·626	0·645	0·664	0·682	0·691	0·699	0·708	0·716	0·725	0·733	0·741	0·749	0·757	0·781
6	0·579	0·641	0·660	0·679	0·698	0·707	0·716	0·724	0·733	0·741	0·750	0·758	0·766	0·775	0·798
7	0·592	0·655	0·675	0·695	0·713	0·722	0·732	0·740	0·749	0·758	0·767	0·775	0·783	0·792	0·816
8	0·605	0·670	0·690	0·710	0·729	0·738	0·747	0·757	0·766	0·774	0·783	0·792	0·800	0·809	0·833
9	0·618	0·684	0·705	0·725	0·744	0·754	0·763	0·772	0·782	0·791	0·799	0·808	0·817	0·826	0·851
10	0·631	0·698	0·719	0·740	0·759	0·769	0·779	0·788	0·797	0·807	0·816	0·825	0·833	0·842	0·868
11	0·643	0·712	0·733	0·754	0·774	0·784	0·794	0·804	0·813	0·822	0·832	0·841	0·850	0·859	0·885
12	0·656	0·726	0·747	0·769	0·789	0·799	0·809	0·819	0·828	0·838	0·847	0·857	0·866	0·875	0·901
13	0·668	0·739	0·761	0·783	0·804	0·814	0·824	0·834	0·844	0·853	0·863	0·872	0·881	0·891	0·917
14	0·680	0·752	0·775	0·797	0·818	0·828	0·838	0·848	0·858	0·868	0·878	0·887	0·897	0·906	0·933
15	0·692	0·765	0·788	0·810	0·832	0·842	0·853	0·863	0·873	0·883	0·893	0·902	0·912	0·921	0·949
16	0·704	0·778	0·801	0·824	0·846	0·856	0·867	0·877	0·887	0·897	0·907	0·917	0·927	0·936	0·964
17	0·715	0·791	0·814	0·837	0·859	0·870	0·880	0·891	0·901	0·911	0·921	0·931	0·941	0·951	0·979
18	0·726	0·803	0·827	0·850	0·872	0·883	0·894	0·904	0·915	0·925	0·935	0·945	0·955	0·965	0·994
19	0·737	0·815	0·839	0·862	0·885	0·896	0·907	0·918	0·928	0·939	0·949	0·959	0·969	0·979	1·008
20	0·748	0·826	0·851	0·874	0·897	0·908	0·919	0·930	0·941	0·952	0·962	0·972	0·982	0·992	1·022
21	0·758	0·838	0·862	0·886	0·909	0·921	0·932	0·943	0·954	0·964	0·975	0·985	0·995	1·005	1·035
22	0·768	0·848	0·873	0·898	0·921	0·932	0·944	0·955	0·966	0·976	0·987	0·998	1·008	1·018	1·048
23	0·778	0·859	0·884	0·909	0·932	0·944	0·955	0·966	0·977	0·988	0·999	1·010	1·020	1·030	1·060
24	0·787	0·869	0·895	0·919	0·943	0·955	0·966	0·977	0·989	1·000	1·010	1·021	1·032	1·042	1·072
25	0·796	0·879	0·905	0·929	0·953	0·965	0·977	0·988	0·999	1·011	1·021	1·032	1·043	1·053	1·084
26	0·805	0·888	0·914	0·939	0·964	0·975	0·987	0·998	1·010	1·021	1·032	1·043	1·053	1·064	1·095
27	0·813	0·897	0·923	0·949	0·973	0·985	0·997	1·008	1·020	1·031	1·042	1·053	1·064	1·074	1·105
28	0·821	0·906	0·932	0·958	0·982	0·994	1·006	1·018	1·029	1·041	1·052	1·063	1·073	1·084	1·115
29	0·829	0·914	0·941	0·966	0·991	1·003	1·015	1·027	1·038	1·050	1·061	1·072	1·083	1·093	1·125
30	0·836	0·922	0·949	0·974	0·999	1·011	1·023	1·035	1·047	1·058	1·069	1·080	1·091	1·102	1·134
31	0·843	0·929	0·956	0·982	1·007	1·019	1·031	1·043	1·055	1·066	1·077	1·089	1·100	1·110	1·142
32	0·849	0·936	0·963	0·989	1·014	1·026	1·039	1·050	1·062	1·074	1·085	1·096	1·107	1·118	1·150
33	0·855	0·943	0·970	0·996	1·021	1·033	1·045	1·057	1·069	1·081	1·092	1·103	1·114	1·125	1·157
34	0·861	0·949	0·976	1·002	1·027	1·040	1·052	1·064	1·076	1·087	1·099	1·110	1·121	1·321	1·164
35	0·866	0·954	0·971	1·008	1·033	1·045	1·058	1·070	1·082	1·093	1·105	1·116	1·127	1·138	1·170

TABLE A9.10

Second apparent ionization constants of carbonic acid (Mehrbach et al., 1973) $(K'_2 \times 10^{10})$

	Salinity (‰)														
	19·00	25·00	27·00	29·00	31·00	32·00	33·00	34·00	35·00	36·00	37·00	38·00	39·00	40·00	43·00
0	1·67	2·44	2·69	2·92	3·14	3·24	3·34	3·44	3·53	3·62	3·71	3·79	3·87	3·94	4·14
1	1·73	2·52	2·76	3·00	3·22	3·32	3·43	3·53	3·62	3·71	3·80	3·88	3·96	4·04	4·24
2	1·81	2·60	2·84	3·08	3·31	3·41	3·52	3·62	3·71	3·81	3·90	3·98	4·06	4·14	4·35
3	1·88	2·68	2·93	3·17	3·40	3·51	3·61	3·72	3·81	3·91	4·00	4·09	4·17	4·25	4·47
4	1·96	2·77	3·03	3·27	3·50	3·61	3·72	3·82	3·92	4·02	4·11	4·20	4·29	4·37	4·60
5	2·05	2·87	3·13	3·37	3·61	3·72	3·83	3·93	4·04	4·14	4·23	4·32	4·41	4·50	4·73
6	2·14	2·97	3·23	3·48	3·72	3·83	3·95	4·05	4·16	4·26	4·36	4·45	4·55	4·63	4·88
7	2·23	3·08	3·34	3·60	3·84	3·96	4·07	4·18	4·29	4·39	4·49	4·59	4·69	4·78	5·03
8	2·34	3·19	3·46	3·72	3·97	4·09	4·20	4·32	4·43	4·53	4·64	4·74	4·84	4·93	5·20
9	2·44	3·31	3·59	3·85	4·10	4·22	4·34	4·46	4·57	4·68	4·79	4·89	4·99	5·09	5·37
10	2·55	3·44	3·71	3·98	4·24	4·36	4·49	4·60	4·72	4·83	4·95	5·05	5·16	5·26	5·56
11	2·67	3·57	3·85	4·12	4·38	4·51	4·64	4·76	4·88	5·00	5·11	5·22	5·33	5·44	5·75
12	2·79	3·70	3·99	4·27	4·54	4·67	4·80	4·92	5·05	5·17	5·29	5·40	5·52	5·63	5·95
13	2·92	3·85	4·14	4·42	4·69	4·83	4·96	5·09	5·22	5·34	5·47	5·59	5·71	5·82	6·16
14	3·06	3·99	4·29	4·58	4·86	5·00	5·13	5·26	5·40	5·53	5·65	5·78	5·90	6·02	6·38
15	3·19	4·15	4·45	4·74	5·03	5·17	5·31	5·45	5·58	5·71	5·85	5·98	6·11	6·23	6·61
16	3·34	4·30	4·61	4·91	5·20	5·35	5·49	5·63	5·77	5·91	6·05	6·18	6·32	6·45	6·84
17	3·49	4·46	4·78	5·08	5·38	5·53	5·68	5·82	5·97	6·11	6·25	6·40	6·54	6·68	7·09
18	3·64	4·63	4·95	5·26	5·57	5·72	5·87	6·02	6·17	6·32	6·47	6·61	6·76	6·91	7·34
19	3·80	4·80	5·12	5·44	5·76	5·91	6·07	6·22	6·38	6·53	6·69	6·84	6·99	7·14	7·60
20	3·96	4·97	5·30	5·62	5·95	6·11	6·27	6·43	6·59	6·75	6·91	7·07	7·23	7·39	7·87
21	4·13	5·15	5·48	5·81	6·14	6·31	6·47	6·64	6·80	6·97	7·13	7·30	7·47	7·63	8·14
22	4·30	5·33	5·66	6·00	6·34	6·51	6·68	6·85	7·02	7·19	7·36	7·54	7·71	7·89	8·42
23	4·47	5·51	5·85	6·19	6·54	6·71	6·89	7·06	7·24	7·42	7·60	7·78	7·96	8·14	8·70
24	4·65	5·69	6·03	6·38	6·74	6·92	7·10	7·28	7·46	7·65	7·83	8·02	8·21	8·40	8·98
25	4·82	5·87	6·22	6·58	6·94	7·12	7·31	7·49	7·68	7·87	8·07	8·26	8·46	8·66	9·27
26	5·00	6·05	6·41	6·77	7·14	7·33	7·52	7·71	7·90	8·10	8·30	8·50	8·71	8·92	9·56
27	5·18	6·23	6·59	6·96	7·34	7·53	7·72	7·92	8·12	8·33	8·54	8·75	8·96	9·18	9·85
28	5·36	6·41	6·77	7·15	7·53	7·73	7·93	8·13	8·34	8·55	8·77	8·99	9·21	9·44	10·15
29	5·53	6·58	6·95	7·33	7·72	7·93	8·13	8·34	8·56	8·77	9·00	9·22	9·46	9·69	10·43
30	5·71	6·75	7·12	7·51	7·91	8·12	8·33	8·54	8·76	8·99	9·22	9·46	9·70	9·94	10·72
31	5·88	6·92	7·29	7·68	8·09	8·30	8·52	8·74	8·97	9·20	9·44	9·68	9·93	10·19	11·00
32	6·04	7·08	7·45	7·85	8·27	8·48	8·70	8·93	9·16	9·40	9·65	9·90	10·16	10·43	11·27
33	6·20	7·23	7·61	8·01	8·43	8·65	8·88	9·11	9·35	9·60	9·85	10·11	10·38	10·66	11·54
34	6·36	7·37	7·75	8·16	8·59	8·81	9·04	9·28	9·53	9·78	10·04	10·31	10·59	10·88	11·79

REFERENCES

Akiyama, T. (1966). *Atlantic Record,* **27**, 2210.
Akiyama, T. (1968). *Oceanogr. Mag.* **20**, 133.
Akiyama, T. (1969a). *Oceanogr. Mag.* **21**, 53.
Akiyama, T. (1969b). *Oceanogr. Mag.* **21**, 129.
Akiyama, T., Sagi, T., Yura, T. and Maeda, Y. (1966). *Oceanogr. Mag.* **18**, 83.
Akiyama, T., Sagi, T., Yura, T. and Maeda, Y. (1968). *Oceanogr. Mag.* **20**, 1.
Almgren, T., Danielsson, L-G., Dyrssen, D., Johansson, T. and Nyquist, G. (1974d). *Thalassia Yugoslavia* (in press).
Almgren, T., Dyrssen, D. and Strandberg, M. (1974a). *Deep-Sea Res.* (in press).
Almgren T., Dyrssen, D. and Strandberg, M. (1974b). *Deep-Sea Res.* (in press).
Almgren, T., Dyrssen, D. and Stranberg, M. (1974c). *Tellus.*
Anderson, D. H. and Robinson, R. J. (1946). *Ind. Eng. Chem. (Anal.),* **18**, 767.
Arnold, J. R. and Anderson, E. C. (1957). *Tellus,* **9**, 28.
Arons, A. B. and Kientzler, C. F. (1954). *Trans. Amer. geophys. Un.* **35**, 722.
Arrhenius, G. (1963). *In,* "The Sea" Vol. 3 (M. N. Hill, ed). Interscience Publishers, New York, 655.
Barnes, H. (1959). "Apparatus and Methods of Oceanography", 341 pp. Interscience Publishers, New York.
Bates, R. G. (1973). "Determination of pH; Theory and Practice". John Wiley and Sons, Inc., New York, 435 pp.
Baxter, M. S. and Walton, A. (1970). *Proc. roy. Soc. Ser. A* **318**, 213.
Baxter, M. S. and Walton, A. (1971). *Proc. roy. Soc. Ser. A,* **321**, 105.
Ben-Yaakov, S. (1970). *Limnol. Oceanogr.* **15**, 326.
Ben-Yaakov, S. (1971). *J. geophys. Res.* **76**, 7417.
Ben-Yaakov, S. (1973). *Limnol. Oceanogr.* **18**, 86.
Ben-Yaakov, S. and Goldhaber, M. B. (1973). *Deep-Sea Res.* **20**, 87.
Ben-Yaakov, S. and Kaplan, I. R. (1968a). *Limnol. Oceanogr.* **13**, 688.
Ben-Yaakov, S. and Kaplan, I. R. (1968b). *Rev. Scient. Instrum.* **39**, 1133.
Ben-Yaakov, S. and Kaplan, I. R. (1969). *Limnol. Oceanogr.* **14**, 874.
Ben-Yaakov, S. and Kaplan, I. R. (1971). *J. geophys. Res.* **76**, 722.
Ben-Yaakov, S. and Kaplan, I. R. (1972). *J. geophys. Res.* **77**, 2735.
Ben-Yaakov, S., Ruth, E. and Kaplan, I. R. (1974). *Deep-Sea Res.* **21**, 229.
Berg, D. and Patterson, A. (1953). *J. Amer. chem. Soc.* **75**, 5197.
Berger, R. and Libby, W. F. (1969). *Science, N.Y.* **164**, 1395.
Berger, W. H. (1967). *Science, N.Y.* **156**, 383.
Berger, W. H. (1968). *Deep-Sea Res.* **15**, 31.
Berger, W. H. (1970). *Marine Geol.* **8**, 111.
Berger, W. H. (1971). *Marine Geol.* **11**, 325.
Berner, R. A. (1965). *Geochim. Cosmoch. Acta,* **29**, 947.
Berner, R. A. (1966). *Science, N.Y.* **153**, 188.
Berner, R. A. (1971). "Principles of Chemical Sedimentology". McGraw Hill, New York, 240 pp.
Berner, R. A. and Morse, J. W. (1974). *Amer. J. Sci.* **274**, 108.
Berner, R. A., Scott, M. R. and Thomlinson, C. (1970). *Limnol. Oceanogr.* **15**, 544.
Berner, R. A. and Wilde, P. (1972). *Amer. J. Sci.* **272**, 826.
Bien, G. S., Rakestraw, N. W. and Suess, H. E. (1960). *Tellus,* **12**, 436.

Bien, G. S., Rakestraw, N. W. and Suess, H. E. (1963a). *Bull. Inst. oceanogr. Monaco*, **61**, No. 1278.

Bien, G. S., Rakestraw, N. W. and Suess, H. E. (1963b). *Nuclear Geophysics Nat. Acad. Sci., Nat. Res. Coun. Publ.* **1075**, 152

Bien, G. S., Rakestraw, N. W. and Suess, H. E. (1965). *Limnol. Oceanogr.* **10**, Suppl. R25–R37.

Bien, G. S. and Suess, H. E. (1967). *Proc. Conf. radioactive dating and methods of low level counting, p. 105. Vienna: Intern. Atomic Energy Agency.*

Bischof, W. (1960). *Tellus,* **12**, 216.

Bischof, W. (1962). *Tellus,* **14**, 87.

Bischof, W. (1965). *Tellus,* **17**, 398.

Bischof, W. (1970). *Tellus,* **22**, 549.

Bischof, W. (1971). *Tellus,* **23**, 558.

Bischof, W. (1973). *Tellus,* **25**, 305.

Bischof, W. and Bolin, B. (1966). *Tellus,* **18**, 155.

Bohr, C. (1899). *Ann. Phys., Chem. Lpz.* **68**, 500.

Bolin, B. (1960). *Tellus,* **12**, 274.

Bolin, B. and Bischof, W. (1970). *Tellus,* **22**, 431.

Bolin, B. and Eriksson, E. (1959). *In* "Rossby Memorial Volume", (B. Bolin, ed.) pp. 130–142. Rockefeller Institute Press, New York.

Bolin, B. and Keeling, C. D. (1963). *J. geophys. Res.* **68**, 3899.

Bradshaw, A. (1973). *Limnol. Oceanogr.* **18**, 95.

Bramlette, M. N. (1961). *In* "Oceanography" (M. Sears, ed.). pp. 345–366, Amer. Assn. adv. Sci. Publ. No. 67, Washington, D.C.

Brander, E. (1932). *Commentat. physico-math, Helsigf.* **6**, 108 pp.

Brannon, H. R., Daughtry, A. C., Perry, D., Whitaker, W. W. and Williams, M. (1957). *Trans. Amer. geophys. Un.* **38**, 643.

Brauner, P., Sillén, L. G. and Whiteker, R. (1969). *Arkiv. Kemi,* **31**, 365.

Bray, D. (1959). *Tellus,* **11**, 220.

van Breemen, N. (1972). *Geochim Cosmochim. Acta,* **37**, 101.

Brindlecombe, P. and Spedding, D. J. (1972). *Nature, Lond.* **236**, 225.

Britton, H. T. S. (1956). "Hydrogen Ions", 476 pp., 4th ed., Vol. I. Van Nostrand, Princeton, New Jersey.

Brodie, J. W. and Burling, R. W. (1958). *Nature, Lond.,* **181**, 107.

Broecker, W. S. (1963). Nuclear Geophysics. *Nat. Acad. Sci., Nat. Res. Coun. Publ.* **1075**, 138.

Broecker, W. S., Ewing, M., Gerard, R., Heezen, B. C. and Kulp, J. C. (1958). *Nat. Acad. Sci., Nat. Res. Coun. Publ.* **572**, 118.

Broecker, W. S. and Gerard, R. (1969). *Limnol. Oceanogr.* **14**, 883.

Broecker, W. S., Gerard, R., Ewing, M. and Heezen, B. C. (1960). *J. geophys. Res.* **65**, 2903.

Broecker, W. S. and Li, Y-H. (1970). *J. geophys. Res.* **75**, 3545.

Broecker, W. S., Li, Y-H. and Peng, T-H. (1971). *In* "Impingement of Man on the Oceans", (D. W. Hood, ed), Wiley-Interscience, New York, pp. 287–324.

Broecker, W. S. and Olson, E. A. (1960). *Science, N.Y.* **132**, 712.

Broecker, W. S. and Olson, E. A. (1961). *Radiocarbon,* **3**, 180.

Broecker, W. S., and Peng, T-H. (1974). *Tellus,* **26**, 21.

Broecker, W. S. and Takahashi, T. (1966). *J. geophys. Res.* **71**, 1575.

Brown, C. W. and Keeling, C. D. (1965). *J. geophys. Res.* **70**, 6077.

Bruneau, L., Jerlov, N. G. and Koczy, F. F. (1953). *Report of the Swedish Deep Sea Expedition*, 1947–48, **3**, 99.
Buch, K. (1929). *Meeresforch.* No. 61.
Buch, K. (1933a). *J. Cons. int. Explor. Mer.* **8**, 309.
Buch, K. (1933b). *Nature, Lond.* **131**, 688.
Buch, K. (1933c). *Rapp. Cons. Explor. Mer.* **85**, 71.
Buch, K. (1938). *Acta Acad. abo.* **11**, No. 5.
Buch, K. (1939a). *Acta Acad. abo.* **11**, No. 9.
Buch, K. (1939b). *Acta Acad. abo.* **11**, No. 12.
Buch, K. (1942). *Ann. Hydrologr., Berl., Jahrg.* **70**, 193.
Buch, K. (1945). *Fennia*, **68**, 208 pp.
Buch, K. (1951). *Meeresforch.* No. 151.
Buch, K., and Gripenberg, S. (1972). *J. Cons. Int. Explor. Mer.* **7**, 233.
Buch, K., Harvey, H. W., Wattenberg, H. and Gripenberg, S. (1932). *Rapp. Cons. Explor. Mer.* **79**, 1.
Buch, K. and Nynas, O. (1939). *Acta Acad. abo.* **12**, No. 3.
Burling, R. W. and Garner, D. M. (1959). *N.Z.J. Geol. Geophys.* **2**, 799.
Butler, J. N. and Huston, R. (1970). *J. phys. Chem.* **74**, 2976.
Byrne, R. H. and Kester, D. R. (1974). *J. mar. Res.* **32**, 119.
Callendar, G. S. (1938). *Quart. J. Roy. met. Soc.* **64**, 223.
Callendar, G. S. (1940). *Quart. J. Roy. met. Soc.* **66**, 395.
Callendar, G. S. (1949). *Weather*, **4**, 310.
Callendar, G. S. (1957). *Tellus*, **9**, 421.
Callendar, G. S. (1958). *Tellus*, **10**, 243.
Chapman, H. W., Gleason, L. S. and Loomis, W. E. (1954). *Plant Physiol.* **29**, 500.
Chave, K. E., Deffeyes, K. S., Weyl, P. K., Garrels, R. M. and Thompson, M. E. (1962). *Science, N.Y.* **137**, 33.
Chave, K. E. and Suess, E. (1967). *Trans. N.Y. Acad. Sci.* **29**, 991.
Chave, K. E. and Suess, E. (1970). *Limnol. Oceanogr.* **15**, 633.
Cloud, P. E. (1962a). *Geochim. Cosmochim. Acta*, **26**, 867.
Cloud, P. E. (1962b). *Prof. pap. U.S. geol. Survey*, 350.
Cox, J. D. and Head, A. J. (1962). *Trans. Faraday Soc.* **58**, 1839.
Craig, H. (1953). *Geochim. Cosmochim. Acta*, **3**, 53.
Craig, H. (1954). *J. Geol.* **62**, 115.
Craig, H. (1957). *Tellus*, **9**, 1.
Craig, H. (1958). *Proc. Second Int. Conf. Peaceful Uses of Atomic Energy*, **18**, 358.
Craig, H. (1963). *In* "Earth Science and Meteoritics" (J. Geiss and E. D. Goldberg, eds.), pp. 103–114. North Holland Publishing Co., Amsterdam.
Craig, H. (1969). *J. geophys. Res.* **74**, 5491.
Craig, H. (1970). *J. geophys. Res.* **75**, 691.
Craig, H. (1971). *J. geophys. Res.* **76**, 5078.
Craig, H. and Weiss, R. F. (1970). *J. geophys. Res.* **75**, 7641.
Culberson, C., Kester, D. R. and Pytkowicz, R. M. (1967). *Science, N.Y.* **157**, 59.
Culberson, C. and Pytkowicz, R. M. (1968). *Limnol. Oceanogr.* **13**, 403.
Culberson, C. and Pytkowicz, R. M. (1970). *J. Oceanogr. Soc. Japan*, **26**, 95.
Culberson, C. and Pytkowicz, R. M. (1973). *Mar. Chem.* **1**, 309.
Culberson, C., Pytkowicz, R. M. and Hawley, J. E. (1970). *J. mar. Res.* **28**, 15.
Culkin, F. (1965). *In* "Chemical Oceanography" (J. P. Riley and G. Skirrow, eds.), Vol. I, pp. 121–161. Academic Press, London.

Curl, H. and Davey, E. W. (1967). *Limnol. Oceanogr,* **12**, 545.
Daly, F. P., Brown, C. W. and Kester, D. R. (1972). *J. phys. Chem.* **76**, 3664.
Danckwerts, P. V. (1970). "Gas Liquid Reactions", McGraw Hill, 276 pp.
Danckwerts, P. V. and Melkersson, K. H. (1962). *Trans. Faraday Soc.* **58**, 1832.
Davies, C. W. (1962). "Ion Association", Butterworths, London, 190 pp.
Denbigh, K. (1971). "The Principles of Chemical Equilibrium". Cambridge University Press, London, 494 pp.
de Vries, H. (1958). *Proc. Acad. Sci. Amst. 61B*, 1.
Deuser, W. G. and Degens, E. T. (1967), *Nature, Lond.* **215**, 1073.
Deuser, W. G. and Hunt, J. M. (1969). *Deep-Sea Res.* **16**, 221.
Distèche, A. (1959). *Rev. scient. Instrum.* **30**, 474.
Distèche, A. (1962). *J. electrochem. Soc.* **109**, 1084.
Distèche, A. (1964). *Bull. Inst. oceanogr. Monaco*, **64**, No. 1320, 10 pp.
Distèche, A. (1972). *In* "Barobiology and Experimental Biology of the Deep Sea" (R. W. Brauer, ed.), pp. 234–265. University of North Carolina.
Distèche, A. (1974). *In* "The Sea" Vol. V. (E. D. Goldberg, ed.). Wiley-Interscience, New York.
Distèche, A. and Distèche, S. (1965). *J. electrochem. Soc.* **112**, 350.
Distèche, A. and Distèche, S. (1967). *J. electrochem. Soc.* **114**, 330.
Distèche, A. and Dubuisson, M. (1960). *Bull. Inst. oceanogr. Monaco*, **57**, No. **1174**, 8 pp.
Dobbins, W. E. (1964). *J. san. eng. Div., Proc. Amer. Soc. Civil Engrs. SA*, **3**, 43.
Dole, M. (1941). "The Glass Electrode", 332 pp. Wiley, New York.
Duedall, I. W. (1972). *Geochim. Cosmochim. Acta*, **36**, 729.
Dyck, W. (1965). Proc. 6th Intern. Conf. radiocarbon and tritium dating. p. 440, Pullman.
Dyrssen, D. (1965). *Acta Chem. Scand.* **19**, 1265.
Dyrssen, D. (1974). unpublished MS.
Dyrssen, D. and Gunderson, K. (1974). *Mar. Chem.* (in press).
Dyrssen, D. and Hansson, I. (1972). *Mar. Chem.* **1**, 137.
Dyrssen, D. and Hansson, I. (1974). Manuscript to be presented at the Dahlem Conference (1975).
Dyrssen, D. and Sillén, L. G. (1967). *Tellus*, **19**, 113.
Edmond, J. M. (1970). *Deep-Sea Res.* **17**, 737.
Edmond, J. M. and Gieskes, J. A. T. M. (1970). *Geochim. Cosmochim. Acta,* **34**, 1261.
Edmond, J. M. (1974). *Deep-Sea Res.* **21**, 455.
Eigen, M., Kustin, K. and Maass, G. (1961). *Z. physik. Chem.* **30**, 130.
Elgquist, B. (1970). *J. Inorg. Nucl. Chem.* **32**, 937.
Enns, T., Scholander, P. E. and Bradstreet, E. D. (1965). *J. Phys. Chem.* **69**, 389.
Fairhall, A. W. (1973a). *Nature. Lond.* **245**, 20.
Fairhall, A. W. (1973b). *Nature, Lond.* **246**, 106.
Fairhall, A. W., Bradford, F., Yang, I. C. and Young, A. W. (1970). *Antarctic J. U.S.* **5**, 190.
Fairhall, A. W., Bradford, F. Yang, I. C. and Young, A. W. (1971). *Antarctic J. U.S.* **6**, 163.
Fairhall, A. W. and Erickson, J. L. (1974). *Nature, Lond.* (in press).
Faurholt, C. (1924). *J. chim. Phys.* **21**, 400.
Fergusson, G. J. (1958), *Proc. roy. Soc.* **A243**, 561.
Fisher, F. H. (1965). *J. acoust. Soc. Amer.* **38**, 805.

Fisher, F. H. (1967). *Science, N.Y.* **157**. 823.
Fisher, F. H. (1972). *Geochim. Cosmochim. Acta,* **36**, 99.
Fonselius, S., Koroleff, F. and Buch, K. (1955). *Tellus,* **7**, 258.
Fonselius, S., Koroleff, F. and Warme, K. E. (1956). *Tellus,* **8**, 176.
Fonselius, S. and Ostlund, H. G. (1959). *Tellus,* **11**, 77.
Gaines, A. G. and Pilson, M. E. Q. (1972). *Limnol. Oceanogr.* **17**, 42.
Garner, D. M. (1958). *Nature, Lond.* **182**, 466.
Garrels, R. M. (1965). *Science, N.Y.* **148**, 69.
Garrels, R. M. and Christ, C. L. (1965). "Solutions, Minerals and Equilibria", Harper Row Inc., New York, 450 pp.
Garrels, R. M. and Thompson, M. E. (1962). *Amer. J. Sci.* **260**, 57.
Garrels, R. M., Thompson, M. E. and Siever, R. (1961). *Amer. J. Sci.* **259**, 24.
Geffken, G. (1904). *Z. phys. Chem.* **49**, 257.
Gieskes, J. M. (1969). *Limnol. Oceanogr.* **14**, 679.
Gieskes, J. M. (1970). *Limnol. Oceanogr.* **15**, 329.
Gieskes, J. M. (1974). *In* "The Sea" Vol. V (E. D. Goldberg, ed.). Wiley-Interscience, New York.
Gibbons, B. H. and Edsall, J. T. (1963). *J. Biol. Chem.* **238**, 3502.
Gold, V. (1956). "pH Measurements", 125 pp. Methuen, London.
Gordon, L. I. and Jones, L. B. (1973). *Mar. Chem.* **1**, 317.
Gordon, L. I., Park, P. K., Kelley, J. J. and Hood, D. W. (1973). *Mar. Chem.* **1**, 191.
Gordon, L. I., Park, P. K., Hager, S. W. and Parsons, T. R. (1971). *J. Oceanogr. Soc. Japan,* **27**, 81.
Gran, G. (1952). *Analyst.* **77**, 661.
Grasshoff, K. (1964). *Cons. perm. int. Explor. Mer.* Bull. No. 123, 7 pp.
Greenberg, G., Moberg, E. G. and Allen, E. (1932). *Industr. Engng. Chem. (Anal.),* **4**, 309.
Greenwald, I. (1941). *J. biol. Chem.* **141**, 789.
Gripenberg, S. (1937). *5th Hydrol. Conf. Helsingfors. Comm. 10B.*
Gripenberg, S. (1960). *J. Cons. int. Explor. Mer.* **26**, 5.
Gunderson, G. and Mountain, C. W. (1973). *Deep-Sea Res.* **20**, 1083.
Hansson, I. (1972). Thesis, University of Göteborg.
Hansson, I. (1973a). *Deep-Sea Res.* **20**, 461.
Hansson, I. (1973b). *Deep-Sea Res.* **20**, 479.
Hansson, I. (1973c). *Acta Chem. Scand.* **27**, 924.
Hansson, I. (1973d). *Acta Chem. Scand.* **27**, 931.
Hansson, I. and Jagner, D. (1973). *Anal. Chim. Acta,* **65**, 363.
Hanya, T. and Ishiwatari, R. (1961). *J. oceanogr. Soc. Japan,* **17**, 28.
Harned, H. S. and Bonner, F. T. (1945). *J. Amer. Chem. Soc.* **67**, 1026.
Harned, H. S. and Davis, R. (1943). *J. Amer. chem. Soc.* **65**, 2030.
Harned, H. S. and Owen, B. B. (1958). "The Physical Chemistry of Electrolyte Solutions" 803 pp. 3rd Ed. Reinhold, New York.
Harned, H. S. and Scholes, S. R. (1941). *J. Amer. chem. Soc.* **63**, 1706.
Harvey, H. W. (1955). "The Chemistry and Fertility of Sea Water". Cambridge University Press, London.
Hawley, J. E. (1973). Thesis, Oregon State University, Corvallis.
Hawley, J. E. and Pytkowicz, R. M. (1969). *Geochim. Cosmochim. Acta,* **33**, 1557.
Hawley, J. E. and Pytkowicz, R. M. (1973). *Mar. Chem.* **1**, 245.

Hayes, F. N., Anderson, E. C. and Arnold, J. R. (1955). *Proc. int. Conf. Peaceful Uses of Atomic Energy, Geneva,* **14**, 188.

Heath, G. R. and Culberson, C. (1970). *Geol. Soc. Amer. Bull.* **81**, 3157.

Herman, H. B. and Rechnitz, G. A. (1974). *Science, N.Y.* **184**, 1074.

Higbie, R. (1935). *Trans. Amer. Inst. Chem. Engrs.* **31**, 365.

Hindman, J. C. (1943). Thesis, U.C.L.A.

Ho, C. and Sturtevant, J. M. (1963). *J. biol. Chem.* **238**, 3499.

Holland, H. D. (1972). *Geochim. Cosmochim. Acta,* **36**, 637.

Hood, D. W., Berkshire, D., Adams, R. and Supernaw, I. (1963). A and M College of Texas, Technical Report, Ref. 63–3D.

Hoover, T. E. (1966). Thesis, A and M College of Texas.

Hoover, T. E. and Berkshire, D. C. (1969). *J. geophys. Res.* **74**, 456.

Houtevmans, J. C., Suess, H. E. and Oeschager, H. (1973). *J. geophys. Res.* **78**, 1897.

Huber, B. (1952). *Forstwiss. Zbl.* **71**, 372.

Hutchinson, G. E. (1954). *In* "The Earth as a Planet" (G. Kuiper, ed.), Chicago University Press, Chicago.

Ibert, E. R. (1963). Ph.D. Thesis. A & M College of Texas. 131 pp.

Ibert, E. R. and Hood, D. W. (1963). A and M College of Texas, Technical Report, Ref. 63–9–T.

Ingle, S. E., Culberson, C., Hawley, J. E. and Pytkowicz, R. M. (1973). *Mar. Chem.* **1**, 295.

Ingri, N. (1963). *Svensk. Kem. Tidskr.* **75**, 3.

Kanwisher, J. (1960). *Tellus,* **12**, 209.

Kanwisher, J. (1963a). *Deep-Sea Res.* **10**, 195.

Kanwisher, J. (1963b). *J. geophys. Res.* **68**, 3921.

Keeling, C. D. (1958). *Geochim. Cosmochim. Acta,* **13**, 322.

Keeling, C. D. (1960). *Tellus,* **12**, 200.

Keeling, C. D. (1961). *Geochim. Cosmochim. Acta,* **24**, 277.

Keeling, C. D. (1968). *J. geophys. Res.* **73**, 4543.

Keeling, C. D. (1973a). *In* "Chemistry of the Lower Atmosphere" (S. I. Rasool, ed.). Plenum Press, New York, 251–329.

Keeling, C. D. (1973b). *Tellus,* **25**, 174.

Keeling, C. D. and Bolin, B. (1967). *Tellus,* **19**, 566.

Keeling, C. D. and Bolin, B. (1968). *Tellus,* **20**, 17.

Keeling, C. D., Harris, T. B. and Wilkins, E. M. (1968). *J. geophys. Res.* **73**, 4511.

Keeling, C. D., Rakestraw, N. W. and Waterman, L. S. (1965). *J. geophys. Res.* **70**, 6088.

Keeling, C. D. and Waterman, L. S. (1968). *J. geophys. Res.* **73**, 4529.

Kelley, J. J. (1968). *Nature, Lond.* **218**, 862.

Kelley, J. J. (1970). *Limnol. Oceanogr.* **15**, 80.

Kelley, J. J. and Hood, D. W. (1971a). University of Alaska College, *Inst. mar. Sci. Rep. No. R71-20,* 9 pp.

Kelley, J. J. and Hood, D. W. (1971b). *Nature, Lond.* **229**, 745.

Kelley, J. J. and Hood, D. W. (1971c). *J. geophys. Res.* **76**, 745.

Kelley, J. J., Longerich, L. L. and Hood, D. W. (1971). *J. geophys. Res.* **76**, 8687.

Kern, D. M. (1960). *J. chem. Educ.* **37**, 14.

Kester, D. R. (1969). Thesis, Oregon State University, Corvallis.

Kester, D. R. and Byrne, R. H. (1972). *In* "Ferromanganese Deposits on the Ocean Floor" (D. R. Horn, ed.). Palisade, New York, pp. 107–116.

Kester, D. R., Duedall, I. W., Connors, D. N. and Pytkowicz, R. M. (1967). *Limnol. Oceanogr.* **12**, 176.
Kester, D. R. and Pytkowicz, R. M. (1967). *Limnol. Oceanogr.* **12**, 243.
Kester, D. R. and Pytkowicz, R. M. (1968). *Limnol. Oceanogr.* **13**, 670.
Kester, D. R. and Pytkowicz, R. M. (1969). *Limnol. Oceanogr.* **14**, 686.
Kester, D. R. and Pytkowicz, R. M. (1970). *Geochim. Cosmoch. Acta*, **34**, 1039.
Kinney, P., Arhelger, M. E. and Burrel, D. C. (1970). *J. geophys. Res.* **75**, 4097.
Kishinevski, M. (1955). *J. Appl. Chem. USSR*, **28**, 881.
Kitano, Y. and Hood, D. W. (1962). *J. Oceanogr. Soc. Japan*, **18**, 35.
Kitano, Y. and Hood, D. W. (1965). *Geochim. Cosmochim. Acta*, **29**, 29.
Kitano, Y., Kanamori, N. and Tokuyama, A. (1970). *Inst. mar. Sci. Occasional Publ. No. 1*, 413.
Knull, J. R. and Richards, F. A. (1969). *Deep-Sea Res.* **16**, 205.
Kramer, J. R. (1958). *Bull. Geol. Soc. Amer.* **69**, 160.
Koczy, F. F. (1956). *Deep-Sea Res.* **3**, 279.
Kolthoff, I. M. and Laitinen, H. A. (1941). "pH and Electrotitrations", 190 pp., 2nd ed. John Wiley and Sons, New York.
Kritchevsky, I. and Iliinskaya, A. (1945). *Acta Physicochim. URSS*, **20**, 327.
Krogh, A. (1904). *Medd. Gronland*, **26**, 331.
Krogh, A. (1910). *Skand. Arch. Physiol.* **23**, 224.
Kroopnick, P. (1974). *Deep-Sea Res.* **21**, 211.
Kroopnick, P., Deuser, W. G. and Craig, H. (1970). *J. geophys. Res.* **75**, 7668.
Ku, T-L., Li, Y-H. and Mathieu, G. (1969). *Antarctic J. U.S.* **4**, 186.
Lafon, G. M. (1970). *Geochim. Cosmochim. Acta*, **34**, 935.
Lal, D. and Rama (1966). *J. geophys. Res.* **71**, 2865.
Leith, H. (1963). *J. geophys. Res.* **68**, 3887.
Lewis, W. K. and Whitman, W. G. (1924). *Ind. Eng. Chem.* **16**, 1.
Leyendekkers, J. V. (1973). *Limnol. Oceanogr.* **18**, 784.
Li, Y-H. (1967). Thesis, Columbia University, New York.
Li, Y-H. and Tsui, T-F. (1971). *J. geophys. Res.* **76**, 4203.
Li, Y-H., Takahashi, T. and Broecker, W. S. (1969). *J. geophys. Res.* **74**, 5507.
Lingfelter, R. E. (1963). *Rev. Geophys.* **1**, 35.
Lisitsin, A. and Petelin, V. P. (1967). *Litol. Poleznye Iskop.* **5**, 50.
Liss, P. (1972). *Nature, Lond.* **233**, 327.
Liss, P. (1973). *Deep-Sea Res.* **20**, 221.
Liss, P and Slater, P. A. (1974). *Nature, Lond.* **247**, 181.
Ljunggren, S. and Lamm, O. (1958). *Acta. chem. Scand.* **12**, 1834.
Lowenstam, H. A. and Epstein, S. (1957). *J. Geol.* **65**, 364.
Lyakhin, Yu-I. (1971). *Okeanologiya*, **11**, 4A.
Lyman, J. (1956). Thesis, University of California, Los Angeles, 196 pp.
Lyman, J. and Fleming, R. H. (1940). *J. mar. Res.* **3**, 134.
Machta, L. (1972). *In* "The Changing Chemistry of the Oceans" (D. Dyrssen and D. Jagner, eds). Almqvist and Wiksell, Stockholm, 121–145.
MacKay, C., Pandow, M., and Wolfgang, R. (1963). *J. geophys. Res.* **68**, 3929.
MacIntyre, W. G. (1965). *Bull. Res. Bd. Can No. 200*, Ottawa, 153 pp.
Markham, A. M. and Kobe, K. A. (1941). *Chem. Rev.* **28**, 519.
Manheim, F. (1961). *Stoch. Control. Geol.* **8**, 27.
Mehrbach, C., Culberson, C. H., Hawley, J. E. and Pytkowicz, R. M. (1973). *Limnol. Oceanogr.* **18**, 897.

Menzel, D. W. and Vaccaro, R. F. (1964). *Limnol. Oceanogr.* **9**, 138.

Millero, F. J. (1969). *Limnol. Oceanogr.* **14**, 376.

Millero, F. J. (1971a). *Geochim. Cosmochim. Acta,* **35**, 1089.

Millero, F. J. (1971b). *Chem. Rev.* **71**, 147.

Millero, F. J. and Berner, R. A. (1972). *Geochim. Cosmochim. Acta,* **36**, 92.

Millero, F. J. and Drost-Hansen, W. (1968). *J. Phys. Chem.* **72**, 1758.

Millero, F. J., Hoff, E. V. and Kahn, L. (1972). *J. Solution Chem.* **1**, 309.

Mills, A. and Urey, H. C. (1940). *J. Amer. chem. Soc.* **62**, 1019.

Miyake, Y. and Saruhashi, K. (1966). *Pap. Met. geophys. Tokio,* **17**, 218.

Miyake, Y. and Sugimura, Y. (1969). *Rec. Oceanogr. Wks. Japan,* **10**, 23.

Moberg, E. G., Greenberg, D. M., Revelle, R. and Allen, E. C. (1934). *Bull. Scripps. Instn. Oceanogr.* **3**, 231.

Mook, W. G. and Koene, B. K. S. (1974) (in press).

Morrison, T. J. and Billett, F. (1952). *J. chem. Soc.* 3819.

Morse, J. W. (1974c). *Amer. J. Sci.* **274**, 97.

Morse, J. W. (1974b). *Amer. J. Sci.* **274**, 638.

Morse, J. W. and Berner, R. A. (1972). *Amer. J. Sci.* **272**, 840.

Mukerjee, P. (1961). *J. phys. Chem.* **65**, 740.

Mukerjee, P. (1966). *J. phys. Chem.* **70**, 2708.

Munnich, K. O. and Roether, W. (1967). *Proc. Conf. Radioactive Dating and Methods of Low Level Counting,* p 93, Vienna; Intern. Atomic Energy Agency.

Munnich, K. O. and Vogel, J. C. (1958). *Naturwissenschaften,* **48**, 327.

Munnich, K. O. and Vogel, J. C. (1959). Paper presented at Radiocarbon Dating Conference, Groningen, Netherlands.

Murray, C. N. (1970). Thesis, University of Liverpool.

Murray, C. N. and Riley, J. P. (1971). *Deep-Sea Res.* **18**, 533.

Murray, J. and Hjort, J. (1912). "The Depths of the Oceans". Macmillan and Co. Ltd., London, 821 pp.

Nasanen, R. (1946). *Acta chem. fenn.* **19B**, 90.

Nasanen, R. (1947). *Acta chem. Scand.* **1**, 204.

Nissenbaum, A., Presley, B. J. and Kaplan, l. R. (1972). *Geochim. Cosmochim. Acta,* **36**, 1007.

Nydal, R. (1963). *Nature, Lond.* **200**, 212.

Nydal, R. (1966). *Tellus,* **18**, 271.

Nydal, R. (1967). *Proc. Conf. Radioactive Dating and Methods of Low Level Counting,* p. 119, Vienna: Intern. Atomic Energy Agency.

Nydal, R. (1968). *J. geophys. Res.* **73**, 3617.

Nydal, R. and Lövseth, K. (1970). *J. geophys. Res.* **75**, 2271.

Oestlund, H. C. (1969). U. S. Atomic Energy Comm. ML 69167, 30 pp. (Chem. abs 1973, 17510).

Olsson, I. U. and Karlen, I. (1963). *Radioactive Dating,* I.A.E.A. 3.

Orr, A. P. (1947). *Proc. roy. Soc. Edinb.* **B63**, 3.

Ostlund, H. G. and Niskin, S. (1970). *J. geophys. Res.* **75**, 7667.

Owen, B. B. and Brinkley, S. R. (1941). *Chem. Rev.* **29**, 461.

Owen, B. B. and King, E. J. (1943). *J. Amer. chem. Soc.* **65**, 1612.

Pales, J. C. and Keeling, C. D. (1965). *J. geophys. Res.* **70**, 6053.

Park, K. (1965). *J. oceanogr. Soc. Japan,* **21**, 54.

Park, K. (1968a). *Deep-Sea Res.* **15**, 721.

Park, K. (1968b). *Deep-Sea Res.* **15**, 171.

Park, K. (1969). *Limnol. Oceanogr.* **14**, 179.

Park, K., Curl, H. C. and Glooschenko, W. A. (1967). *Nature, Lond.* **215**, 380.

Park, K. and Hood, D. W. (1963). *Limnol. Oceanogr.* **8**, 287.

Park, K., Hood, D. W. and Odum, H. T. (1958). *Publ. Inst. mar. Sci. University of Texas*, **5**, 47.

Park, K., Kennedy, G. H. and Dobson, H. H. (1964). *Anal. Chem.* **36**, 1686.

Park, K., Webster, G. R. and Yamamoto, R. (1969). *Limnol. Oceanogr.* **14**, 559.

Pearman, G. I. and Garratt, J. R. (1973). *Tellus*, **25**, 309.

Peterson, M. N. A. (1966). *Science, N.Y.* **154**, 1542.

Pinsent, B. W. R., Pearson, L. and Roughton, F. J. W. (1956). *Trans. Faraday Soc.* **52**, 1512

Plass, G. N. (1956). *Tellus*, **8**, 140.

Pond. S., Pytkowicz, R. M. and Hawley, J. E. (1971). *Deep-Sea Res.* **18**, 1135.

Postma, H. (1964). *Netherlands J. Sea. Res.* **2**, 258.

Presley, B. J. and Kaplan, I. R. (1968). *Geochim. Cosmochim. Acta*, **32**, 1037.

Pytkowicz, R. M. (1963). *Deep-Sea Res.* **10**, 633.

Pytkowicz, R. M. (1965a). *J. Geol.* **73**, 196.

Pytkowicz, R. M. (1965b). *Limnol. Oceanogr.* **10**, 220.

Pytkowicz, R. M. (1967). *Geochim. Cosmochim. Acta*, **31**, 63.

Pytkowicz, R. M. (1968). *In* "Oceanogr. mar. biol. Ann. Rev", Vol. 6 (H. Barnes, ed.). Allen and Unwin Ltd., London, p. 83.

Pytkowicz, R. M. (1969a). *Amer. Zoologist*, **9**, 673.

Pytkowicz, R. M. (1969b). *Geochem. J.* **3**, 181.

Pytkowicz, R. M. (1970). *Geochim. Cosmochim. Acta*, **34**, 836.

Pytkowicz, R. M. (1971a). Tech. Rep. 214 (71-20) Dept. Oceanogr. Oregon State University, Corvallis, Oregon.

Pytkowicz, R. M. (1971b). *Geochim. Cosmochim. Acta*, **35**, 509.

Pytkowicz, R. M. (1972a). *Geochim. Cosmochim. Acta*, **36**, 631.

Pytkowicz, R. M. (1972b). *J. geophys. Res.* **77**, 2733.

Pytkowicz, R. M. (1972c). *Comm. on Earth Sciences: Geophysics*, **3**, 15.

Pytkowicz, R. M. (1972d). *Limnol. Oceanogr.* **17**, 958.

Pytkowicz, R. M. (1973a). *Amer. J. Sci.* **273**, 515.

Pytkowicz, R. M. (1973b). *Schweiz. Z. Hydrol.* **35**, 8.

Pytkowicz, R. M. and Connors, D. N. (1964). *Science, N.Y.* **144**, 840.

Pytkowicz, R. M., Disteche, A. and Disteche, S. (1967). *Earth Planet Sci. Letters*, **2**, 430.

Pytkowicz, R. M. and Fowler, G. A. (1967). *Geochem. J.* **1**, 169.

Pytkowicz, R. M. and Gates, R. (1968). *Science, N.Y.* **161**, 690.

Pytkowicz, R. M. and Hawley, J. E. (1974). *Limnol. Oceanogr.* **19**, 223

Pytkowicz, R. M., Ingle, S. E. and Mehrbach, C. (1974). *Limnol. Oceanogr.* **19**, 665

Pytkowicz, R. M., Kester, D. R. and Burgener, B. C. (1966). *Limnol. Oceanogr.* **11**, 417.

Pytkowicz, R. M. and Kester, D. R. (1969). *Amer. J. Sci.* **267**, 217.

Pytkowicz, R. M. and Kester, D. R. (1971). *In* "Oceanographic Marine Biology Annual Revue" (H. Barnes, ed.) Vol. 9. p. 11. Allen and Unwin, London.

Quinn, J. A. and Otto, N. C. (1971). *J. geophys. Res.* **76**, 1539.

Rafter, T. A. (1955). *N.Z. J. Tech. Bull.* **37**, 20.

Rafter, T. A. and Fergusson, G. J. (1958a). *N.Z.J. Tech. Bull.* **38**, 871.

Rafter, T. A. and Fergusson, G. J. (1958b). Proc. 2nd Conf. Peaceful uses of Atomic Energy, **18**, 526.

Rafter, T. A. and O'Brien, B. J. (1970). *In* "Radiocarbon Variations and Absolute Chronology" (I. U. Olsson, ed.) pp. 355–377. Wiley Interscience, New York.

Redfield, A. C., Ketchum, B. H. and Richards, F. A. (1963). *In* "The Sea; Ideas and Observations" (M. N. Hill, ed.) Vol. 2, p. 26. Interscience, N.Y., 554 pp.

Reid, J. L. (1962). *Limnol. Oceanogr.* 7, 287.

Reusmann, G. (1968). *Kiel. Meeresforsch.* 241, 14.

Revelle, R. (1965). *In* "Restoring the Quality of our Environment", Reports of Environment Pollution Panel, Presidents Science Advisory Committee, White House, pp. 111–113.

Revelle, R. and Suess, H. E. (1957). *Tellus,* 9, 18.

Richards, F. A. (1965). *In* "Chemical Oceanography" (J. P. Riley and G. Skirrow, eds.), Academic Press, London, Vol. I.

Richards, F. A., Cline, J. D., Broenkow, W. W. and Atkinson, L. P. (1965). *Limnol. Oceanogr.* 10, R185–R201.

Robinson, R. A. (1954). *J. mar. biol Ass. U.K.* 33, 449.

Rotty, R. M. (1973). *Tellus,* 25, 508.

Rotschi, H. (1965). *Prog. Oceanogr.* 3, 301.

Roughton, F. J. W. (1941). *J. Amer. chem. Soc.* 63, 2930.

Roughton, F. J. W. and Booth, U. H. (1937). *Chem. Ind.* 56, 1053.

Rulolf, W. (1971). *Meteor Forschungerget Reihe* No 6, 12.

Sackett, W. M., Eckelmann, W. R., Bender, M. L. and Bel, A. W. H. (1965). *Science, N.Y.* 148, 235.

Sackett, W. M. and Moore, W. S. (1966). *Chem. Geol.* 1, 323.

Saruhashi, K. (1953). *Pap. Met. Geophys., Tokyo,* 3, 202.

Saruhashi, K. (1955). *Pap. Met. Geophys., Tokyo,* 6, 38.

Schmalz, R. F. and Swanson, F. J. (1969). *J. Sediment. Petrol.* 39, 255.

Seibold, E. (1962). *Sedimentology,* 1, 50.

Sharma, M. M. and Danckwerts, P. (1963). *Chem. Eng. Sci.* 18, 729.

Shedlovsky, T. and MacInnes, D. A. (1935). *J. Amer. chem. Soc.* 57, 1705.

Shishkin, O. V. (1958). *Okeanol. Akad. Nauk SSSR.* 33, 178.

Siever, R., Beck, K. C. and Berner, R. A. (1965). *J. Geol.* 73, 39.

Sillén, L. G. (1961). The Physical Chemistry of Sea Water. *In* "Oceanography" (M. Sears, ed.). Amer. Assn. for adv. of Sci. Publ. 67. Washington, pp 549–581.

Sillén, L. G. (1963). *Svensk. Kem. Tidskr.* 75, 161.

Sillén, L. G. (1967). *Science, N.Y.* 156, 1189.

Sillén, L. G. and Martell, A. E. (1964). Chem. Soc. spec. publ. No 17 754 pp.

Simkiss, D. (1964). *J. Cons. int. Explor. Mer.* 29, 6.

Simpson, H. J. and Broecker, W. S. (1973). *Limnol. Oceanogr.* 18, 426.

Sirs, J. A. (1958). *Trans. Faraday Soc.* 54, 201, 207.

Skirrow, G. and Whitfield, M. (1974). *Limnol. Oceanogr.* (In press).

Slocum, G. (1955). *Monthly Weather Rev.* Oct. 225.

Smallwood, P. (1972). Ph.D. Thesis, University of Liverpool.

Smith, C. L. (1940). *J. mar. Res.* 3, 171.

Smith, C. L. (1941). *J. mar. biol. Ass. U.K.* 25, 235.

Smith, V. N. (1953). *Instruments,* 26, 421.

Smith, W. H. and Hood, D. W. (1964). *In* "Ken Sugawara Festival Volume, Recent Researches in the fields of the Hydrosphere, Atmosphere and Nuclear Geochemistry" (Y. Miyake and T. Koyama, eds.). Maruzen Co., Tokyo pp. 185–202.

Spedding, D. J. (1972). *Atmos. Environ. New York,* 6, 583.

Spencer, C. P. (1965). *Oceanogr. mar. biol. Ann. Rev.* **3**, 31.

Spiro, T. G., Lee, J. and Revesz, A. (1968). *J. Amer. chem. Soc.* **90**, 4000.

Strickland, J. D. H. and Parsons, T. R. (1960). *Bull. Fish Res. Bd. Can.* **125**, 185.

Stuiver, M. (1965). *Science, N.Y.* **149**, 533.

Stumm, W., and Belinski, H. (1972). *Proc. Water Poll. Res.*

Stumm, W., and Morgan, J. J. (1970). "Aquatic Chemistry". Wiley, New York, 583 pp.

Suess, H. E. (1955). *Science, N.Y.* **122**, 415.

Suess, H. E. (1973). *Endeavour*, **32**, 34.

Suess, E. (1970). *Geochim. Cosmochim. Acta*, **34**, 157.

Suess, E. (1973). *Geochim. Cosmochim. Acta.* **37**, 2435.

Sugiyura, Y. (1964). *In* "Recent Researches in the Fields of the Hydrosphere, Atmosphere and Nuclear Geochemistry". (Y. Miyake and T. Koyama, eds). Maruzen Co., Tokyo.

Sugiura, Y. (1965a). *La Mer. (Bull. Soc. franco-japonaise d'oceanographie)*, **2**, 87.

Sugiura, Y. (1965b). *Pap. Met. Geophys. Tokyo*, **15**, 208.

Sugiura, Y. (1969a). *Oceanogr. Mag.* **21**, 39.

Sugiura, Y. (1969b). *Kagaku No Ryoiki*, **23**, 396.

Sugiura, Y. and Yoshimura, H. (1964). *J. oceanogr. Soc. Japan*, **20**, 14.

Sugiura, Y., Ibert, E. R. and Hood, D. W. (1963). *J. mar. Res.* **21**, 11.

Sverdrup, H. U., Johnson, M. W. and Fleming, R. H. (1942). "The Oceans", 1087 pp. Prentice Hall, New York.

Swinnerton, J. W., Linnenbom, V. J. and Cheek, H. (1962). *Anal. Chem.* **34**, 483.

Szabo, B. J., Koczy, F. F. and Östlund, G. (1967). *Earth Planet Sci. Letters*, **3**, 51.

Takahashi, T. (1961). *J. geophys. Res.* **66**, 477.

Takahashi, T. (1974). *J. foraminif. Res.* (in press).

Takahashi, T., Weiss, R. F., Culberson, C. H., Edmond, J. M., Hammond, D. E., Wong, C. S., Li, Y-H. and Bainbridge, A. E. (1970). *J. geophys. Res.* **75**, 7648.

Teal, J. J. and Kanwisher, J. (1966). *J. mar. Res.* **24**, 4.

Thompson, M. E. and Ross, J. W. (1966). *Science, N.Y.* **154**, 1643.

Thompson, T. G. and Anderson, D. H. (1940). *J. mar. Res.* **3**, 224.

Thompson, M. E. (1966) *Science. N.Y.* **153**, 866.

Thorstenson, D. C. (1970). *Geochim. Cosmochim. Acta*, **34**, 745.

Turekian, K. K. (1965). *In* "Chemical Oceanography" Vol. 2 (J. P. Riley and G. Skirrow, eds.). Academic Press, London.

Turekian, K. K., Bowen, P. M. and Woodburn, J. (1971). *Atlantic Res. Ser.* **15**, 315.

van Slyke, D. D. (1922). *J. biol. Chem.* **52**, 525.

van Slyke, D. D. and Neill, J. M. (1924). *J. biol. Chem.* **61**, 523.

van Slyke, D. D. and Sendroy, J. (1927). *J. biol. Chem.* **73**, 127.

Walker, A. C., Bray, V. B. and Johnston, J. (1927). *J. Amer. chem. Soc.* **49**, 1235.

Walton, A. and Baxter, M. S. (1969). *Nature, Lond.* **220**, 475.

Walton, A., Ergin, M. and Harkness, D. D. (1970). *J. geophys. Res.* **75**, 3089.

Wangersky, P. J. (1972a). *Limnol. Oceanogr.* **17**, 1.

Wangersky, P. J. (1972b). *Limnol. Oceanogr.* **17**, 960.

Waterman, L. S. (1965). *Nature, Lond.* **205**, 1099.

Wattenberg, H. (1933). *Wiss. Erg.* **8**, 122.

Wattenberg, H. (1936). *Fortschr. Mineral.* **20**,

Wattenberg, H. and Timmerman, E. (1936). *Ann. Hydrogr. Mar. Mat.* **64**, 23.

Weast, R. C. (1965). Handbook of Chemistry and Physics, Chemical Rubber Publishing Company.

Weber, J. N. and Woodhead, P. M. J. (1971). *Geochim. Cosmochim. Acta,* **35**, 891.
Weiss, R. F. (1974). In press.
Weiss, R. F. and Craig, H. (1968). *Trans. Amer. geophys. Union,* **49**, 216.
Weiss, R. F. and Craig, H. (1973). *Deep-Sea Res.* **20**, 291.
Wells, A. J. and Illing, L. V. (1964). *In* "Developments in Sedimentology", Vol I.
   (L. van Straaten, ed.) pp. 429–435. Elsevier Publishing Co., New York.
West, L. E. and Robinson, R. J. (1970). *J. mar. Res.* **4**, 38.
Weyl, P. K. (1961). *J. Geol.* **69**, 33
Whitfield, M. (1972). *J. Chem. Eng. Data,* **17**, 124.
Whitfield. M. (1974a). *Limnol. Oceanogr.* **19**, 235.
Whitfield, M. (1974b). *Nature, Lond.* **247**, 523.
Whitfield, M. (1974c). *Nature, Lond.* **249**, 818.
Whitfield, M. (1974d). *Nature, Lond.* (in press).
Williams, R. M. and Gordon, L. I. (1970). *Deep-Sea Res.* **17**, 19.
Williams, P. M. and Strack, P. M. (1966). *Limnol. Oceanogr.* **11**, 401.
Wissburn, K. F., French, D. M. and Patterson, A. (1954). *J. Phys. Chem.* **58**, 697.
Wong, C. S. (1970). *Deep-Sea Res.* **17**, 9.
Worthington, L. V. (1954). *Deep-Sea Res.* **1**, 244.
Wyatt, B., Gilbert, W., Gordon, L. and Barstow, D. (1970). Oregon State University:
   Dept. Oceanogr. Data Rep. 42 155 pp.
Young, J. A. (1967). Thesis, University of Washington, Seattle.
Young, J. A. and Fairhall, A. W. (1968). *J. geophys. Res.* **73**, 1185.
Zimen, K. E. and Altenhein, F. K. (1973a). *Naturwiss.* **60**, 198.
Zimen, K. E. and Altenhein, F. K. (1973b). *Naturforschung,* **28a**, 1747.
Zirino, A. and Healey, M. L. (1970). *Limnol. Oceanogr.* **15**, 956.
Zirino, A. and Yamamoto, S. (1972). *Limnol. Oceanogr.* **17**, 661.

Chapter 10

Chemistry of the Sea Surface Microlayer

PETER S. LISS

School of Environmental Sciences, University of East Anglia,
Norwich, NR4 7TJ, England

10.1. INTRODUCTION

Although it has been recognized for many years that the properties of the surface layer of the sea differ from those of the bulk water, it is only in the last decade that attempts have been made to study this layer. The earlier lack of research in this field stems in large measure from the difficulty of collecting a representative sample of the layer, and also, in part, from the lack of analytical techniques of sufficient sensitivity for the examination of many of its components. Even now the former difficulty (the collection of a realistically thin layer) has not been overcome, and consequently many of the results which have been published are equivocal. Nevertheless, some real progress has been made, and although selected aspects of the subject have been reviewed

193

by MacIntyre (1974) and by Parker and Barsom (1970), no general review of the subject has appeared. It is the aim of this chapter to provide a general introduction to the topic, concentrating as equally as possible on its organic, physical and inorganic facets, some attention also being given to the experimental aspects. In the selection of results for inclusion, a conscious effort has been made to choose those obtained using real, rather than artificial, sea water, and to make use of data collected with a view to preserving *in situ* sea conditions rather than those generated in laboratory simulation experiments.

10.2. Origin of Microlayer Material

Figure 10.1, which is freely adapted from Garrett (1970), illustrates some of the routes of entry and removal of material to and from the sea surface, and also the internal transformations of such material within the microlayer. Most of the pathways and processes shown apply equally to both natural and anthropogenic materials. Spillage at sea refers to input directly to the sea surface of man-made products from ships and other craft.

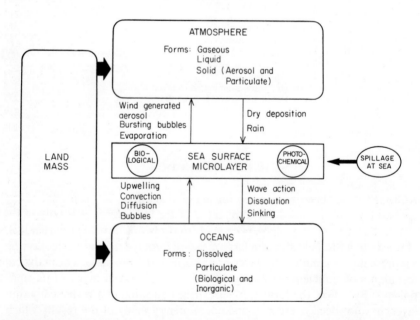

FIG. 10.1. Sources and sinks of natural and man-made materials for the sea surface microlayer, including internal processes within the layer.

Material may be brought to the sea surface from the main body of the oceans by diffusion, convection, upwelling and rising gas bubbles. The last of these is probably the most important. It is thought that the rising bubble adsorbs dissolved and particulate material (both organic and inorganic) during its passage through the water. When the bubble bursts at the interface only a part of the adsorbed material is ejected into the atmosphere, the remainder being left behind at the sea surface (Blanchard, 1964). From laboratory studies on the rate of accumulation of surface active material at the air–sea interface Jarvis (1967) has shown that the rate increases markedly if a stream of air bubbles is passed through the water. Further examples of the importance of bubbles in microlayer processes will appear as the chapter proceeds. The vexed question of the role of bubbles in producing organic particulates at the surface will also be discussed. An additional source of material for the microlayer is the atmosphere. Material transported in this way can reach the interface either as rainfall or by dry deposition (gaseous, aerosol and particulate depositions).

Microlayer material may be dispersed into the main body of the water by, amongst other things, sinking of particles which may or may not have adsorbed molecules on their surfaces. Another mechanism is the dissolution of water soluble molecules, a process which is greatly aided by compression of the sea surface by wind, waves, currents, etc. Transport from the sea surface to the atmosphere can be by the formation of wind generated aerosols or via the bursting of bubbles (Blanchard, 1964). An additional route is by evaporation of volatile compounds such as iodomethane and dimethyl sulphide (Lovelock et al., 1972, 1973).

Processes operating on materials within the microlayer itself may be either biological surface- or photo-chemical. The products of such processes may remain in the layer or be transported to the atmosphere or subsurface water. Results to be presented in a later section show that the microlayer is exceptionally rich in bacteria, plankton, etc., relative to subsurface water. Such productive populations will both consume and produce a wide variety of organic (and inorganic) substances. For instance, Wilson and Collier (1972) have shown that various marine phytoplankton are capable of producing surface active material, although not all species studied had equal ability in this respect. Pilpel (1968) has suggested that microorganisms may be important in the oxidative breakdown of petroleum products accidentally introduced into the oceans. The existence of processes of this type, coupled with the high microorganism density at the sea surface, argue for the importance of biological activity in the microlayer in bringing about the transformation of both natural and man-made materials in the marine environment. The role of photochemical reactions in the breakdown of organic molecules at the air–

water interface has been the subject of a variety of laboratory studies (Timmons, 1962; Baier, 1972; Wheeler, 1972). The results of these show that radiation in the U.V. region of the spectrum can bring about oxidative breakdown of many of those types of organic molecules which are found at the sea surface. It is difficult to know to what extent this information can be applied to the interpretation of conditions at the sea surface, but since the microlayer receives U.V. radiation unattenuated by any appreciable thickness of water, it seems likely that it will be a site for considerable photochemical activity. All the experiments to date have been confined to photochemically induced transformations of dissolved and particulate organic material. It is not known whether U.V. irradiation affects the inorganic material in the microlayer to any significant extent. However, the enhanced biological and photochemical activity in the microlayer means that the surface film is a microenvironment which is likely to be very different from bulk sea water with respect to such properties as pH, pE and speciation.

10.3. Thickness of the Sea Surface Layer—I

The question of the thickness of the sea surface layer (also referred to as the surface film and sea surface microlayer) is an important one. This matter will be side-stepped for the moment, because much of the information presently available on the chemical nature of the sea surface layer has been obtained from studies in which a portion of the surface water has been directly sampled and subsequently analysed for both chemical composition and physical properties. A discussion of the thickness of the skin of surface water sampled by these devices and its relationship to surface properties will follow at end of this chapter (Section 10.6). This means that for the moment the sea surface layer is defined as that water which is collected by the various surface sampling devices. Details of these devices are given in the next section.

10.4. Method of Collecting Surface Material

Garrett (1965) has described a 16-mesh screen (75 × 60 cm) made from 0·14 mm diameter Monel metal wire for collecting surface material. The screen is immersed and withdrawn horizontally through the water surface and the water which adheres to it is drained into a sample bottle. Material from the top 150 μm of the water is sampled with an efficiency of about 75%. The less than 100% efficiency is due to irreversible adsorption of some microlayer material onto the mesh when it is first immersed in the water. This material does not drain into the sample bottle and deactivates a fraction of the screen for subsequent immersions. By using the screen 200–250 times a considerable volume (~20 litres) of sample may be collected. A sample of this size corresponds to collection of approximately $100 \, m^2$ of surface water. Sieburth (1965) used a screen of similar design, but with a stainless steel mesh, to sample bacteria from the ocean surface. Microlayer samples for trace metal analysis have been collected using a 20-mesh polythene screen by Piotrowicz et al. (1972). Samplers of this type have been used in most of the studies of sea surface material so far conducted. A deficiency of the screen sampler is that it takes a long time to get a reasonable volume of water and there is some doubt whether a sufficiently thin and unmixed layer is selectively sampled. Furthermore, subsurface objects such as weed, plankton and rope fragments may adhere to the mesh and so contaminate the sample.

To try to overcome these problems Harvey (1966) has designed a microlayer collector which consists of a rotating stainless steel drum (38 cm diameter, 60 cm long) with a ceramic coating, as shown in Fig. 10.2. The surface layer (60–100 μm) adheres to the drum and is removed by the wiper. The thickness of the layer sampled depends on the speed of rotation of the drum and the water temperature. Under average operating conditions the sampling rate is about one third of a litre per minute and, as with the screen sampler, a considerable volume can be collected. The principal difficulty with this sampler is that it will work effectively only under calm conditions.

More recently, Harvey and Burzell (1972) have described what must be the simplest method so far for the collection of microlayer material. A clean glass plate (20 × 20 cm and 0·4 cm thick) is immersed vertically through the surface and then withdrawn vertically at a rate of $20 \, cm \, s^{-1}$. A surface layer approximately 60–100 μm thick remains on the plate and is removed from both sides using a neoprene wiper blade. The technique has been used successfully for natural organic surface films, but not for oil layers significantly thicker than a monolayer.

Harvey and Burzell (1972) have made a comparison between the ceramic drum and glass plate microlayer collectors. Surface layer material was harvested from a large laboratory tank of sea water using both methods simultaneously. A sample from 10 cm depth was collected for comparison. The results are given in Table 10.1.

The concentration of various classes of microorganisms and typical fluorescent compounds was measured in all three samples. The results show reasonably good agreement between the two microlayer collectors. They also illustrate the very considerable enrichment of the surface layer, relative to a depth of 10 cm, with respect to both organisms and fluorescent compounds.

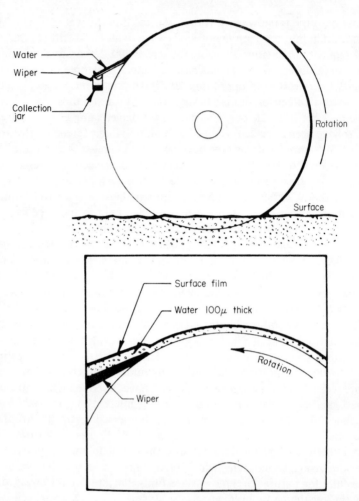

Fɪɢ. 10.2. Schematic diagram of the rotating drum microlayer sampler designed by Harvey (1966). (Reprinted with permission from American Society of Limnology and Oceanography).

Larsson *et al.* (1974) describe the use of a teflon plate perforated with a large number of conical holes to sample the microlayer. As teflon is strongly hydrophobic this device will preferentially sample lipids and other highly surface active compounds. A collection efficiency for lipids of 70–100% is claimed, the thickness of the layer sampled being between 50 and 100 µm.

Baier (1970) has given details of the use of internal reflection prisms made of optically polished germanium for sampling and infrared analysis of surface layer material. The prism (5 × 2 × 0·1 cm) is immersed and retracted, with

TABLE 10.1

Comparison of drum and plate microlayer collectors (Harvey and Burzell, 1972).

	10 cm depth	Ceramic drum	Glass plate
	(Number of particles cm^{-3})		
Detritus	3,640	54,400	47,600
Ciliates	322	2,210	3,140
Colourless flagellates	296	1,700	5,250
Chain diatoms	112	568	389
	Fluorometer reading—relative values		
Tryptophan	2·17	31·50	29·80
Tyrosine	2·00	22·00	19·40
Chlorophyll *a*	0·19	2·03	2·35

its main axis vertical, through the water surface and the I.R. spectrum of the adhering layer of surface material is analysed by multiple attenuated internal reflection (MAIR). It was claimed that a single dip in and out of the water samples only a monomolecular film from the interface. By allowing the prism plus adsorbed material to dry between dips, a multilayer of mono-molecular film can be collected. It is also possible to determine the surface tension, thickness and electrical potential of the material adsorbed on the prism. A most valuable property of this device is that the collection and analysis are carried out without transferring the material from the sampler. Against this must be set the fact that I.R. spectroscopy will yield only information on the major classes of organic groups present. It has been used most elegantly to study the input and fate of organic material on the surface of an inland lake (Baier, 1970) and in laboratory experiments to elucidate the factors leading to the breakdown of surface contaminants on fresh waters (Baier, 1972). The prism dip technique is reported to have been used at sea (Baier and Goupil, 1973), but at the time of writing very few results from these studies are available.

A novel approach to sampling the sea surface, called a "bubble micro-tome", has been proposed by MacIntyre (1968). He suggested that rising gas bubbles may be very efficient collectors of surface material. When the bubble breaks at the interface jet droplets of surface material are ejected into the air from where they may be harvested by a variety of techniques. The relation-ship between bubble diameter and thickness of the microtome cut is not well established, but MacIntyre (1974) has argued that it is likely to be roughly 0·05% of the bubble diameter. Fasching et al. (1973) have reported the use of the bubble microtome in coastal sea water and have estimated that the top micrometre of the water is sampled. At present, the results of these measurements are not available.

The screen, rotating drum and plate techniques take samples from the water surface of thickness $100 \pm 50\,\mu m$. In contrast, the prism dip and bubble microtome methods sample much thinner layers; about $1\mu m$ for the bubble microtome and of order $10^{-2}\,\mu m$ for the prism. This last result is the mean of the values found by Baier (1970) for the thickness of the film adhering to the prism when used in the single dip (monomolecular layer) mode on Lake Chautauqua in the United States. It is to be expected that the chemical and physical properties of the surface layer will differ considerably depending on their thickness. However, in the absence of results for the bubble microtome and prism sampler this prediction cannot be tested. The majority of results to be presented in subsequent sections have been obtained using the samplers which collect a slice of surface material approximately $100\,\mu m$ thick. In many cases this surface layer shows enhanced properties relative to a more conventional surface (i.e. bucket) or deep sample. It may well be that if similar measurements were available for material collected by the bubble microtome or prism the enhancement would be even more striking; however, for the present, such a conclusion is speculative.

10.5. PROPERTIES OF THE SEA SURFACE LAYER

10.5.1. ORGANIC CHEMICAL COMPOSITION

The most extensive published information on the chemical composition of surface material is that given by Garrett (1967a). A brief description of the methods used in his work together with the results obtained are given below.

Surface and subsurface water samples were collected using the screen technique (Garrett, 1965) and a 16 l PVC Van Dorn bottle, respectively. After collection, samples (unfiltered) were transferred to clean polythene carboys which had previously been rinsed with excess sample water to reduce adsorption losses. The organic material in each sample was concentrated by adding iron(III) to the water, the resulting iron(III) hydroxide precipitate being allowed to settle and the supernatant liquid siphoned off. The concentrate was frozen and returned to the laboratory for analysis. On receipt, samples were thawed, acidified to pH 2 and the organic material extracted into purified chloroform. The extract was analysed by gas chromatography both before and after esterification.

Using the above procedure it is difficult to know whether there is any appreciable adsorption of organic material onto the walls of the polythene carboy. Some precautions were taken to avoid adsorption, but no quantitative estimate of their efficiency was given. The co-precipitation with iron(III) hydroxide was shown to be at least 90% efficient for surface active molecules. However, because coprecipitation works well only for surface active compounds, the analysis may not represent all the organic material at the sea surface. Unfiltered samples were used on the grounds that much of the organic material at the interface occurs in combination with organisms or particles which would be removed by filtration. Furthermore, filtration may lead to loss of surface active material from solution onto the filter.

Prior to the work of Garrett (1967a) no serious attempt appears to have been made to characterize the chemical nature of the organic matter at the ocean surface. The samples analysed by Garrett are listed in Table 10.2; they were collected from a number of coastal areas in the Atlantic and Pacific oceans, but also included some samples from the open Atlantic. Water from areas likely to be contaminated by ships or other pollution sources was avoided. The results from gas chromatography of the esterified chloroform extracts are given in Table 10.3. The extracts contained free and combined fatty acids having from 11 to 22 carbon atoms, Of the higher aliphatic alcohols, dodecanol occurred in most samples, hexadecanol in ten, of which only three were from the surface; oleyl and stearyl alcohols being essentially absent. The average carbon chain length for the surface samples was 16·8, and for the subsurface samples 16·0.

Free fatty acids were determined in some samples by gas chromatography of the chloroform extract without esterification. The results indicated that up to 25% of the extract might have consisted of free rather than esterified fatty acids and, furthermore, that some samples contained lower molecular weight fatty acids having from 8 to 10 carbon atoms. This latter finding probably implies that the esterification procedure used to determine free and combined acids did not efficiently esterify acids of low molecular weight.

Barger and Garrett (1970) have analysed organic material collected from the air above the sea surface. They were able to identify five fatty acids (carbon atoms: double bonds, 14:0, 16:0, 16:1, 18:0, 18:1) in the chloroform extracts. The relative proportions of these were in good agreement with the ratios for these compounds in sea surface material found by Garrett (1967a). However, these five components, which for the surface samples shown in Table 10.3 make up, on average, 66% of the material analysed, account for only about 5% of the organic material in the chloroform extracts from the air samples.

As shown above, the organic molecules from the subsurface water appear to have an average carbon chain length somewhat shorter than does material from the sea surface. The point is shown very clearly by samples N-3 and N-4

TABLE 10.2

Details of samples analysed by Garrett (1967a).

Sample designation	Date	Condition of surface	Sampling depth (m)	Sea conditions	Location
N-3	June 62	75% slick-covered	surface	calm	19 mi S., Block Island, R.I.
N-4	June 62	75% slick-covered	near surface (bucket)	calm	19 mi S., Block Island, R.I.
B-2	July 62	no visible slicks	surface	moderate swell	North Atlantic, 56°30'N, 51°W.
A-1	Aug. 62	no visible slicks	surface	choppy	200 mi E., Atlantic City, N.J.
A-2	Aug. 62	no visible slicks	surface	choppy	200 mi E., Atlantic City, N.J.
A-3	Aug. 62	no visible slicks	surface	calm	200 mi E., Atlantic City, N.J.
A-4	Aug. 62	no visible slicks	surface	calm	200 mi E., Atlantic City, N.J.
B-3	Oct. 62	no visible slicks	surface		Mid Atlantic, 47°W., 35°N.
P-11	Nov. 62	20% slick covered	surface	swell, no chop	8 mi W., San Diego, Calif.
P-12	Nov. 62	20% slick-covered	surface	swell, no chop	8 mi W., San Diego, Calif.
G-10	Nov. 62	slicks sampled	surface	calm	Gulf of Lower California
G-20	Nov. 62	no visible slicks	surface	choppy	30 mi SW., Key West, Fla.
G-21	Nov. 62	no visible slicks	surface	choppy	30 mi SW., Key West, Fla.
G-22	Nov. 62	no visible slicks	surface	choppy	30 mi SW., Key West, Fla.
G-23	Nov. 62	no visible slicks	surface	choppy	30 mi SW., Key West, Fla.
G-9	Jan. 63	slick associated with planktonic bloom			8 mi SE., Galveston, Tex.
P-20	Apr. 63	slicks present	4	calm	10–15 mi S., Panama City, Pan. Bay of Panama
P-21	Apr. 63	slicks present	12	calm	10–15 mi S., Panama City, Pan. Bay of Panama
P-22	Apr. 63	no visible slicks	4	choppy	10–15 mi S., Panama City, Pan. Bay of Panama
P-23	Apr. 63	no visible slicks	8	choppy	10–15 mi S., Panama City, Pan. Bay of Panama

TABLE 10.2—*cont.*

Sample designation	Date	Condition of surface	Sampling depth (m)	Sea conditions	Location
P-25	Apr. 63	slicks present	8	choppy	10–15 mi S., Panama City, Pan. Bay of Panama
P-26	Apr. 63	slicks present	surface	choppy	10–15 mi S., Panama City, Pan. Bay of Panama
W-4	June 63	no slicks	surface	choppy	7 mi E., Wilmington, N.C.
P-31	Dec. 63	20% slick-covered	12	choppy, well mixed	15–20 mi S., Panama, Bay of Panama
P-32	Dec. 63	20% slick-covered	25	choppy, well mixed	15–20 mi S., Panama, Bay of Panama
P-33	Dec. 63	20% slick-covered	surface	choopy, well mixed	15–20 mi S., Panama, Bay of Panama
P-39	Dec. 63	20% slick-covered	25	choppy, well mixed	10 mi S., Panama, Bay of Panama
P-40	Dec. 63	20% slick-covered	12	choppy, well mixed	10 mi S., Panama, Bay of Panama

1 mi = 1 nautical mile = 1·85 km

TABLE 10.3

Methyl esters and alcohols obtained from chloroform extra

Alcohols or methyl esters of fatty acids (carbon atoms: double bonds)	Relative quantities* (%)									
	N-3	N-4	B-2	A-1, A-3	A-2, A-4	B-3	P-11	P-12	G-10	G-20
11:0		0·5	0·5							
12:0			1·0	0·5	1·0	1·0	0·5		1·5	
Dodecanol	12·5	54·5	2·0	2·5	10·5	4·0	2·0	1·0	3·0	
13:0								2·0		
14:0	16·5	2·5		2·0	5·5	10·0			13·5	3·5
14:1							2·0			
	4·0	1·0	1·0		2·0	1·5				
15:0						2·5			2·5	
16:0	26·0	14·5	26·5	29·5	37·0	31·5	8·0	27·0	35·0	38·5
16:1		2·5	1·0		3·0	9·5	1·0	1·0	10·5	8·0
Cetyl alcohol		0·5						1·0		3·5
17:0	6·0	1·0	1·0			4·0			2·5	8·5
	11·0		0·5				5·0			
18:0	11·0	10·0	6·0	19·0	20·5	19·0		4·5	12·0	22·5
18:1	4·5	3·5	19·0	6·5	10·5	13·5		21·0	7·0	15·5
18:2	1·0	2·5	38·5	40·0	4·0	1·0		42·5	1·5	
Stearyl alcohol										
19:0										
18:3	3·0	1·5	1·0						2·0	
									1·0	
20:0	2·0	1·5							1·5	
20:4						2·5				
	2·5	3·0	2·0		6·0		81·5		4·0	
22:0										
Average carbon chain length	16·0	14·8	17·2	17·2	16·5	16·2	19·3	17·3	17·2	16·8

he samples listed in Table 10.2 (Garrett, 1967a).

Relative quantities* (%)												
G-22	G-23	G-9	P-20	P-21	P-22	P-23	P-25	P-26	W-4	P-39	P-40	P-31, 32, 33
		0·5		0·5	1·0	2·0		1·0	1·5			
	3·0	2·0		19·5	7·0	15·0	11·0	2·5	10·5	6·0	11·0	
		1·0						0·5				
3·0	3·5		11·0	4·5	4·5	3·5	8·0	4·0	9·0	4·5	5·5	
				1·5	1·5							
			5·0						6·0			
	3·0		1·5	1·5	0·5		4·0	2·5		2·5	2·5	
				1·5								
			6·0	8·0	5·5							
	29·0	25·0	20·0	30·00	21·0		33·0	31·0	16·0	34·5	30·0	NO PEAKS OBSERVED
{48·5						{32·5						
	9·0	1·5	10·0	3·5	1·0		10·0	3·0	8·0	9·5	2·0	
											1·5	
	3·0		1·0		1·0	2·0	1·0					
										{3·0		
5·5	1·5		2·5			7·5	5·0	1·5	1·5		2·0	
										1·0		
	30·5	3·0	17·5	17·5	27·0	18·5	19·0	29·5	9·5	16·5	18·0	
{28·5												
	17·5	18·5	20·0	8·5	21·5	14·0	9·0	15·0	12·5	5·5		
		48·5	5·5	3·5	4·0	3·0		5·5	25·5	1·5	1·5	
										0·5		
										{2·0	{0·5	
					2·5	2·0		4·0				
					2·0							
										2·5	3·0	
										10·5	22·5	
14·5												
15·6	16·7	17·3	16·5	15·6	16·9	16·1	16·0	16·9	16·2	16·7	15·7	

er cent of total chromatogram area occupied by a particular peak. A few minor peaks were
ted as unimportant. Values calculated to nearest 0·5%.

in Table 10.3. In this instance the shorter chain length is the result of the considerably larger amount of dodecanol in the subsurface relative to the surface sample. A possible mechanism for such effects is displacement of soluble and/or non-polar substances by more strongly surface active molecules during compression of the surface layer by wind, surface currents, etc.

It has already been mentioned that the procedure used by Garrett (1967a) probably measures only a fraction of the total organic matter in the sample. There are a number of reasons for this. (i) It is not certain that all organic compounds are quantitatively coprecipitated with iron(III) hydroxide. (ii) The esterification procedure is not fully effective for short chain (8–10 carbon atoms) fatty acids. (iii) In a number of samples (P-31, 32, 33 in Tables 10.2 and 10.3) neither free nor combined fatty acids were detected. However, these samples, which originated from the surface and various depths in an area of high biological production, contained considerable amounts ($\sim 5 \, \text{mg} \, l^{-1}$) of chloroform-soluble organic material, very little of which was surface active. This material is probably composed of long chain hydrocarbons, and from elemental analysis a possible molecular formula of $C_{29}H_{52}$ has been suggested. (iv) Only 5% of the material in the chloroform extract of the samples of airborne organic matter was identifiable by gas chromatography. (v) Although water-insoluble compounds are likely to occur preferentially at the sea surface, it seems probable that a variable amount of water-soluble organic material will also be found there. From the data given by Williams (1967) for waters off Peru and California it is possible to calculate average figures for dissolved (DOC), particulate (POC) and total (TOC = DOC + POC) organic carbon in the sea surface layer in these areas. The values obtained are, DOC = $2 \cdot 6 \, \text{mg} \, l^{-1}$, POC = $0 \cdot 9 \, \text{mg} \, l^{-1}$, hence TOC = $3 \cdot 5 \, \text{mg} \, l^{-1}$. The figure for total organic carbon may now be compared with the range of values for chloroform-soluble organic matter given by Garrett (1967a) for his samples ($0 \cdot 2$–$1 \cdot 0 \, \text{mg} \, l^{-1}$, ignoring the samples discussed in (iii) above). Even if it is assumed that the gas chromatographic technique estimates all the components in the chloroform extract, the technique used by Garrett appears to analyse at most 25% of the total organic carbon. It seems likely that much of the unanalysed material is composed of water-soluble organic compounds. MacIntyre (1974) has argued strongly for the existence of polypeptide (and polysaccharide) components in the water at the interface.

More recently, Quinn and Wade (1972) have collected and analysed surface samples using techniques similar to those employed by Garrett (1967a). They collected microlayer material using the screen technique, and also subsurface water from a depth of 20 cm from Narragansett Bay, New York Bight and the North Atlantic. Results for fatty acids are generally similar to those found by Garrett (1967a), except that there is not such an obvious difference in carbon chain length between those in the microlayer and those in subsurface water. This may be because Garrett's subsurface samples came from depths considerably greater than the 20 cm used by Quinn and Wade. The fatty acids show a pronounced increase in concentration between the surface and subsurface, values of the enrichment factor

$$\left[\frac{\text{microlayer concentration}}{\text{subsurface concentration}} - 1 \right]$$

covering the range $-0{\cdot}1 - 7{\cdot}9$ and averaging approximately $2{\cdot}0$. Similar enrichments of organic materials in the microlayer have been reported by Goering and Menzel (1965) for DOC, by Williams (1967) and Nishizawa (1971) for DOC and POC, and by Barker and Zeitlin (1972) for TOC. Observed concentration gradients for POC are generally substantially higher than those for DOC. In the study made by Quinn and Wade (1972), hydrocarbons showed no significant enrichment in the surface layer and occurred at considerably lower concentrations than did fatty acids. They found that the average carbon chain length for hydrocarbons was close to 23. Once again these results for fatty acids and hydrocarbons probably represent only a partial analysis of the total dissolved organic material in the microlayer. Thus, at present, our knowledge of the organic chemical composition of the sea surface layer is restricted to a detailed description of only a small part of the material, the majority being almost completely uncharacterized. The portion analysed in detail is probably the most highly surface-active part, and it could be argued that as the interface is approached this analysis will describe an increasingly large fraction of the total dissolved organic material in it. However, the preliminary report of the findings of Baier and Goupil (1973) using the prism dip sampler ($\sim 10^{-2}$ μm layer collected) indicates that protein and polysaccharide are the dominant components very close to the interface. It has been reported (MacIntyre 1974) that Baier and Goupil found lipid material only in polluted waters, and they proposed that the lipids found by other workers was extracted from organisms in the samples. If this indeed proves to be the case it will be necessary to reappraise completely our concepts of the nature of the materials present in the sea surface.

10.5.2. BACTERIA AND OTHER MICROORGANISMS

The first reported investigation of bacteria in the marine surface layer is that by Sieburth (1963). Using a screen sampler having an zseptic stainless steel mesh (Sieburth, 1965) he found considerably higher numbers of bacteria in the surface layer than in the immediate sub-surface water. At 22 offshore stations between San Diego (California) and Balboa (Canal Zone) the mean bacterial population was 2398 organisms ml^{-1} in the microlayer, compared with 8 and 26 organisms ml^{-1} at 1 and 10 m respectively. Similar results have been obtained for the Caribbean and the Atlantic (Sieburth, 1971). Enhanced bacteria levels in the microlayer have also been reported by Harvey (1966) for nearshore samples collected off La Jolla, California and by Morita and Burton (1970) who studied water samples from various stations in the Arctic.

A rather different technique for sampling bacteria from the sea surface has been used by Bezdek and Carlucci (1972). This involves passing a stream of nitrogen bubbles through a natural sea water sample in the laboratory and collecting the jet drops formed as the bubbles burst at the interface. The technique is said to sample the top 10 µm of the water surface. Results obtained for two samples collected from Scripps Pier, California show concentration factors (ratio of the population density of bacteria in drops to that in the bulk liquid) of between 33 and 250. Because of the large spread of results both between water samples and type of bacteria, it is not possible to compare in detail the findings obtained by bubbling and screen techniques. The results do show that in the microlayer high concentrations of bacteria are regularly observed; enhancements of the order 10^3 are not uncommon.

Similarly, large enhancements are also found in the surface layer for microorganisms other than bacteria. Harvey (1966) used a rotating drum sampler at a station 11 km off La Jolla, California to measure the numbers of various classes of microorganisms at the surface and in sub-surface water. His results are shown in Table 10.4. Similar observations which have been

TABLE 10.4

Relative proportions (numbers litre^{-1}) of living microorganisms in rotating drum and bucket samples collected 11 km off La Jolla, California in May 1964 (Harvey, 1966).

Type of sample	Unidentified flagellates < 15 µm	Unidentified flagellates > 15 µm	Dino-flagellates	Ciliates	Diatoms
Rotating drum	4,470	30	31,270	330	930
Bucket	0	0	3,900	370	3,770
13 m	0	0	1,100	100	16,100

made by Harvey and Burzell (1972) are given in Table 10.1. It is apparent that, except for the diatoms, all the organisms investigated were markedly enriched in the surface layer. Diatoms present a somewhat conflicting picture (see Tables 10.1 and 10.4). The results are difficult to reconcile, but from Table 10.4 it will be seen that if enhancement is found, its magnitude is considerably smaller than that observed for other microorganisms. Maynard (1968) has reported high concentrations of diatoms in natural surface foams collected off the coast of Florida. In contrast, the data in Table 10.4 seems to show that diatoms can be markedly depleted in the microlayer and near surface, the highest concentrations occurring at the maximum depth sampled. As suggested by MacIntyre (1974), it is possible that the light is too intense near the surface for optimum growth of diatoms. An alternative explanation is that the surface layer contains insufficient dissolved silicon for prolific diatom growth. This view is supported by the work of Goering and Menzel (1965) who found no enrichment of dissolved silicon in the surface layer at a station in the Tropical Atlantic. This could arise if dissolved silicon, unlike dissolved and particulate organic matter, is not transported to the interface by rising bubbles etc. Lack of vertical mixing coupled with some biological utilization could thus lead to depletion of silicon near the surface.

The foregoing results imply that bacteria and many other types of micro-organism occur preferentially at the sea surface. It has already been established (Section 10.5.1) that there is an enhancement of both dissolved and particulate organic matter in the microlayer. However, it is not clear whether the high microorganism population exists because of the abundant food provided by the organic matter, or *vice versa*. Possibly both the organisms and the organic matter are enriched at the surface because both are concentrated there by similar mechanisms. Thus, laboratory studies by a number of workers (Carlucci and Williams, 1965; Blanchard and Syzdek, 1972; Wallace *et al.*, 1972) have shown that bubbling will produce very substantial enrichments of bacteria and phytoplankton at the surface. There is considerable evidence that bubbling will likewise scavenge organic matter from the bulk of the fluid and transfer it to the interface. However, despite the considerable effort which has been directed to the study of this phenomenon, there is still a great deal of controversy (see Section 12.4.1).

Sutcliffe *et al.* (1963) have shown that organic particles can be produced by bubbling air through filtered sea water (see also Riley *et al.*, 1964, 1965)*. The operative mechanism is thought to be adsorption of the dissolved organic matter from the main body of the fluid onto the rising gas bubbles. When the bubbles break at the air–water interface some of the organic matter is ejected into the atmosphere, the rest remains at the sea surface in particulate form. Menzel (1966) considered that these experiments were defective in two respects. Firstly, using an apparatus similar to that employed by Sutcliffe *et al.* (1963), he found that bubbling for 3 hours increased the dissolved organic carbon in the water samples by factors of between 2 and 4. This was shown to be the result of the introduction of organic material from the

* See also Vol. 2, Chapter 13, Section 13.4.

unpurified air stream and could be eliminated by careful purification of the air (including precombustion at 600°C). Secondly, using the purified air stream and sea water which had been prefiltered 3 times through 0·8 μm silver filters, Menzel found that in both bubbled (test) and unbubbled (control) duplicates of the samples small quantities of organic particulates were present, but that there was no significant difference between tests and controls. He concluded that bubbling is an unlikely mechanism for the production of organic particulates. Barber (1966) found that the presence of bacteria is essential for particles to be formed in the bubbling process. The lack of agreement between the above results led Batoosingh et al. (1969) to conduct a careful review of the existing data and to perform additional experiments. They found that small particles (0·22–1·2 μm) are very important because they act as nuclei for the growth of particles during the bubbling process. The discrepancies between the previous results probably arose because different modes of filtration were used to pretreat the samples. This led to a difference in the distribution of particle sizes in the water. The absence of organic particulate formation in Menzel's experiments was a consequence of the efficient filtration processes which he employed for the air stream and the sea water. Evidently, bacteria were no more effective than other particles in promoting organic particle formation.

There are probably sufficient particles in the sea to ensure that natural bubbling processes will harvest organic matter and transport it to the micro-layer. It might be noted in passing, that Wallace and Wilson (1969) have proposed the use of scavenging by rising bubbles ('foam separation') as a laboratory procedure for isolating dissolved organic molecules from sea water samples. In conclusion, it seems likely that although organisms utilize and hence recycle the organic matter close to the interface, the real cause of the enrichment of organic matter and organisms in the microlayer is the scavenging effect associated with the rising bubbles produced by wind and wave action.

10.5.3. HALOGENATED HYDROCARBONS

The very few available measurements indicate that both PCBs and halo-genated hydrocarbon pesticides, such as DDT, can occur at concentrations in the surface microlayer which are enhanced with respect to those in the immediate underlying water. Seba and Corcoran (1969) have reported measurement of various pesticides in surface slicks collected in Biscayne Bay, Florida. The surface water was collected by holding the tip of the sample bottle just below the water surface; this procedure almost certainly sampled a rather thick slice of the surface layer. Subsurface water was also collected. Table 10.5 shows the amounts of various pesticides found in the water. Subsurface samples showed no detectable amounts of pesticide (less than $0.001 \, \mu g \, l^{-1}$). Surface samples No. 1–6 were all taken within approximately two miles of the shore and No. 7 was from the main Florida Current

TABLE 10.5

Concentration of various pesticides in surface slick samples from Biscayne Bay, Florida (Seba and Corcoran, 1969).

Surface Sample No.	Pesticide residues in $\mu g \, l^{-1}$					
	p,p'– DDT	p,p'– DDE	Dieldrin	o,p– DDT	Aldrin	Total
1	0·049	0·124	0·003	0·012	0·018	0·206
2	0·090	0·266	0·004	0·013	0·015	0·388
3	3·460	9·250	0·022	0·005	0·017	12·750
4	1·460	2·880	0·035	0·081	0·025	4·480
5	0·114	0·266	0·010	0·002	0·034	0·426
6	0·140	0·178	0·021	0·014	0·005	0·358
7	0·017	0·061	0·002	0·002	0·011	0·093

about 20 miles offshore. The nearshore samples are obviously profoundly affected by run-off from the mainland, which is an agricultural area drained by canals discharging directly into the sea. Furthermore, the sampling was carried out during a period of extremely heavy rainfall when land drainage would have been maximal. Their single offshore sample showed very much lower levels of all the pesticides. The fact that pesticides were not detectable in any of the subsurface samples indicates that they occur preferentially in the surface layer.

At two stations in Narragansett Bay, Duce *et al.* (1972a) collected surface layer (using the screen technique) and subsurface (20 cm) water which was analysed for PCBs and halogenated pesticides. The latter were not found, presumably because in these samples they occurred at concentrations below the detection limit of the analytical procedure. The Narragansett Bay PCB results are shown in Table 10.6. Sample 1 was collected from a heavy frothy slick and Sample 2 from a less pronounced slick. Both showed substantial enrichment of PCB in the microlayer.

These two studies indicate that where PCBs and pesticides occur in measurable amounts they tend to be found preferentially at the sea surface. This is probably because both groups of compounds are readily soluble in surface active lipid type molecules which occur at enhanced levels at the air–sea interface (Section 10.5.1). It would seem that most of the techniques used so far have been unable to detect PCBs and halogenated pesticides in

TABLE 10.6

Polychlorinated biphenyls in two samples collected from Narragansett Bay, Rhode Island (Duce et al., 1972a).

	Surface*	Subsurface*	Enrichment factor†
Sample 1	4·2	0·15	27
Sample 2	0·45	≤0·05	≥ 8

* PCB concentrations in $\mu g \, l^{-1}$ expressed as Aroclor 1254.
† Enrichment Factor = (Surface concentration/Subsurface concentration) − 1.

solution in the marine environment, other than in coastal waters, and then only in close proximity to a source of such compounds. However, in a recent study in which the sensitivity of the analysis appears to be considerably greater than that achieved previously, Bidleman and Olney (1974) reported measurable enrichment of chlorinated hydrocarbons in the microlayer in samples collected in the Sargasso Sea between 80 and 320 km south of Bermuda. The observed enhancements are considerably greater for PCBs than for DDT residues.

10.5.4. TRACE METALS

Using a screen sampler fabricated from polythene, Piotrowicz *et al.* (1972) have collected surface layer and subsurface (20 cm) materials from Narragansett Bay, New York Bight and the North Atlantic and analysed them for a variety of trace metals. Samples were divided into up to three fractions: (i) "particulate"—retained by a 0·45 μm membrane filter; (ii) "chloroform extractable"—that portion of the filtrate extractable into chloroform; (iii) "inorganic"—that portion of the filtrate remaining after the chloroform extraction. Samples from Narragansett Bay were split into these three fractions and each was analysed for Cu, Fe, Ni and Pb. New York Bight samples were divided into particulate and chloroform extractable fractions and these were analysed for Al, Cu, Fe, Pb, Mn and V. As the North Atlantic samples were very low in particulate matter, only the chloroform extractable fraction was analysed, in this instance for Al, Cu, Fe, Mn, Pb and V. Table 10.7 gives the range of enrichment factors observed in each sampling area. The results presented in the table together with the much larger body of data given by Piotrowicz *et al.* (1972) may be summarized as follows. In Narragansett Bay no significant enhancement was found, for any of the elements studied, in the inorganic fraction (which was the major fraction for Cu, Ni and Pb). Enrichment of the particulate and chloroform fractions with Cu, Fe and Ni

TABLE 10.7

Range of trace metal enrichment factors* for various fractions of samples from Narragansett Bay, New York Bight and the North Atlantic (Piotrowicz et al., 1972).

| Element | Narragansett Bay | | | New York Bight | | N. Atlantic |
	Particulate	CHCl$_3$-extractable	Inorganic	Particulate	CHCl$_3$-extractable	CHCl$_3$-extractable
Cu	0–35	–0·1–28	–0·1–1·7	–0·4–3·1	0–4·8	–0·5–48
Fe	–0·3–28	–0·9–5·2	–0·1–1·7	–0·2–6·0	–0·2–2·6	–0·4–57
Ni	–0·6–49	–0·7–9	–0·3–0·3			
Pb	0·1–15	–0·1–7·5	–0·4–1·0	0·6–5·5	–0·3–5·6	–0·8–14
Al				–0·2–6·0	0·1–2·3	0·6–7·5
Mn				–0·5–0·5	–0·8–3·5	0–5·0
V				–0·3–4·3	4·3–4·5	0·5–3·0

* Enrichment factor = (surface concentration/subsurface concentration) − 1.

was very variable ranging from negligible to 50-fold. Pb was also variable in the same fractions, but was not so strongly fractionated. There appeared to be a correlation between enrichment in the surface layer and the presence of visible slicks on the sea surface. When the New York Bight and open ocean samples were collected there was no visible slick material at the interface. In spite of this, significant but highly variable enhancement of the concentrations of all metals analysed (except Mn) was observed for most samples from these two areas. Manganese exhibited measurable enrichment in only one sample.

Barker and Zeitlin (1972) collected surface layer samples (using the plate technique) at a number of sites off Hawaii and analysed them for Cu, Fe and Zn. Their findings are in general agreement with those of Piotrowicz *et al.* (1972). Barker and Zeitlin found that the average enrichment factors for the organic fraction were 5·7 for Cu, 6·1 for Fe and 5·1 for Zn; high enrichment factors were associated with visible slicks. The ratio of organically extractable to total metal was found to be greater in microlayer samples than in those taken 0·6 m below the surface. These data confirm that there is no enrichment of inorganic Cu and Zn in the microlayer, although an enrichment factor of about two can be calculated for iron.

Szekielda *et al.* (1972) have reported analyses for a large number of elements in surface slick and foam material collected at frontal convergences in Delaware Bay. Concentrations of numerous trace metals in chloroform extracts were given and it was observed that Cr, Cu, Fe, Pb and Zn were enriched in the surface film by up to 10^4 times relative to sea water. A chloroform extract was found to contain an amount of mercury equivalent to 0·3 ppm in the original sample which would thus appear to be enriched by three orders of magnitude relative to sea water. This result contrasts strongly with the findings of Fitzgerald and Hunt (1973) who reported concentrations of 32 ng organic $Hg\,l^{-1}$ in the water of Long Island Sound, but no significant quantity of organic mercury in the surface layer of the open ocean. In the western Sargasso Sea they could find no fractionation of inorganic Hg ($<15\,ng\,l^{-1}$) between surface and subsurface water. Enrichment of the order reported by Szekielda *et al.* (1972) seems very unlikely, but paucity of information on the sampling, separation or analytical methods employed precludes further comment.

The results surveyed in this section indicate that enrichment is generally observed only in the 'organic' fraction of the surface layer material. All the metals studied, with the exception of manganese, show an enrichment in the chloroform extract, although the magnitude is highly variable. Except for iron in the samples from Hawaii analysed by Barker and Zeitlin (1972), there are no reports of enrichment of the inorganic forms of trace metals in the surface layer. The prevalence of enhancement in the chloroform extracts suggests that the metals enriched at the interface are associated with organic groups.

A number of authors (Piotrowicz et al., 1972; Barker and Zeitlin, 1972) have discussed the possibility that enrichment factors observed with micro-layer material, collected using any of the devices which take a slice of the surface approximately 100 μm thick, may be only minimum values. According to these groups of workers, organic compounds at the sea surface really occur as either a monomolecular film ($\sim 2 \times 10^{-3}$ μm thick, Piotrowicz et al. 1972) or in a layer approximately 1 μm thick (Barker and Zeitlin, 1972). If it is assumed that the trace metals are associated with such relatively thin organic layers, then analysis of the material contained in a 100 μm slice of the surface will underestimate the concentration of the interfacial material by a factor of between 10^2 and 5×10^4. Szekielda et al. (1972) have used this argument in reverse in order to explain the very large microlayer enrichments of trace metals which they found in Delaware Bay. The implication is that their surface sampler is able to collect virtually monomolecular layers, although the brief description of the sampler which they used ('a V-shaped plastic tube that was sealed at both ends') makes objective comment difficult. The possibility that measured enrichments are low because of the limitations of sampling methods used is one which should be borne in mind. Experimental investigation of this problem is desirable.

Regardless of the degree of enrichment of the microlayer in trace metals with respect to the subsurface water it is important to establish the mechanism by which enrichment occurs. In Section 10.5.2 the importance of bubbling as a concentration mechanism for dissolved and particulate material was discussed. The technique of ion flotation as described by Sebba (1962) suggests that rising bubbles containing surface active organic molecules can scavenge trace metals from a large volume of water. This mechanism, coupled with the existence of enrichments for both trace metals and organic matter in the microlayer, provides considerable evidence for the importance of bubbling in the trace metal enrichment process. Another possibility is that the high concentration of organic molecules at the interface brings about the *in situ* solubilization of metal-rich particulate matter via the formation of soluble metal–organic complexes. The particulate material may come from the water body or, more likely, from airborne dust settling on the sea surface. If such a process occurs, it has important implications for the fate of trace metals (and other chemicals) introduced into the atmosphere by man's activities. Such solubilization of airborne particulates at the sea surface will lead to their rapid introduction into marine food chains via the high concentrations of microorganisms found in the microlayer (see Section 10.5.2).

10.5.5. NUTRIENT ELEMENTS

The concentration of the nutrient elements phosphorus, nitrogen and silicon in the sea surface layer has been studied by a small number of workers. In view of the fundamental importance of these nutrients to phytoplankton growth, and hence to the whole marine food chain, the paucity of field observations is surprising. Some attention has been devoted to the natural mechanisms by which surface enhancement of nutrients might be produced. Most of this effort has been put into laboratory studies to try to ascertain the role of bubbles in scavenging nutrients, especially phosphorus, and hence producing high levels at the interface. In this section the field data will be presented for each of the three nutrient elements in turn, along with some discussion of the possible causes of the *in situ* findings.

10.5.5.1. *Phosphorus*

The first workers to measure dissolved phosphorus levels at the sea surface did not have microlayer collectors of the type described in Section 10.4 at their disposal. Goering and Menzel (1965) used a polythene sampler to collect water from a depth of about 3 mm in the tropical Atlantic. Comparison of the water from 3 mm and 1 m depths showed that the surface water was significantly enhanced in dissolved inorganic phosphate (mean difference $0.8 \,\mu\text{mol}\,l^{-1}$). Somewhat different results were obtained by Goering and Wallen (1967) who measured phosphate in samples pumped from depths of 1, 2, 5, 10, 15, 20, 30 and 50 cm in the southeast Pacific. They found two types of phosphate distribution with depth. At some stations highest concentrations occurred at the 1 cm level, whereas other stations showed distinct maxima at lower depths (generally 10–30 cm). Perhaps the most detailed investigation to date into nutrient concentrations at the sea surface is that by Williams (1967). He used a screen microlayer collector at stations off Peru and California. Subsurface water was collected at the same time, and samples were analyzed for dissolved (and in some cases particulate) nutrients, including phosphorus. Positions of stations sampled in this study together with other data are given in Table 10.8. Results of the analyses for dissolved inorganic and dissolved organic phosphorus are shown in Table 10.9. It can be seen that for most stations enrichment in the surface is observed for both dissolved phosphate and organic phosphorus. The mean enrichment factor for phosphate is 2.5, considerably larger than the value of 0.9 found for dissolved organic phosphorus.

In conclusion, the field observations to date show that there is an enhancement of both organic and inorganic phosphorus dissolved in the sea surface layer relative to subsurface water. The laboratory studies on the

TABLE 10.8

Station locations and related data for samples analysed by Williams (1967).

Date	Station number (Peru, 1966)	Station position	Bottom depth (m)	Distance from coast (miles)*	Secchi Disc reading (m)	Sea State (Beaufort)	Surface condition
27 March	22	09°04'S 83°37'W	4300	110	13	3	No visible slicks
1 April	29	13°32'S 77°20·5'W	2000	60	10	3	No visible slicks
2 April	30	14°05'S 78°11'W	4000	106	17	3	No visible slicks
3 April	32	13°09'S 76°47'W	140	18	3	1	Heavy slick (diatom "bloom")
4 April	33	14°50'S 79°19'W	4600	187	18	3	No visible slicks
5 April	34	16°13'S 80°50'W	4000	318	17	3	No visible slicks
	(Southern California, 1965)						
11 November	1	32°37'N 117°27·3'W	640	12	22	2	Moderate slick
11 November	2	32°38'N 117°21·3'W	140	7	15	2	Moderate slick
11 November	3	32°39.2'N 117°21·3'W	15	2	5	1	Very heavy slick, (kelp beds)

* 1 sea mile = 1·85 km.

TABLE 10.9

Results for phosphate and dissolved organic phosphorus (DOP) in samples analysed by Williams (1967)

Station No.*	Depth	PO_4^{3-} μ mol l^{-1}	DOP
22	S	1·5	0·6
	SS	0·2[25]	0·3[25]
29	S	1·6	0·7
	SS	0·6[20]	0·4[20]
30	S	1·4	0·6
	SS	0·4[25]	0·4[25]
32	S	1·5	1·0
	SS	2·2[25]	0·5[25]
33	S	2·2	0·8
	SS	0·5[35]	0·5[35]
34	S	1·9	0·6
	SS	0·5[25]	0·5[25]
1	S	1·3	0·4
	SS	0·4[10]	1·0[10]
2	S	1·5	0·4
	SS	0·6[10]	0·1[10]
3	S	1·8	0·5
	SS	0·6[10]	0·2[10]

* For details see Table 10.8.

S = Surface microlayer sample (Screen sampler).

SS = Subsurface sample (depth in metres indicated by superscript).

relationship between bubbling and dissolved phosphorus in sea water have not necessarily had as their main objective the explanation of the surface layer enrichments observed in the field. Indeed, some of the laboratory studies pre-date the work at sea. Nonetheless, the laboratory work gives important clues to the mechanism by which enhanced surface phosphorus levels may occur in the marine environment and will be briefly reviewed at this point.

Baylor *et al.* (1962) conducted laboratory experiments in which a stream of small bubbles was passed through sea water. They found that for both filtered and unfiltered sea water 90% of the dissolved inorganic phosphate was removed after 24 hours bubbling. Aerosol particles collected above the bubbled samples appeared to contain only about 25% of the phosphate removed. The authors suggested that this may have been due to inefficiency of the collector used (a sloping glass plate) with respect to very small droplets, biological absorption, or to formation of particulate phosphate which was not measured in the aerosol samples. Later laboratory experiments by Sutcliffe *et al.* (1963) showed that much of the phosphate is removed from solution as particles in the spray droplets. Artificial sea water and solutions of disodium phosphate in distilled water bubbled in a similar way showed no measurable loss of phosphate. Extraction of sea water with xylene prior to bubbling decreased the rate of phosphate loss by 25%. Baylor *et al.* (1962) concluded that because removal occurs in both filtered and unfiltered sea water, biological activity cannot be responsible. The results point to the existence in sea water, but not in synthetic aqueous media, of surface active molecules able to bond with phosphate anions. These findings are consistent with the hypothesis that bubbling removes surface active material together with its associated phosphate ions from the body of the water and concentrates it in the microlayer and/or aerosol droplets when the bubbles burst.

These results are in general agreement with laboratory studies reported by MacIntyre and Winchester (1969) and by MacIntyre (1970) in which solutions containing ^{22}Na and phosphate labelled with ^{32}P were bubbled and the ratios $^{32}P:^{22}Na$ in both the aerosol and the solution were compared. The enrichment factor $\{[(PO_4^{3-}/Na^+)_{aerosol}/(PO_4^{3-}/Na^+)_{solution}] - 1\}$ in these experiments was always positive and varied considerably with drop size, the mean value lying somewhere near three. It is argued that the scavenging mechanism involves selective adsorption of phosphate as counterions on micelles of surface active molecules, the high ionic potential (large charge per surface area) of PO_4^{3-} making it particularly suitable in this context.

In general, the laboratory experiments show that phosphate levels in the bulk fluid can be readily reduced when solutions containing surface active molecules and dissolved inorganic phosphorus are bubbled. This may occur frequently in the upper few metres of the oceans, and it seems very likely that bubbling produced by the action of wind and waves plays a large part in producing the enrichments for various forms of phosphorus observed in the microlayer at the sea surface.

10.5.5.2. *Nitrogen*

The first workers to examine the concentrations of dissolved nitrogen species at the sea surface were Goering and Menzel (1965). Using a polythene sampler to collect material from the top 3 mm in the tropical Atlantic they found enhancement, relative to water at a depth of 1 m, for ammonia (significant at the 1% level) and to a smaller extent nitrite (significant at the 5% level). The samples (both surface and 1 m) were also analysed for nitrate, but at most stations levels were below the detection limit of the method employed. In contrast to the results for the tropical Atlantic, nitrite was found by Goering and Wallen (1967) to be generally uniform in the top 30 cm in the South-east Pacific.

Williams (1967) measured particulate and a number of dissolved forms of nitrogen in the microlayer samples collected off Peru and California, details of which are given in Table 10.8. Results for the various nitrogen forms analysed by Williams are shown in Table 10.10. Nitrite was enhanced in the

TABLE 10.10

Results for ammonia, nitrite, nitrate, dissolved organic nitrogen (DON), and particulate organic nitrogen (PON) in samples analysed by Williams (1967).

Station No.*	Depth	NH_3	NO_2^-	NO_3^- $\mu \, mol \, l^{-1}$	DON†	PON
22	S	8·5	0·4	6·3	27·2	15·0
	SS	1·7[25]	0·1[25]	1·2[25]	6·7[25]	1·7
29	S	11·8	0·1	3·2	31·9	11·7
	SS	0·4[20]	0·2[20]	2·5[20]	6·9[20]	0·9
30	S	8·2	1·0	1·7	23·8	9·6
	SS	0·6[25]	0·1[25]	0·1[25]	7·6[25]	1·3
32	S	13·7	0·0	1·6	39·2	60·0
	SS	2·2[25]	1·2[25]	21·5[25]	2·5[25]	4·6
33	S	10·0	0·1	2·4	31·2	14·6
	SS	0·4[35]	0·1[35]	1·3[35]	6·1[35]	3·7
34	S	14·4	0·1	0·9	37·7	20·9
	SS	0·5[25]	0·0[25]	0·1[25]	6·9[25]	0·4
1	S	n.d.	n.d.	1·2	8·7	6·2
	SS	n.d.	n.d.	0·1[10]	5·4[10]	n.d.
2	S	n.d.	n.d.	0·2	9·1	2·1
	SS	n.d.	n.d.	0·1[10]	5·1[10]	n.d.
3	S	n.d.	n.d.	0·5	9·1	13·5
	SS	n.d.	n.d.	0·1[10]	5·3[10]	n.d.

* For details see Table 10.8.
† Includes NH_3 and labile amino acids.
S = Surface microlayer sample (Screen sampler).
SS = Subsurface sample (depth in metres indicated by superscript).
n.d. = not determined.

microlayer at three stations, but the opposite occurred at the other three stations. This supports the results of Goering and Menzel (1965) and Goering and Wallen (1967) reported above, who found small or non-existent enrichments for nitrite. In marked contrast, NH_3 and PON, in the samples analysed by Williams (1967), show pronounced enhancement in the surface layer. All samples analysed showed enhancement, the mean enrichment factor for NH_3 being 17·0 and for PON 15·4. Nitrate and DON exhibit somewhat lower enhancements, with enrichment factors of 4·9 and 3·8 respectively. More recently, Nishizawa (1971) used a screen sampler to collect surface material at a station on the equator at 155° W. In comparison with water from a depth of 5 cm, the following enrichment factors were obtained: NH_3, 3·6; NO_2^-, 0·0; NO_3^-, 0·2; PON, 5·9. The lack of enhancement for nitrite confirms previous work. Enrichments for other nitrogen forms appear significant, although substantially less than those found by Williams (1967). The results presented in this section indicate that surface layer enrichment is general, but of variable magnitude, for particulate and dissolved forms of nitrogen, with the exception of nitrite; the order of enrichment is approximately $NH_3 \sim PON > NO_3^- \sim DON$.

In contrast with phosphorus, it is difficult with nitrogen species to be at all sure of the role played by bubble scavenging in producing the observed surface layer enrichments. This is partly because no laboratory experiments, analogous to those for phosphorus, have been performed. Furthermore, it can be argued that the low ionic potential of most dissolved forms of nitrogen will militate against their successful adsorption by surface active material during bubbling. Alternative mechanisms are equally problematical. Hamilton (1964) has conducted field and laboratory studies into the possibility of photochemical transformations of inorganic nitrogen forms in surface waters. He concluded that photo-oxidation of ammonia is of no significance. Photo-reduction of nitrate was observed, but the reduction appeared to go only as far as nitrite. The field results reported above show that of all the nitrogen species investigated, nitrite is the only one not exhibiting enrichment in the surface microlayer. On present evidence photochemical processes seem to be of little help in explaining the enhancements which have been found. The possibility of input of fixed nitrogen in rainwater producing the observed enrichments of nitrogen forms in the sea surface layer has been discussed by a number of workers (see for example Goering and Menzel, 1965). In contrast, other authors consider the sea to be a source of fixed nitrogen to the atmosphere (Wilson, 1959; Bloch and Luecke, 1970). Williams (1967) concluded that *in situ* bacterial nitrification is the most likely mechanism responsible for nitrate enrichment in the microlayer. In summary, the existence of enhanced levels for a number of nitrogen forms at the sea surface is well established, but our present knowledge of the mechanisms producing such enrichment leaves much to be desired.

10.5.5.3. *Silicon*

The only reported measurements of dissolved silicon at the sea surface are those of Goering and Menzel (1965) for the tropical Atlantic. Compared with water from a depth of one metre the surface material showed no significant enhancement with respect to silicon in solution.

10.5.6. FRACTIONATION

Much effort in studies of atmospheric chemistry has been devoted to an examination of the fractionation that elements may undergo in moving from the sea to the atmosphere. MacIntyre (1974) has recently given a critical account of this topic and for this reason only a general discussion will be presented here.

In accordance with the recommendations of the 'Working Symposium on Sea–Air Chemistry' (Duce *et al.*, 1972b) fractionation is defined by:

$$F_{Na}(X) = (X/Na)_{atm}/(X/Na)_{sea\ water}$$

where, $F_{Na}(X)$ is the observed fractionation for component X, normalized to sodium as reference element; $(X/Na)_{atm}$ is the weight ratio of element X to sodium in an atmospheric sample (rain, particulate material etc.); $(X/Na)_{sea\ water}$ is the weight ratio of element X to sodium in sea water. Enrichment (E) is defined:

$$E_{Na}(X) = F_{Na}(X) - 1$$

It will be seen that E is analogous to the measure already used for reporting microlayer/subsurface water enrichments.

As defined above, fractionation concerns changes in element ratios with respect to transfer from bulk sea water to the atmosphere. It has already been shown that the concentrations of many dissolved and particulate species differ considerably in the microlayer from that in the bulk fluid, and it does not seem unreasonable to presume that fractionation and processes in the microlayer are closely interrelated. For this reason, an article on the sea surface microlayer must include a review of the mechanisms which have been invoked to explain fractionation and must also assess the observed fractionations for a number of elements in the marine environment.

10.5.6.1. *Mechanisms of Fractionation*

A variety of mechanisms have been proposed to explain the fractionation of elements which occurs on crossing the air–sea interface; four of these will be discussed here.

(a) *Gibbsian (equilibrium) Adsorption.* A solute which lowers the free energy of the water surface will be preferentially adsorbed there and, conversely, one which raises it will tend to be rejected. Provided that the surface is at thermal equilibrium the equilibrium uptake is expressed by the Gibbs Adsorption Equation (Glasstone, 1968),

$$\tau = -\frac{a}{RT} \cdot \frac{d\gamma}{da}$$

where τ is the excess concentration of the solute per unit area of interface,

a is the activity of the solute, γ is the surface tension and R and T have their usual meanings.

It can be seen that if $d\gamma/da$ is negative (i.e. the surface tension decreases with increase in solute concentration) then τ will be positive and the concentration of solute will be greater in the water close to the interface than in the bulk aqueous phase. The opposite will apply if $d\gamma/da$ is positive. Two main difficulties are encountered when attempts are made to apply this to the sea surface: (i) the assumption of thermal equilibrium is unlikely to hold at natural air–water interfaces, (ii) the adsorption/rejection can be shown to be operative over depths of, at most, 10 Å (MacIntyre 1972, 1974). For fractionation at the sea surface to be explained in this way there must be some mechanism which skims off a layer of solution which is only a few molecules thick.

(b) *Thermal Diffusion.* Because of evaporative cooling, the sea surface is seldom at thermal equilibrium. Field measurements using radiometric techniques show temperature gradients in the oceanic surface layer of several degrees Celsius per cm (Ewing and McAlister, 1960; McAlister and McLeish, 1970; McAlister *et al.*, 1971). Komabayasi (1962) has suggested that thermal diffusion along such temperature gradients (Ludwig–Soret effect) might cause fractionation at the sea surface. He has argued that heavier ions will be enriched in the cooler surface layer, and that this would qualitatively explain the approximate increase in fractionation with atomic weight observed in aerosols, rainwater, snow etc. However, both Komabayasi (1962) and MacIntyre (1974) have calculated that the concentration differences to be expected with the observed temperature gradients at the sea surface are too small (approximately $10^{-8}\,mol\,\mu m^{-1}$) to account for the fractionations observed.

(c) *Selective Adsorption by Surfactants.* It has been shown previously (Section 10.5.1) that the sea surface layer is generally enriched in organic material relative to subsurface water. If such surface material adsorbs ions from the bulk fluid during passage to the interface then this constitutes a possible mechanism for fractionation. The adsorbed species may act as counter ions on colloidal micelles of surface active material, and thus, ions exhibiting high ionic potential should show the greatest enrichment (MacIntyre, 1970). Such a mechanism has already been invoked to explain phosphate enrichment in aerosols produced by bubbling solutions in the laboratory (Section 10.5.5.1). MacIntyre (1972) concluded that selective adsorption by surfactants is the most likely of those mechanism which have been proposed so far to explain surface enrichment.

All three mechanisms discussed above basically account for enhancement in the surface layer. The way in which inter-phase fractionation subsequently occurs is usually thought to be by ejection of sea surface material into the atmosphere during bubble rupture. In the fourth mechanism (described below) it is not necessary to postulate bubbling in the fractionation process.

(d) *Gaseous Transfer.* Where an element can exist in a gaseous form at the sea surface there is the possibility of a net flux of the gas across the interface. The magnitude and direction of the flux will depend on the state of saturation of the sea surface with respect to the atmospheric concentration of the gas concerned. If the flux is from the sea to the air, then atmospheric samples will be enriched in the element concerned relative to those other elements for which gaseous transport from sea to atmosphere does not occur. Gaseous transfer has been proposed as a possibility for a small number of elements (B, I, and S). These will be discussed when fractionation of these elements is described below.

10.5.6.2. *Observed Fractionations in the Marine Environment*

(a) *Group I and II Ions.* Table 10.11, which summarizes the results of recent investigations into the fractionation of alkali and alkaline earth elements, has been adapted from Table X of MacIntyre (1974) with some additional information. In summary, the field data provide little or no evidence of enhancement of Mg and Sr. The available information indicates that, relative to sodium, strontium is in fact depleted in the marine aerosol. The results for K and Ca, are somewhat equivocal, although all the studies listed in Table 10.11 show enhancement of potassium. However, it is difficult to be sure of even this apparently clear-cut result for K, because the two sets of data for which standard deviations are available indicate that the observed fractionation is within the error of the measurements.

(b) *Group VII Elements.* (i) *Fluorine.* The only workers who seem to have measured fluorine in marine aerosols are Wilkniss and Bressan (1972). They found considerable depletion of F relative to Na in aerosol samples from the major oceans, the mean fractionation being 0·38.

(ii) *Chlorine.* Modern field studies seem to indicate little or no significant fractionation of chlorine at the air–sea interface. For instance, Lazrus *et al.* (1970) found that water collected from clouds over Puerto Rico had a fractionation for Cl of 1·05 ± 0·14. The mean fractionation found by Wilkniss and Bressan (1972) for Cl in marine aerosol particles of size ⩾ 2 μm collected over the Atlantic and Pacific Oceans was 1·1. Approximately 400 rainwater samples collected by Seto *et al.* (1969) on Hawaii showed no significant difference in their Na/Cl ratios in comparison with sea water for which the ratio is 0·56. This is illustrated in Fig. 10.3 which is adapted from Seto *et al.*

TABLE 10.11

Observed in situ *fractionations* (relative to Na as reference element) for Group I and II Ions.

K^+	Mg^{2+}	Ca^{2+}	Sr^{2+}	Sample	Reference
2·42	*1·04 ± 0·01	†no real enrichment		Average of variety of air, rain and snow samples	Bloch et al. (1966)
				Orographic cloud	Lazrus et al. (1970)
1·69 ± 0·96	0·997 ± 0·195	1·72 ± 0·51	0·546 ± 0·271	Hawaiian aerosol	Barker and Zeitlin (1972)
1·35	1·06	1·42		Antarctic aerosol	Chesselet et al. (1972)
1·05 ± 0·36	0·98 ± 0·03	0·97 ± 0·17	0·88 ± 0·28	Hawaiian aerosol	Hoffman and Duce (1972)
1·4				Various, major ocean	Wilkniss and Bressan (1972)

* Relative to Cl as reference element.
† Apparent fractionation in favour of Ca was observed, but all of it was ascribed to non-marine sources.

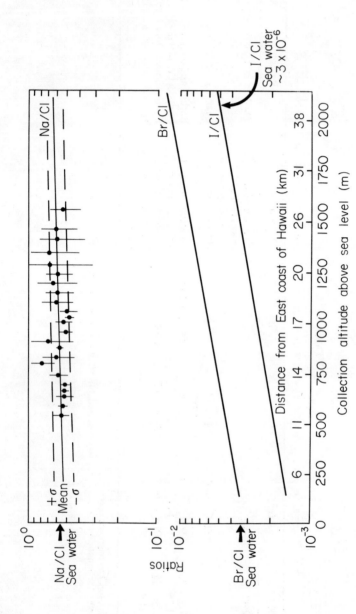

FIG. 10.3. Variation of the Na/Cl, Br/Cl and I/Cl ratios in Hawaiian rainwater samples as a function of collection altitude and distance inland (Seto *et al.*, 1969). Sea water ratios are as indicated. Reprinted with permission of the American Geophysical Union.

(1969). There seems to be very little change in the Na/Cl ratio with height (i.e. with distance inland from the east coast of the island), in marked contrast to the variations found for Br and I (see below).

MacIntyre (1974) has reviewed earlier work. Some of this indicates substantial depletion of Cl relative to Na in aerosol samples, volatilization of gaseous chlorine compounds being the mechanism usually suggested to explain such depletions. In contrast, other workers consider that Cl is enriched in atmospheric samples.

The ambivalence of the older results, coupled with the small or insignificant fractionations found in more modern studies, must lead to the conclusion that significant fractionation for Cl in crossing from the sea to the atmosphere has yet to be demonstrated.

(iii) *Bromine*. Measurements on freshly produced marine aerosol or sea spray appear to show no real fractionation effect for Br between sea water and the atmosphere. Most workers report their results for Br with Cl as the reference element, the Br/Cl ratio in sea water being 3.4×10^{-3}. Dean (1963) collected sea spray in Titahi Bay, New Zealand and found it to have a Br/Cl ratio of 3.3×10^{-3}. Duce et al. (1963) reported that samples collected in Hawaii showed Br/Cl ratios which average $(2.4 \pm 0.6) \times 10^{-3}$ for aerosols and $(5.4 \pm 1.0) \times 10^{-3}$ for rain waters. Later measurements (Duce et al., 1965 and Seto et al., 1969, illustrated in Fig. 10.3) indicated a Br/Cl ratio near sea level of about $(3.5 \pm 2.4) \times 10^{-3}$. Furthermore, analysis of sea salt particles collected from approximately 4 m above mean sea level at a coastal site in Hawaii showed no significant variation in the Br/Cl ratio with particle size, the value of the ratio being essentially the same as that for sea water (Duce and Woodcock, 1971). Thus, there is considerable evidence for non-fractionation of Br relative to Cl in transfer across the air–sea interface.

Once spray has entered the atmosphere the Br/Cl ratio may show quite considerable changes. For example, sea salt particles collected 20 m above sea level may show a depletion of Br relative to Cl (Duce and Woodcock, 1971; Moyers and Duce, 1972). At greater heights, the Br/Cl ratio appears to gradually increase with increasing altitude (Duce et al., 1965 and Seto et al., 1969; see Fig. 10.3). Bloch et al. (1966) reported a very substantial enrichment of Br relative to Cl in rainwater and snow, a result which is not supported by the data of Dean (1963). In view of the absence of any change in the Br/Cl ratio as the air–sea interface is crossed, any changes in the ratio well above sea level must be ascribed to processes in the free atmosphere and will not be pursued further here where the main emphasis is on microlayer phenomena.

(iv) *Iodine*. There is abundant and unequivocal evidence that I/Cl ratios in aerosols, rain and snow are very markedly higher than the same ratio in sea water (Komabayasi, 1962; Dean, 1963; Miyake and Tsunogai, 1963;

Duce *et al.*, 1963, 1965; Duce and Woodcock, 1971; see also Fig. 10.3). The magnitude of the enhancement, as reported by a number of workers for different types of samples, is generally in the range 10^2 to 10^3. As it has already been shown that there is little or no fractionation for Cl in crossing the air–sea interface, the observed enhancement of I/Cl in atmospheric and freshwater samples compared to sea water must be due to preferential transfer of I across the interface. Two general mechanisms have been proposed to explain this effect.

Dean (1963) observed that about 40% of the I in rainwater collected in New Zealand was in an organic form and, furthermore, the water contained much organic debris which he identified as disintegrated marine algae and plankton. From this he concluded that high I levels in freshwater samples are the result of transport of fragments of marine microorganisms by sea spray. At present, it is not possible to assess quantitatively the importance of this process on a global scale.

In the second suggested mechanism the preferential exchange is thought to be brought about by escape of iodine gas molecules from the sea surface. From laboratory studies using ^{131}I, Miyake and Tsunogai (1963) suggested that molecular iodine was produced by photochemical oxidation of iodide in the water, according to the following reaction:

$$2I^- + \tfrac{1}{2}O_2 + H_2O \xrightarrow{h\nu} I_2 + 2OH^-$$

Further laboratory simulations of conditions at the interface (Martens and Harriss, 1970; Seto and Duce, 1972) give some support to this idea, but also raise a number of problems; for example temperature appears to be at least as important as U.V. radiation in promoting emission of $^{131}I_2$; some production of $^{131}I_2$ occurs in the absence of U.V. light; natural sea water (no ^{131}I added) failed to release I_2 on irradiation. It seems likely that laboratory studies have so far failed to model conditions at natural air–water interfaces with sufficient accuracy to produce clear-cut results.

Other objections can be raised to the concept of the volatilization of I_2 from the sea surface. The I_2 concentration in natural sea water has not yet been measured and, even if I_2 was produced, it would be likely to react rapidly with the high concentration of organic material found in the microlayer (Truesdale, 1974). A more likely explanation is that preferential transfer of I relative to Cl or Br across the air-sea interface is through the mediation of a volatile organo-iodine compound such as iodomethane. This compound has been detected in sea water by Lovelock *et al.* (1973) who found a mean surface water concentration in the Atlantic Ocean of 135×10^{-12} cm^3 CH$_3$I cm^{-3}; atmospheric samples collected simultaneously contained much less

CH_3I, probably because of its photolysis under the influence of sunlight. Liss and Slater (1974) have used a two-film model of the air–sea interface, together with this observed concentration differential to calculate that the total flux of CH_3I from the oceans to the atmosphere is 2.4×10^{11} g I yr^{-1}. Miyake and Tsunogai (1963) constructed a global budget for iodine and found that in order to achieve a balance, 5×10^{11} g I yr^{-1} must cross the sea–air interface; this is about twice the flux calculated by Liss and Slater (1974). The reasonable agreement between these estimates lends weight to the idea that CH_3I rather than I_2 is the species involved in the preferential transfer of iodine from the ocean to the atmosphere.

(c) *Sulphur.* There is little or no evidence for fractionation of SO_4^{2-} between the atmosphere and the oceans (MacIntyre, 1974). However, in most budgets of sulphur at the earth's surface there is generally an imbalance which is usually made up by invoking transfer of a volatile sulphur compound (often H_2S) from the land and/or oceans to the atmosphere. As far as the air–sea interface is concerned, H_2S is an unlikely transfer species because, except in very specialized (anoxic) areas, it is not possible to detect it in surface sea water. Lovelock *et al.* (1972) have suggested that a more likely vehicle for the transport is dimethyl sulphide (DMS). They report average concentrations in surface water from the Atlantic of 1.2×10^{-11} g cm^{-3}. DMS is not detectable in marine air, presumably because of its rapid breakdown in the atmosphere. Liss and Slater (1974) have calculated the sea-air flux of DMS (3.7×10^{12} g S yr^{-1}) implied by the above concentration difference, which is only about 4 % of the amount needed to balance the sulphur budget. Recently, Lovelock (1974) has reported that sea water in equilibrium with the alga *Polysiphonia fastigiata* contains 10^5 times more DMS than does water from the open Atlantic. At present it is impossible to assess the quantitative importance of such DMS 'hot spots' to the global sulphur budget. Carbon disulphide has also been detected in sea water (Lovelock, 1974), but its low concentration (5.2×10^{-13} g CS$_2$ cm^{-3} in the open Atlantic) means that it cannot be an important component in the sulphur cycle.

Present evidence thus indicates that processes at the sea surface are quantitatively unable to provide sufficient volatile sulphur compounds to balance the global budget for this element. The presence of methylated compounds such as DMS and iodomethane in sea water is surprising, and their existence is often ascribed to biological activity. It is possible to envisage alternative mechanisms involving interaction between dissolved organic molecules and iodine or sulphur containing species under the action of U.V. radiation. Such processes, as well as biological methylation, are likely to occur preferentially in the extreme surface layers of the ocean.

(d) *Boron.* There is considerable evidence that the B/Cl ratio in rain water and other freshwaters is substantially higher than that in sea water (Gast and Thompson, 1959; Brujewicz and Korzh, 1971). From laboratory experiments Gast and Thompson (1959) concluded that boric acid can evaporate from sea water, and attributed the excess boron in atmospheric samples to this source. In contrast, Nishimura and Tanaka (1972), using a very similar laboratory set-up, found that boric acid in the atmosphere is supersaturated with respect to sea water and so, far from being a source, the ocean should be regarded as a sink for boron.

In trying to resolve these contradictory results, MacIntyre (1974) has pointed out that in the experiments of Gast and Thompson the air temperature (25°C) was 7°C lower than that of the water, but that Nishimura and Tanaka used a system at thermal equilibrium, both air and water being at 25°C. For this reason the two sets of laboratory experiments are not comparable and this highlights the difficulty of producing from laboratory studies results which can be meaningfully interpreted in terms of *in situ* air–sea processes. If the sea turns out not to be the source of the excess boron in the atmosphere, then input from volcanic activity is another possibility (Horn and Adams, 1966).

10.5.6.3. *Difficulties of field and laboratory studies on fractionation*

(a) *Field studies.* From the data presented in Section 10.5.6.2 it would seem that it is hard to find clear-cut evidence for air-sea fractionation, especially for the major dissolved constituents of sea water. One is forced to this conclusion in spite of many reports in the literature of substantial fractionation. Many of these studies suffer from deficiencies which make it hard to accept their conclusions, and highlight the great difficulties involved in studies of fractionation. Factors which must be taken into consideration include natural geographical variability, the need to collect a statistically significant number of samples, attention to problems of analytical and sampling errors and contamination and the necessity for collecting samples from regions free from terrigenous dust and man-made pollution.

(b) *Laboratory studies.* Some of the difficulties of simulating fractionation at the sea-air interface in the laboratory are apparent from the previous examples. The importance of adsorption of species onto surface active material has already been stressed. MacIntyre (1975) has detailed the difficulties of obtaining laboratory solutions free from, or containing reproducible amounts of, surface active materials, and the lack of agreement between different workers apparently conducting similar experiments may in large part be attributable to this cause.

10.5.7. GAS EXCHANGE

The transfer of gas across a clean air–water interface is subject to control (resistance) by processes in the liquid and/or gas phases. For most gases the resistance of one phase tends to predominate; this means that transfer may generally be treated as being under the control of processes within the phase which is rate determining for the gas concerned. Gases which are sparingly soluble in water (e.g. N_2, O_2, inert gases, etc.) are invariably subject to liquid phase control. In contrast, the exchange of very soluble gases (e.g. H_2O, SO_2, NH_3), which also often exhibit marked chemical reactivity in the water, is controlled by processes in the gas phase (Liss, 1971, 1973). The presence of impurities/contaminants (surface active molecules, oil films, etc.) may affect gas exchange across the interface. In order to understand the relevant processes it will be helpful to consider gas and liquid phase control separately.

10.5.7.1. *Gases whose exchange is subject to gas phase resistance*

The most important gas whose air–sea exchange is subject to gas phase control is water vapour. In considering the evaporation of water molecules from a clean surface it is usually argued that the air immediately in contact with the aqueous surface is saturated with water vapour. Exchange between this saturated air and the free atmosphere is then by diffusional processes in the gas phase. Hence, for a contaminant free interface, liquid phase resistance may be ignored for water vapour (Whitney and Vivian, 1949).

The presence of a continuous film of surface active material will add a liquid phase component to the existing gas phase resistance, and very substantial decreases in evaporation rates would be expected in these circumstances. Such reduction in evaporation has been substantiated in various laboratory studies (e.g. Jarvis *et al.*, 1962; Frenkiel, 1965; La Mer and Healy, 1965; Garrett, 1971). As pointed out by Jarvis *et al.* (1962), the molecules of the monolayer must be linear and capable of close packing in a compressed film in order to be effective. The presence of double bonds, ionised polar groups, or branching of the hydrocarbon chain greatly reduces the ability of the film to reduce evaporation. The presence of 1% of non-linear impurities in the monolayer can lower its ability to reduce evaporation by 90–99% (La Mer, 1962). The mixed nature of the organic compounds found at the sea surface indicates that the occurrence of areas where significant evaporation retardation can take place is likely to be rare. In this context it is important to note that the chloroform extracts analysed by Garrett (1967a) contained a number of unsaturated compounds, even for samples taken from slicks.

Similarly, Garrett (1972) has argued that monolayers on the sea surface arising from anthropogenic input of hydrocarbons will not significantly affect gas exchange rates. The mixed chemical nature of petroleum products will ensure that they cannot form a compressed monomolecular film. In contrast, thick ($>5\,\mu$m) multi-layer films of oil, etc. can considerably decrease evaporation rates. Such films generally exist only in the vicinity of major oils spills (and also possibly of shipping lanes), and will rapidly be thinned and dispersed under the action of wind and waves.

10.5.7.2. *Gases whose exchange is subject to liquid phase resistance*

Here the resistance to gas transfer is normally envisaged as being caused by diffusional processes in the water close to the interface (Kanwisher, 1963). The effect of aqueous phase contaminants will be to add an additional diffusion resistance to the one naturally occurring at the surface of the liquid. Moderate decreases in the rate of gas exchange might be expected provided that the impurity forms a continuous layer at the interface. Such effects have been demonstrated for a number of gases in laboratory experiments using a variety of surface active substances, e.g. long chain fatty acids and alcohols and detergents (Downing *et al.*, 1957; Hawke and Alexander, 1962; Mancy and Okun, 1965; Boyd and Marchello, 1966; Sada and Himmelblau, 1967). A number of workers have noted that an appreciable decrease in the gas exchange rate is observed only under regimes of moderate stirring of the liquid phase (Mancy and Okun, 1965; Downing, 1962). Under calm conditions the natural resistance of the aqueous phase masks any effect of the contaminant film. With very turbulent mixing the surface film is broken up and transfer is dominated by the vigorous stirring of the liquid phase. Hawke and Alexander (1962) and Garrett (1972) have pointed out that the film must be in a close packed condition in order to offer any appreciable resistance to passage of gas. As already discussed, monolayers (whether natural or from oil spills) at the sea surface are unlikely to fulfil this condition, and thus under normal marine conditions gas exchange should not be inhibited. Thick oil layers will increase the liquid phase resistance and so may decrease the rate of gas exchange (Downing and Truesdale, 1955). However, Garrett (1972) has pointed out that, except for really large oil spills or possibly in the vicinity of shipping lanes, such man-made multi-layered films are unlikely to have substantial lifetimes in the marine environment.

10.5.8. SURFACE TENSION

The surface tension of clean sea water is approximately $75 \times 10^{-3}\,\mathrm{N\,m^{-1}}$ at a salinity of 35‰ and temperature $10°C$.* This value is little affected by change in salinity, but temperature has a much more pronounced effect (Cox, 1965).

* See Appendix Table 24.

The presence of surface films will lower the surface tension, the magnitude of the decrease being equal to the pressure exerted by the film. The most widely used method for measuring such film pressures in the field is the spreading drop technique devised by Adam (1937). It has been used on fresh water bodies by Goldacre (1949), who found measurable film pressures for all waters examined, with a few values in excess of $30 \times 10^{-3}\,\mathrm{N\,m^{-1}}$.

Measurements in the marine environment indicate that films at the sea surface exert considerably smaller pressures. Lumby and Folkard (1956) measured surface tension depressions in Monaco Bay which were consistently between 2 and $3 \times 10^{-3}\,\mathrm{N\,m^{-1}}$ in ruffled water, but in slick covered water values sometimes exceeded $22 \times 10^{-3}\,\mathrm{N\,m^{-1}}$. These very high film pressures are thought to result from input of land-derived material rather than from natural marine surface film. For highly productive areas of the Sargasso Sea, Sieburth and Conover (1965) found film pressures from less than 1 to just over $3 \times 10^{-3}\,\mathrm{N\,m^{-1}}$; values in rippled water were somewhat less than those in slick covered regions. From measurements on a slick in Chesapeake Bay, Garrett (1965) found film pressures in the range $3\text{–}12 \times 10^{-3}\,\mathrm{N\,m^{-1}}$, with an average value within the slick of $5 \times 10^{-3}\,\mathrm{N\,m^{-1}}$. Sturdy and Fischer (1966) measured surface tensions in and around kelp beds just off the Californian coast. Outside the area of the kelp beds surface tensions were near 70×10^{-3} $\mathrm{N\,m^{-1}}$ but decreased to between 50 and $60 \times 10^{-3}\,\mathrm{N\,m^{-1}}$ directly over the kelp beds. The most comprehensive measurements of film pressures using the spreading drop technique are those reported by Barger *et al.* (1974). They made over 170 measurements of surface tension over a period of 9 days at a station 0·7 miles offshore from Mission Beach, San Diego. More than 150 of the observations were in clean water, and nearly all of these indicated film pressures of $1 \times 10^{-3}\,\mathrm{N\,m^{-1}}$ or less. Where visible slicks were present film pressures were higher, the greatest value recorded being $23 \times 10^{-3}\,\mathrm{N\,m^{-1}}$.

All the film pressure observations reported in this section have been made in coastal waters or other areas of high biological production. Even under these conditions film pressures in excess of $1 \times 10^{-3}\,\mathrm{N\,m^{-1}}$ are uncommon, except in the presence of visible slicks. Although there are no measurements from typical open ocean situations, it would appear that film pressures in such areas are likely to be considerably less than $1 \times 10^{-3}\,\mathrm{N\,m^{-1}}$. In order to obtain data to confirm this conclusion it will be necessary to devise a method of measuring *in situ* film pressures (under all sea states, if possible) with a sensitivity greater than that obtainable with the spreading drop technique ($1 \times 10^{-3}\,\mathrm{N\,m^{-1}}$).

10.5.9. FORCE–AREA BEHAVIOUR

Jarvis *et al.* (1967) have determined force-area curves for the chloroform extracts of some of the surface and near-surface sea water samples listed in Table 10.2. Most of the samples were found to give curves of similar shape. The reproducibility was good providing that measurements were made during the initial compression stage of duplicate runs. This was necessary because the expansion curves did not retrace the compression curves; continued compression and expansion cycles led to displacement of the curves to smaller areas. It is apparent that compression of the water surface brings about downward displacement of less surface active material, implying that many samples collected from the surface contain appreciable amounts of material which is only weakly surface active. Typical force-area curves from such experiments are shown in Fig. 10.4. On compression the surface sample (N-3 in Tables 10.2 and 10.3) behaves in a similar way to the highly surface active oleic acid. In contrast, the subsurface sample (N-4 in Tables 10.2 and 10.3) gave a film several molecules thick, which tended to dissolve and was hard to reproduce. At a film pressure of $19 \times 10^{-3} \, \text{N m}^{-1}$ the area per unit weight for films N-3 and N-4 were 74% and 26% respectively of that for the oleic acid film. The lack of surface activity shown by sample N-4 is probably due to the large amount of dodecanol which it contains; this is more soluble than are the longer chain fatty acids of high surface activity which constitute the bulk of sample N-3. These experiments were all carried out using the chloroform soluble fraction of the organic material.

Force-area plots for untreated surface layer samples have been obtained by Barger *et al.* (1974). Results of the experiments, which were performed on microlayer material collected in the Pacific just offshore from San Diego, are shown in Fig. 10.5. Samples were taken from both rippled and slicked water surfaces, and the similarity of the force-area curves indicates that the chemical nature of the surface material did not change appreciably between these two sea states. The shape of the curves is very different from that exhibited by the chloroform extracts shown in Fig. 10.4. With the untreated microlayer material considerable compression of the film must take place before there is any dramatic increase in film pressure. In contrast, the chloroform extracts show sharp increases in film pressure as the surface material is compressed. This may be further evidence that chloroform extraction preferentially selects the more surface active material and so gives a poor indication of the full range of substances in the microlayer. The low slope of the force-area curve of the untreated samples at high film areas probably represents a squeezing out of the less surface active molecules as the film is compressed. As this takes place the concentration of surface molecules increases until they form a close packed layer. Once in a close packed condition, further decrease in area leads to a very rapid increase in film pressure. This change in behaviour occurs when the film pressure is approximately $1 \times 10^{-3} \, \text{N m}^{-1}$. As will

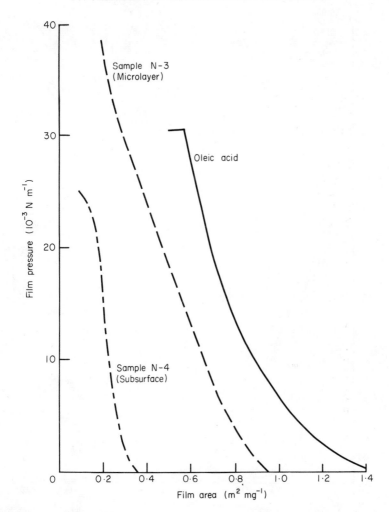

Fig. 10.4. Force-area curves for chloroform extracts of samples N-3 and N-4 (for details see Tables 10.2 and 10.3) from Jarvis *et al.* (1967) and for oleic acid, from Garrett (1970).

be shown in Section 10.5.10 this is near the optimum film pressure for the damping of capillary waves. The conclusion of Section 10.5.8, that open ocean film pressures are probably less than $1 \times 10^{-3}\,\mathrm{N\,m^{-1}}$, lends weight to the contention, discussed in Section 10.5.7 on the effect of microlayers on gas exchange, that close packed films are unlikely to be common at the sea surface.

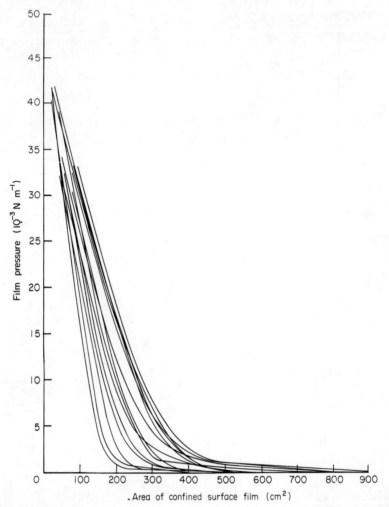

FIG. 10.5. Force-area curves for untreated microlayer samples collected 0·7 miles offshore from Mission beach, San Diego, California (Barger *et al.*, 1974). Reprinted with permission of Pergamon Press Ltd.

10.5.10. CAPILLARY WAVE DAMPING AND SLICK FORMATION

In order for insoluble monomolecular films to damp out capillary waves (wavelength less than 1·7 cm) at the sea surface a film pressure in the region of $1 \times 10^{-3}\,\mathrm{N\,m^{-1}}$ or more is required (Garrett and Bultman, 1963; Garrett, 1967b). Fig. 10.6 shows how the damping coefficient varies with film pressure. There is little damping at low film pressures, and the curve passes through

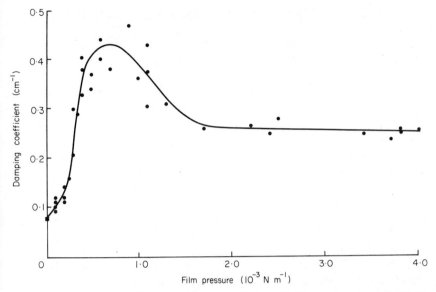

FIG. 10.6. Variation of capillary wave damping coefficient with film pressure, Garrett (1967b). The damping coefficient (k) is defined by the relationship $a = a_0 e^{-kx}$, where a_0 is the wave amplitude at its source and a is the wave amplitude at distance x from the source.

a maximum when the film pressure is about $1 \times 10^{-3} \, N \, m^{-1}$. As discussed in Section 10.5.9 this is the pressure at which the film seems to achieve a close packed condition. Films having sufficient pressure to damp capillary waves give rise to a change in light reflection properties of the water, such areas being known as "slicks".

From laboratory studies on the rate of formation of films on sea water samples collected from the Bay of Panama, Jarvis (1967) concluded that most samples contained insufficient surface active material to form a close packed film, except when the molecules are compressed under the action of light winds or convergences of surface water. Plainly, slick formation is dependent on the availability of enough surface active material and a water circulation which will compress the film sufficiently to damp out capillary waves without the water being so turbulent that the slick is broken up. Slicks are thus unlikely in the open ocean both because of the paucity of surface active molecules and because conditions are generally too rough. They may occur in shipping lanes or regions where oil has been spilled; such slicks are caused by the presence of petroleum hydrocarbons in the form of multi-molecular layers. Natural slicks are most commonly found in coastal or other waters where there is a high rate of biological production of surface active material and only light winds (Ewing, 1950; Garrett, 1970).

I

10.6. THICKNESS OF THE SEA SURFACE LAYER—II

Most of the results presented in Section 10.5 have been obtained with samples collected from the sea surface using collectors which harvest approximately the top $100 \pm 50\,\mu m$ of the water. Since the diameter of a water molecule is about $3 \times 10^{-4}\,\mu m$ and the length of a typical surface active molecule oriented normally to the surface is ca. $3 \times 10^{-3}\,\mu m$, $100\,\mu m$ appears to be a rather thick slice of the surface. For this reason, a number of workers (Piotrowicz et al., 1972; Barker and Zeitlin, 1972; Szekielda et al., 1972) have argued that measured microlayer enhancements for trace metals based on samples from the top $100\,\mu m$ give estimates which are low, because of dilution of "true" surface material by subsurface water (Section 10.5.4). As a result of the enormous practical difficulties of sampling in the micrometre, let alone the nanometre, range such suggestions still await experimental test.

As an alternative to considering the surface to be a monomolecular layer, it is sometimes helpful to study the depth to which the water molecules are perturbed from the normal distribution which they have in the bulk of the liquid. Horne (1969) has reviewed theories in this field; these fall into two categories. In the first, the preferential orientation of water molecules is thought to fall off exponentially with distance from the interface, and the surface zone can be considered to be effectively 2–3 nm in thickness (Fletcher, 1962). In the second (that preferred by Horne), the surface forces in liquids are assumed to be very much greater than those for individual molecules so that the perturbation extends several hundred nanometres into the liquid.

The discussion so far has been based on ideas concerning the nature and structure of molecules close to the interface. Estimates of the surface layer thickness may be inferred from the results of experimental studies at air–water interfaces aimed at measuring phenomena such as optical properties of the surface, thermal layering and air–sea gas exchange rates.

Kinosita and Yokota (1965) measured the elliptical polarization produced in plane polarized light after it has been incident on a water surface at the Brewsterian angle. Such ellipticity is thought to be a consequence of a transition layer at the water surface, the degree of elliptical polarization being related to the layer thickness. The measurements were performed over a range of temperatures, and at 20°C the layer thickness was found to be in the range 0·5–1 nm.

It was pointed out in Section 10.5.6.1b that considerable temperature gradients exist at the sea surface. Under calm conditions such cooled surface layers appear to have a thickness of about $100\,\mu m$, with a considerable thinning as the degree of turbulence increases (Ewing and McAlister, 1960). The thickness of the interfacial layering implied by such thermal data is substantially greater than that deduced from any of the other approaches described hitherto. Surface ordering of this magnitude is also implied if gas exchange rates at the air–sea interface are interpreted in terms of the film model (Whitman, 1923). In this approach, transfer of a sparingly soluble gas across the interface is controlled by molecular diffusion through a laminar water layer at the surface (Kanwisher, 1963; Hoover and Berkshire, 1969). By dividing the appropriate coefficient of molecular diffusion by the observed gas exchange constant (flux of gas per unit concentration gradient) the thickness of the film may be calculated. It follows that, as the rate of gas exchange increases, so the apparent film thickness decreases, and vice versa. Gas exchange rates measured at sea have been summarized by Liss (1973) and Broecker and Peng (1974) and a mean value for the exchange constant of 20 cm hr^{-1} seems reasonable. Assuming a value for the coefficient of molecular diffusion in the laminar layer of $1 \cdot 5 \times 10^{-5} \text{ cm}^2 \text{ sec}^{-1}$, this exchange constant corresponds to a film thickness of approximately 30 μm.

Holley (1973) has compared layer thickness from thermal and gas exchange measurements in laboratory experiments in which a thermistor was used to define the near surface temperature profile and the rate of oxygen exchange was monitored. He found that, under identical conditions of turbulence in the liquid phase, the layer thickness from the oxygen measurements was twenty times less than that deduced from the temperature profiles, this difference being attributed to change in the Schmidt or Prandtl Number (viscosity/diffusivity) for mass and heat.

It is perhaps hardly surprising that these different approaches produce thicknesses for the "surface" water layer which cover six orders of magnitude. However, when discussing microlayer phenomena, it is important to know that the layer thickness is very dependent on the parameter used to define or measure it. To date, most microlayer measurements have been on layers whose thicknesses are about 100 μm. An important objective for future research must be the development of techniques to sample thinner slices of the surface water in order to determine the extent to which they differ in chemical, physical and biological properties from the comparatively thick layers so far investigated.

ACKNOWLEDGEMENTS

The author is greatly indebted to Dr. Ferren MacIntyre for a pre-publication copy of his article in "The Sea" Vol. V and to Dr. Fred Culkin and Dr. Mike Rhead for helpful discussions.

REFERENCES

Adam, N. K. (1937). *Proc. R. Soc.* **B122**, 134–139.

Baier, R. E. (1970). *Proc. 13th Conf. Great Lakes Res. Internat. Assoc. Great Lakes Res.* 114–127.

Baier, R. E. (1972). *J. geophys. Res.* **77**, 5062–5075.

Baier, R. E. and Goupil, D. W. (1973). Abstract A-1, Int. Symp. on Chemistry of Sea/Air Particulate Exchange Processes, Nice.

Barber, R. T. (1966). *Nature, Lond.* **211**, 257–258.

Barger, W. R. and Garrett, W. D. (1970). *J. geophys. Res.* **75**, 4561–4566.

Barger, W. R., Daniel, W. H. and Garrett, W. D. (1974). *Deep Sea Res.* **21**, 83–89.

Barker, D. R. and Zeitlin, H. (1972). *J. geophys. Res.* **77**, 5076–5086.

Batoosingh, E., Riley, G. A. and Keshwar, B. (1969). *Deep Sea Res.* **16**, 213–219.

Baylor, E. R., Sutcliffe, W. H. and Hirschfeld, D. S. (1962). *Deep Sea Res.* **9**, 120–124.

Bezdek, H. F. and Carlucci, A. F. (1972). *Limnol. Oceanogr.* **17**, 566–569.

Bidleman, T. F. and Olney, C. E. (1974). *Science, N.Y.* **183**, 516–518.

Blanchard, D. C. (1964). *Science, N.Y.* **146**, 396–397.

Blanchard, D. C. and Syzdek, L. D. (1972). *J. geophys, Res.* **77**, 5087–5099.

Bloch, M. R. and Luecke, W. (1970). *Israel J. Earth Sci.* **19**, 41–49.

Bloch, M. R., Kaplan, D., Kertes, V. and Schnerb, J. (1966). *Nature, Lond.* **209**, 802–803.

Boyd, D. P. and Marchello, J. M. (1966). *Chem. Engng Sci.* **21**, 769–776.

Broecker, W. S. and Peng, T. H. (1974). *Tellus,* **26**, 21–35.

Brujewicz, S. W. and Korzh, V. D. (1971). *Okeanologiia* **11**, 414–422.

Carlucci, A. F. and Williams, P. M. (1965). *J. Cons. perm. int. Explor. Mer* **30**, 28–33.

Chesselet, R., Morelli, J. and Buat-Menard, P. (1972). *J. geophys. Res.* **77**, 5116–5131.

Cox, R. A. (1965). *In* "Chemical Oceanography" (J. P. Riley and G. Skirrow, eds.) Vol 1, pp. 73–120. Academic Press, New York and London.

Dean, G. A. (1963). *N.Z. Jl Sci.* **6**, 208–214.

Downing, A. L. (1962). *In* "River Pollution" (L. Klein, ed.) Vol 2, pp. 224–253. Butterworths, London.

Downing, A. L. and Truesdale, G. A. (1955). *J. appl. Chem., Lond.* **5**, 570–581.

Downing, A. L., Melbourne, K. V. and Bruce, A. M. (1957). *J. appl. Chem., Lond.* **7**, 590–596.

Duce, R. A. and Woodcock, A. H. (1971). *Tellus* **23**, 427–435.

Duce, R. A., Wasson, J. T., Winchester, J. W. and Burns, F., (1963). *J. geophys. Res.* **68**, 3943–3947.

Duce, R. A., Winchester, J. W. and Van Nahl, T. W. (1965). *J. Geophys. Res.* **70**, 1775–1799.

Duce, R. A., Quinn, J. G., Olney, C. E., Piotrowicz, S. R., Ray, B. J. and Wade, T. L. (1972a). *Science, N.Y.* **176**, 161–163.

Duce, R. A., Stumm, W., and Prospero, J. M. (1972b). *J. geophys. Res.* **77**, 5059–5061.

Ewing, G. (1950). *J. mar. Res.* **9**, 161–187.

Ewing, G. and McAlister, E. D. (1960). *Science, N.Y.* **131**, 1374–1376.

Fasching, J. L., Courant, R. A. and Duce, R. A. (1973). Abstract A-4, Int. Symp. on Chemistry of Sea/Air Particulate Exchange Processes, Nice.

Fitzgerald, W. F. and Hunt, C. D. (1973). Abstract A-5, Int. Symp. on Chemistry of Sea/Air Particulate Exchange Processes, Nice.

Fletcher, N. H. (1962). *Phil. Mag.* **7**, 255–269.

Frenkiel, J. (1965). "Evaporation Reduction". UNESCO, Paris.
Garrett, W. D. (1965). *Limnol. Oceanogr.* **10**, 602–605.
Garrett, W. D. (1967a). *Deep Sea Res.* **14**, 221–227.
Garrett, W. D. (1967b). *J. mar. Res.* **25**, 279–291.
Garrett, W. D. (1970). *In* "Organic Matter in Natural Waters" (D. W. Hood, ed.) pp. 469–477. Institute of Marine Science, University of Alaska.
Garrett, W. D. (1971). *J. Atm. Sci.* **28**, 816–819.
Garrett, W. D. (1972). *In* "The Changing Chemistry of the Oceans" (D. Dyrssen and D. Jagner, eds.) pp. 75–91. Wiley & Sons, New York.
Garrett, W. D. and Bultman, J. D. (1963). *J. Colloid Sci.* **18**, 798–801.
Gast, J. A. and Thompson, T. G. (1959). *Tellus* **11**, 344–347.
Glasstone, S. (1968). "Textbook of Physical Chemistry". Macmillan, London.
Goering, J. J. and Menzel, D. W. (1965). *Deep Sea Res.* **12**, 839–843.
Goering, J. J. and Wallen, D. (1967). *Deep Sea Res.* **14** 29–33.
Goldacre, R. J. (1949). *J. Anim. Ecol.* **18**, 36–39.
Hamilton, R. D. (1964). *Limnol. Oceanogr.* **9**, 107–111.
Harvey, G. W. (1966). *Limnol. Oceanogr.* **11**, 608–613.
Harvey, G. W. and Burzell, L. A. (1972). *Limnol. Oceanogr.* **17**, 156–157.
Hawke, J. G. and Alexander, A. E. (1962). *In* "Retardation of Evaporation by Monolayers: Transport Processes" (V. K. La Mer, ed.) pp. 67–73. Academic Press, New York and London.
Hoffman, G. L. and Duce, R. A. (1972). *J. geophys. Res.* **77**, 5161–5169.
Holley, E. R. (1973). *Wat. Res.* **7**, 559–573.
Hoover, T. E. and Berkshire, D. C. (1969). *J. geophys. Res.* **74**, 456–464.
Horn, M. K. and Adams, J. A. S. (1966). *Geochim. cosmochim. Acta* **30**, 279–297.
Horne, R. A. (1969). "Marine Chemistry". Wiley & Sons, New York.
Jarvis, N. L. (1967). *Limnol. Oceanogr.* **12**, 213–221.
Jarvis, N. L., Timmons, C. O. and Zisman, W. A. (1962). *In* "Retardation of Evaporation by Monolayers: Transport Processes" (V. K. La Mer, ed.) pp. 41–58. Academic Press, New York and London.
Jarvis, N. L., Garrett, W. D., Scheiman, M. A. and Timmons, C. O. (1967). *Limnol. Oceanogr.* **12**, 88–96.
Kanwisher, J. (1963). *Deep Sea Res.* **10**, 195–207.
Kinosita, K. and Yokota, H. (1965). *J. phys. Soc. Japan* **20**, 1086
Komabayasi, M. (1962). *J. met. Soc. Japan* **40**, 25–38.
La Mer, V. K. (1962). *In* "Retardation of Evaporation by Monolayers: Transport Processes" (V. K. La Mer, ed.) pp vii–xvii. Academic Press, New York and London.
La Mer, V. K. and Healy, T. W. (1965). *Science, N.Y.* **148**, 36–42.
Larsson, K., Odham, G. and Södergren, A. (1974). *Marine Chemistry* **2**, 49–57.
Lazrus, A. L., Baynton, H. W. and Lodge, J. P. (1970). *Tellus* **22**, 107–113.
Liss, P. S. (1971). *Nature, Lond.* **233**, 327–329.
Liss, P. S. (1973). *Deep Sea Res.* **20**, 221–238.
Liss, P. S. and Slater, P. G. (1974). *Nature, Lond.* **247**, 181–184.
Lovelock, J. E. (1974). *Nature, Lond.* **248**, 625–626.
Lovelock, J. E., Maggs, R. J. and Rasmussen, R. A. (1972). *Nature, Lond.* **237**, 452–453.
Lovelock, J. E., Maggs, R. J. and Wade, R. J. (1973). *Nature, Lond.* **241**, 194–196.
Lumby, J. R. and Folkard, A. R. (1956). *Bull. Inst. oceanogr. Monaco* No. 1080, 1–19.
MacIntyre, F. (1968). *J. phys. Chem., Ithaca* **72** 589–592.
MacIntyre, F. (1970). *Tellus* **22**, 451–462.

MacIntyre, F. (1972). *J. geophys. Res.* **77**, 5211–5228.

MacIntyre, F. (1974). *In* "The Sea" (E. D. Goldberg, ed.) Vol. 5, pp 245–299 Wiley & Sons, New York.

MacIntyre, F. (1975). *J. Rech. Atmos.* In the press.

MacIntyre, F. and Winchester, J. W. (1969). *J. phys. Chem., Ithaca* **73**, 2163–2169.

McAlister, E. D. and McLeish, W. (1970). *Appl. Opt.* **9**, 2697–2705.

McAlister, E. D., McLeish, W. and Corduan, E. A. (1971). *J. geophys. Res.* **76**, 4172–4180.

Mancy, K. H. and Okun, D. A. (1965). *J. Wat. Pollut. Control Fed.* **37**, 212–227.

Martens, C. S. and Harriss, R. C. (1970). *In* "Precipitation Scavenging (1970)" (R. J. Engelmann and W. G. N. Slinn, eds.) pp. 319–324. U.S. Atomic Energy Commission.

Maynard, N. G. (1968). *Z. allg. Mikrobiol.* **8**, 119–126.

Menzel, D. W. (1966). *Deep Sea Res.* **13**, 963–966.

Miyake, Y. and Tsunogai, S. (1963). *J. geophys. Res.* **68**, 3989–3993.

Morita, R. Y. and Burton, S. H. (1970). *In* "Organic Matter in Natural Waters" (D. W. Hood, ed.) pp. 275–285. Institute of Marine Science, University of Alaska.

Moyers, J. L. and Duce, R. A. (1972). *J. geophys. Res.* **77**, 5330–5338.

Nishimura, M. and Tanaka, K. (1972). *J. geophys. Res.* **77**, 5239–5242.

Nishizawa, S. (1971). *Bull. plankt. Soc. Japan* **18**, 42–44.

Parker, B. and Barsom, G. (1970). *Bioscience* **20**, 87–93.

Pilpel, N. (1968). *Endeavour* **27**, 11–13.

Piotrowicz, S. R., Ray, B. J., Hoffman, G. L. and Duce, R. A. (1972). *J. geophys. Res.* **77**, 5243–5254.

Quinn, J. G. and Wade, T. L. (1972). *In* "Baseline Studies of Pollutants in the Marine Environment" (E. D. Goldberg, ed.) pp. 633–663. National Science Foundation.

Riley, G. A., Wangersky, P. J. and Van Hemert, D. (1964). *Limnol. Oceanogr.* **9**, 546–550.

Riley, G. A., Van Hemert, D. and Wangersky, P. J. (1965). *Limnol. Oceanogr.* **10**, 354–363.

Sada, E. and Himmelblau, D. M. (1967). *A.I.Ch.E. Jl* **13**, 860–865.

Seba, D. B. and Corcoran, E. F. (1969). *Pestic. Monit. Jl* **3**, 190–193.

Sebba, F. (1962). "Ion Flotation". Elsevier, New York.

Seto, F. Y. B. and Duce, R. A. (1972). *J. geophys. Res.* **77**, 5339–5349.

Seto, F. Y. B., Duce, R. A. and Woodcock, A. H. (1969). *J. geophys. Res.* **74**, 1101–1103.

Sieburth, J. McN. (1963). Abstract A8, Am. Soc. Microbiol. 63rd Annual Meeting, Cleveland.

Sieburth, J. McN. (1965). Trans. Joint. Conf. Ocean Sci. Ocean Engng, Washington, D.C. 1064–1068.

Sieburth, J. McN. (1971). *Deep Sea Res.* **18**, 1111–1121.

Sieburth, J. McN. and Conover, J. T. (1965). *Nature. Lond.* **205**, 830–831.

Sturdy, G. and Fischer, W. H. (1966). *Nature, Lond.* **211**, 951–952.

Sutcliffe, W. H., Baylor, E. R. and Menzel, D. W. (1963). *Deep Sea Res.* **10**, 233–243.

Szekielda, K. H., Kupferman, S. L., Klemas, V. and Polis, D. F. (1972). *J. geophys. Res.* **77**, 5278–5282.

Timmons, C. O. (1962). U.S. Naval Research Laboratory Report No. 5774, 1–8.

Truesdale, V. W. (1974). *Deep Sea Res.* **21**, 761–766.

Wallace, G. T. and Wilson, D. F. (1969). U.S. Naval Research Laboratory Report No. 6958, 1–22.

Wallace, G. T., Loeb, G. I. and Wilson, D. F. (1972). *J. geophys. Res.* **77**, 5293–5301.

Wheeler, J. (1972). *J. geophys. Res.* **77**, 5302–5306.

Whitman, W. G. (1923). *Chem. metall. Engng.* **29**, 146–148.
Whitney, R. P. and Vivian, J. E. (1949). *Chem. Engng Prog.* **45**, 323–337.
Wilkniss, P. E. and Bressan, D. J. (1972). *J. geophys. Res.* **77**, 5307–5315.
Williams, P. M. (1967). *Deep Sea Res.* **14**, 791–800.
Wilson, A. T. (1959). *Nature, Lond.* **184**, 99–101.
Wilson, W. B. and Collier, A. (1972). *J. mar. Res.* **30**, 15–26.

Chapter 11

The Micronutrient Elements

C. P. SPENCER

Marine Science Laboratories, University College of North Wales,
Menai Bridge, Anglesey, North Wales

11.1. Introduction

A nutrient element is one which is functionally involved in the processes of living organisms. Traditionally, in chemical oceanography the term has been applied almost exclusively to silicon, phosphorus and inorganic nitrogen, but strictly a number of the major constituents of sea water, together with a large number of the essential trace metals, are also nutrient elements. In comparison with the major constituents, silicon, phosphorus and inorganic nitrogen

245

occur in sea water in low concentrations, and in certain localities living organisms are responsible for the removal or excretion of sufficiently large amounts of these elements, relative to the total quantity present, for their effects to be easily detectable. Large amounts of some of the major constituents of sea water are also cycled by biological systems, but for these elements the total amounts available are so great that the action of living organisms produces only small or undectable changes in their concentrations. Many of the other essential nutrient elements are also present in sea water in very small amounts, but only minute quantities of these are required by living organisms; geochemical reactions are often more important in the control of the concentrations of these elements. Other elements which are not functionally involved are sometimes adventitiously incorporated into living organisms, and biological action can play an important part in the distribution of these elements in the sea. Equally, some nonfunctional incorporation of essential trace elements may also occur, and in such instances their distribution is also markedly affected by the action of living organisms.

This chapter is confined to a consideration of the chemistry of silicon, phosphorus and inorganic nitrogen in the sea. It is particularly for these elements that functional involvement in living organisms is a dominant process in the control of their concentration and distribution. It is also true that the availability of these elements plays an important role in controlling the growth of marine plants and animals, but consideration of these aspects is outside the scope of this Chapter.

11.2. GENERAL CONSIDERATION OF THE SPATIAL AND TEMPORAL VARIATIONS OF MICRONUTRIENT ELEMENTS IN THE SEA

The primary processes which influence the concentrations of the nutrient and other elements in the sea are the geophysical and geochemical processes which control the addition to sea water of these elements and those responsible for their dispersion and removal. The products of rock weathering and of the decay of organic material, together with discarded wastes, are the major sources of most forms of the nutrient elements to the sea, to which they are usually carried by terrestrial drainage. Contributions also arise from submarine weathering, glacial action at the poles, and volcanic and other geothermal activity. There is also the possibility of contributions as a result of transport from the atmosphere; this may occur by direct gaseous uptake by the sea, by addition of rain water and by the deposition of solid particles.

The geochemical processes which control the removal of elements from solution in sea water involve essentially the formation of solid phase material and sedimentation. Such geochemical removal processes might, *a priori*, be expected to include the formation of authigenic minerals and sorption reactions onto suspended material. In addition, there is the possibility of loss of elements from the sea by droplet transport into the atmosphere.

In conjunction with the mixing and advective processes at work in the seas and oceans, the location of the sources and the sites at which an element is removed from solution will cause gradients in concentration and produce particular patterns of distribution. Depending on the interplay between the supply and the removal processes and the efficiency of mixing in the oceans as a whole, the localized concentration of an element in solution in sea water will tend towards a steady state.

The incorporation of silicon, nitrogen and phosphorus into the cells, tissues and extracellular structures of living organisms and the excretion and regeneration of these elements in solution will impose additional mechanisms of addition, removal and transport on the geophysical and geochemical processes considered above. The site of these biologically activated mechanisms will depend upon the ecology of the organisms responsible, and this will set up quite different patterns of distribution of the elements from those produced by abiological processes. Biological activity will, therefore, tend to modify the movements towards geophysical and geochemical equilibrium. These effects will also be apparent in the distribution of any element which is incorporated adventitiously into living organisms in sufficiently large quantities relative to its concentration in sea water. Barium is an example of this (Chow and Goldberg, 1960*), since, although as far as is known it is not functionally involved in living processes it does show distribution patterns which have much in common with those of the nutrient elements.

The primary agencies for the biological removal of inorganic nitrogen, phosphorus and silicon from solution in sea water are the unicellular algae of the phytoplankton. On the coasts, the littoral and benthic algae also remove these elements from sea water, but the total amounts involved are probably small in relation to the supply of these elements in the seas and oceans as a whole. Many phytoplanktonic organisms can, under certain circumstances, assimilate all these elements to a limited extent in the dark, but ultimately incorporation into algal cells is dependent on the products of photosynthesis. The removal of these elements from solution in sea water by biological action is, therefore, limited to the upper, well illuminated, layers of the sea. The primary and other consumers of the marine food web act as agencies, directly or indirectly, for the regeneration of these elements in soluble form. Bacteria are particularly involved in these regeneration processes.

The incorporation of the nutrient elements into the cells and extracellular structures of living organisms may be a reversible process, the elements being

* See also Vol. 1, Section 7.6.2.

returned in solution to the oceans in their original chemical forms. Alternatively, they may ultimately be released in a different chemical state. When this occurs the regeneration of the elements in their most stable inorganic forms is often activated by bacteria. Often however, the incorporation of a nutrient element by living organisms is an irreversible process, and a proportion of the element may enter the sediments and be lost from the sea water phase.

The removal of phosphorus and combined inorganic nitrogen from the upper layers of the sea as a result of the growth of the phytoplankton is a well established phenomenon. If the phytoplankton crop includes diatoms or silicoflagellates, silicon will also be removed from solution. The amount removed and the rate of removal will be controlled by the many interdependent factors which regulate phytoplankton growth. In shallow well mixed water, the depletion in the concentration of these elements is often uniform throughout the water column because even if photosynthesis is limited to the upper layers, vertical mixing will disperse the effects throughout the water column. If, however, thermal stratification develops, the light intensity below the thermocline may be insufficient to sustain net photosynthesis, and further removal of the nutrient elements from solution will be limited to the water above the thermocline. It is not uncommon in temperate latitudes for the concentration of combined inorganic nitrogen, phosphorus or silicon in solution in the surface waters to fall to very low values during the summer months. In areas of the ocean where a permanent thermocline exists, very low concentrations of these elements are maintained throughout the year. In other regions, for example the Antarctic and some coastal areas, although very considerable removal of dissolved nutrients occurs, the supply is so abundant that the growth of the phytoplankton is ultimately limited by other factors with the result that considerable concentrations of the nutrient elements are always available in the water. Most of the detailed data on the temporal variations of the nutrient elements which are available relate to easily accessible inshore waters, but such information as there is makes it reasonable to believe that at a particular latitude, the pattern of events in the surface water further offshore will follow the same general pattern as that in the adjacent coastal water.

A small proportion of the phytoplankton cells which have grown in the upper layers of the sea may become senescent and sink into deeper water. Autolysis and bacterial action will result in the regeneration of the nitrogen and phosphorus from such cells. The bulk of the cells of the phytoplankton will probably be ingested by herbivores and a proportion of their nitrogen and phosphorus will be excreted in solution or as faecal pellets into the waters of the euphotic zone. The soluble products of excretion will be available for successive crops of phytoplankton and a quasi-steady state will result, the balance between supply and removal processes maintaining low concentrations of the elements. A proportion of the nutrient elements ingested by herbivores will be released as excretory products by animals at other trophic levels in the food web. Many of these animals will have migrated to the water below the euphotic zone and their excretory products will join the detrital matter and faecal pellets which have sedimented from the upper layers. All this material will ultimately be remineralized. These processes will combine to transport the elements down the water column. In stratified shallow water this sometimes causes an increase in the concentration of the nutrient elements in the bottom water during the summer months, but in the absence of other data it is often impossible to interpret positively such changes in terms of nutrient regeneration. In the deeper water of the oceans this downward transport of material causes the well known increases in the concentration of the nutrient elements with depth. Here there is the superimposition of two distinct transport mechanisms; firstly the normal advective process and secondly the process which is the consequence of assimilation of dissolved material by living organisms. This is not unique to the nutrient elements but the distribution with depth which it produces is indicative of the intervention of living organisms.

The replenishment of the supply of the nutrient elements in the surface water depends upon the action of physical processes. In shallow water in temperate latitudes, the breakdown of thermal stratification as a result of seasonal cooling allows wind generated mixing of the whole water column. In the deeper ocean water of similar latitude, the wind mixed surface layer increases with depth in the latter part of the year and regenerated nutrients are brought to the surface. Some nutrient regeneration continues throughout the water column during the autumn and winter. In regions of the oceans where there is no marked annual variation in the depth of the surface mixed layer, the downward transport of the nutrient elements is not reversed seasonally, and the surface waters become denuded of nutrients. Under these circumstances if there is to be a continuous supply of these elements there must be lateral advective movements, which cause some exchange of the surface water, and also contributions from the atmosphere. In other areas of the oceans, the upwelling of deep water at the continental margins reverses the downward transport of material.

The rates of accumulation of inorganic nitrogen, phosphorus and silicon in deep water are controlled by the biological productivity of the surface layers, by the rates of regeneration of the different elements and by the vertical eddy diffusivity and lateral advective movements of the deep water which act by dispersing the accumulating solutes. Differences in the rates of regeneration of the nutrient elements can cause fractionation effects and may result in differences in the relative concentrations of these elements in different water masses. Differences in the circulation patterns of the deep water in the various oceans will cause characteristic differences in the distribution of the nutrient elements. When the horizontal movement and exchange of deeper water is restricted or prevented, as in fiords, deep ocean basins and trenches, the concentrations of nutrient elements may reach higher values than in the oceanic deep water as a whole.

Particular examples of many of the general processes which affect the distribution of inorganic nitrogen, phosphorus and silicon in the sea will be considered in the following, more detailed, accounts.

11.3. SILICON

11.3.1. ASPECTS OF THE CHEMISTRY OF SILICON IN SEA WATER

Silicon is present in sea water in solution and as solid material in suspension. The particulate silicon consists of the extracellular structures of diatoms, silicoflagellates and radiolarians, both living and dead, and of inorganic silicon in various mineralogical forms.

The silicon in solution in sea water is probably in the form of silicic acid $Si(OH)_4$. The value of the first ionization constant of silicic acid in $0.5 M$ solutions of sodium chloride at $25°C$ is 3.9×10^{-10} (Ingri, 1959) and it therefore seems likely that unionized silicic acid is the major silicon containing species present in sea water at pH 8.2. There is still some doubt about the solubility of amorphous silica in sea water, but at $22-27°C$ the value is probably about $1600-1800$ μg-at $Si\,l^{-1}$ (Krauskopf, 1956). The solubility probably decreases with decreasing temperatures, and at $0°C$ may be about $1000-1200$ μg-at $Si\,l^{-1}$. The maximum concentrations of silicic acid found in sea water are always lower than those at which spontaneous polymerization of silicic acid occurs and, in addition, experimental evidence indicates that polymers of silicic acid are unstable in sea water (Burton et al., 1970). The methods of determination of dissolved silicon in sea water depend upon the formation of a silicomolybdate complex. Since only ionized or molecularly dispersed silicic acid will react it is likely that current analytical methods measure all the dissolved species of silicic acid. Because natural sea water is always undersaturated with respect to silicic acid, the concentration of this element in the sea is clearly not controlled by precipitation of hydrated forms of silica. The processes which may control the amounts of dissolved silicon in the oceans are discussed later in this Chapter.

In addition to that fraction of the suspended particulate silicon which consists of the extracellular structures of diatoms and other siliceous organisms, the particulate silicon in the sea also includes a wide range of minerals (Atkins *et al.*, 1954; Lowndes, 1955; Gunnerson and Emery, 1962; Lisitsin, 1961; Ishii and Ishikawa, 1964.). Much of this material will have been transported in suspension in river water, and some of the degraded clay minerals may occur in the sea perhaps in modified form. Other siliceous particles enter the sea from the atmosphere. These comprise wind-borne particles from desert regions as well as materials of volcanic and cosmic origin (Radcezewski, 1937, 1939; Rex and Goldberg, 1958; Murray and Renard, 1891.).

The ultimate fate of much of the particulate silicon in suspension is to enter the sediments. The mineralogical composition of sedimentary material depends on the source of the material and on any modifications which it has undergone during transport, suspension in sea water or in the sediment itself. Some of the minerals found in marine sediments, such as illite and chlorite, are relatively stable in sea water and seem to be largely of terrestrial origin (Weaver, 1958; Hurley *et al.*, 1959; Griffin and Goldberg, 1963). In contrast, some minerals, e.g. kaolinite, seem to be less stable in sea water (Mackenzie and Garrels, 1965). The particulate remains of siliceous organisms will also enter the sediments in some circumstances, and may form a major fraction of diatomaceous deposits in the polar regions.

11.3.2. SPATIAL AND TEMPORAL VARIATIONS OF SILICON IN THE SEA

11.3.2.1. Spatial Variation

The concentration of dissolved silicon in solution in the sea varies more than that of any other element. Armstrong (1965a) has provided a general account of the global distribution of this element in the sea. Surface waters of the open ocean often contain very low and sometimes undetectable amounts, but high concentrations are found at the surface in areas of upwelling and up to $50 \, \mu$g-at $Si \, l^{-1}$ have been recorded for the South Atlantic. The concentrations of silicic acid in deep ocean waters reaches $100 \, \mu$g-at $Si \, l^{-1}$ in the Atlantic and higher values of up to $175 \, \mu$g-at $Si \, l^{-1}$ in the Pacific Ocean. In common with many other elements which are incorporated into living organisms, the concentration of dissolved silicic acid increases with depth (Fig. 11.1). The circulation and exchange of the deep water differs from ocean to ocean, and this clearly affects the accumulation and dispersion of the silicic acid. The increase in concentration of silicic acid with depth in the sea is not always regular, and maxima occur in its vertical distribution in some regions (Fig. 11.2). The intermediate water layers which contain high concentrations of silicic acid are presumably major sites of solution of particulate silicon.

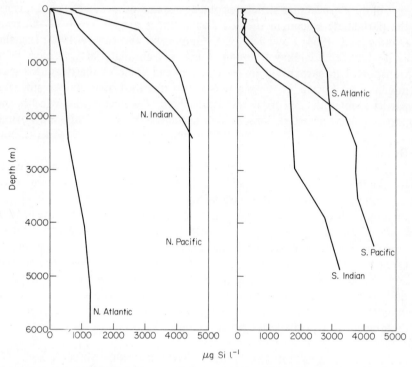

FIG. 11.1. Typical depth profiles of the concentrations of dissolved silicic acid in the oceans of the world (From Armstrong, 1965a).

Richards (1958) and Metcalf (1969) have suggested that the high concentrations of dissolved silicic acid in the water occupying basins and trenches may be the result of restrictions on the exchange of this water. Often there may be subsurface utilization of dissolved silicic acid by radiolarian populations, and this and the subsequent re-solution of this silicon also affects its vertical distribution. River water usually contains much higher concentrations of dissolved silicic acid than does sea water and, in consequence, the dissolved silicon concentration in coastal regions which are appreciably affected by land drainage is higher than that of open ocean surface waters (Stefánsson and Richards, (1963). When this is so, an inverse silicic acid–salinity relationship often applies. The distribution in estuaries can apparently sometimes be more complex, and some examples of this will be considered later in this chapter.

The quantity of particulate silicon found in suspension in sea water also shows considerable variation. Armstrong (1958) found between 54 and 520 $\mu g \ SiO_2 \ l^{-1}$ in the surface waters about 20 miles off the coast in the English

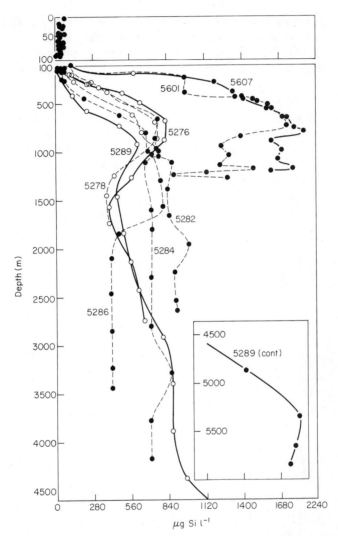

FIG. 11.2. Depth profiles of the concentrations of silicic acid with depth at various stations in the Antilles Arc region. 5601 and 5607, Cariaco Trench. 5282 and 5284, Venezuelan Basin. 5276 and 5278, Puerto Rica Trench. 5286, Jungfern Passage region (From Richards, 1958).

Channel, the lowest concentrations being recorded during the summer months. Except in one instance the amounts of particulate silicon exceeded the quantity in solution. In the surface waters of the eastern Atlantic Ocean the amounts of suspended silicon ranged from 50 to 350 µg SiO_2 l^{-1}, these amounts also being in excess of the concentrations of silicon in solution.

The amounts of particulate silicon did not vary with depth in any systematic way, but because the concentrations of dissolved silicic acid increased with depth, the particulate silicon was always a small proportion of the total silicon in the deeper water.

11.3.2.2. Seasonal variation

In those regions of the oceans which support a seasonal growth of diatoms, the concentrations of silicic acid in the surface waters decreases as the bloom of diatoms develops. This seasonal consumption of silicon in temperate waters is a well documented and regular occurrence (Fig. 11.3). If the water column is shallow and well mixed the effects of silicon consumption are dispersed to all depths. Observations have been continued for many years at a station in the English Channel by workers from the Plymouth Laboratory. Armstrong and Butler (1968) have provided extensive references to this work,

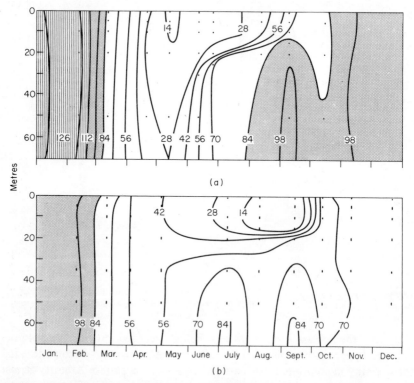

FIG. 11.3. Variations in the vertical distribution of silicic acid (in μg Si l^{-1}) in the English Channel (50° 04′ N, 4° 22′ W). (a) 1960; (b) 1961 (From Armstrong and Butler, 1960 and 1962).

and Armstrong (1965a) has described other long-term surveys. Similar records have also been made at a station off the Isle of Man in the Irish Sea (see Slinn, 1958 and subsequent Annual Reports). From such records a common pattern of changes is evident although the net amount of silicon removed from solution often shows marked variation from year to year. This may be the result of variations in diatom growth and, especially in inshore regions, of variations in advection and vertical mixing. Evidence of similar seasonal variations in the concentrations of dissolved silicic acid is available for Antarctic waters (Clowes, 1938). The silicic acid concentration in the surface water in the tropics is usually very low (e.g. $0.3–1.8$ μg-at l^{-1} in the Sargasso Sea; Menzel and Ryther, 1960). In areas where diatoms are not as ecologically important, the annual variation in the concentration of dissolved silicic acid is not as marked, and in some tropical inshore waters any consumption of silicon which occurs is sometimes masked by seasonal increases in the additions of silicon to the sea from the enhanced river discharge which occurs during the wet season (Ganapati et al., 1956; Krishnamurthy, 1967).

The development of thermal stratification in shallow water may prevent consumption of silicic acid by diatoms in the poorly illuminated bottom layers (Fig. 11.3) and under such conditions there are sometimes increases in the concentration of this constituent in the deeper waters during the summer months. This may be the result of regeneration of silicic acid, but advective transport can also produce similar effects. Temporal variations in the concentration of silicic acid in deep oceanic water have been recorded (Phelps, 1937). Although these are often the result of advective transport, it might be expected that seasonal changes in the additions of siliceous detritus would cause temporal changes in the concentration of dissolved silicic acid in deeper water in some areas.

11.3.3. THE CYCLING AND REGENERATION OF SILICON IN THE SEA

Diatoms assimilate silicon from solution in the form of dissolved silicic acid (Lewin, 1955). Their ability to do this seems to be very dependent upon the integrity of the cell surface (Lewin and Chin-Hon Chen, 1968). This can be impaired by washing the cells, and the resulting inhibition can be removed by the addition of the wash water or by the addition of aspartic or glutamic acid. Earlier reports had suggested that in some species at least, the effects of washing the cells can also be reversed by the addition of divalent sulphur (Lewin, 1954). Although the silicon is deposited in an extracellular structure, there seems to be an obligate link between some of the cellular metabolic processes and the assimilation and deposition of silicon in diatoms. Thus, Darley and Volcani (1969) have reported that net synthesis of DNA in a marine diatom was blocked as soon as silicon deficiency developed, and this suggests that the frustule is more functionally linked to diatom metabolism than had been assumed previously.

The physical and chemical nature of the silicon in the frustules of diatoms is poorly understood. Kamatani (1971) has presented evidence that a proportion of the silicon in these structures is amorphous, but there are considerable differences from species to species. Rogall (1939) has reported the presence of both iron and aluminium in association with the silicon in diatom frustules.

Earlier reports on the dissolution of the silicon from the frustules of diatoms gave variable and inconsistent results. Atkins (1945) was unable to dissolve diatom frustules even using solutions of relatively high pH. Harvey (1955) has cited an experiment in which cells of the marine diatom *Ditylum brightwelli*, which had been killed by warming, released 50% of the silicon in their frustules into sea water within 2 months. Jørgensen (1955) investigated the rate of solution of silicon in water buffered with bicarbonate–carbonate mixtures, or acidified with hydrochloric acid, using two species of diatoms. He found the rates of solution to be variable, but to be only low at pH 8·0. Lewin (1961) was able to show that the silicon in the frustule of the fresh water diatom *Navicula pelliculosa* tended to dissolve in Tris buffer of pH 8·0 if the cells were first killed. Removal of the organic matter by digestion of frustules with nitric acid also resulted in more rapid solution. Solution of the silicon was retarded by various cations, notably iron and aluminium, and the protective action of these ions was removed by treating the frustules with a chelating agent. The silicon in the frustules of living diatoms seems to be protected against solution by some vital process, but digestive action in the gut of herbivorous zooplankton coupled with mechanical damage of the frustules is likely to assist solution of the silica in the sea. Lewin also reported that the silicon in fossil diatoms can be solubilized by treatment with chelating agents. Kamatani (1971) found considerable variability in the solubility of the silicon of various species of marine diatoms and suggested that this is related to the proportions of amorphous and pseudo-opaline silica present. Kamatani also confirmed Lewin's observation that the presence of sodium chloride increased the rate of solution of silicon from diatom frustules.

The recognizable remains of radiolarians form a large proportion of the sediments over wide areas of the ocean bed at lower latitudes. This suggests that considerable amounts of silicon are removed from solution at intermediate depths by these organisms and that a proportion of this silicon is lost to the sediments. Berger (1968) has reported *in situ* measurements of the solution of silicon from radiolarian skeletons. The rate of solution of silicon from this material decreased with increasing depth, and Berger ascribed this to the lower temperatures and the higher concentrations of silicic acid in solution in the deeper water. Both these factors might encourage the formation of diatomaceous sediments in the polar regions.

Grill (1970) has developed a mathematical model to describe the vertical distribution of dissolved silicic acid in the north and equatorial Pacific Ocean and the equatorial Atlantic Ocean. The model takes account of the flux of silicic acid caused by diatom growth in the upper layers and their dissolution at depth and of the similar processes activated by radiolarians. On the basis of this model, an equation can be derived for the depth profiles of silicic acid observed at a number of stations. The model also permits an expression to be derived for the ratio of the rate of increase of concentration of silicic acid to the coefficient of vertical eddy diffusivity at a particular point in the sea. Absolute values of this ratio cannot be easily calculated, but its sign indicates addition or removal of silicic acid. The variation of the ratio with depth at various stations implies a net generation of silicic acid in solution below the surface mixed layer at those stations where radiolarians are not ecologically important. At those stations where sedimentological evidence points to the presence in the intermediate water of large populations of radiolarians, the negative sign of the ratio suggests that there is significant assimilation of silicic acid by these organisms. The model satisfactorily describes the vertical distribution of silicic acid at the stations included in the study in terms of the activity of diatoms and radiolarians, and suggests that the latter organisms can be important in controlling the vertical distribution of silicic acid in some areas of the oceans.

There seems no reason to doubt that at least a proportion of the silicon in diatom frustules and the skeletons of radiolarians dissolves in the deeper water of the oceans. The sedimentary material derived from diatoms and radiolarians is presumably the structural material containing relative large amounts of the forms of silica which are resistant to solution in sea water. The low temperatures of deep ocean water will probably decrease the rate of solution of amorphous silica considerably, but it seems doubtful if the highest concentrations of silicic acid which occur in the sea are generally sufficiently close to saturation values to have a significant mass effect on the rates of solution of this material. Association with iron and aluminium may also be important in the stabilization of the solid phase material.

The concentration of silicic acid at depth in the oceans can, in some regions, be affected by hydrographical factors (Cooper, 1952; Richards, 1958). The model proposed by Grill (1970) took no account of horizontal advection; since this model gives a satisfactory description of the observed vertical profiles it seems likely that horizontal advection plays little part in determining these profiles. However, it is likely that a combination of both biological and hydrographical factors will be involved in some areas.

11.3.4. GEOCHEMICAL AND BIOLOGICAL CONTROL OF THE CONCENTRATION OF
SILICIC ACID IN THE SEA

There is no doubt that biological activity is an important factor in the control
and distribution of the dissolved silicon in the sea and that a considerable
amount of silicon enters the sediments as a result of irreversible incorporation
into siliceous organisms. The possible importance of geochemical removal
processes has also been widely considered. Sillén (1967) assumed alumino-
silicate clay minerals to be important phases in his model of sea water and a
similar suggestion was made by Garrels (1965). Mackenzie and Garrels (1965)
demonstrated that various amounts of silicic acid dissolved from a range of
minerals when these were suspended in sea water, and later Mackenzie *et al.,*
(1967) also showed that silicic acid was removed by the same minerals from
sea water enriched with $25 \, mg \, SiO_2 \, 1^{-1}$. Since little uptake of dissolved
silicic acid occurred from distilled water, it is likely that cations are required
and are involved in the formation of some ill defined aluminosilicates. Similar
results were reported by Siever (1968). Helgerson and Mackenzie (1970)
refined some aspects of the equilibrium treatment used by Sillén by using more
recent thermodynamic data, and in their treatment postulated that aluminosilicates play an important role in poising the concentration of silicic acid in
the sea. Although certain aspects of this treatment have been criticized by
Perry (1971), there seems no doubt that some aluminosilicates are reactive in
sea water and the above suggestions are consistent with present understanding of the chemistry of these materials.

The importance of geochemical reactions in the distribution and control of
the amounts of silicic acid in the sea at the present time are difficult to assess.
Various attempts have been made to draw up a budget for the silicon in the
oceans by making estimates of the total amounts delivered via continental
drainage and other sources, and the amounts removed by various routes.
(Harris, 1966; Gregor, 1968; Calvert, 1968; Burton and Liss, 1968). If
geochemical processes are not important in removing silicic acid from sea
water and if the total amount of silicon in solution in sea water is in a steady
state, the additions from river drainage and other sources should balance the
removal which occurs as a result of irreversible incorporation into siliceous
organisms. There are clearly uncertainties in the estimates of the rates of both
silicon addition to the oceans and its sedimentation. Most workers have
assumed that river drainage is the most important route of its addition to the
sea, but the identity of the sources of supply and the amounts supplied by
them are in dispute. Whether or not irreversible biological removal of silicic
acid from solution balances the annual additions of silicon to the oceans cannot be decided with certainty at the present time, and the possibility of abiological removal by clay minerals or by other mechanisms cannot be excluded.

The most likely sites of removal of silicic acid from solution by reaction with clay minerals seem to be deep oceanic waters and the margins of the oceans which receive terrestial drainage. The aluminosilicates in oceanic water must have been in suspension in sea water for a considerable time and equilibrium might be expected. However, sedimentation into waters containing high concentrations of silicic acid might allow further reaction leading to the inclusion of silicon into solid phases as a result of processes similar to those described by Mackenzie *et al.* (1967). Terrestrial drainage also supplies to inshore waters aluminosilicates which have not equilibrated with sea water.

Bien *et al.* (1958) reported anomalies in the silicate: chlorinity ratios of the waters in the Mississippi delta and suggested that abiological removal of silicic acid was occurring. Similar results have been reported for the River Conway in North Wales by Liss and Spencer (1970). In this instance (Fig. 11.4), removal of silicic acid from solution was recorded under conditions such that biological uptake seems unlikely. It is possible that some fraction of the dissolved silicic acid in the Vella estuary in southern India is also removed by a similar process (Burton, 1970). In contrast, Makimoto *et al.* (1955) and

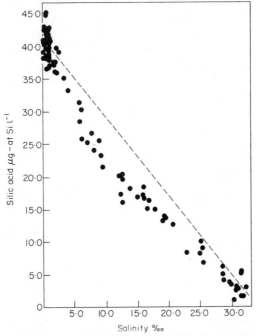

FIG. 11.4. Relationship between dissolved silicic acid and salinity in samples collected over a tidal regime at four stations covering a 9 km reach of the River Conway, North Wales (From Liss and Spencer, 1970).

Maeda and Takesue (1961) have reported a linear relationship between the chlorinity and the silicic acid concentration in the estuaries of two rivers in Japan. Similarly, Stefánsson and Richards (1963) could find no evidence of abiological removal of silicic acid in the plume of the Columbia River off Washington, and Ryther *et al.* (1967) reported similar findings for that of the River Amazon off Brazil.

Some of these studies were restricted to water within a rather narrow range of salinity and this might account for the absence of any abiological removal process. On all occasions when evidence of abiological removal has been obtained, the water has carried a heavy load of suspended material, and this may be important if appreciable removal is to occur. Schink (1967) has drawn up a budget for the silicon in the Mediterranean Sea and his results suggest that the amount of silicon added by rivers draining into this sea is balanced by that leaving in the flow of water through the Straits of Gibraltar. Schink interprets this result as evidence that there is no abiological removal of silicic acid from solution when fresh water drainage enters the Mediterranean Sea. Wollast and De Broeu (1971) reported large departures from a linear inverse relationship between salinity and silicic acid in a 100 km reach of the River Scheldt. Because these workers failed to demonstrate removal of silicic acid from these waters in model experiments using pre-treated kaolin, they considered that most, if not all, of the silicic acid was removed by biological activity. Evidence which suggests abiological removal of silicic acid from the inshore waters of the eastern Irish Sea has been reported by Ewins and Spencer (1967). In this area the concentration of silicic acid in solution seemed to be controlled at a maximum value of about 7·0 µg-at Si l^{-1} during the winter months, although this is a period when the input of fresh water drainage is at its maximum and the continued rise in the concentration of nitrate–nitrogen throughout the period suggests that minimal biological removal was occurring.

At the present time, the knowledge available of the geochemical processes which might remove silicic acid from solution in sea water is unsatisfactory. Some of the data available is contradictory; further investigations are required before the complex processes involved can be adequately described and satisfactorily understood.

11.4. PHOSPHORUS

11.4.1. ASPECTS OF THE CHEMISTRY OF PHOSPHORUS IN SEA WATER

Phosphorus exists in solution in sea water in a variety of forms, and these may be fractionated arbitrarily by suitable analytical techniques. Strickland and Austin (1960) recognized four fractions, the total phosphorus present first being separated by filtration into a soluble and an insoluble fraction. The soluble material can be divided into a further two fractions—one which will react with molybdate under the conditions used in the analytical procedure and another which will react only after oxidation of the organic matter present. The particulate fraction can also be divided into a fraction which is removed from the solid material in reactive form under the conditions used for analysis and one that becomes reactive only after oxidation. None of these fractions can be strictly defined in chemical terms. The soluble reactive fraction is likely to be largely orthophosphate, and the soluble unreactive fraction is probably mainly phosphorus in organic combination. If very labile esters occur in samples they will be included in the soluble reactive phosphate fraction. The particulate phosphorus which reacts directly with molybdate will include any inorganic phosphate that dissolves under these conditions. The additional phosphorus produced if the particulate material is first oxidized may also include any inorganic material which dissolves during the oxidation process and any organically combined phosphorus present.

11.4.1.1. Inorganic phosphorus in solution

Phosphorus occurs in solution in sea water as orthophosphate; in this medium the apparent ionization constants at 20°C in 33‰ S are $K'_1 = 2.35 \times 10^{-3}$, $K'_2 = 8.8 \times 10^{-7}$, and $K'_3 = 1.37 \times 10^{-10}$ (Kester and Pytkowicz, 1967). In sea water at pH 8.0 about 1 % of the orthophosphate is present as $H_2PO_4^-$, 87 % as HPO_4^{2-} and 12 % as PO_4^{3-}. The values of the apparent ionization constants differ considerably in solutions of sodium chloride and sea water as a result of specific ion effects. 96 % of the PO_4^{3-} and 44 % of the HPO_4^{2-} are apparently present in sea water as ion pairs probably with calcium and magnesium (Fig. 11.5). See also Vol. 1, Section 3.4.3.

Calcium phosphate is more soluble in sea water than in distilled water because of the effects of complex and ion pair formation on the activities of both calcium and phosphate ions. Pytkowicz and Kester (1967) have given a value for its solubility product in sea water at 20°C of $K'_{sp} = 4.0 \times 10^{-20}$— three orders of magnitude higher than earlier estimates. The effects of temperature and pressure on the solubility of calcium phosphate are not known, but it is likely that in areas where phosphorites are found in the sediments, the sea water is more highly saturated with calcium phosphate than in other areas. The greater degree of saturation probably results from higher pH values, which increase the activity of the triply charged orthophosphate ion rather than from higher concentrations of orthophosphate. Cooper (1948) has suggested that the solubility product of ferric phosphate in sea water is about

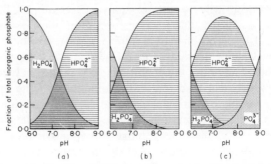

Fig. 11.5. Variations with pH of the relative proportions of orthophosphoric acid at 20°C in (A) pure water, (B) 0·68 M NaCl and (C) artificial sea water, 33‰S (From Kester and Pytkowicz, 1967).

$10^{-31·5}$, but it seems likely that ferric phosphate will not be stable at the pH of sea water (Armstrong and Harvey, 1950).

A considerable proportion of the inorganic phosphate associated with the clay minerals in sediments will exchange with dissolved phosphate and this may enrich estuaries (Rochford, 1951) and inner continental shelf waters (Cooper, 1951) with phosphate. Carritt and Goodgal (1954) have suggested that these exchange reactions in the sediments could act as a buffer for the concentration of phosphate in the overlying water. Pomeroy et al. (1965) examined the exchange of ^{32}P between bottom sediments and water, and confirmed that sorption reactions took place. At equilibrium, concentrations of phosphate of between 0·7 and 1·0 μg-at P l^{-1} were present in the water. Evidently, material in suspension in sea water may exert a controlling influence on the concentration of dissolved phosphate, and it is perhaps significant that the maximum concentrations of phosphate in inshore waters carrying a heavy load of suspended inorganic material are often within this range of concentrations.

Large amounts of the inorganic phosphorus in algae occurs as polyphosphates which will not react with molybdate. Acid hydrolysis of sea water samples will also hydrolyse any esterified phosphate which is labile under the conditions used, and such esters could not be distinguished by this method from condensed inorganic polyphosphates. These latter compounds are not hydrolysed by the photo-oxidation method for the determination of organically combined phosphorus in solution. Solórzano and Strickland (1968) did not detect significant quantities of inorganic polyphosphates in offshore or oceanic samples, although appreciable quantities of these inorganic phosphorus compounds were found in samples of harbour water and in a sample taken from a "red tide" bloom. Consequently, the increase in reactive phosphate in filtered water samples which results from the use of an appropriate oxidation procedure is usually probably a measure of organically combined phosphorus.

11.4.1.2. Organic phosphorus in solution

The detailed nature of the organically combined phosphorus dissolved in sea water is unknown. Using a sample of sea water which had been incubated for three days, Watt and Hayes (1963) separated the dissolved organic phosphorus into six main fractions. Some, but not all, of the organic phosphorus released by unicellular algae was hydrolysed by an alkaline phosphatase thus indicating the presence of monophosphate esters in this material. However, Strickland and Solórzano (1966) found that only a small proportion of the dissolved phosphorus in natural sea water samples was hydrolysed by a monophosphoesterase. It seems likely that aminophosphonic acids which occur in many marine organisms (Quin, 1965; Kittredge et al., 1967; Kittredge et al., 1969) will make up part of the dissolved organic phosphorus in sea water together with compounds such as phosphonucleotides.

11.4.1.3. Particulate phosphorus

The particulate material in suspension in the sea contains both inorganic phosphate and organic phosphorus. It is not possible to separate these different fractions satisfactorily, and most of the data available relates to the total phosphorus in suspension. Some inorganic phosphate is associated with the clay minerals suspended in inshore sea water (Ewins, 1964) and solid calcium and ferric phosphate may also be present (Cooper, 1948). Phosphate is also likely to be adsorbed on to organic matter, and the organic detritus present in suspension in the sea will contain a wide range of the phosphorus compounds which occur in living tissue.

Correll (1965) fractionated particulate material from the surface of the Antarctic Ocean and found that between 30 and 60% of the phosphorus was orthophosphate. Small amounts (2 to 10%) were in the form of oligopolyanions. RNA-polyphosphates, DNA and some phosphoproteins were the main components of the organic phosphorus fraction which made up 25 to 57% of the total particulate phosphorus. Between 2 and 15% of the material was present in an acid soluble organic phosphorus fraction, and some 4 to 7% occurred as lipid phosphorus.

11.4.2. SPATIAL AND TEMPORAL VARIATIONS OF PHOSPHORUS IN THE SEA

11.4.2.1. Spatial variation

Inorganic phosphate was one of the first micronutrients in sea water for which convenient analytical methods were available, and there is probably more information available about the distribution and variation of ortho-phosphate than there is for any other constituent of sea water. The concentration of dissolved inorganic phosphate in the surface waters of the oceans is variable, but over large areas the maximum concentrations are in the range of 0·5 to 1·0 µg-at P l^{-1}. Lower concentrations occur in the surface water of the tropics where maximum concentrations of 0·1 to 0·2 µg-at P l^{-1} are common. Significantly higher concentrations of phosphate have been demonstrated to be present in the top few cm of the sea than at a depth of 1 m (Goering and Menzel, 1965). Larger amounts of orthophosphate occur in the surface waters of the North Pacific Ocean (up to 2·0 µg-at P l^{-1}), in the Antarctic Ocean and in areas of upwelling. Reddy *et al.* (1968) have reported very high concentrations of inorganic phosphate (12 µg-at P l^{-1}) at all depths in the waters around the Andaman Islands in the Bay of Bengal, but it is not clear from the published report if all the phosphate is in solution. High concentrations of orthophosphate (> 3·0 µg-at l^{-1}) are also found in anoxic basins (Richards, 1960). The concentration of phosphate in coastal waters is often higher than in the water further off shore, and this may sometimes be ascribed to enrichment by freshwater drainage. Certainly, in enclosed areas, input via terrestrial drainage may cause higher concentrations of the nutrient elements in general; where this occurs there is invariably a reduction in salinity. In less enclosed coastal areas the volume of fresh water is small in relation to the body of sea water which receives it and this, coupled with continual exchange, maintains high salinity and appreciable enrichments of orthophosphate do not occur. In the particular instance of inorganic phosphate, river water usually contains quite low concentrations, and Riley (1967) has shown that the enrichment of inshore water with nutrients can be satisfactorily described in terms of a two layered system in which differences in the effects of horizontal advection and vertical mixing of the deeper nutrient rich water combine with the effects of biological uptake, regeneration and transport. The interplay of these processes seem likely to be generally important for maintaining relatively high concentrations of nutrients in coastal waters.

In common with the other nutrient elements, the concentration of inorganic phosphate increases with depth. The vertical distribution of phosphate in the oceans generally shows a gradual increase from low concentrations at the surface to a maximum value at between 500 and 2000 m. The phosphate maximum layer is most marked in the Atlantic Ocean (Fig. 11.6). The layers below this usually have fairly constant, or slightly decreasing, concentrations with increasing depth. The highest concentrations in deep water occur in the North Pacific and Indian Oceans (over 3·0 µg-at P l^{-1}). In both the Pacific and the Indian Ocean the maximum concentrations are found north of the equa-

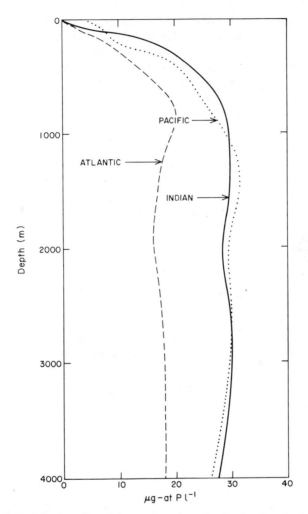

FIG. 11.6. Typical depth profiles of the concentrations of orthophosphate in the oceans of the world (From Sverdrup *et al.*, 1942).

tor. In contrast, the circulation of the deep water in the Atlantic Ocean produces a general decrease in the concentration of phosphate–phosphorus
toward the north. Riley (1951) has developed a model which describes the
effects of various physical and biological processes on the distribution of phosphorus and other nutrient elements and which predicts some of the major
features of the observed distribution of phosphorus in the Atlantic Ocean.
Higher concentrations of inorganic phosphate are found in anoxic basis and

trenches (Richards and Vaccaro, 1956). Armstrong (1965b) has given a more extensive account of the general features of the global distribution of phosphate in the oceans, and Rochford (1967) has more recently described the distribution of phosphate in the major currents of the Indian Ocean.

The concentration of organically combined phosphorus in surface water also varies greatly, and although there are reports of low concentrations (Rochford, 1967), concentrations of up to 1.3 μg-at P l^{-1} in the presence of roughly equal amounts of inorganic phosphate occur in the surface waters of the North-east Pacific Ocean (Strickland and Austin, 1960). The surface water in the tropics sometimes contains appreciable quantities of organic phosphorus in solution although inorganic phosphate may be undetectable (Kuenzler et al., 1963).

The high concentrations of inorganic phosphate in deep water make it difficult to detect low concentrations of organic phosphorus, but Ketchum et al. (1955) showed that the frequency of samples from the equatorial Atlantic Ocean which showed significant concentrations of organically combined phosphorus decreased with depth, and that no measurable amounts could be detected in any samples taken from depths greater than 1000 m. Similar observations have been reported by Armstrong (1965b) (see also Fig. 11.7). However, the concentrations of organically bound phosphorus are significant in the Atlantic deep water near the North American slope (McGill et al., (1964). High concentrations of organic phosphorus in solution for samples taken

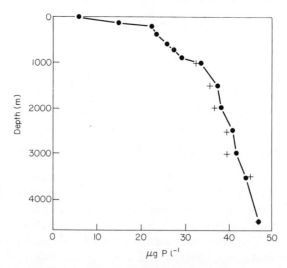

Fig. 11.7. Distribution of phosphorus with depth in the eastern North Atlantic (46° 28′ N, 8° 06′ W) in September, 1952. ● Orthophosphate, + total phosphate (From Harvey, 1955).

from between 1000 and 2000 m in the North East Pacific Ocean have also been reported by Strickland and Austin (1960).

Kalle (1937) detected between 40 and 50 $\mu g \, l^{-1}$ of particulate phosphorus in the turbid inshore waters of the North Sea, but found much smaller concentrations further off shore. Armstrong and Harvey (1950) reported quantities of between 2 and 3 $\mu g \, P \, l^{-1}$ in the suspended material in the English Channel during the early summer months. A considerable proportion of this phosphorus was in the cells of phytoplankton. The detrital phosphorus in the particulate material probably accounted for about 50% of the total phosphorus throughout most of the year. Vaccaro (1963) reported maximum concentrations of about 5 $\mu g \, l^{-1}$ of particulate phosphorus in the Atlantic Ocean off New England. Menzel and Ryther (1960) found between 0·6 and 6·0 $\mu g \, P \, l^{-1}$ of particulate phosphorus in the Western Atlantic. Newell and Kerr (1968) reported smaller amounts ranging from 0·4 $\mu g \, P \, l^{-1}$ at the surface to 0·16 $\mu g \, P \, l^{-1}$ in the deep water of the south eastern Indian Ocean. Szekielda (1963) found between 2 and 10 $\mu g \, P \, l^{-1}$ of particulate phosphorus in the shallow water of Kiel Bay and Correll (1965) reported between 1 and 4 $\mu g \, P \, l^{-1}$ in suspension in the surface water of the Antarctic Ocean. Yoshimasa and Okabe (1967) recorded quantities of particulate phosphorus in the range of 15 to 60 $\mu g \, P \, l^{-1}$ in inshore sea water samples, but in oceanic samples the concentrations were usually lower than 15 $\mu g \, P \, l^{-1}$.

11.4.2.2. Seasonal variation

The seasonal variations of the concentrations of dissolved inorganic phosphate in the surface water in temperate latitudes have been extensively studied, and a prolonged series of observations has been made at a station in the English Channel by workers from the Plymouth Laboratory. These observations were initiated by Atkins (1923), and apart from short breaks have been continued by later workers. Armstrong and Butler (1968) have provided extensive references to this work. Slinn has also reported records which have been continued for 20 years from a station off the Isle of Man in the Irish Sea (see Slinn, 1956, and subsequent Annual Reports). The pattern of events observed is common to all temperate latitudes, although variations in the magnitude and the timing of the changes occur from year to year and from place to place (Fig. 11.8). Although removal of phosphate is accompanied by the removal of other nutrient elements, synoptic records sometimes show changes in the concentration of one element which are are not accompanied by similar changes in another. This is often the result of advective movements of water which contains different relative amounts of the nutrient elements (Fig. 11.9). The utilization of phosphate without concommittant utilization of silicon can be the result of the growth of nonsiliceous algae. The surface water

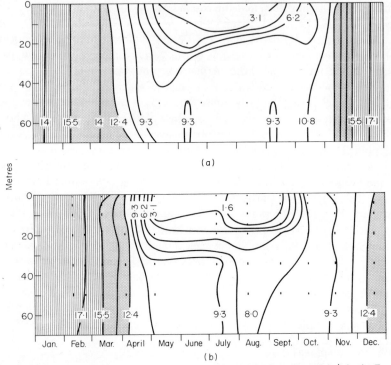

FIG. 11.8. Variations in the vertical distributions of orthophosphate (in μg P l^{-1}) in the English Channel (50° 04′ N, 4° 22′ W). (a) 1959; (b) 1960 (From Armstrong and Butler, 1960, 1962).

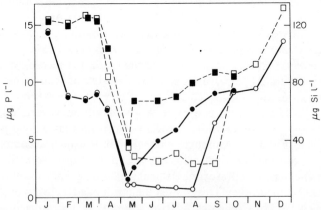

FIG. 11.9. Seasonal variations of orthophosphate and silicic acid at 10 and 50 m in the English Channel (50° 04′ N, 4° 22′ W) during 1955 − − − − orthophosphate; □ m; ■, 50 m; ——— silicic acid; ○, 10 m; ●, 50 m (From Armstrong 1955).

in the tropics usually contains very low concentrations of inorganic phosphate $(0.02$ to $0.16 \, \mu g\text{-at P } l^{-1})$, and little seasonal variation occurs (Menzel and Ryther, 1960). Seasonal consumption of phosphate occurs in the nutrient rich waters in the Antarctic region, but it is probable that the concentration of phosphate in these waters is always high (Deacon, 1933; Bienati and Comes, 1971). The phosphate concentration in the water of the Atlantic intermediate current shows a series of maxima along the direction of the current (Clowes, 1938). This water is recruited from the surface Antarctic water and the yearly cycle of phosphorus in the sub-Antarctic and Antarctic regions seems to be the main factor controlling the spatial variation in the concentration of phosphate in the water in this current.

The annual variation in the concentration of organic phosphorus in solution has been studied by several workers. Redfield et al (1937) showed that over 90% of the phosphorus in the surface waters of the Gulf of Maine was present as inorganic orthophosphate during the winter months. During the summer, large quantities of organic phosphorus appeared in solution at all depths, and this was converted to inorganic phosphate during the early winter. Armstrong and Harvey (1950) have reported a similar phenomenon in the English Channel, and this pattern of events has been recorded over a period of several years for these waters (Armstrong, 1954, 1955, 1957; Armstrong and Butler, 1959). A similar sequence has been recorded by Strickland and Austin (1960) for the waters of Departure Bay, British Columbia. Ketchum and Corwin (1965) studied the cycling of phosphorus during a phytoplankton bloom in the Gulf of Maine and also showed an increase in dissolved organic phosphorus over a period of 10 days. Ketchum and Keen (1948) reported greatly increased concentrations of organic phosphorus in solution during a bloom of *Gymnodinium* off the Florida coast, and Ramamurthy and Seshadri (1966) found concentrations of up to $10.0 \, \mu g\text{-at } l^{-1}$ of total phosphorus during a bloom of *Trichodesmium* near Porte Novo. Solórzano and Strickland (1968) showed the presence of appreciable quantities of inorganic polyphosphate during a "red tide" bloom. This suggests that on some occasions the increases recorded in the total phosphorus concentration during the growth of phytoplankton may be, at least partially, the result of the production of inorganic polyphosphates. Clearly, the term organic phosphorus in connection with results obtained under conditions of high productivity must be used with caution. The annual changes in the concentration of dissolved organic phosphorus are sometimes less marked in inshore waters (Ewins and Spencer, 1967) where the effects of terrestial drainage may mask the usual pattern of annual variation which accompanies phytoplankton growth.

The amount of particulate phosphorus in the surface waters of the sea may be expected to reflect the changes in the standing crop of phytoplankton and

K

in the amounts of organic detritus in suspension. Armstrong and Harvey (1950) found about $0.5\,\mu g\,l^{-1}$ of particulate phosphorus in suspension in the English Channel during the winter months, and this increased to between 2 and $3\,\mu g\,l^{-1}$ during the early summer. The detrital fraction accounted for about 50% of the particulate phosphorus throughout most of the year.

11.4.3. THE CYCLING AND REGENERATION OF PHOSPHORUS IN THE SEA

The principal route of removal of inorganic phosphate from the sea is assimilation by the phytoplankton. Senescent or damaged cells release ortho-phosphate as a result of autolysis (Vaccaro, 1963). Inorganic phosphate is rapidly regenerated by microbial action from natural plankton and detritus when they are suspended in sea water in the dark (Cooper, 1935b), but Johannes (1965) has suggested that protozoa may be more important in the re-generation of phosphate from detritus than are bacteria. Gardener (1937) measured the excretion of phosphorus by living zooplankton, and the subject which subsequently received considerable attention, has been reviewed by Corner and Davies (1971). Harris (1959) measured the rate of excretion of phosphate by zooplankton, and reported rates of $11\,\mu g\,P\,mg^{-1}$ (dry wt.) day^{-1}. It is clear, however, that the rate of phosphate excretion varies con-siderably with the physical conditions, the activity of the animals (Hargrave and Geen, 1968) and the feeding regime (Marshall and Orr, 1961). The basal rates of excretion by herbivores are likely to be of the order of 2 or $3\,\mu g\,P\,mg^{-1}$ (dry wt.) day^{-1}. Beers (1964) has recorded a similar rate of phosphorus excre-tion by a carnivorous zooplankton organism. The higher rates of excretion sometimes reported may often be the result of inefficient assimilation under conditions of luxury feeding, but even allowing for this, the turnover rates of the body phosphorus of zooplankton are high, and many of these organisms may on occasion excrete between 10 and 40% of their body phosphorus per day. Johannes (1965) has pointed out that, per unit weight, the rates of excretion of phosphate by protozoa are one or two orders of magnitude greater than those achieved by crustacea. Other animals in the marine food web will also excrete inorganic phosphate. Whitledge and Packard (1971) have measured the excretion rates of phosphate by the Peruvian anchovy and report values of $90\,\mu g\,g^{-1}$ (dry wt.) day^{-1}.

Often the excretion of phosphate by zooplankton may be the most impor-tant route of regeneration of phosphorus, but when the standing population of zooplankton is low and the biomass of fish is high, excretion by larger organisms will be important. A proportion of the phosphate excreted by animals will be in solid form, and this material will join the particulate detritus from other sources. The relative importance of bacteria in the regeneration of phosphate in the upper layers of the sea is difficult to assess. A large proportion of marine bacteria are capable of liberating inorganic phosphate from a range of organic compounds in laboratory experiments, and the oxidation of such material by bacteria for their metabolic and material requirements must, on occasion, result in the liberation of some inorganic phosphate. The total biomass of bacteria in the sea is small, but against this must be counted the very high metabolic rates of these organisms. Johannes (1965) showed that a mixed population of bacteria and protozoa caused higher rates of regeneration of inorganic phosphate than do either bacteria or axenic protozoa alone. It is probable that in this system, assimilation of detrital organic phosphorus by bacteria and digestion of the bacterial tissue by protozoa is more important than direct mineralization by the bacteria themselves. In nature, such particulate material will presumably be continuously worked by a succession of particle feeders, and animals other than protozoa may thus be involved in a similar way. In the experiments described by Johannes, continuous grazing of the bacteria seemed to increase the activity of the population, perhaps by preventing too dense colonization of the surfaces. It is also possible that some of the excretion products of the animals in such a system may stimulate bacterial growth. The rates of mineralization of phosphorus from particulate organic matter which are measured in laboratory experiments when bacteria are the sole agency may, therefore, be considerably lower than those achieved in nature. A number of estimates of the proportion of the phosphate which is required by phytoplankton in the euphotic zone and which is supplied by zooplankton excretion are listed by Johannes (1968). These vary from $\sim 100\%$ in winter to $\sim 10\%$ during very active phytoplankton growth. In deeper water, bacterial activity is, however, likely to be more important and must be the dominant process under anoxic conditions.

The relationship between the sea water concentration of dissolved organic phosphorus in temperate latitudes and the annual cycle of biological activity has been discussed above. Organic phosphorus is released into solution by healthy phytoplankton (Kuenzler and Ketchum, 1962; Watt and Hayes, 1963; Johannes, 1964; Kuenzler, 1970) and is also released from senescent or damaged cells. A considerable proportion of the phosphorus excreted by animals is also organically combined (Pomeroy et al., 1963; Hargrave and Geen, 1968). Many algae are able to reassimilate the dissolved organic phosphorus excreted either by themselves or by other algae, although species apparently differ in the quantities or types of organic phosphorus compounds which they are able to assimilate (Kuenzler, 1970). It is probable that both phytoplankton and animals are equally important in the production of dissolved organic phosphorus. The contribution from phytoplankton will be particularly important in the brightly illuminated immediate surface layers where high light intensities will favour the excretion of extra-cellular meta-bolites. It is also possible that the phytoplankton may be important ·in recycling this fraction of the organic phosphorus. In deeper waters, bac-terial action is likely to be the dominant initial process because, as the varia-tion with depth of dissolved organic phosphorus shows, there is usually evidence of complete mineralization of this material. The high concentrations of organic phosphorus in the deep water of the north eastern Pacific Ocean (Strickland and Austin, 1960) is not easily explained, but this water is of greater age than the bottom water in many other oceans and has received contributions of organic detritus for a considerable period of time. It is possible that in this instance the capacity for rapid oxidation of organic material has been exceeded and that bacterial mineralization processes are retarded.

Bacteria may play an important part in the regeneration of soluble phosphate from the solid calcium phosphate of bone, and bacteria capable of carrying out this process are known to occur in the marine environment (Ayyakkannu and Chandramohan, 1971). Their activity is probably very important in sediments.

Ketchum and Corwin (1965) studied various aspects of the phosphorus cycle over a period of 10 days during a phytoplankton bloom in the Gulf of Maine. From their results they were able to make estimates of the minimum uptake and production of the various fractions of phosphorus in the water column. There were no changes below 136 m, but inorganic phosphate was consumed in the upper 50 m and particulate and dissolved organic phosphorus increased. There was a net loss in the upper layers of about 5·5 µg-at $P\,l^{-1}$ and a similar increase in the deepest layers. The accumulation below 50 m in-cluded a net increase in organic phosphate and particulate phosphorus, and a decrease in dissolved organic phosphorus at a rate of $4·0 \times 10^{-3}$ µg-at l^{-1} day^{-1}. Inorganic phosphate was lost from the water column at a considerably greater rate, and under these conditions regeneration will ultimately limit the rate of primary production. These regeneration rates are, in any case, higher than other estimates which have been calculated for deep water (Riley, 1951; McGill et al., 1964).

Using the extensive data available from Long Island Sound, Riley (1956) estimated the net rates of regeneration of phosphate on the assumption that the effects of horizontal diffusion and advection could be neglected. The rate of change in the concentration of phosphate at a particular depth was equated to the sum of biological effects and of vertical eddy diffusion. There was a net utilization of phosphate at a rate of about 0·025 µg-at $P\,l^{-1}$ day^{-1} in the upper 2·5 m of the water column, and the rate was fairly constant throughout the year. Between depths of 2·5 and 17·5 m, utilization rates varied seasonally, reaching a rate of over $0.1\ \mu\text{g-at P l}^{-1}\ \text{day}^{-1}$ on one occasion in the water between 2·5 and 7·5 m. There were some instances when a net regeneration of phosphate occurred in mid-water, but the bulk of the regeneration (rates between 0.01 and $0.125\ \mu\text{g-at P l}^{-1}\ \text{day}^{-1}$ occurred in the water below 17·5 m, the highest rates being recorded during the summer months following phytoplankton blooms in the upper layers.

Postma (1971) has considered the general distribution of nutrients in the oceans as a whole with special reference to phosphorus. He has calculated that the various advective movements of water transport to the warm surface layers about half the amount of phosphorus that is required to maintain the estimated global primary production. This suggests that, on average, an atom of phosphorus is cycled twice through a cycle of assimilation and regeneration before it is lost to deep water by sedimentation. It is to be expected, however, that recycling will vary in its importance in different regions and will be particularly important in the tropics.

11.5. Nitrogen

11.5.1. ASPECTS OF THE CHEMISTRY OF NITROGEN IN SEA WATER

The most abundant form of nitrogen in the sea is elemental nitrogen (see Chapter 8). Sea water also contains much lower concentrations of a wide range of inorganic and organic nitrogen compounds in which the nitrogen occurs in many of the nine different oxidation states from -3 to $+5$. The sea also contains organic nitrogen in living and detrital particulate matter.

The development of satisfactory analytical techniques for the determination of some of the forms of combined inorganic nitrogen in sea water has been slow. Accurate and precise methods for the determination of the concentration of nitrite–nitrogen have been available for many years, but it was not until the introduction of cadmium reductors that the concentration of nitrate–nitrogen in sea water could be measured reliably. No completely satisfactory method is available today for the determination of ammonium–nitrogen in sea water, but some of the methods proposed recently provide opportunities for greater study of the chemistry of this species in the sea (see Chapter 19). The lack of satisfactory techniques has resulted in much less information being available about the chemistry of nitrogen in the sea than has been accumulated for that of phosphorus and silicon.

Combined inorganic nitrogen is added to the oceans by terrestrial drainage, and most river waters contain substantially greater amounts of combined nitrogen than does sea water. It has been known for many years that rain water contains nutrient salts, and the rates of addition of combined nitrogen from this source (8×10^7 ton yr^{-1}, Clarke, 1924; Hutchinson, 1944) exceeds the amount removed from the sea as a result of incorporation in to the sediments $(0.9 \times 10^7$ ton yr^{-1}, Trask, 1939; Emery and Rittenberg, 1952). Biological processes are dominant in the interconversion of the various forms of nitrogen in the sea and, in turn, all marine organisms are dependent, directly or indirectly, upon the store of inorganic nitrogen in the sea for the synthesis of many of their essential components.

11.5.1.1. *Elemental nitrogen* (see also Chapter 8)

Exchange with the atmosphere maintains the concentrations of elemental nitrogen in solution in sea water near saturation values. The small variations in the *in situ* saturation values are usually within the range that could result from variations in the atmospheric pressure at the surface when equilibration occurred (Carritt, 1954). Benson and Parker (1961) showed that the ratios of dissolved argon to nitrogen in the North and South Atlantic Ocean were within 1 % of those expected from solution of the atmospheric gases. In these areas of the oceans therefore, elemental nitrogen behaves as a conservative element. Elemental nitrogen is certainly removed and produced by biological action in some areas of the marine environment. The impact of these processes on the total amounts of elemental nitrogen in the oceans as a whole is unknown, but it does not seem likely to be very great. There is some doubt if the slowness of the oxidation of elemental nitrogen to nitrate–nitrogen can be the only reason for the apparent geochemical stability of elemental nitrogen, and it has been suggested that there may be some unrecognised process which converts combined nitrogen into its elemental form (L. G. Sillén, quoted by Riley and Chester, 1971).

11.5.1.2. *Nitrate nitrogen*

Nitrate–nitrogen is the thermodynamically stable form of combined inorganic nitrogen in well oxygenated sea water, and variation in the concentration of this species and of the more reduced inorganic nitrogenous compounds is predominantly the result of biologically activated reactions. The general preponderance of nitrate–nitrogen in deep well oxygenated water is, therefore, to be expected on energetic grounds.

Rakestraw and Hollaender (1936) reported that the photo-reduction of nitrate occurred readily in sea water, and this has been confirmed more recently by Hamilton (1964). This reaction is unlikely to be very important in nature, but the latter worker showed that appreciable reduction of nitrate–nitrogen occurred *in situ* when samples of sea water were enclosed in plastic bags which had optical properties similar to those of the water. One sample of sea water enriched with 15 µg-at NO_3-N l^{-1} was found to have had 4 % of its nitrate–nitrogen converted to nitrite–nitrogen over a period of 24 hours, and the process cannot therefore be completely ignored in the immediate surface layers of the sea in areas which receive large amounts of incident radiation.

11.5.1.3. *Nitrite–nitrogen*

Inorganic nitrogen in the form of nitrite–nitrogen is usually present in lower concentrations in the sea than are the other forms of combined inorganic nitrogen. The ionization constant of nitrous acid is about $4·5 \times 10^{-5}$ and the concentration of unionized nitrous acid in solution will, therefore be minute. Nitrite–nitrogen is the intermediate oxidation state between ammonium and nitrate–nitrogen, and as such it can appear as a transient in both the oxidation of ammonium–nitrogen and the reduction of nitrate–nitrogen. Both of these processes are predominantly activated in the sea by biological agents, although as noted above, some photochemical reduction of nitrate ions cannot be totally excluded. It does not appear, however, that the nitrite ion is susceptible to further photoreduction in sea water under natural conditions.

11.5.1.4. *Ammonium–nitrogen*

The lack of a convenient method of analysis for ammonium–nitrogen has hindered research, and the chemistry of this species in sea water has been little explored. The ionization constant of the equilibrium $NH_{3(aq.)} + H_2O \rightleftharpoons NH_4^+ + OH^-$ is about 10^{-5} and if this value is accepted, about $\frac{1}{10}$ of the combined inorganic nitrogen at this oxidation level is present in the sea as hydrated ammonia.* The dominant species is, however, likely to be cationic, and it seems justifiable to refer to this form of combined inorganic nitrogen in sea water as ammonium–nitrogen.

It is probable that both hydrated ammonia and the ammonium cations will adsorb readily on to many types of surfaces and will interact with clay mineral material. There is evidence that adsorption assists the microbial activation of ammonium–nitrogen in some circumstances (Lees and Quastel, 1946), and it has been suggested that adsorption may also affect both abiological (Cooper, 1948) and photochemical activations (Zobell, 1933).

The possibility of the photochemical oxidation of ammonium–nitrogen has been investigated by several workers. Many years ago Zobell (1933) and Rakestraw and Hollaender (1936) suggested that ultra-violet light might activate this reaction in sea water. More recently Hamilton (1964) reinvestigated the process, but was unable to demonstrate appreciable oxidation of ammonium–nitrogen in sea water and suggested that the reaction may require catalysis. Zobell (1933) reported that the autoclaving of sea water inhibited photochemical oxidation, but the nature of the catalytic species is not known. Joussot-Dubien and Kadri (1970) have shown that singlet oxygen produced

* For a recent investigation of the hydrolysis of ammonium ions in sea water see Whitfield (1974).

in sea water as a result of the photosensitization of suitable dye molecules by visible light will oxidize ammonium to nitrite–nitrogen. The presence of appreciable quantities of suitable sensitizing molecules in natural sea water is unproven, but this process could be important at greater depths than those at which direct short wavelength photochemical oxidation of ammonium–nitrogen might occur. The possibility of photochemical oxidation of ammonium–nitrogen in the sea cannot be discounted, but the knowledge at present available does not suggest that it is an important process in the transformations of inorganic nitrogen compounds in the sea.

11.5.1.5. Organic nitrogen in solution

Natural sea waters contain a wide range of free amino acids in solution. Duursma (1965) has given an extensive tabulation of the earlier analyses of amino acids in sea water. Alanine, glycine, serine and ornithine are often the most abundant, and relatively high concentrations of threonine, valine and leucine also occur (Pocklington, 1971). Significantly, arginine is often not detectable (Siegel and Degens, 1966). The greatest proportion of the amino acids in solution in coastal sea water is present in combined form as peptides (Riley and Segar, 1970). Urea (Newell *et al.*, 1967; Remsen, 1971), and presumably other soluble nitrogenous excretory products of aminals and a wide range of compounds which have dissolved from nitrogenous detritus are also present (see also Chapter 13).

12.5.1.6. Particulate nitrogen (see also Chapter 14)

The tissues of living micro-organisms together with detrital material are included in the particulate nitrogen fraction. Both algae and bacteria will, therefore, contribute a wide range of nitrogenous substances to the particulate nitrogen in suspension in the sea. Cowey and Corner (1963) showed that the amino acid composition of the suspended material from the inshore water off Plymouth was very similar to that of phytoplankton. Nitrogenous organic matter in solution will also be expected to be adsorbed by detritus, particularly clay minerals, and this may result in some modifications to the organic matter (Degens, 1970). Inorganic nitrogen, particularly in the form of ammonium–nitrogen, is also likely to occur in association with clay minerals in suspension.

11.5.2. SPATIAL AND TEMPORAL VARIATIONS IN THE AMOUNTS OF NITROGEN COMPOUNDS IN THE SEA

11.5.2.1. Spatial variation

The total combined inorganic nitrogen in the surface layers of the oceans shows a similar spatial variation to that of the other micronutrient elements. Typical maximum concentrations in temperate latitudes at the surface are in the range 8–15 μg-at $N l^{-1}$, but higher concentrations occur in the surface waters of the North-east Pacific, in the Antarctic Ocean and on coastal margins where upwelling occurs. Maximum concentrations of about 40 μg-at $N dl^{-1}$ were reported by Deacon (1933) in the Antarctic. Total inorganic nitrogen concentrations are usually low in the surface waters in tropical regions, and the maximum nitrate–nitrogen plus ammonium–nitrogen concentrations are often about 1 or 2 μg-at $N l^{-1}$ in such areas (Menzel and Ryther, 1960; Thomas, 1966a).

The concentrations of inorganic nitrogen increase with depth (Fig. 11.10), and reaches typical concentrations of about 20 μg-at $N l^{-1}$ in the deep water of the Atlantic Ocean, 38 μg-at $N l^{-1}$ in the North Pacific Ocean and up to 42 μg-at $N l^{-1}$ in the Indian Ocean. These distributions are produced by an interplay of similar processes to those which set up the analogous distributions of phosphorus in the deep waters of the oceans of the world. The vertical distributions of combined inorganic nitrogen show pronounced maxima (typically about 30 μg-at $N l^{-1}$) at intermediate depths in the Atlantic Ocean which parallel those of the phosphorus profiles. Vaccaro (1965) has provided a detailed account of the global distribution of nitrate–nitrogen in the oceans.

Nitrate–nitrogen is the most abundant of the three forms of combined inorganic nitrogen in surface water when biological activity is at a minimum, and in deep, well oxygenated, water nitrite and ammonium–nitrogen if detectable, form a negligible proportion of the total inorganic nitrogen present. The latter two forms of inorganic nitrogen attain their highest concentrations in surface and sub-surface water, but they are often evanescent and intimately linked to biological processes. For this reason they are most appropriately considered in the discussion of the temporal variations of inorganic nitrogen compounds which follows later in this Chapter. In anoxic water in enclosed basins and trenches, nitrate–nitrogen is utilized and, under these extreme conditions, will disappear from the water.

Early measurements of the amounts of dissolved organic nitrogen compounds in sea water were made by Krogh (1934) and Robinson and Wirth, (1934). The concentration of dissolved inorganic nitrogen in the sea has more recently been extensively investigated by Duursma (1961) in a series of studies of the Wadden Sea, the Norwegian Sea and the North Atlantic Ocean. He found concentrations of between 0·02 and 0·44 mg organic-N kg^{-1} in solution, but reported no general systematic change in concentration with depth in the North Atlantic. Fraga (1966, 1969) reported between 3 and 8 μg-at l^{-1} of dissolved organic nitrogen in the waters of the western Indian Ocean, the highest values occurring in surface samples. The surface water of

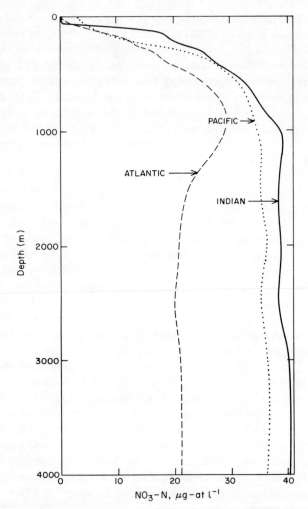

FIG. 11.10. Typical depth profiles of the concentrations of nitrate–nitrogen in the oceans of the world (From Svedrup et al., 1942).

the tropical eastern Pacific Ocean contains between 3 and $14\,\mu\text{g-at}\,l^{-1}$ of dissolved organic nitrogen (Thomas *et al.*, 1971). Pocklington (1971) has reported concentrations of individual amino acids in the North Atlantic ranging from 6 to $47\,\mu\text{g}\,l^{-1}$, with a mean value of $22\,\mu\text{g}\,l^{-1}$.

Vaccaro (1963) found up to $42\,\mu\text{g}\,l^{-1}$ of particulate nitrogen in the Atlantic Ocean off New England, and Ryther and Menzel (1964) reported that the western North Atlantic Ocean carried between 7 and $70\,\mu\text{g}\,l^{-1}$ of particulate nitrogen. Its concentration showed little systematic variation with depth below about 200 m. That of particulate nitrogen in the western Indian Ocean (Fraga, 1966) was somewhat lower (0·4 to 30 $\mu g\,l^{-1}$, and decreased with depth. Dugdale and Goering (1967) found maximum concentrations of 2·8 $\mu g\,l^{-1}$ in the surface water of the Sargasso Sea. Newell and Kerr (1968) using a ninhydrin method have recorded a significant decrease with depth in the concentration of particulate peptide nitrogen. For the Gothland Deep Ehrhardt (1969) reported typical values of 36 $\mu g\,l^{-1}$ of particulate nitrogen at the surface. decreasing to 14 $\mu g\,l^{-1}$ at a depth of 75 m. About 30% of the particulate nitrogen in surface samples in this area was protein nitrogen. Flemmer and Biggs (1971) quote values of 100 $\mu g\,l^{-1}$ of particulate nitrogen as being typical of inshore waters.

11.5.2.2. Seasonal variations

Inorganic nitrogen is removed from the surface waters during the growth of phytoplankton. In culture, many unicellular algae utilize ammonium–nitrogen in preference to nitrate–nitrogen (Harvey, 1955), and under some conditions will grow more rapidly using ammonium–nitrogen as a nitrogen source than they do with nitrate–nitrogen (Paasche, 1971). Dugdale and Goering (1967) have measured the rate of uptake of ^{15}N-labelled ammonium and nitrate–nitrogen by phytoplankton during incubation of natural water samples. They found that, in general, the rates of uptake of ammonium–nitrogen were higher than those of nitrate–nitrogen, the differences being most marked in tropical waters. Phytoplankton are also able to use nitrite–nitrogen (Eppley et al., 1969).

The utilization of inorganic nitrogen in the well illuminated surface layers of the sea in temperate latitudes results in the concentrations of these species falling dramatically in the spring and early summer. Very low or undetectable concentrations of nitrate–nitrogen often persist throughout the summer months (Fig. 11.11). If thermal stratification develops in the water column, further decreases in the concentration of inorganic nitrogen may be limited to the upper well illuminated layers. Some increase in the combined inorganic nitrogen often occurs in the deeper water as a result of regeneration processes. The decrease in the intensity of assimilatory biological activity in the autumn allows a more general increase in the concentration of the inorganic nitrogen compounds.

In areas where the inorganic–nitrogen concentrations in the surface waters are high, such as the Antarctic Ocean, appreciable utilization of nitrate–nitrogen also occurs. Deacon (1933) reported a decrease from concentrations of 40 μg-at $N\,l^{-1}$ in October and November to 27 μg-at $N\,l^{-1}$ during the summer in these regions. The concentration of nitrate–nitrogen in the surface

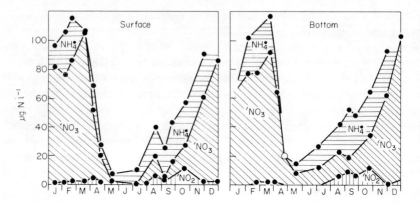

FIG. 11.11. Seasonal variations in the concentrations of ammonium-, nitrite- and nitrate-nitrogen in the upper layers and near the bottom at a position in the English Channel during 1931 (After Cooper, 1933).

waters of tropical regions varies only slightly (undetectable to $1{\cdot}8$ µg-at $N\,l^{-1}$; Menzel and Ryther, 1960). Dugdale and MacIsaac (1971) have developed a model for nitrate–nitrogen assimilation which is based on Michaelis–Menton kinetics and relates uptake rate to the NO_3^- concentration and to the incident light energy. The incident light intensity decreases exponentially with depth. Predictions based on this model for the integrated effects of these two factors in the Peru upwelling region agreed well with the depth profiles observed. The concentration of nitrate–nitrogen is often increased in inshore locations by land drainage, but as for phosphate, hydrographical factors can also produce similar effects (Riley, 1967).

There is little information about annual changes in the concentration of dissolved organic nitrogen in the sea. Riley and Segar (1970) showed that marked variations occurred in the concentrations of combined amino acids in the Irish Sea. The concentrations of free amino acids showed less total variation, but were related to the seasonal growth of the phytoplankton.

11.5.3. THE CYCLING AND REGENERATION OF NITROGEN IN THE SEA

11.5.3.1. *The formation of ammonium nitrogen*

The concentration of ammonium–nitrogen in the sea shows considerable variations and can change rapidly. Menzel and Spaeth (1962) showed a correlation between the concentration of ammonium–nitrogen in the surface waters of the Sargasso Sea and rainfall (Fig. 11.12). However, Dugdale and Goering (1967) have suggested that the average daily rates of addition of ammonium–nitrogen to the euphotic zone in this area are insignificant in comparison with the daily requirements of the phytoplankton. Ammonium–nitrogen can often be the most abundant form of combined inorganic nitrogen in the surface water in temperate latitudes when phytoplankton growth has removed the greater part of the nitrate–nitrogen (Fig. 11.11) and it is generally the dominant form of combined inorganic nitrogen in the surface water in the tropics (Menzel and Spaeth, 1962; Thomas, 1966a). The vertical distribution of ammonium–nitrogen in the sea sometimes shows a sub-surface maximum (Fig. 11.13). Ammonium–nitrogen is the dominant form of combined inorganic nitrogen when anoxic conditions have developed to the stage when all the nitrate–nitrogen has been reduced, and if the system becomes anoxic, the concentration of ammonium–nitrogen can become high.

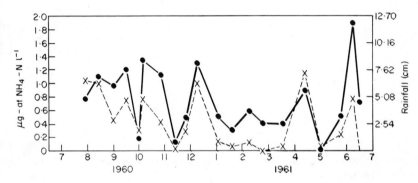

FIG. 11.12. The variations in ammonium–nitrogen concentrations in the surface water near Bermuda (———) and the total rainfall for the 5 days previous to each determination (------) (From Menzel and Spaeth, 1962).

Dead or senescent algal cells will autolyse, and this effect coupled with bacterial action will release as ammonium–nitrogen a large proportion of the organic nitrogen originally bound in particulate form (von Brand et al., 1937, 1939, 1941, 1942; Vaccaro, 1965). Most of the cellular organic nitrogen synthesized by the phytoplankton will, however, probably be ingested by herbivores and enter the marine food web. Nitrogen is also excreted in soluble form by animals, and for aquatic animals ammonium–nitrogen is usually the main soluble inorganic nitrogenous excretion product. Urea may also be excreted (Corner and Newell, 1967; Jawed, 1969), and the excretion of amino acids by zooplankton has also been reported (Johannes and Webb, 1965; Webb and Johannes, 1967; Jawed, 1969). The excretion of trimethylamine oxide and other nitrogenous compounds has been reviewed by Nichol (1960). The rate of excretion of nitrogen and the relative proportions of the various nitrogen compounds are, as with phosphorus, very dependent upon the

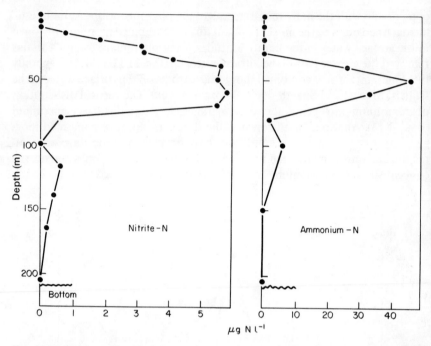

Fig. 11.13. Depth profiles of nitrite- and ammonium–nitrogen in the Gulf of Maine (From Redfield and Keys, 1938).

physical conditions, the food supply, the body weight of the animals and other factors. Corner and Davis (1971) have listed nitrogen excretion rates for zooplankton ranging from 0·3 to 43 μg N mg^{-1} (dry wt.) day^{-1} which represents from about 0·5 to 48% of the body nitrogen of the animals. Although very variable results have been recorded, the rates of excretion of nitrogen by zooplankton are high relative to those of other animals.

Marine algae also excrete organic nitrogen under some circumstances (Hellebust, 1965). Thomas et al. (1971) have suggested that the organic nitrogen in the sea is not generally available to phytoplankton, and Johannes and Webb (1970) concluded that, although uptake of dissolved amino acids by animals does occur, there is usually a net flux of dissolved organic nitrogen into the water. It is probable therefore that much of this material in the sea must be recycled, directly or indirectly, by bacteria. Many marine bacteria possess the potential to use a wide variety of nitrogenous substances, and the liberation of ammonium–nitrogen is readily observed in cultures. Utilization by bacteria in the sea of proteinaceous organic matter as a source of energy probably causes some liberation of ammonium–nitrogen as a result of oxidation–deamination reactions. It is, however, difficult to assess how much of the ammonium–nitrogen is liberated directly into the sea by bacterial action and how much of it is recycled as a result of assimilation by bacteria and subsequent ingestion of the bacterial cells by particle feeders; often this latter process must be of importance. Estimates of the proportion of the nitrogen requirements of the phytoplankton which are supplied by zooplankton excretion have been made by Johannes (1968); it is probable that, as for phosphorus, in winter more is made available than is required. However, during active phytoplankton growth a smaller fraction of the nitrogen requirements than that of phosphorus is supplied in this way. Bacterial assimilation of the naturally occurring low concentrations of dissolved materials will be greatly assisted if these materials are first adsorbed onto clay minerals and other surfaces. This will also facilitate subsequent ingestion of the bacteria by particle feeders. Both the particulate organic nitrogen voided by animals and other nitrogenous detrital material which occur in suspension in sea water must ultimately be regenerated by similar mechanisms. Subsurface maxima in the vertical distribution of ammonium–nitrogen are presumably often indicative of the sites of very active regeneration of organic nitrogen (Fig. 11.13).

11.5.3.2. The oxidation of ammonium–nitrogen

In the absence of efficient competitive removal of ammonium–nitrogen from solution by phytoplankton, the inorganic nitrogen is transformed to nitrate–nitrogen (+5 oxidation state) Bacterial nitrification is a well recognized process in many non-marine environments, but the *in situ* process has proved particularly difficult to investigate. Early work met with limited success because oxidation of ammonium–nitrogen could be demonstrated only in inshore water samples and sediments but not in water collected further from the land (see Zobell 1946). Spencer (1956) reported similar results and showed that in some instances the addition of iron to sea water samples stimulated the production of nitrite–nitrogen. Of the earlier workers Thomsen (1910) was the one to succeed in isolating nitrifying bacteria from the sea. However, he was able to detect these organisms only in inshore waters.

There was, nevertheless, considerable evidence which suggested that nitrification proceeds in oceanic water and that it is a particularly active process in the mid-depths between 250 and 800 m in the Atlantic Ocean (Riley 1951). Redfield and Keys (1938) found that the maximum concentrations of both ammonium and nitrite–nitrogen in the water column of the Gulf of Maine (Fig. 11.13) occurred in the sub-surface water. Brandhorst (1959) reported the presence of a nitrite–nitrogen maximum in the upper 100 m over large areas of the eastern tropical Pacific Ocean. In several instances the nitrite-maxima were found in water containing quite high concentrations of oxygen, which suggests that they may be, at least in part, attributable to nitrification. Wada and Hattori (1971) have produced evidence that the oxidation of ammonium–nitrogen makes an important contribution to the concentration of nitrate–nitrogen at depths of 50 to 80 m in the North East Pacific Ocean. There is usually a transitory but significant increase in the concentration of nitrite–nitrogen in the autumn in temperate latitudes. In shallow inshore water this is often dispersed throughout the water column and occurs at the time when the concentrations of nitrate–nitrogen are increasing (Fig. 11.11).

Rakestraw (1936) has pointed out that there are many instances when the accumulation of nitrite–nitrogen could be explained satisfactorily either in terms of the oxidation of ammonium–nitrogen, or of the reduction of nitrate–nitrogen. The latter process is discussed later in this Chapter. However, when there is evidence for the formation of nitrite–nitrogen under well oxygenated conditions, especially if there is a concurrent increase in nitrate–nitrogen, nitrification is often the most likely process. Photochemical oxidation which has been discussed earlier cannot be totally excluded, and there are other processes (see below) which can also cause the formation of nitrite–nitrogen.

Vaccaro (1962) was able to demonstrate the formation of nitrite–nitrogen in both coastal and oceanic water samples after 48 hours incubation in the dark. The rates were often sufficiently high to suggest that chemical as well as bacterial oxidation was occurring. Watson (1962, 1963, 1965) reported the isolation of a marine bacterium, *Nitrocystis oceanus* which, although morphologically different from terrestial nitrifying bacteria, was similar physiologically. Similar organisms were shown to be present in the continental shelf waters of the Atlantic Ocean and in the Sargasso Sea. Carlucci and Strickland (1968) isolated a number of bacteria from the surface waters of the Pacific Ocean which were capable of oxidizing ammonium–nitrogen. In addition, Hattori and Wada (1971) found evidence for the presence of nitrifying bacteria in the upper 125 m over a wide area in the Pacific Ocean, but they were detected only occasionally in samples from deeper water. From these and other results (Kimata *et al.*, 1961, 1963a, b; Vargues and Brisou, 1963) there is now good evidence that nitrifying bacteria are widely distributed in the sea. The rates of these bacterially induced oxidation processes observed in culture are generally too low to account satisfactorily for the rates of production of nitrite and nitrate–nitrogen found for the sea. The bacterial oxidation of marine ammonium–nitrogen is not limited to well oxygenated water since Carlucci and MacNally (1969) have demonstrated that this process can occur in poorly oxygenated environments.

Carlucci et al. (1970) have made a further examination of the rate at which ammonium–nitrogen is formed during the decay of phytoplankton cells and faecal pellets. In contrast to the approach used in most earlier studies, these workers used low concentrations of particulate nitrogen, and their results showed that, in the absence of nitrifying bacteria, ammonium–nitrogen accumulated in the water. However, when inocula of nitrifying bacteria were added, nitrite–nitrogen accumulated concomitantly with the release of ammonium–nitrogen. The sequential production of ammonium–nitrogen and nitrite–nitrogen observed by the earlier workers was presumably the result of the small populations of nitrifying bacteria present during the early stages of the process. This pattern of events does not appear to happen in nature. The rates of oxidation of the ammonium–nitrogen derived from the particulate material observed by Carlucci et al., were greater than those achieved under comparable conditions by cultures of the nitrifying bacteria when supplied with preformed ammonium–nitrogen. It seems likely therefore that the bacterially promoted oxidation of ammonium–nitrogen is very dependent upon surface and other physico-chemical factors. At the present time knowledge is insufficient to allow either a fully satisfactory description of the process of nitrification in the sea or an evaluation of its importance.

11.5.3.3. *The reduction of nitrate–nitrogen*

A proportion of the heterotrophic bacteria in the sea is able to use nitrate–nitrogen as a source of nitrogen. Thus, in common with most of the phytoplanktonic algae they have the ability of assimilatory nitrate reduction. In this type of process the nitrogen is normally used solely for the formation of cellular nitrogenous substances. However, Vaccaro and Ryther (1960) have shown that nitrite–nitrogen is produced by marine unicellular algae under conditions of low incident light intensity if nitrate–nitrogen is present in sufficiently large amounts. Such conditions exist in temperate coastal waters in the autumn and in the lower part of the euphotic zone in tropical oceanic waters. Increased concentrations of nitrite–nitrogen are, in fact, frequently found at, and below, the compensation depth (Fig. 11.14). Carlucci et al. (1970) showed that nitrite–nitrogen was also produced by *Skeletonema costatum* under low light intensities, and in addition demonstrated that vitamin B_{12} deficient cells of this organism also released nitrite–nitrogen. Thus, it seems that phytoplankton may produce nitrite–nitrogen in the sea as a result of assimilatory nitrate reduction in excess of the requirements of the cells. This process may therefore sometimes contribute to the formation of nitrite–nitrogen in the sub-surface layers of the oceans.

Some marine heterotrophic bacteria are also capable of using nitrate–nitrogen as an alternative respiratory electron acceptor in place of oxygen,

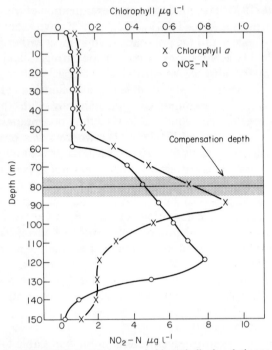

Fɪɢ. 11.14. Depth profiles of nitrite–nitrogen and chlorophyll a in relation to the compensation depth in the Sargasso Sea (From Vaccaro and Ryther, 1960).

particularly when the dissolved oxygen concentration is low. The nitrite–nitrogen so formed can be further reduced to ammonium–nitrogen by some species of facultative anaerobes. There is a further pathway by which the oxidized forms of inorganic nitrogen can be used as respiratory electron acceptors by some facultative anaerobic bacteria. In this way, nitrite–nitrogen can be reduced to nitrous oxide or even to elemental nitrogen. It is important to differentiate between the three types of nitrate reduction, and to confine the term denitrification to the process which leads to the production of nitrous oxide or elemental nitrogen.

It has long been recognized that the bacterial reduction of nitrate–nitrogen in the marine environment involves endergonic reactions and that an abundant supply of organic matter is necessary. For this reason it was often suggested that, except in highly localized environments, the process was not likely to be of great significance in the sea (Zobell, 1946). However, Carlucci and Schubert (1969) have recently reported the isolation of marine bacteria capable of reducing nitrate–nitrogen under anaerobic conditions when supplied low concentrations of glucose. Thus, the occurrence of these processes in oceanic regions is unlikely to be restricted because of limitations in the supply of organic matter.

Respiratory reduction of nitrate–nitrogen and denitrification are faculta-tive abilities, and nitrate–nitrogen and oxygen are interchangeable as res-piratory electron acceptors. Skerman and MacRae (1957) showed that nitrate reduction was inhibited by oxygen concentrations above about $0.15 \text{ cm}^3 \text{ l}^{-1}$. Although concentrations of oxygen lower than this allow respira-tory reduction to occur, sometimes with the formation of ammonium–nitrogen, a greater degree of anoxia is generally considered to be required for denitrifi-cation (Kluyver, 1953).

A number of species of bacteria which are able to reduce nitrate–nitrogen can be isolated from the marine environment (Zobell, 1946). This ability is widely distributed among a number of bacterial genera, but considerable variation probably exists between species in their ability to bring about the various types of nitrate reduction. In addition, some organisms are probably capable of both respiratory reduction and denitrification depending upon the prevailing oxygen concentration. Sreenivasan and Venkataraman (1956) isolated 32 strains of nitrate reducing bacteria from sea water from off the Indian coast. More recently, Hattori and Wada (1971) were able to detect large numbers of nitrate reducers in the tropical South Pacific Ocean, but were able to demonstrate denitrifying ability (release of elemental nitrogen) only in a small proportion of the water samples. Wada and Hattori (1971) have reported similar results for other regions of the South Pacific Ocean and for the coastal waters off Japan.

Brandhorst's (1959) studies of the eastern tropical Pacific Ocean indicated nitrite–nitrogen production to occur in the upper 100 m of water (Fig. 11.15).

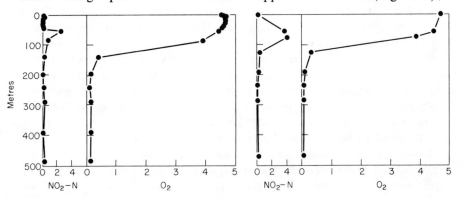

FIG. 11.15. Depth profiles of dissolved oxygen ($\text{cm}^3 \text{ l}^{-1}$) and nitrite–nitrogen ($\mu$g-at l^{-1}) in the eastern tropical Pacific Ocean showing nitrite–nitrogen maxima in well-oxygenated water (From Brandhorst, 1959).

The concentration of oxygen at some of these points was sufficiently high to
make respiratory reduction of nitrate–nitrogen unlikely, and possibly
interrupted assimilatory reduction by phytoplankton or nitrification occurs.
At other stations, Brandhorst found secondary nitrite maxima at greater
depth (Fig. 11.16). The oxygen concentrations at these points were very low

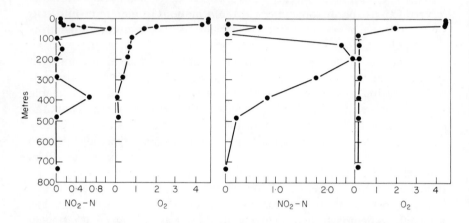

FIG. 11.16. Depth profiles of dissolved oxygen (cm^3 l^{-1}) and nitrite–nitrogen (μg-at l^{-1}) in the
eastern Pacific Ocean showing nitrite–nitrogen maxima in the presence of low concentrations of
dissolved oxygen (From Brandhorst, 1959).

(about 0.1 cm^3 l^{-1}). Wooster et al. (1965) have reported similar secondary
maxima coincident with low dissolved oxygen concentrations (< 0.2 cm^3 l^{-1})
for water lying below the Peru Current. Wooster (1967) also observed that
nitrite–nitrogen concentration maxima were associated with amounts of
dissolved oxygen of 0.15 cm^3 l^{-1} or less in the tropical Pacific Ocean. Thomas
(1966b) confirmed the occurrence of a nitrite-rich layer in the deep water of
the Pacific Ocean north of the equator. The expected nitrate–nitrogen concentrations calculated on the basis of the observed concentrations of dissolved
phosphate exceeded those actually found.

When they examined the data collected on several cruises in the eastern
tropical Pacific, Fiadero and Strickland (1968) observed that there were
marked differences between the shapes of the plots of nitrate–nitrogen or
phosphate against apparent oxygen utilization (see section 11.5). These
differences seemed to be the result of a loss of combined nitrogen rather than
interrupted mineralization and oxidation of organic nitrogen. No accumulation of nitrite–nitrogen was noted in water containing more than about

$0.2 \, cm^3 \, l^{-1}$ oxygen, but some samples which contained greater amounts of oxygen than this also showed an apparent deficiency in the amount of nitrate–nitrogen present. Only a small proportion of the deficiency of nitrate–nitrogen was accounted for by the amounts of nitrite–nitrogen present, and denitrification is probably the explanation. However, it is unlikely that respiratory nitrate reduction or denitrification occurs at oxygen concentrations greater than about $0.2 \, cm^3 \, l^{-1}$. The apparent loss of nitrate under more oxygenated conditions could be the result of the oxygen concentrations being raised by mixing with another more highly oxygenated water mass after nitrate reduction had occurred. The occurrence of nitrite-rich water in association with relatively high concentrations of dissolved oxygen in other areas can be explained in the same way.

Craig and Gordon (1963) have shown that some of the waters containing high concentrations of nitrite–nitrogen are enriched with nitrous oxide; this suggests the occurrence of denitrification. Goering and Dugdale (1966) found that incubated ^{15}N-nitrate enriched samples taken from the deep water off the Peruvian coast gave no direct evidence of denitrification; however, samples containing high concentrations of nitrite–nitrogen collected off the west coast of Mexico did yield some elemental ^{15}N (Goering, 1968). Goering and Cline (1970) have demonstrated the formation of nitrite–nitrogen and a net loss of combined inorganic nitrogen from solution during incubation of water samples taken from an oxygen minimum layer. Little ammonium nitrogen was formed in these samples, and no nitrous oxide could be detected. Similar results were reported by Barbaree and Payne (1967) who studied the products of denitrification produced by marine bacteria grown anaerobically under helium. Although elemental nitrogen was liberated, nitrous oxide was formed in only trace quantities when the organisms were grown in the presence of high concentrations of nitrate–nitrogen.

Direct evidence of denitrification in anoxic waters containing sulphide has been obtained by Richards and Benson (1961) by measurements of $N_2:Ar$ ratios. There is no doubt that the bacterial flora of many samples of sea water have the potential to bring about denitrification, and there seems no reason to doubt that respiratory reduction of nitrate–nitrogen may occur when the oxygen concentration is low. The concentration of oxygen in sea water in which nitrite–nitrogen accumulates is often similar to the critical concentrations of oxygen which Skerman and MacRae (1957) have found to be optimal for nitrate reduction. The evidence for denitrification in such waters is mainly indirect; however, most of the relevant observations, particularly the low concentrations of ammonium–nitrogen generally found in these waters certainly suggests that any reduction of nitrite–nitrogen which occurs proceeds by dentrification. On the other hand, nitrification is apparently not excluded by the low concentrations of oxygen. It is perhaps rather surprising that denitrification occurs appreciably in a water containing the low but finite concentrations of dissolved oxygen typical of many of these situations. The experimental evidence that the process is inhibited until lower concentrations of oxygen are reached was obtained in the presence of high concentrations of organic matter and nitrate–nitrogen. It is possible that the marine bacteria which are responsible for denitrification differ in their responses to low concentrations of oxygen, or that the balance between the further reduction of nitrite–nitrogen to ammonium–nitrogen and denitrification is affected by the low concentrations of organic matter available in the sea. Denitrification, which seems likely to place a lower demand on electron donor systems than does the reduction of nitrite–nitrogen to ammonium–nitrogen, might therefore be favoured.

If appreciable denitrification does occur (e.g. in the oxygen depleted waters of the Pacific) the associated regions may be important sites for the cycling of elemental nitrogen. As noted above, appreciable formation of elementary nitrogen from its combined forms may be necessary in order to maintain a geochemical steady state with respect to this element.

11.5.3.4. *Nitrogen fixation*

In general, the marine environment contains few species of organisms which are known to fix elemental nitrogen. The terrestrial bacteria, or their marine counterparts, which possess this ability are not widely distributed in the sea, and the abundant supply of organic matter which such organisms require is also generally absent (Zobell, 1946). The blue-green algae also have a limited distribution in the oceans. However, the availability of new techniques for detecting nitrogen fixation has recently stimulated research on this topic.

Many species of marine blue-green algae have been tested for their ability to fix elementary nitrogen, but with the exception of species of *Trichodesmium*, they are all littoral or coastal species. It is, however, the planktonic species which occur in the open sea which seem likely to be most important in the nitrogen economy of the oceans. Stewart (1971) has recently reviewed the distribution of these organisms in the sea.

Trichodesmium spp. usually occur in quantity only in thermally stratified water which is depleted of nutrients and which is sparsely populated with other planktonic algae. Nitrogen fixation in natural samples by *Trichodesmium* was first demonstrated by incubation experiments (Dugdale *et al.*, 1961), and in the course of subsequent work (Dugdale *et al.*, 1964; Goering *et al.*, 1966) high rates of nitrogen fixation were recorded for samples of the algae taken from the tropical Atlantic Ocean and the Arabian Sea. Dugdale and Goering (1967) found that the rate of assimilation of elemental nitrogen by this alga could equal or even exceed that of ammonium–nitrogen assimilation found for tropical water samples. *Trichodesmium* assimilates ammonium–nitrogen more rapidly than elemental nitrogen or nitrate–nitrogen. Nitrogen fixation can apparently occur concurrently with assimilation of other forms of combined inorganic nitrogen. The fact that the presence of other nitrogen sources does not preclude nitrogen fixation strengthens the view that the contribution made by *Trichodesmium* to the supply of combined inorganic nitrogen may have a profound influence on the nitrogen budget of the tropical oceans. The ability of *Trichodesmium* to grow so successfully in nutrient depleted water raises interesting questions about the role of other nutrients in controlling nitrogen fixation. Dugdale *et al.* (1961) reported that nitrogen fixation by this organism was inhibited when incubated samples were enriched with phosphate, vitamins and trace metals, and it is likely that appreciable concentrations of such nutrients may be more important in inhibiting nitrogen fixation by *Trichodesmium* than is the presence of combined inorganic nitrogen.

The nitrogen fixed by blue-green algae may be made available to other organisms not only in the form of cellular nitrogen but also in the form of excreted extra-cellular nitrogenous compounds. It is not known if *Trichodesmium* excretes appreciable quantities of such material, but it seems likely as many species of blue-green algae do so (Stewart, 1971).

A number of the bacterial species which are found in organic-rich sediments are known to be able to fix elementary nitrogen. Brooks *et al.* (1971) have demonstrated *in situ* nitrogen fixation in an estuarine sediment and were able to isolate a *Clostridium* sp. which was shown to be capable of nitrogen fixation when grown in a medium free from fixed nitrogen. The bulk of the nitrogen-fixing activity in this sediment was located in the subsurface layer of consolidated material and little activity occurred either in the unconsolidated surface layer or in the deeper layers.

11.6. STOICHIOMETRY OF THE UPTAKE AND REGENERATION OF THE NUTRIENT ELEMENTS AND OF OXYGEN

The fact that nitrate–nitrogen and phosphate are removed from sea water in constant proportions to each other as a result of phytoplankton growth was noted by Harvey (1926). Redfield (1934) examined the data then available on the concentrations of these two constituents in the sea, and showed that water from different localities and depths also contained these two nutrients in approximately constant proportions. Cooper (1937, 1938b) confirmed these observations, and after correcting the earlier values for a "salt error" involved in the determination of the inorganic phosphate, reported a ratio of N:P (by atoms) of 15:1. It was recognized at this time that, although this ratio held for many water masses, variations occurred in some areas. For example, anomalously high values were found in the Mediterranean Sea, and low ratios occurred in water from the South Atlantic. Subsequently, Phiefer and Thompson (1937) reported lower values for the N:P ratio (mean value 12·7:1) for the water in the San Juan Channel, and values ranging between 15·6:1 and 10·5:1 have been reported by a number of workers (Ketchum et al., 1958; Megia and Lao, 1955; Armstrong and Butler, 1960; McAllister et al., 1960; Stefánsson and Richards, 1963, Stefánsson, 1968). N:P ratios of between 5:1 and 8:1 are common in inshore waters (Riley and Conover, 1965; Stefánsson and Richards; 1963; Pratt, 1965) and in surface waters in which the ratio may approach zero after phytoplankton growth has removed the greater part of the nitrate. However, values as high as 25:1 have also been reported for some inshore waters (Ewins and Spencer, 1967). Ketchum et al. have pointed out that the N:P ratio of zooplankton is higher than that of phytoplankton, and the soluble excretion products of these animals and others in the food web must, therefore, have a relatively low N:P ratio. These effects may make some contribution to the low N:P ratios often observed during the summer in surface waters. Some of the low N:P ratios reported may be partially the result of ammonium–nitrogen not being included when assessing the nitrogen. Furthermore, the methods used for the determination of nitrate–nitrogen in the earlier studies were not very satisfactory. Nevertheless there are undoubtedly real variations in the N:P ratios of various water masses and Cooper (1937) urged a search for the underlying cause. Stefánsson (1968) has recently demonstrated the usefulness of the N:P ratio as a means of identifying various water masses in the Irminger Sea.

The earlier analyses of the elemental composition of plankton (Redfield, 1934; Wakesman et al., 1937; Cooper, 1935b; von Brand, 1938) suggested that the ratio of N:P in this material is also about 16:1. More recent analyses have shown that the N:P ratio in the cells of cultured phytoplankton can vary from 17·4:1 to 5·5:1 (Parsons et al., 1961), and the elemental composition of individual zooplankton species also varies very considerably (Corner and Davies, 1971). Values for coastal phytoplankton of between 16·5:1 and 13·5:1 were obtained by Antia et al. (1963), and McAllister et al. (1960) reported values of 17:1 for an oceanic population. Ketchum (1939) has noted that Nitzschia closterium (Phaeodactylum tricornutum) initially assimilated nitrogen and phosphorus with a normal ratio of 15:1. However, when the medium becomes impoverished in one nutrient, the cells assimilate one nutrient preferentially. It seems likely therefore that the anomalous values reported from time to time for N:P ratios in phytoplankton may reflect a relatively transient

state; on a sufficiently long time-scale, these organisms will remove ɪ
and phosphorus from sea water in the approximate ratio 16:1. Certaɪnɪy, ᴜɪᴇ
ratio in which these elements are removed from sea water by the growth of the
phytoplankton in nature is often close to this value (Ketchum et al., 1958;
Stefánsson and Richards, 1963). This presumably reflects the integrated
effects of phytoplankton growth better than does the instantaneous relative
amounts of these elements in a standing population. It also seems, in practice,
that the assumption that nitrogen and phosphorus generally occur in these
proportions in marine organic matter provides a satisfactory basis for the
calculation of the oxygen demand of this material during regeneration.

The concept of apparent oxygen utilization was introduced by Redfield
(1942) (see also Chapter 8). In this concept it is assumed that the oxygen
concentration in a water mass which is isolated from the surface will be
changed only by biological activity which, in deep water, will be limited to the
oxygen consumption involved in remineralization of organic matter. Assuming that it was originally in equilibrium with the atmosphere, the original
oxygen concentration of the water can be calculated from a knowledge of its
salinity and temperature. Any anomaly between the assumed and measured
oxygen concentrations is ascribed to photosynthesis or to the biological
oxidation of organic matter. The apparent utilization or production of oxygen
should, therefore, be related to the nutrient changes which accompany these
processes; in turn, these should be related to the relative amounts of the
nutrient elements and to the carbon in the organic material. Assuming $C:N:P$
ratios of $106:16:1$, remineralization can be represented by the equation
(Richards, 1965)

$$(CH_2O)_{106}(NH_3)_{16}H_3PO_4 + 138O_2 = $$
$$106CO_2 + 122H_2O + 16HNO_3 + H_3PO_4$$

Riley (1951 and 1956) has discussed the effects of variations in the composition of plankton in relation to the oxygen demands for regeneration and
came to the conclusion that between 250 and 300 atoms of oxygen might be
required for each atom of phosphorus mineralized. For large masses of water
and over a sufficiently long time scale ratios of $106:16:1$ are usually adopted.

Pytkowicz (1971) has pointed out that the apparent oxygen utilization in a
mixture of two different water masses is related in a simple linear fashion to
the proportion of each in the mixture. A change in the oxygen utilization
at a point does not necessarily indicate oxygen utilization, but can be the
result of the mixing of waters having different apparent oxygen utilization
values. However, the resultant apparent oxygen utilization is a true reflection

of the magnitude of the oxidative changes which have occurred in each water mass and of the relative proportions of the two which make up the mixture. Oxygen solubility is not a linear function of temperature, but if the temperatures of the two water masses do not differ by more than a few degrees, the uncertainties which this causes can be neglected (Pytkowicz and Kester,1966).

Effective use has been made of these ideas for relating nutrient uptake and photosynthesis in surface waters (Stefánsson and Richards, 1964; Ketchum and Corwin, 1965). The apparent oxygen utilization in deeper waters is related to the amounts of inorganic nitrogen and phosphate released. The total amounts of inorganic nitrogen and phosphate present will include any pre-formed material in addition to the nutrient salts of oxidative origin. Plots of the concentration of nutrients against apparent oxygen utilization should be linear and the intercept at zero apparent oxygen utilization will correspond to the concentration of pre-formed nutrients. The pre-formed nutrient concentrations should be conservative and be of potential use for hydrographical studies. Treatments based on these assumptions often agree well with the observed distributions of the nutrients and of dissolved oxygen, and such an agreement has been observed for deep waters of the Atlantic and Pacific Oceans (Redfield et al., 1963). These concepts have recently been increasingly adopted as valuable parameters for use in hydrographical investigations (Pytkowicz, 1968; Pytkowicz and Kester, 1966, Culbertson and Pytkowicz, 1970) and other investigations (Fiadero and Strickland, 1968).

Silicon would not be expected to show a close stoichiometric relationship to the other nutrient elements and to the oxygen consumption because the proportions of diatoms in phytoplankton crops and their degree of silicification varies considerably. However, Richards (1958) reported ratios of ΔSi: ΔN:ΔP of 16:16:1 for the Atlantic. Stefánsson (1968) also found ratios of ΔO:ΔSi:ΔN:ΔP of 276:16:14·3:1 for the Irminger Sea, but, more recently Schott and Ehrhardt (1969) reported ratios of ΔO:ΔSi:$\Delta N \Delta P$ of 276:7·2: 14·7:1·2 from the analyses of over 400 samples taken from the North Sea. It appears that the extent of regeneration of silicic acid can vary considerably. Grill and Richards (1964) found ΔO:ΔSi ratios of 276:23 during the decomposition of diatoms in laboratory experiments. Only a proportion of the silicic acid regeneration seemed to be closely associated with the oxidation and regeneration of the nitrogen and phosphorus of the organic matter. Park (1967) has commented on the apparent oxygen utilization-silicon plots for samples taken off the Oregon coast. At lower concentrations of silicic acid the regression agreed well with the value reported by Grill and Richards. The data reported by Park suggested that at depths below 1200 m in this area, increases in the concentration of silicic acid are not related to oxygen demand and are presumably the result of the solution of particulate silicon.

REFERENCES

Antia, N. J., McAllister, C. D. Parsons, T. R., Stephens, K. and Strickland, J. D. H. (1963). *Limnol. Oceanogr.* **8**, 166.
Armstrong, F. A. J. (1954). *J. mar. biol. Ass. U.K.* **33**, 381.
Armstrong, F. A. J. (1955). *J. mar. biol. Ass. U.K.* **34**, 223.
Armstrong, F. A. J. (1957). *J. mar. biol. Ass. U.K.* **36**, 317.
Armstrong, F. A. J. (1958). *J. mar. Res.* **17**, 23.
Armstrong, F. A. J. (1965a). In "Chemical Oceanography" (J. P. Riley and G. Skirrow, eds.), Vol 1, pp 409–432. Academic Press, London and New York.
Armstrong, F. A. J. (1965b). In "Chemical Oceanography" (J. P. Riley and G. Skirrow, eds.), Vol. 1, pp 323–364. Academic Press, London and New York.
Armstrong, F. A. J., and Butler, E. I. (1959). *J. mar. biol. Ass. U.K.* **38**, 41.
Armstrong, F. A. J., and Butler, E. I. (1960). *J. mar. biol. Ass. U.K.* **39**, 525.
Armstrong, F. A. J., and Butler, E. I. (1962). *J. mar. biol. Ass. U.K.* **42**, 253.
Armstrong, F. A. J., and Butler, E. I. (1968). *J. mar. biol. Ass. U.K.* **48**, 153.
Armstrong, F. A. J., and Harvey, H. W. (1950). *J. mar. biol. Ass. U.K.* **29**, 145.
Atkins, W. R. G. (1923). *J. mar. biol. Ass. U.K.* **13**, 119.
Atkins, W. R. G. (1945). *Nature, Lond.* **156**, 446.
Atkins, W. R. G., Jenkins, P. G. and Warren, F. T. (1954). *J. mar. biol. Ass. U.K.* **33**, 497.
Ayyakkannu, K., and Chandramohan, D. (1971). *Mar. Biol.* **11**, 201
Barbaree, J. M., and Payne, W. J. (1967). *Mar. Biol.* **1**, 136.
Berger, W. H. (1968). *Science* **159**, 1237.
Beers, J. R. (1964). *J. Cons. int. Explor. Mer.* **28**, 123.
Benson, B. B., and Parker, P. D. M. (1961). *Deep-Sea Res.* **7**, 237.
Bien, G. S., Contois, D. E. and Thomas, W. H. (1958). *Geochim. Cosmochim. Acta* **14**, 35.
Bienati, N. L., and Comes, R. A. (1971). In "Fertility of the Sea" (J. D. Costlow, ed.) pp 51–70. Gordon and Breach, New York, London and Paris.
von Brand, T. (1938). *J. Cons. int. Explor. Mer.* **31**, 187.
von Brand, T., Rakestraw, N. W. and Renn, C. E. (1937). *Biol. Bull. Woods Hole* **72**, 165.
von Brand, T., Rakestraw, N. W. and Renn, C. (1939). *Biol. Bull. Woods Hole* **77**, 285.
von Brand, T. and Rakestraw, N. W. (1941). *Biol. Bull. Woods Hole* **81**, 63.
von Brand, T., Rakestraw, N. W. and Zabor, J. W. (1942). *Biol. Bull. Woods Hole* **83**, 273.
Brandhorst, W. (1959). *J. Cons. int. Explor. Mer.* **25**, 3.
Brooks, R. H., Brezonik, P. L., Putman, H. D. and Keirn, M. A. (1971). *Limnol. Oceanogr.* **16**, 701.
Burton, J. D. (1970). *J. Cons. int. Explor. Mer.* **33**, 141.
Burton, J. D., Leatherhead, T. M. and Liss P. S. (1970). *Limnol. Oceanogr.* **15**, 473.
Burton, J. D. and Liss, P. S. (1968). *Nature, Lond.* **220**, 905.
Calvert, S. E. (1968) *Nature, Lond.* **219**, 919.
Carlucci, A. F. and McNally, P. M. (1969). *Limnol. Oceanogr.* **14**, 736.
Carlucci, A. F. and Schubert, H. R. (1969). *Limnol. Oceanogr.* **14**, 187.
Carlucci, A. F. and Strickland, J. D. H. (1968). *J. exp. mar. Biol. Ecol.* **2**, 156.
Carlucci, A. F., Hartwig, E. O. and Bowes, P. M. (1970). *Mar. Biol.* **7**, 161.
Carritt, D. E. (1954). *Deep-Sea Res.* **2**, 59.

Carritt, D. E. and Goodgal. S. (1954). *Deep-Sea Res.* 1, 224.
Chow, T. J. and Goldberg, E. D. (1960). *Geochim. Cosmochim. Acta,* 20, 192.
Clarke, F. W. (1924). *U.S. geol. Surv. Bull.* 770, 841.
Clowes, A. J. (1938). *"Discovery" Rep.* 19, 1.
Cooper, L. H. N. (1933). *J. mar. biol. Ass. U.K.* 18, 677.
Cooper, L. H. N. (1935a). *J. mar. biol. Ass. U.K.* 19, 755.
Cooper, L. H. N. (1935b). *J. mar. biol. Ass. U.K.* 20, 197.
Cooper, L. H. N. (1937). *J. mar. biol. Ass. U.K.* 22, 177.
Cooper, L. H. N. (1938a). *J. mar. biol. Ass. U.K.* 22, 183.
Cooper, L. H. N. (1938b). *J. mar. biol. Ass. U.K.* 23, 179.
Cooper, L. H. N. (1948). *J. mar. biol. Ass. U.K.* 27, 322.
Cooper, L. H. N. (1951). *J. mar. biol. Ass. U.K.* 30, 21.
Cooper, L. H. N. (1952). *J. mar. biol. Ass. U.K.* 30, 511.
Corner, E. D. S. and Davies, A. G. (1971). *Adv. mar. Biol.* 9, 101.
Corner, E. D. S. and Newell, B. S. (1967). *J. mar. biol. Ass. U.K.* 47, 131.
Correll, D. L. (1965). *Limnol. Oceanogr.* 10, 364.
Cowey, C. B. and Corner, E. D. S. (1963). *J. mar. biol. Ass. U.K.* 43, 495.
Craig, H. and Gordon, L. I. (1963). *Geochim. Cosmochim. Acta.* 27, 949.
Culberson, C. and Pytkowicz, R. M. (1970). *J. oceanogr. Soc. Japan* 26, 95.
Darley, W. M. and Volcani, B. E. (1969). *Exptl. Cell Res.* 58, 334.
Deacon, G. E. R. (1933). *"Discovery" Rep.* 7, 173.
Degens, E. T. (1970). *In* "Organic Matter in Natural Waters" (D. W. Hood, ed.)
　　pp 77–106. Inst. Marine Science, Alaska. Occ. Publ. No. 1.
Dugdale, R. C. and Goering, J. J. (1967). *Limnol. Oceanogr.* 12, 196.
Dugdale, R. C. and MacIsaac, J. J. (1971). *Investigación pesq.* 35, 299.
Dugdale, R. C., Menzel, D. W. and Ryther, J. H. (1961). *Deep-Sea Res.* 7, 298.
Dugdale, R. C., Goering, J. J. and Ryther, J. H. (1964). *Limnol. Oceanogr.* 9, 507.
Duursma, E. K. (1961). *Netherl. J. Sea Res.* 1, 1.
Duursma, E. K. (1965). *In* "Chemical Oceanography" (J. P. Riley and G. Skirrow,
　　eds.), Vol I, pp 433–475. Academic Press, London and New York.
Ehrhardt, M. (1969). *Kieler Meersforch.* 25, 71.
Emery, K. O. and Rittenberg, S. C. (1952). *Bull. Amer. Ass. petrol. Geol.* 36, 735.
Eppley, R. W., Rogers, J. N. and McCarthy, J. J. (1969). *Limnol. Oceanogr.* 14, 912.
Ewins, P. A. (1964). M.Sc. Thesis, University of Wales.
Ewins, P. A. and Spencer, C. P. (1967). *J. mar. biol. Ass. U.K.* 47, 533.
Fiadero, M. and Strickland, J. D. H. (1968). *J. mar. Res.* 26, 187.
Flemer, D. A. and Biggs, R. B. (1971). *J. Fish. Res. Bd. Can.* 28, 911.
Fraga, F. (1966). *Deep-Sea Res.* 13, 413.
Fraga, F. (1969). *Investigación pesq.* 33, 163.
Ganapati, P. N., La Fond, E. C. and Bhavanarayana, P. V. (1956). *Ind. Acad. Sci.* 44,
　　68.
Gardener, A. C. (1937). *J. Cons. int. Explor. Mer.* 12, 144.
Garrells, R. M. (1965). *Science.* 148, 69.
Goering, J. J. (1968). *Deep-Sea Res.* 15, 157.
Goering, J. J. and Cline, J. D. (1970). *Limnol. Oceanogr.* 15, 306
Goering, J. J. and Dugdale, R. C. (1966). *Science* 154, 505.
Goering, J. J. and Menzel, D. W. (1965) *Deep-Sea Res.* 12, 839.
Goering, J. J., Dugdale, R. C. and Menzel, D. W. (1966). *Limnol. Oceanogr.* 11, 614.
Gregor, B. (1968). *Nature, Lond.* 219, 360.

Griffin, J. J. and Goldberg, E. D. (1963). *In* "The Sea" (M. N. Hill, ed.) Vol. III, pp. 728–741. Interscience, New York.
Grill, E. V. (1970). *Deep-Sea Res.* **17**, 245.
Grill, E. V. and Richards, F. A. (1964). *J. mar. Res.* **22**, 51.
Gunnerson, C. G. and Emery, K. O. (1962). *Limnol. Oceanogr.* **7**, 14.
Hamilton, R. D. (1964). *Limnol. Oceanogr.* **9**, 107.
Hargrave, B. T. and Geen, G. H. (1968). *Limnol. Oceanogr.* **13**, 332.
Harris, R. C. (1966). *Nature, Lond.* **212**, 275
Harris, E. (1959). *Bull. Bingham oceanogr. Coll.* **17**, 31.
Harvey, H. W. (1926). *J. mar. biol. Ass. U.K.* **14**, 71.
Harvey, H. W. (1955). "The Chemistry and Fertility of Sea Water", 224 pp. Cambridge University Press, Cambridge, U.K.
Hattori, A. and Wada, E. (1971). *Deep-Sea Res.* **18**, 557.
Helgerson, H. C. and Mackenzie, F. T. (1970). *Deep-Sea Res.* **17**, 877.
Hellebust, J. A. (1965). *Limnol. Oceanogr.* **10**, 192.
Hurley, P. M., Hart, S.M., Pinson, W. H. and Fairbain, H. W. (1959). *Bull. geol. Soc. Amer.* **70**, 1622.
Hutchinson, G. E. (1944). *Amer. Scient.* **32**, 178.
Ingri, N. (1959). *Acta chem. Scand.* **13**, 758.
Ishii, J. and Ishikawa, T. (1964). *In* "Studies in Oceanography" (Hidaka Commemorative Volume), pp 288. Tokyo.
Jawed, M. (1969). *Limnol. Oceanogr.* **14**, 748.
Johannes, R. E. (1964). *Limnol. Oceanogr.* **9**, 192.
Johannes, R. E. (1965). *Limnol. Oceanogr.* **10**, 434.
Johannes, R. E. (1968). *Adv. Microbiol. Sea,* **1**, 203.
Johannes, R. E. and Webb, K. L. (1965). *Science,* **150**, 7677.
Johannes, R. E. and Webb, K. L. (1970). *In* "Organic Matter in Natural Waters" (D. W. Hood, ed.), pp 257–273. Inst. Marine Science, Alaska. Occ. Publ. No. 1.
Jørgensen, E. G. (1955). *Physiol. Plant.* **8**, 846.
Joussot-Dubien, J. and Abdelmajid Kadiri (1970). *Nature, Lond.* **227**, 700.
Kalle, K. (1937). *Ann. Hydrogr. Berlin,* **65**, 1.
Kamatani, A. (1971). *Mar. Biol.* **8**, 89.
Kester, D. R. and Pytkowicz, R. M. (1967). *Limnol. Oceanogr.* **12**, 243.
Ketchum, B. H. and Corwin, N. (1965). *Limnol. Oceanogr. Suppl.* **10**, R 148.
Ketchum, B. H. and Keen, J. (1948). *J. mar. Res.* **7**, 17.
Ketchum, B. H., Corwin, N. and Keen, D. J. (1955). *Deep-Sea Res.* **2**, 172.
Ketchum, B. H., Vaccaro, R. F. and Corwin, N. (1958) *J. mar. Res.* **17**, 282.
Kimata, M. A., Kawai and Yoshida, Y. (1961). *Bull. Jap. Soc. scient. Fish.* **27**, 593.
Kimata, M. A., Kawai, A. and Yoshida, Y. (1963a). *Bull. Jap. Soc. scient. Fish.* **29**, 1027.
Kimata, M. A., Kawai, A. and Yoshida, Y. (1963b). *Bull. Jap. Soc. scient. Fish.* **29**, 1031.
Kittredge, J. S., Isabell, A. F. and Hughes, R. R. (1967). *Biochemistry* **6**, 289.
Kittredge, J. S., Horiguchi, M. and Williams, P. M. (1969). *Comp. Biochem. Physiol.* **29**, 859.
Kluyver, A. J. (1953). *In* VI Int. Cong. Microbiol. Rome, pp 71–91. Fondazione Emanuele Paterno Viale Regina Margherita, Roma. Italy.
Krauskopf, K. B. (1956). *Geochim. Cosmochim. Acta* **10**, 1.
Krishnamurthy, K. (1967). *Int. Rev. ges. Hydrobiol.* **52**, 427.

Krogh, A. (1934). *Ecol. Monogr.* **4**, 430

Kuenzler, E. J. (1970). *J. Phycol.* **6**, 7.

Kuenzler, E. J. and Ketchum, B. H. (1962). *Biol. Bull. Woods Hole* **123**, 134.

Kuenzler, E. J., Guillard, R. L. and Corwin, N. (1963). *Deep-Sea Res.* **10**, 749.

Lees, H. and Quastel, J. H. (1946). *Biochem. J.* **40**, 815.

Lewin, J. C. (1954). *J. Gen. Physiol.* **37**, 589.

Lewin, J. C. (1955). *J. Gen. Physiol.* **39**, 1.

Lewin, J. C. (1961). *Geochim. Cosmochim. Acta* **21**, 182.

Lewin, J. C., and Chin-Hon Chen, (1968). *J. Phycol.* **4**, 161.

Lisitsin, A. P. (1961). *Okeanol. Issled. Akad. Nauk. S.S.S.R.* **3**, 42.

Liss, P. S. and Spencer. C. P. (1970). *Geochim. Cosmochim. Acta* **34**, 1073.

Lowndes, A. G. (1955). *Sci. Progress* **171**, 434.

Mackenzie, F. T. and Garrels, R. M. (1965). *Science* **150**, 57.

Mackenzie, F. T., Garrels, R. M., Bricker, O. P. and Brickley, F. (1967). *Science* **155**, 1404.

Maeda, H. and Takesue, K. (1961). *Rec. oceanogr. Wks. Japan* **6**, 112.

Makimoto, H., Maeda, H. and Era, S. (1955). *Rec. oceanogr. Wks. Japan* **2**, 106.

Marshall, S. M. and Orr, A. P. (1961). *J. mar. biol. Ass. U.K.* **41**, 463.

McAllister, C. D., Parsons, T. R. and Strickland, J. D. H. (1960). *J. Cons. int. Explor. Mer.* **25**, 240.

McGill, D. A., Corwin, N. and Ketchum, B. H. (1964). *Limnol. Oceanogr.* **9**, 27.

Megia, T. G. and Lao, R. G. (1955). *Philippine J. Fish.* **3**, 55.

Menzel, D. W. and Ryther, J. H. (1960). *Deep-Sea Res.* **6**, 351.

Menzel, D. W. and Ryther, J. H. (1964). *Limnol. Oceanogr.* **9**, 179.

Menzel, D. W. and Spaeth, J. P. (1962). *Limnol. Oceanogr.* **7**, 159.

Metcalf, W. G. (1969). *Deep-Sea Res.* **16** (Suppl.), 139.

Murray, J. and Renard, A. F. (1891). *Voyage of H.M.S. Challenger, 1872–76. Report of the Deep Sea Deposits.* H.M.S.O. London.

Newell, B. S. and Kerr, J. D. (1968). *Aust. J. mar. Freshwater Res.* **19**, 129.

Newell, B. S., Morgan, B. and Cundy, J. (1967). *J. mar. Res.* **25**, 201.

Nichol, J. A. C. (1960). *The Biology of Marine Animals,* pp 669, Pitman, London.

Paasche, E. (1971). *Physiol. Plant.* **25**, 294.

Park, K. (1967). *Limnol. Oceanogr.* **12**, 353.

Parsons, T. R., Stephens, K. and Strickland, J. D. H. (1961). *J. Fish. Res. Bd. Can.* **18**, 1001.

Perry, E. A. (1971). *Deep-Sea Res.* **18**, 921.

Phelps, A. (1937). *Trans. Amer. Phil. Soc.* **29**, 153.

Phiefer, L. D. and Thompson, T. G. (1937). *J. mar. Res.* **1**, 34.

Pocklington, R. (1971). *Nature, Lond.* **230**, 374.

Pomeroy, L. R., Mathews, H. M. and Min, H. S. (1963). *Limnol. Oceanogr.* **8**, 50.

Pomeroy, L., Smith, E. E. and Grant, C. M. (1965). *Limnol. Oceanogr.* **10**, 167.

Postma, H. (1971). *In* "Fertility of the Sea" (J. D. Costlow, ed.) pp 337–349. Gordon Breach, New York, London and Paris.

Pratt, D. M. (1965). *Limnol. Oceanogr.* **10**, 173.

Pytkowicz, R. M. *J. oceanogr. Soc. Japan.* **24**, 21.

Pytkowicz, R. M. (1971). *Limnol. Oceanogr.* **16**, 39.

Pytkowicz, R. M. and Kester, D. R. (1966). *Deep-Sea Res.* **13**, 373.

Pytkowicz, R. M. and Kester, D. R. (1967). *Limnol. Oceanogr.* **12**, 714.

Quin, L. D. (1965). *Biochemistry* **4**, 324.

Radcezewski, O. E. (1937). *Wiss. Erg. dtsch. atlant. Exped.* 'Meteor' **3**, 262.

Radcezewski, O. E. (1939). *In* "Recent Marine Sediments" (Trask eds), pp 496–502. American Association of Petroleum Geologists, Thomas Murby and Co., London.

Radkestraw, N. M. (1936). *Biol. Bull. Woods Hole* **71**, 133.

Rakestraw, N. W. and Hollaender, A. (1936). *Science* **84**, 442.

Ramamurthy, V. D. and Seshadri, R. (1966). *Current Sci.* **35**, 100.

Reddy, C. V., Murty, P. S. N. and Sankaranarayanan, V. N. (1968) *Current Sci.* **37**, 17.

Redfield, A. C. (1934). *In* "James Johnstone Memorial Volume" (R. J. Daniel, ed.), pp 176–192. University of Liverpool Press, Liverpool, U.K.

Redfield, A. C. (1942). *Pap. phys. Oceanogr. Met.* **9**, 1.

Redfield, A. C. and Keys, A. B. (1938). *Biol. Bull. Woods Hole* **74**, 83.

Redfield, A. C., Smith, H. P. and Ketchum, B. H. (1937). *Biol. Bull. Woods Hole* **73**, 421.

Redfield, A. C., Ketchum, B. H. and Richards, F. A. (1963). *In* "The Sea", (M. N. Hill, ed.), pp 26–87. Interscience, New York.

Remsen, C. C. (1971). *Limnol. Oceanogr.* **16**, 732.

Rex, R. W. and Goldberg, E. D. (1958). *Tellus*, **10**, 153.

Richards, F. A. (1958). *J. mar. Res.* **17**, 449.

Richards, F. A. (1960). *Deep-Sea Res.* **7**, 163.

Richards, F. A. (1965). *In* "Chemical Oceanography" (J. P. Riley and G. Skirrow, eds.), Vol I, pp 611–645. Academic Press, London.

Richards, F. A. and Benson, B. B. (1961). *Deep-Sea Res.* **7**, 254.

Richards, F. A. and Vaccaro, R. F. (1956). *Deep-Sea Res.* **3**, 214.

Riley, G. A. (1951). *Bull. Bingham oceanogr. Coll.* **13**, 1.

Riley, G. A. (1956). *Bull. Bingham oceanogr. Coll.* **15**, 324.

Riley, G. A. (1967). *Bull. Bingham oceanogr. Coll.* **19**, 72.

Riley, G. A. and Conover, S. A. M. (1956). *Bull. Bingham oceanogr. Coll.* **15**, 47.

Riley, J. P. and Chester, R. (1971). "Introduction to Marine Chemistry", 465 pp. Academic Press, London.

Riley, J. P. and Segar, D. (1970). *J. mar. biol. Ass. U.K.* **50**, 713.

Robinson, R. J. and Wirth, H. E. (1934). *J. Cons. int. Explor. Mer.* **9**, 15.

Rochford, D. J. (1951). *Aust. J. mar. Freshwater Res.* **2**, 1.

Rochford, D. J. (1967). *Aust. J. mar. Freshwater Res.* **18**, 1.

Rogall, E. (1939). *Planta* **29**, 279.

Ryther, J. H., Menzel, D. W. and Corwin, N. (1967). *J. mar. Res.* **25**, 69.

Schink, D. R. (1967). *Geochim. Cosmochim. Acta* **31**, 987.

Schott, F. and Ehrhardt, M. (1969). *Kieler Meeresforch.* **25**, 272.

Siegel, A. and Degens, E. T. (1966). *Science* **148**, 235.

Siever, R. (1968). *Earth planet. Sci. Letters* **5**, 106.

Sillén, L. G. (1967). *Science* **156**, 1189.

Skerman, V. D. B. and MacRae, I. C. (1957). *Can. J. Microbiol.* **3**, 506.

Slinn, D. J. (1956). *Ann. Rep. mar. biol. St. Port Erin* **69**, 29.

Slinn, D. J. (1958). *Ann. Rep. mar. biol. St. Port Erin* **71**, 24.

Solórzano, L. and Strickland, J D. H. (1968). *Limnol. Oceanogr.* **13**, 515.

Spencer, C. P. (1956). *J. mar. biol. Ass. U.K.* **35**, 621.

Sreenivasan, A. and Venkataraman, R. (1956). *J. Gen. Microbiol.* **15**, 241.

Stefánsson, U. (1968). *J. Cons. int. Explor. Mer.* **32**, 188.

Stefánsson, U. and Richards, F. A. (1963). *Limnol. Oceanogr.* **8**, 394.

300   C. P. SPENCER

Stefánsson, U. and Richards, F. A. (1964). *Deep-Sea Res.* **11**, 353.
Stewart, W. D. P. (1971). *In* "Fertility of the Sea" (J. D. Costlow, ed.), pp 537–564. Gordon and Breach, New York, London and Paris.
Strickland, J. D. H. and Austin, K. H. (1960). *J. Fish. Res. Bd. Can.* **17**, 337.
Strickland, J. D. H. and Solórzano, L. (1966). *In* "Some Contemporary Studies in Marine Science" (H. Barnes, ed.), pp 655–674. George Allen and Unwin Ltd., London.
Svedrup, H. U., Johnson, M. W. and Fleming, R. H. (1942). "The Oceans", pp 1087, Prentice Hall Inc. New York.
Szekielda, K. H. (1963). *Kieler Meeresforch* **19**, 16.
Thomas, W. R. (1966a). *Limnol. Oceanogr.* **11**, 393.
Thomas, W. H. (1966b). *Deep-Sea Res.* **13**, 1109.
Thomas, W. H., Renger, E. H. and Dodson, A. N. (1971). *Deep-Sea Res.* **18**, 65.
Thomsen, P. (1910). *Wiss. Meeresunters Abt. Kiel* **11**, 1.
Trask, P. D. (1939). *In* "Recent Marine Sediments" (P. D. Trask ed.), American Association of Petroleum Geologists, pp 428–453. Thomas Murley, Tulsa, U.S.A.
Vaccaro, R. F. (1962). *J. Cons. int. Explor. Mer.* **27**, 1.
Vaccaro, R. F. (1963). *J. mar. Res.* **21**, 284.
Vaccaro, R. F. (1965). *In* "Chemical Oceanography" (J. P. Riley and G. Skirrow, eds.), Vol I, pp 356–408. Academic Press, London.
Vaccarro, R. F. and Ryther, J. E. (1960). *J. Cons. int. Explor. Mer.* **25**, 260.
Vargues, H. and Brisou, J. (1963). *In* "Marine Microbiology" (C. H. Oppenheimer, ed.), pp 415–426. Charles C. Thomas, Springfield, U.S.A.
Wada, E. and Hattori, A. (1971). *Limnol. Oceanogr.* **16**, 766.
Waksman, S. A., Stokes, J. L. and Butler, M. R. (1937). *J. mar. biol. Ass. U.K.* **22**, 359.
Watson, S. W. (1962). Abst. 8th Int. Congr. Microbiol. Montreal, Nov. 1962.
Watson, S. W. (1963). *In* "Marine Microbiology" (C. H. Oppenheimer, ed.), pp 73–84. Charles C. Thomas, Springfield, U.S.A.
Watson, S. W. (1965). *Limnol. Oceanogr.* Suppl. **10**, R274.
Watt, W. D. and Hayes, F. R. (1963). *Limnol. Oceanogr.* **8**, 276.
Weaver, C. E. (1958). *Bull. Amer. Ass. petrol. Geol.* **42**, 254.
Webb, K. L. and Johannes, R. E. (1967). *Limnol. Oceanogr.* **12**, 376.
Whitfield, M. (1974) *J. Mar. Biol. Ass. U.K.* **54**, 565.
Whitledge, T. E. and Packard, T. T. (1971). *Investigación pesq.* **35**, 243.
Wooster, W. S. (1967). *J. mar. Res.* **25**, 154.
Wooster, W. S., Chow, T. J. and Barrett, I. (1965). *J. mar. Res.* **23**, 210.
Wollast, R. and De Broeu, F. (1971). *Geochim. Cosmochim. Acta* **35**, 613.
Yoshimasa, T. and Okabe, S. (1967). *J. oceanogr. Soc. Japan.* **23**, 1.
Zobell, C. E. (1933). *Science* **77**, 27.
Zobell, C. E. (1946). "Marine Microbiology" 240 pp. Chronica Botanica, Waltham, U.S.A.

Chapter 12

Biological and Chemical Aspects of Dissolved Organic Material in Sea Water

P. J. le B. WILLIAMS

Department of Oceanography,
University of Southampton, Southampton SO9 5NH, England

12.1. INTRODUCTION

The earliest studies of dissolved organic material in sea water appear to be those of Natterer and of Pütter at the end of the last century; Natterer attempted to determine the gross organic content of sea water; Pütter, at about the same time, was examining the role of dissolved organic material as an alternative food supply for planktonic animals. Pütter came to the conclusion that organic material in solution must be a significant source of food for such organisms. These early workers were severely limited by the analytical

L

methods available to them. The study of the organic chemistry of sea water really began in the 1930's with the careful work of Krogh and subsequently of Datsko on the dissolved organic carbon content of sea water and with Redfield's work on dissolved organic phosphorus.

After this initial start, no great advance was made until the late fifties. Thereafter, methods were devised for the semi-routine analysis of sea-water for dissolved organic carbon, nitrogen and phosphorus as well as for the analysis of sea water for individual compounds, such as vitamins, and intermediates of decomposition, e.g. sugars and amino acids. More recently, as a result of the growing awareness of the impact of man's activity on the environment, work has commenced on the determination of the concentrations and distributions of compounds such as hydrocarbons and their chlorinated derivatives in the oceans. In recent years, the research effort has increased as a result of the development of sensitive and specific physico-chemical analytical techniques. The subject of dissolved organic material in sea water is now being studied over a broad front, but despite this mounting effort, many aspects of the subject are clearly still in a rudimentary state.

The importance of dissolved organic material in sea water is now recognized in a variety of marine sciences and this topic acts as a strong link between subjects which otherwise have very little in common. It is a central problem in marine microbiology, an important consideration in the study of phytoplankton and zooplankton, and it is now recognized that dissolved organic compounds may influence the state of inorganic substances in sea water and vice versa. In this and other ways dissolved organic compounds may affect sedimentary processes in the oceans.

The subject of the organic chemistry of sea water has been reviewed by various authors. The accounts given by Vallentyne (1957), Hood (1963) and Riley (1965) cover the analytical aspects of the subject, whereas more general reviews of the chemistry and biology of organic matter in sea water have been published by Duursma (1961, 1965), Skopintsev (1971) and in Volume 5 of "The Sea" (Goldberg, 1974), the chapter by Menzel in the latter is particularly valuable. Two recent symposia have dealt in detail with many aspects of the organic material in sea water (Hood, 1970; Faust and Hunter, 1971).

12.2. SOURCES OF SUPPLY

Geochemists recognize that for inorganic compounds the sea must be considered as an open system in which the terrestrial inputs may be an important part of the budget. As will be discussed in subsequent sections, the gross input of organic material from external sources is probably normally of minor importance. However, in recent years the effect of man's activity on the oceans has led to instances in which the external inputs may be a significant or indeed the sole, source of supply. Accordingly, the sources of organic material in the sea will be considered under two general headings : external and internal.

12.2.1. EXTERNAL SOURCES

Three external sources may be recognized: the rivers, the atmosphere and the marine sediments. The input from the sediments of either soluble organic material released via decay processes, or oil from seepages is very poorly understood. Blumer (1972) has argued that natural oil seepages probably make only a minor contribution to the input of oil into the sea in comparison with that spilt and released into the sea by man. Only the inputs from the rivers and the atmosphere will be considered here in any detail.

12.2.1.1. *Input from the atmosphere*

There is now good evidence that there is a significant input of organic material to the oceans via the atmosphere. Williams (1971) estimated that dissolved organic material brought into the oceans by precipitation amounts to $2 \cdot 2 \times 10^{14} g \, yr^{-1}$, an amount comparable with the input from rivers (Table 12.1) and equivalent to about 1 % of the total net primary production. The input of dissolved organic material into the North Sea via precipitation has been estimated to be $10000 \, kg \, km^{-2} \, yr^{-1}$ (i.e. $10 \, g \, m^{-2} \, yr^{-1}$); on an annual basis this is equivalent to about 5% of both the standing amount of dissolved organic material and the annual production (Goldberg, 1971).

TABLE 12.1

Inputs, reservoirs and losses of organic material in the sea

Reservoirs	
Dissolved organic carbon (assuming 700 µg C l^{-1})	1×10^{18} g C
Particulate organic carbon (assuming 20 µg C l^{-1})	3×10^{16} g C
Plankton	5×10^{14} g C
Annual inputs	
Net primary productivity (assuming 100 g C fixed $m^{-2} \, yr^{-1}$)	$3 \cdot 6 \times 10^{16}$ g C
Rain (assuming 1 mg C l^{-1})	$2 \cdot 2 \times 10^{14}$ g C
Rivers (assuming 5 mg C l^{-1})	$1 \cdot 8 \times 10^{14}$ g C
Possible annual inputs into the dissolved fraction	
Phytoplankton excretions (10 % of production, see text)	$3 \cdot 6 \times 10^{15}$ g C
Resistant material from phytoplankton (5% of production, see text)	$1 \cdot 8 \times 10^{16}$ g C
Annual losses by sedimentation	
Nearshore sedimentation	$2 \cdot 7 \times 10^{12}$ g C
Pelagic sedimentation	$9 \cdot 2 \times 10^{13}$ g C

Data partially from Williams (1971) and Skopintsev (1971)

The most striking instance of the input of organic material into the sea from the atmosphere is the chlorine containing pesticides. It is now believed (National Academy of Sciences, 1971) that the major route of DDT and its derivatives into the sea is via the atmosphere. Only this mode of transport can account for the widespread distribution of DDT and its derivatives in plankton. Like many of the other pesticides of this type it has a low solubility but is comparatively volatile. Lloyd-Jones (1971) has calculated the evaporation rate to be equivalent to about 2 lb acre^{-1} yr^{-1} (0·25 g m^{-2} yr^{-1}), which is equivalent to about half its yearly application in the U.K. It is also dispersed into the atmosphere during spraying from the air, either when it vaporizes from the talc carrier or when the talc itself remains in the atmosphere as dust. Risebrough *et al.* (1968) found that the amounts of DDT and its derivatives present in the N.E. trade winds off Barbados averaged 7·8 × 10^{-14} g m^{-3}. They found that samples taken at La Jolla, southern California, contained about a thousand times more (average 7·0 × 10^{-11} g m^{-3}), and they calculated that over the area influenced by the N.E. trade winds (1·94 × 10^7 km^2), 600 kg of insecticides would be deposited per year. This rate of deposition (3 × 10^{-5} kg km^{-2} yr^{-1}) is much less than that estimated for the North Sea, into which 300 tons are thought to be deposited annually (Goldberg, 1971). This is equal to a deposition rate of 0·5 kg km^{-2} yr^{-1}. In a report prepared for the American National Academy of Sciences (National Academy of Sciences, 1971) it was assumed that the average DDT residue content of rain was 80 parts in 10^{12}, from this it was calculated that the total annual input into the oceans would be 2·4 × 10^7 kg (24000 tons); this is equal to 25% of the estimated world annual production.

Compounds of the polychlorinated biphenyl (PCB) group are also widely found in the marine environment. These compounds are less likely to enter the atmosphere than are those of the DDT group. Firstly they are less volatile and more water-soluble than DDT. Secondly, their use is almost exclusively industrial, and they thus have less opportunity than does DDT to enter the atmosphere. As a result of this, manufactured PCB compounds are more likely to enter the sea in river water, effluents or sewage sludge. However, the distribution of PCB compounds in the marine environment is more widespread than would be expected if sewage and rivers were the only sources of these compounds to the sea. PCB compounds can be produced photochemically in the vapour phase from DDT and its derivatives, and it has been suggested (see Maugh, 1973) that this may be a contributing reason for the widespread distribution of PCB compounds in the environment.

12.2.1.2. *Input by rivers*

The contribution to the DOC pool by the rivers is probably comparatively small on a global scale. Duursma (1961, p. 73) noted no seaward fall in the DOC content of the oceans; Williams and Gordon (1970) determined the $^{13}C/^{12}C$ ratio of the dissolved organic material in ocean water and found the $\delta^{13}C$ to be in the range -21 to $-24\cdot4\%_0$ (see Table 12.11 and Section 12.5.2.1), quite distinct from, and typically less negative than, that of dissolved organic material from fresh water. The DOC of the Amazon River, for example, was reported to have a $\delta^{13}C$ of $-28\cdot5$.

Undoubtedly, in regions where great rivers flow into the sea, there will be an enhanced DOC content of the adjacent waters. For example, Fredericks and Sackett (1970) have estimated the average DOC content of the river run-off into the Gulf of Mexico to be 5 mg C 1^{-1}, which amounts to $0\cdot6 \times 10^{13}$ g C annually. This is comparable to the primary production of the whole of the Gulf of Mexico. Williams (1971) and Skopintsev (1971) have both made estimates of the total amount of organic material brought into the sea annually along with river water. Williams gave an estimate of $3\cdot1 \times 10^{13}$ g C yr^{-1}, whereas Skopintsev calculated a higher one: 18×10^{13} g C yr^{-1}. In part, the disparity between the two estimates stems from the very real problem of ascribing a meaningful average for the organic content of river water. Skopintsev's estimate is given in Table 12.1, although this is not meant to imply that there is any reason to believe it is more accurate than that of Williams. Precise estimates of the contribution of riverborne organics to the overall organic budget of the oceans probably have little meaning, for it is known that, on mixing with seawater, a portion of the DOM in river water precipitates (Skopintsev, 1947; Sieburth and Jensen, 1968) and is lost by sedimentation.

Chlorinated hydrocarbon pesticides enter the rivers along with run off from agricultural land and precipitation. Risebrough et al. (1968) have calculated the inputs by two American rivers—the San Joaquin and the Mississippi —and calculated that they transported respectively 1900 and 10000 kg yr^{-1} of these compounds. It has been estimated (National Academy of Sciences, 1971) that only about $0\cdot1\%$ of the world DDT production reaches the oceans via surface runoff; this is small compared with the amount which is estimated to enter the sea via the atmosphere. Thus, the rivers would appear to be a less important source of these compounds than is the atmosphere. This reflects the relatively high volatility and low solubility of the chlorinated pesticides.

Polychlorinated biphenyls are used in a variety of industries and are often discharged into sewers. Depending upon the method used for the treatment of sewage, these compounds may enter the seas directly along with the effluent or along with sewage sludge. PCB may be present in sewage sludge in significant amounts; Holden (1970) found that the sludge which he analysed contained from 1–185 µg of PCB per gram dry weight. He calculated that approximately 1 ton of PCB was dumped annually into the Firth of Clyde along with sewage sludge from Glasgow treatment plants. He estimated that comparable amounts were dumped along with the wastes from two other British cities.

12.2.2. INTERNAL SOURCES

Organic material may be produced in the sea by chemical or biological processes. It is probable that chemical formation, as opposed to chemical transformation, of organic material under the action of radiation is no longer of any great significance, although it is thought (Bernal, 1961) to have played the dominant part in the origin of life. Phytoplankton are the main producers of organic material in the oceans as a whole. It has been estimated (Ryther, 1963) that benthic algae account for one-tenth of total production in the sea. The amount of bacterial primary production in the sea by autotrophs such as the photosynthetic and nitrifying bacteria may be regarded as insignificant.

There are a variety of routes whereby the products of plankton photosynthesis may enter the DOM fraction. Algae release some of their photosynthetic products from the cell. Zooplankton release dissolved and particulate organic material as excretion products; during grazing there may be damage of algal cells resulting in the leaking of soluble cell components into the environment. Finally, after death, the autolytic and decay processes will result in the production of soluble compounds.

12.2.2.1. *Excretion by phytoplankton* (*see also Section* 14.2.4.)

Experimental evidence now strongly suggests that healthy algal cells excrete organic material during growth. Hellebust (1965) studied 22 species of marine algae from a representative range of families and found that during logarithmic growth in axenic culture, 3–6% of the photosynthetic products were usually released from the cell; a few species excreted as much as 10–25%. He, and other workers (e.g. Watt, 1966), observed that high light intensities caused significant increases in the rate of excretion.

Hellebust made some studies with concentrated natural plankton and found 4–16% excretion of photosynthetic products. Only recently has data become available for natural unconcentrated plankton. Anderson and Zeutschel (1970), who have made a series of measurements in the North-east Pacific Ocean, noted that although the rate of excretion was dependent upon the photosynthetic rate, the ratio was not constant, and the proportion of photosynthetically fixed carbon excreted increased as the photosynthetic rate decreased. Watt (1966), working with freshwater algae observed a similar phenomenon, i.e. that the fraction excreted increased as the plankton concentration diminished. Anderson and Zeutschel found that at an oligotrophic station 26% was excreted on average throughout the euphotic zone whereas at a eutrophic station only 7% was excreted. Thomas (1971) found that the proportion excreted increased in a seaward direction from the east coast of the United States. For estuaries he recorded less than 7% excretion; for coastal water it was generally less than 13%, and in the Sargasso Sea it was up to 44%. This, again, represents an increase in the fraction excreted with a decrease in photosynthetic rate. However, the vertical profiles at the individual stations suggest that, in fact, light intensity, rather than photosynthetic rate, is the important factor controlling excretion. From the above work it appears that 10% is probably a representative figure for the fraction of the photosynthetic products excreted by offshore plankton; on a global scale this figure would amount to 3.6×10^{15} g C yr^{-1}, using the estimate for total primary production given in Table 12.1. Khaylov and Burlakova (1969), using light extinction at 260 nm as a measure of organic material, reported that shore weeds excreted about 30% of their gross photosynthetic production. A similar figure was quoted by Sieburth (1969) for *Fucus vesiculosus*. This implies that, in rocky coastal regions, attached algae are probably an important source of dissolved organic material.

The chemical composition of the excreted material has been examined by several workers, one of the most comprehensive studies with marine algae being that by Hellebust (1965). One important finding was that there was no constancy in the composition of the excretion products within the algae as a whole or even within families: therefore it is difficult to make generalizations. The protein content of the excretion products, determined by trichloracetic acid precipitation, amounted to 0.2–5.9% of the excretion products; the total amount of amino acids in the excretion products exceeded that associated with the protein, and Hellebust concluded, as did Jones and Stewart (1969), that low molecular weight peptides accounted for a significant fraction of the nitrogen excreted. Tolbert and Zill (1956) observed that glycolic acid is an important excretion product of freshwater species of *Chlorella*, and it was found by Hellebust to account for 10% or more of the excretion products of a few algae (e.g. *Skeletonema* and *Chaetoceros*). However, in the great majority of species it accounted for <5% of their gross photosynthetic production. Carbohydrate appears to be a common component of the excreted material. Hellebust detected mannitol, glycerol and, to a lesser extent, arabinose and glucose in the excretion products of algal cultures. Sieburth (1969) observed that carbohydrate was the most abundant fraction in the excretion products of *Fucus vesiculosus*. Carlucci and Bowes (1970a) found that the excretion products of algae included thiamine, biotin and vitamin B_{12} and they further showed (Carlucci and Bowes, 1970b) that in certain instances these vitamins could be used by vitamin-requiring species inoculated into the medium. For example, *Coccolithus huxleyi*, a thiamine–requiring alga, would grow in thiamine deficient medium in the presence of a thiamine–excreting alga such as *Dunaliella tertiolecta* or *Skeletonema costatum*.

In making studies of algal excretion, it is usually necessary to filter off the cells in order to analyse the extracellular environment, and this approach is open to criticism on the grounds that damage to algal cells could occur during filtration and lead to the release of organic material. Fogg (1971) takes the view that little cell damage normally occurs during filtration. Arthur and Rigler (1967) on the other hand presented data that they interpreted as evidence for cell damage during filtration. At present, the matter remains unsettled. In view of this uncertainty, the work of Carlucci and Bowes (1970b), cited above, is particularly interesting, because excretion was demonstrated without the necessity for filtration.

12.2.2.2. *Other internal sources.*

Information available about other inputs is scattered and very incomplete. When zooplankton feed on algal cells there is a possibility of breakage of the cell and leakage of the soluble cellular contents into the external environment. Although it is generally recognized that this process occurs, it is difficult to estimate its quantitative importance as a source of DOM. Corner *et al.* (1972) have recently made a detailed study of the copepod *Calanus helgolandicus* feeding on a large diatom *Biddulphia sinensis*. They concluded that although the copepod loses a substantial fraction of this food as faeces, there was no evidence of extensive release of soluble organic material attributable to damage to algal cells during feeding.

Zooplankton may be responsible for the production of DOM in other ways. Firstly, they release soluble organic excretions from their bodies as a result of their internal metabolism; secondly, some soluble compounds will be released along with the faeces; finally, the faeces themselves will contain residual digestive enzymes as well as bacteria from the gut, and these will act on the faecal material to produce soluble organic material. It is, in practice, usually difficult to distinguish between these various processes.

Johannes (1966) showed that for the benthic amphipod *Lembos intermedius* the amount of dissolved organic phosphorus released from the body and that associated with the faeces were comparable, each amounting to 4% of the phosphorus in the diet. Butler *et al.* (1970) studied the release of organic phosphorus and nitrogen by the planktonic copepod *Calanus finmarchicus*. They found that during the periods when the algal cells in the water exceeded about 10^6 cells l^{-1} the proportion of organic phosphorus excreted reached as much as 70%. However, they found no associated rise in the excretion of organic nitrogen, which amounted to less than 12% of the total nitrogen excreted. There is, at present, no general agreement about the amount of organic nitrogen excreted by zooplankton. Johannes and Webb (1965) measured excretion of free amino acids by zooplankton. From these data, obtained with net zooplankton, they calculated that the amount of amino acid excretion was equivalent to 22% and 25% of plankton production in the Sargasso Sea and Gulf Stream respectively. This extensive release of amino acids may be an artifact of the experimental conditions (Corner and Newell, 1967), but the matter remains unsettled (Webb and Johannes, 1969).

After death, decomposition of plants and animals is initiated by autolytic enzymes (cathepsins) present in the tissue, but subsequent colonization by bacteria and other micro-organisms is almost certainly essential for complete breakdown and mineralization. Dissolved organic compounds will be intermediate products of the decay process, and those not taken up by the attendant microbes will diffuse out into the surrounding environment.

Studies on decomposing algal cells indicate that, in addition to biologically labile material such as sugars and amino acids, biologically resistant material is also produced. Skopintsev (1960) has reported that about 20–30% of plankton resists decay over a period of 6–7 months. Essentially similar results were obtained by Otsuki and Hanya (1968, 1972a, b), who followed the decay of a culture of freshwater alga, *Scenedesmus*, over a period of 200 days. About 5–10% of the original cellular organic material was found to remain as resistant dissolved organic material; the atomic C/N ratio of the soluble organic fraction decreased with time from 11 to 5·9 during the process of decay. If the lower end (5%) of the range given above is used to estimate the amount of resistant soluble material accumulating annually, then a figure of $1·8 \times 10^{15}$ g C yr^{-1} is obtained. It is not known if, when material is eaten rather than allowed to decay, the same amount of resistant material is produced. In its passage through the gut the food will pick up members of the gut microbial flora and these may exhibit a greater metabolic versatility than does the resident population of sea water. This is an important point as Harvey (1950) considered that the normal fate of phytoplankton cells in the sea is to be eaten; few apparently simply die.

Faecal material will decompose by basically the same steps as do the dead organisms described above. However, in this instance residual digestive enzymes will be present in faeces (instead of the autolytic ones present in dead organisms); more important, the faeces will contain bacteria from the gut (Frankenberg et al., 1967). There is a growing awareness of the importance of the faecal pellets as a source of organic material, particularly in coastal regions. Frankenberg *et al.* (1967) have studied and discussed the more biological aspects of the role of faecal pellets in the food web.

12.3. THE ORGANIC COMPOSITION OF SEA WATER

There are two fundamental approaches to the study of the organic content and composition of sea water. Either analyses may be made of some feature related to the gross organic content of sea water or, alternatively, measurements may be made of a specific compound or class of compounds. The latter approach is often carried out as part of microbiological or biological studies.

The determination of the gross organic content of sea water is a problem that has long concerned marine chemists, and the problem is still not satisfactorily solved. It is not possible to analyse sea water for total dissolved organic material as such, but it is conventional to oxidise the organic material and subsequently to determine one of its constituent elements, usually carbon, nitrogen or phosphorus. The results are then expressed as organic carbon, nitrogen, etc. Physical methods, usually based on optical properties of the sea water, have been used to gain an indication of the gross levels of organic material. Such methods have the advantage that they are readily automated for use *in situ* or in continuous monitoring systems. Earlier procedures (see Kalle, 1966) were based on light extinction at the blue end of the visible spectrum. More recently, the ultra violet region of the spectrum has been used. Ogura and Hanya (1967), who worked using radiation of 220 nm, found that only 20 % of the light extinction was due to organic material, most of the absorption at this wavelength was caused by bromide. Foster and Morris (1968) studied the relationship between the integrated light absorption between 250 and 350 nm and the measured organic carbon content, and found that, although there was a functional relationship between the two, the degree of correlation was very poor. Despite the practical simplicity of optical methods, they do not seem to give a good measure of organic material.

12.3.1. ELEMENTAL COMPOSITION

Analysis for the three elements: carbon, nitrogen and phosphorus, probably accounts for the major part of the work on the dissolved organic content of sea water. As in many other branches of the organic chemistry of sea water, progress has been severely hampered by lack of satisfactory methods. Few, if any, of the existing methods for the measurement of dissolved organic carbon, nitrogen or phosphorus content are widely accepted as being accurate.

Carbon is the major component of organic material and, accordingly, it will be considered first. Two approaches can be used for measuring the dissolved organic carbon content of sea water (see Chapter 19). Soviet workers have evolved methods based on the evaporation of the sample and a subsequent dry combustion at high temperature (e.g. Skopintsev, 1960). Most of the work outside of the Soviet Union has been based on wet oxidation of sea water, originally using acid dichromate (Krogh, 1934; Duursma, 1961), but more recently peroxydisulphuric acid has been used (Menzel and Vaccaro, 1964). The dry combustion methods quite clearly give results which are about double those obtained by the wet combustion procedures (see Table 12.2). Recently two papers have appeared (Sharp, 1973; Gordon and Sutcliffe, 1973) which contain analyses that relate directly to the wet combustion–dry combustion problem; data from both papers are included in Table 12.2. The above authors concluded that the wet combustion method gives an incomplete analysis of sea water; however, there are still some details yet to be resolved (see Chapter 19, Section 19.2.4.1). Much of the discussion in the subsequent sections will be based on the results obtained by the peroxydisulphate method, and if this turns out to be seriously in error then it will be necessary to reassess the conclusions drawn from these analyses.

In contrast to the conclusions of Krogh (1934), most workers have found that the concentrations of dissolved organic carbon are generally highest in the upper 100 m of the water column in the oceans. A typical value for this zone would be $0.6–1.0$ mg C l^{-1} for a wet oxidation analysis and $1.5–2.0$ mg C l^{-1} for a dry combustion analysis. The values for deep water are with both methods typically about two-thirds of the surface ones. Table 12.2 gives a summary of the results obtained by different workers for various locations, and indicates the analytical technique employed. A typical profile of DOC obtained with a wet combustion method is given in Fig. 12.1, along with profiles of DON, DOP, their inorganic counterparts and vitamin B_{12}.

Until recent years the determination of dissolved organic nitrogen has been a matter of considerable difficulty (see Chapter 19), and the results obtained by different methods are usually not very consistent. It is, therefore, difficult to generalize on the results of organic nitrogen analyses of sea water. Most of the analyses fall within the range of $2–20$ μg-at N l^{-1} (Table 12.3). How much of this spread is caused by the analytical method, as opposed to real variations in the composition of sea water is not clear. Some workers (e.g. Fraga, 1966; Holm–Hansen et al., 1966, Banoub and Williams, 1972) have reported values in the range $2–8$ μg-at N l^{-1}; typically concentrations are $5–8$ μg-at N l^{-1} in the upper 100 m, falling to $2–5$ μg-at N l^{-1} in deep ocean. Duursma (1961) and Skopintsev et al. (1967) who analysed samples from the North Atlantic and the Black Sea respectively reported much higher values (typically $10–20$ μg-at N l^{-1}) with no general decrease with depth.

TABLE 12.2

Average and/or typical range of dissolved organic carbon analyses for various oceanic areas (expressed as mg C l⁻¹)

Reference	Krogh (1934)	Duursma (1961)	Holm-Hansen et al. (1966)	Skopintsev et al. (1966)	Skopintsev et al. (1967)	Menzel and Ryther (1968, 1970)	Fredericks and Sackett (1970)
Location	Temperate N. Atlantic	N. Atlantic	N. E. Pacific	Tropical Atlantic	Black Sea	N. and S. Atlantic	Gulf of Mexico
Oxidation method	Dichromate oxidation	Dichromate oxidation	Peroxydisulphate oxidation	Dry combusion	Dry combustion	Peroxydisulphate oxidation	Peroxydisulphate oxidation
Depth (m)							
0–100	2·4	0·6 (0·3–1·2)	0·7 (0·5–0·8)	2·0 (1·7–2·2)	3·0	0·8 (0·6–1·0)	0·8 (0·6–1·1)
100–300	—	0·5 (0·3–1·0)	0·6 (0·5–0·6)	1·7 (1·5–1·8)	2·8	0·7 (0·4–0·8)	0·6 (0·4–0·8)
300–1000	—	0·4 (0·3–0·6)	0·5 (0·4–0·6)	1·5 (1·4–1·7)	2·5	0·5 (0·4–0·6)	0·5 (0·3–0·7)
1000–2000	2·35	0·4 (0·3–0·6)	0·5 (0·4–0·6)	1·5 (1·4–1·6)	—	0·5 (0·4–0·6)	0·5 (0·3–0·6)
2000+	2·33	0·4 (0·3–0·6)	—	1·5 (1·4–1·6)	—	0·5 (0·4–0·6)	0·5 (0·4–0·6)

Reference	Starikova (1970)			Banoub and Williams (1972)	Sharp (1973)	Gordon and Sutcliffe (1973)
Location	Pacific Ocean	Atlantic Ocean	Indian Ocean	W. Mediterranean	N. Atlantic	N. Atlantic
Oxidation method	Dry combustion	Dry combustion	Dry combustion	Peroxydisulphate oxidation	Peroxydisulphate oxidation	Dry combustion
Depth (m)						
0–100	1·79 (1·49–1·98)	1·72 (1·30–1·86)	1·73 (1·57–1·99)	0·6 (0·37–0·82)	1·0–1·3	1·5–2·5
100–300	1·45 (1·11–1·70)	1·53 (1·08–1·66)	1·57 (1·45–1·67)	0·36 (0·29–0·43)	0·7–1·0	1·5–2·0
300–1000				0·34 (0·23–0·55)	0·6–0·8	1·2–2·0
1000–2000	1·26 (1·09–1·31)	1·36 (1·13–1·50)	1·35 (1·30–1·47)	0·28 (0·22–0·34)	0·6–0·8	1·2–1·5
2000+				0·24 (0·21–0·27)		1·2–1·5

TABLE 12.3

Average and/or typical range of dissolved organic nitrogen analyses for various oceanic areas (expressed as μg-at N l⁻¹)

Reference	Krogh (1934)	Duursma (1961)	Armstrong et al. (1966)	Holm-Hansen et al. (1966)	Fraga (1966)	Skopintsev et al. (1967)	Banoub and Williams (1972)
Location	Temperate N. Atlantic	N. Atlantic	N.E. Pacific	N.E. Pacific	Indian Ocean	Black Sea	W. Mediterranean
Oxidation method	Dry combustion	Dry combustion	U.V. irradiation	U.V. irradiation*	Not given	Kjeldahl digestion	U.V. irradiation*
Depth (m)							
0–100	18·4	15	6·0	6·0 (5–7·5)	7·5 (6–8)	17	5·2 (3·5–6·1)
100–300	—	15	5·0	5·0 (4–6)	7 (6·5–8)	15	3·3 (2·0–4·5)
300–1000	17·7	15	4·0	3·5 (2–5)	6 (5·5–6·7)	16	3·3 (2·6–4·6)
1000–2000	—	10	4·0	3·5 (2–4)	5 —	15	4·8 (1·4–6·7)
2000+	17·1 (16–18)	10	—	——	4·3 —	—	—

* Values include ammonia.

TABLE 12.4

Average and/or typical range of dissolved organic phosphorus analyses for various oceanic areas (expressed as µg-at P l^{-1})

Reference	Holm-Hansen et al. (1966)	Armstrong etal. (1966)	Skopintsev et al. (1967)	Banoub and Williams (1972)
Location	N. E. Pacific	N. E. Pacific	Black Sea	W. Mediterranean
Oxidation method	U. V. irradiation	U.V. irradiation	H_2SO_4 hydrolysis	U. V. irradiation
Depth (m)				
0–100	0·3 (0·2–0·4)	0·3 (0·2–0·4)	0·25	0·07 (0·0–0·15)
100–300	0·15 (0·1–0·2)	0·2 (0·15–0·25)	0·80	0·05 (0·03–0·06)
300–1000	0·15 (0·1–0·2)	0·2 (0·15–0·25)	0·90	0·05 (0·02–0·08)
1000–2000	0·15 (0·1–0·2)	—	1·0	0·06 (0·0–0·14)
2000 +	—	—	—	0·04 (0·03 and 0·05)

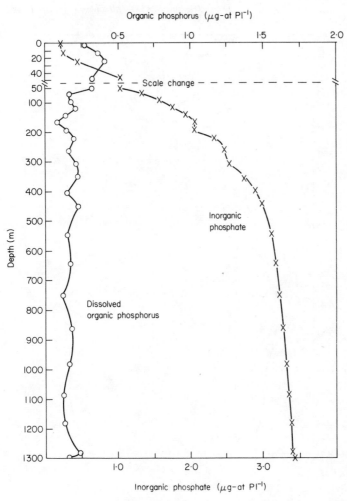

Fig. 12.1a

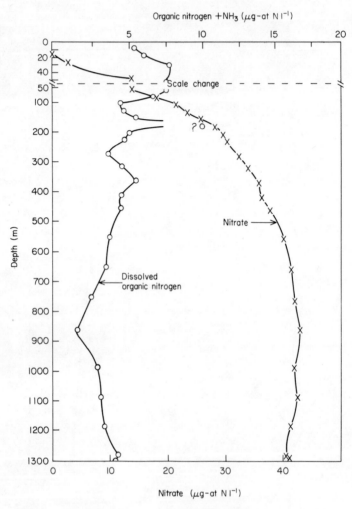

FIG. 12.1b

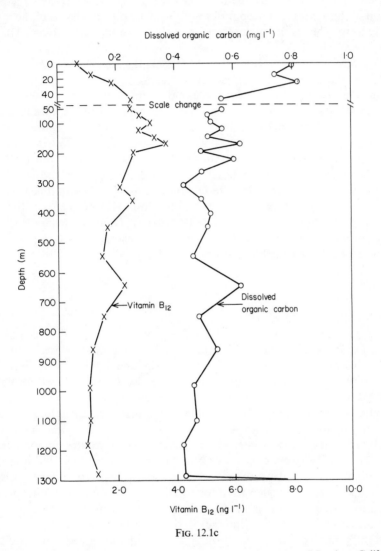

FIG. 12.1c

FIGS. 12.1a, b, c. Profiles of organic and inorganic material at a station off Southern California. (From Holm–Hansen *et al.*, 1966).

The concentration of dissolved organic phosphorus is generally very low, and in common with some methods for dissolved organic nitrogen, the precision of the analyses in deep water is reduced by the high levels of the inorganic form. The results of analyses typically fall in the range 0·05–0·2 μg-at P l^{-1} (see Table 12.4).

For the above reasons it is hardly surprising that, in general, there is not a high correlation between the dissolved organic C, N and P content of sea water. Duursma (1965), Fraga (1966) and Holm-Hansen *et al.* (1966) have presented data for both C/N ratios and their change with depth. Duursma expected an increase in C/N ratio with depth, but found no evidence for this, nor is one apparent from the results of Holm-Hansen *et al*; Fraga found an indication of a decrease in C/N ratio with depth. Otsuki and Hanya (1968) observed that when algal material was allowed to decay the atomic C/N ratio of the dissolved organic material produced decreased with time from 11 to 5·9. The average atomic ratio reported by Holm-Hansen *et al.* was 10·5 : 1; most of those observed by Fraga were close to 9 : 1; Duursma generally found lower ratios, often as low as 2·8 : 1. Until analytical methods improve, C/N ratios probably provide little information of any value.

12.3.2. DETAILED COMPOSITION

The gross analysis of sea water for either organic C, N or P gives only a limited amount of information, and for a variety of reasons analyses are made for individual compounds or classes of compounds. It has been clear for some time that the compounds for which one can devise analyses normally account for only a small fraction of the total organic material in the sea and it is unlikely that it will ever be possible to produce a complete organic description of sea water. With few exceptions, analyses have been made for compounds of biological interest: fatty acids, carbohydrates, amino acids, vitamins etc. The recent concern with pollution has generated interest in the concentrations of hydrocarbons, pesticides and similar compounds in sea water, and it has also drawn attention to the fact that in this, and other areas, of marine chemistry there is a paucity of data on the normal, or baseline, concentrations of individual compounds in ocean and coastal waters.

At this point it is well to recognize that contamination is a serious problem in this type of work and in the ensuing discussion it has been necessary to ignore data which are suspect. It is often true that low values for a particular compound are more likely to be correct than high ones. The reason for this is that low values most likely result from an incomplete recovery or determination which, in many instances may be easily checked; high values can result from contamination during sampling, storage and analysis: this is always difficult to monitor and control.

12.3.2.1. *Lipids*

The term lipids embraces a heterogeneous group of biological compounds soluble in non-polar or semi-polar solvent systems. It includes hydrocarbons, although in the present account these compounds have been considered separately. In view of the ease with which lipids can be extracted from sea

TABLE 12.5

Fatty acid content of sea water. Data abstracted from Williams (1965).

Fatty acid	Mean concentration (μg methyl ester l^{-1})	Depth (m)	Total fatty acid content (μg methyl ester l^{-1})
$C_{14:0}$	0·65	20	4·02
$C_{14:1}$	0·07	73	3·26
$C_{14:2}$	0·07	405	3·98
$C_{16:0}$	1·17	480	4·01
$C_{16:1}$	0·68	891	8·84
$C_{16:2}$	0·17	1826	2·90
$C_{16:3}$	0·09	2310	3·12
$C_{18:0}$	0·51		
$C_{18:1}$	0·33		
$C_{18:2}$	0·43		
$C_{18:3}$	0·11		
$C_{20:1}$	0·73		

water compared with other classes of compounds, it is surprising that there are so few quantitative data for them. The lipid composition of sea water has been reviewed by Jeffrey (1970) who has given data for the amount of organic material extracted from sea water with chloroform; this ranged from 0·15 to 1·59 mg l^{-1} for a series of samples from the coastal water of the Gulf of Mexico. In two profiles in the central and western regions of the Gulf of Mexico, the concentration of extractable lipid was found to be 0·15–0·31 mg l^{-1}, equivalent to 13·5 to 21 % of the total dissolved organic material, as determined by wet combustion analysis. Over the depth range 10 to 3200 m there was no evidence for a decrease in extractable material with depth.

Williams (1965) carried out a detailed analysis of the fatty acids of the lipids extracted from filtered sea water taken from coastal and offshore regions of the N.E. Pacific; both free fatty acids as well as those combined as triglycerides would be determined by the method he used. He reported individual fatty acid concentrations ranging from undetectable levels to 1·93 µg l^{-1}, the total fatty acid content varied from 1·25 to 8·8 µg l^{-1}, expressed in terms of methyl esters. A selection of his data is given in Table 12.5. The predominant fatty acids were: $C_{14:0}$; $C_{16:0}$; $C_{16:1}$; $C_{18:0}$; $C_{18:1}$; $C_{18:2}$; and $C_{20:1}$. These findings are similar to those of Garrett (1967b) who analysed material from the sea surface, and Jeffrey (1970) who analysed samples from the Gulf of Mexico. These three workers reported that uneven carbon numbered fatty acids ($C_{15:0}$, $C_{17:0}$ and $C_{19:0}$) were present as a small but not insignificant fraction of the total fatty acids. Quinn and Wade (1972) have given concentrations of total fatty acid for samples from the North Atlantic as 13–60 µg l^{-1}. Williams's (1965) results strongly suggest that the distribution of fatty acids is not influenced by factors such as standing crop, depth or distance offshore; similarly Stauffer and MacIntyre (1970) found little systematic quantitative difference between samples from an estuary and those from oceanic water.

12.3.2.2. *Hydrocarbons and their halogenated derivatives*

This group includes three of the important organic pollutants of the sea: petroleum hydrocarbons, the chlorinated hydrocarbon pesticides (the DDT group) and the polychlorinated biphenyls (PCB). Except where there is acute pollution, data is still scarce for all of the groups. However, this situation is likely to change over the next few years.

The unmodified hydrocarbons differ from the other two classes of compounds because they may have a recent planktonic origin as well as a source in seepages or losses of oil in the course of exploitation. With such a group of compounds there is need and an urgency to establish baseline concentrations in the ocean if, indeed, it is still possible to do so. Parker *et al.* (1972) have given analyses of the *n*-paraffin content of three samples from offshore, two from the Caribbean and one from the Gulf of Mexico. The total *n*-paraffin content ranged from 0·78–1·1 µg l^{-1}, the predominant paraffins being C_{28} to C_{31}. They regarded the *n*-paraffin concentration of 0·78 µg l^{-1} in a sample from a depth of 2000 metres in the Caribbean as being a typical baseline value.

Quinn and Wade (1972) and Barbier *et al.* (1973) have given results of total hydrocarbon analyses of ocean water from the North and the Equatorial Atlantic; the range of concentrations was 9–56 µg l^{-1}. Even and uneven chain number hydrocarbons were present in similar amounts.

Very interesting data are now accumulating for the gaseous hydrocarbons in sea water. Linnenbom and Swinnerton (1970) and Sackett *et al.* (1972) have reported analyses of a variety of low molecular weight hydrocarbons in ocean water. The distribution of these shows a series of interesting patterns.

TABLE 12.6

Concentration of hydrocarbons in ocean water

Compounds	Location	Depth (m)	Concentration (ml l^{-1})	Concentration (ng l^{-1})	Reference
Methane	N. Atlantic	Sub-surface	$4\text{-}5 \times 10^{-5}$	30–35	Linnenbom and Swinnerton (1970)
Methane		deep	$0\cdot6 \times 10^{-5}$	4·3	Linnenbom and Swinnerton (1970)
Methane	Black Sea	1000+	15×10^{-2}	100000	Linnenbom and Swinnerton (1970)
Ethane	Black Sea	0–100	5×10^{-7}	0·6	Linnenbom and Swinnerton (1970)
Ethane + Ethylene	Black Sea	1000+	400×10^{-7}	50	Linnenbom and Swinnerton (1970)
Ethylene	Black Sea	0–300	5×10^{-6}	6	Linnenbom and Swinnerton (1970)
Propylene	Black Sea	0–300	1×10^{-6}	2	Linnenbom and Swinnerton (1970)
Propane	Gulf of Mexico	0–1000	1×10^{-7}	0·2	Sackett et al. (1972)
Propane	Cariaco Trench	0–100	$<10 \times 10^{-7}$		Linnenbom and Swinnerton (1970)
Propane	Cariaco Trench	300+	$20\text{-}50 \times 10^{-7}$		Linnenbom and Swinnerton (1970)
Total n-paraffins	Gulf of Mexico	2000		780	Parker et al. (1972)
Total hydrocarbons	N. Atlantic	Sub-surface		9000–56000	Quinn and Wade (1972)
Total hydrocarbons	Equatorial Atlantic	50		43000	Barbier et al (1973)
Total hydrocarbons		500–4500		10000–37000	Barbier et al. (1973)
PCB	N. Atlantic	Sub-surface		35	Harvey et al. (1973)
PCB	N. Atlantic	200		10	Harvey et al. (1973)
DDT residues	Coastal E. Pacific	Sub-surface		2·3–5·6	Cox (1971)

Typical concentrations for the gases methane, ethylene, ethane, and propylene are given in Table 12.6. Linnenbom and Swinnerton found that the methane content of surface water in the N.W. Atlantic Ocean is close to equilibrium with that in the atmosphere; this contrasts with the distribution of carbon monoxide which is supersaturated in the sea implying that the sea is a source of carbon monoxide (see Vol. 1, Section 8.7.3).

In the vertical profiles given by the above workers, the ethylene concentratration is high in the upper 100 m but below this it decreases to about the same value as those of ethane and propylene ($5-10 \times 10^{-7}$ ml 1^{-1}); these latter two gases show a much less pronounced concentration decrease with depth. The high levels of ethylene in surface water may be interpreted as a sign of input from the atmosphere, but could also result from photochemical production; Wilson et al. (1970) discovered that carbon monoxide, ethylene and propylene are produced by photochemical decomposition of organic material originating from algae.

In deoxygenated waters, such as those of the Black Sea or the Cariaco Trench, the paraffinic hydrocarbon gases, methane, ethane and propane, show pronounced concentration increases in the anoxic zone, but this does not occur with the olefinic gases ethylene and propylene. Methane shows the greatest increase in anoxic waters and it is well known to be a product of microbial anaerobic metabolism.

The presence of the DDT group of insecticides and their residues in marine plankton is now well established. Cox (1971) analysed unfiltered sea water from the Californian coast and found concentrations of DDT residues ranging from $2\cdot3-5\cdot6$ ng 1^{-1}; surprisingly only about 10% of this was sorbed onto particles having a size of 1 μm or greater. The remainder of the DDT was believed to be sorbed onto smaller particles. It is interesting to note that the above concentrations are well below the solubility of DDT (ca. 1 μg 1^{-1}; Bowman et al., 1960). Thus, present information would suggest that solubility is not a factor controlling the distribution of DDT in ocean water. Harvey et al. (1973, 1974) have published data on polychlorinated biphenyls in North Atlantic surface water. Those for 1972 averaged 35 ng 1^{-1} at the surface to 10 ng 1^{-1} at 100 m; however, values ca. 40 times lower were found in 1973. This reduction was considered to be the result of the decrease in industrial use of these compounds after 1970. R. Dawson (private communication) has obtained similar results for dissolved chlorinated hydrocarbons in subsurface waters (1–2 m deep) of the English Channel and from the Atlantic off the west coast of Scotland in 1973. Ranges and averages (in parentheses) for PCBs, total DDT residues and dieldrin were found to be (ng 1^{-1}) $0\cdot1-0\cdot24$ ($0\cdot16$); $0\cdot02-0\cdot05$ ($0\cdot03$); $0\cdot01-0\cdot09$ ($0\cdot02$) respectively. Appreciably higher concentrations of PCBs and DDT residues were found in inshore waters

TABLE 12.7

Concentrations of carbohydrates in sea water

Method	Sampling area	Carbohydrate	Depth of sampling (m)	Range of results (µg l⁻¹)	Reference
Anthrone and n-ethyl carbazole	Pacific Ocean	Total	Not specified	700	Lewis and Rakestraw (1955)
Phenol-sulphuric acid	Indian Ocean	Total	0–2400	200–600	Handa (1966)
Anthrone	Sargasso Sea	Total	0–2100	200–800	Walsh and Douglas (1966)
Extraction and paper chromatography	E. Pacific Ocean	Galactose	0–3720	1·3–5·4	Degens et al. (1964)
		Glucose	0–3720	5·4–15·5	Degens et al. (1964)
		Mannose	0–3720	2·7–17·8	Degens et al. (1964)
Bioassay	S. W. Atlantic	Glucose	10–100	0–60	Vaccaro and Jannasch (1966)
Bioassay	N. Atlantic	Glucose	0–100	0–190	Vaccaro et al. (1968)
Enzyme assay	N. Atlantic	Glucose	0–100	0–150	Vaccaro et al. (1968)
Extraction and enzyme assay	English Channel	Glucose	10 and 50	0·4–5·7	Andrews and Williams (1971)

waters, such as the Irish sea, in one area of which $1 \, \text{ng} \, 1^{-1}$ of PCB and $0.24 \, \text{ng}$ 1^{-1} of DDT residues were found.

12.3.2.3. *Carbohydrates*

Two approaches have been used to study dissolved carbohydrates in the sea. One of these is to measure the so-called total carbohydrate content of sea water by treating the sample with a strong dehydrating agent and determining the furfural derivatives produced by a colorimetric method, such as the anthrone procedure. The other is to determine individual sugars either chemically, microbiologically or biochemically.

Measurements of total carbohydrate have been made by Lewis and Rakestraw (1955) for the Pacific Ocean, Handa (1966, 1967) for the Indian Ocean and the Kuroshio Current, and Walsh and Douglas (1966) for the Sargasso Sea and are summarised in Table 12.7. The concentrations reported are typically 200–$600 \, \mu\text{g} \, 1^{-1}$. However, the details of the vertical distribution vary. Walsh (1965) has made an interesting study of a coastal pond, in which he found a diurnal change in carbohydrate concentration; the increase during the day was found to range from 14–45% of the measured phytoplankton production, implying extensive excretion of polysaccharides by phytoplankton.

The concentrations of individual sugars in sea water have been measured by various workers (Degens *et al.*, 1964; Vaccaro and Jannasch, 1966; Vaccaro *et al.*, 1968, and Andrews and Williams, 1971) using a variety of methods and a summary of their results is given in Table 12.7. Not surprisingly perhaps, the range is considerable. Vaccaro *et al.* found good agreement between two essentially independent methods with samples exhibiting a wide range of glucose concentrations, showing that the variations do not necessarily result from analytical error. Hobbie and Wright (1965) found concentrations of glucose in freshwater samples to range from 10–$59 \, \mu\text{g} \, 1^{-1}$, similar to those in sea water. As far as can be judged, individual sugar concentrations normally fall in the range undetectable to $20 \, \mu\text{g} \, 1^{-1}$. Degens *et al.* (1964) found no decrease in concentration with depth.

12.3.2.4. *Amino acids and polypeptides*

Both amino acids and sugars are probably important in the turnover of organic material by heterotrophic micro-organisms and in supplying them with energy. For these reasons the concentration and flux of these compounds are of interest to the microbial ecologist. The concentrations of individual free amino acids usually fall within the range <1–$10 \, \mu\text{g} \, 1^{-1}$, only exceptionally reaching $20 \, \mu\text{g} \, 1^{-1}$ or more (Pocklington, 1971; Degens *et al.*, 1964; Starikova and Korzhikova, 1969; Bohling, 1970, 1972; Riley and Segar,

1970; Andrews and Williams, 1971). Table 12.8 gives a selection of data from the above analyses. The low molecular weight amino acids (glycine, serine and frequently also alanine) are generally dominant among the free amino acids in sea water. The reason for this is not known, but it appears to be a general phenomenon for it is also observed for fresh water systems.

Pocklington's (1971) paper provides the most extensive data on the distribution of amino acids in ocean water. He found that the total amino acid concentration ranged from 6–47 µg l^{-1}, with a mean value of 22 µg l^{-1}. The amino acid concentration in the upper part of the water column was generally above the mean, although there was a great deal of scatter in the results. The difference between the concentration in upper and deep water was generally no more than twofold. Riley and Segar (1970) and Andrews and Williams (1971) made studies of the seasonal change in the free amino acid concentrations of the water at stations on the continental shelf; both groups observed no pronounced systematic seasonal change in free amino acid concentrations.

TABLE 12.8

Amino acid analyses of sea water

Location	Total amino acid content (µg l^{-1})	Individual amino acid content (µg l^{-1})	Reference
E. Pacific Ocean	16–124	0–38	Degens *et al.* (1964)
Black Sea	2·3–37	0–8·8	Starikova and Korzhikova (1969)
Irish Sea	4·5–31	0–8·1	Riley and Segar (1970)
Heligoland Sound	8–70*	0–15	Bohling (1970)
N. Atlantic Ocean	6–47	0–16	Pocklington (1971)
English Channel	9·8–80	0–8·1	Andrews and Williams (1971)

*Calculated assuming a mean molecular weight of 100.

Riley and Segar (1970) also determined the dissolved combined amino acids and found pronounced variations in this fraction with concentrations varying from 2·1 to 120 µg l^{-1}. Similar results were obtained by Bohling (1972), and the general conclusion is that whereas the combined amino acids concentrations may show increases that may be related to plankton activity, such changes are not normally encountered with the free amino acids. Short term fluctuations may occur. Bohling (1970) made an intensive study over a period of two months (April to May) of the free amino acid concentrations of water near Heligoland in the North Sea. He observed what appear to be real changes in amino acid concentration with time; this was not confined to a single amino acid, but was characteristic of the whole group and was associated with an algal bloom. Thus, although it would appear that there are increases in amino acid concentration in the sea, presumably as a result of plankton activity, they appear not to persist for longer than a few weeks since they are not seen in less intensive sampling programmes. This is no doubt a consequence of the comparatively rapid growth of heterotrophic micro-organisms in response to increases in the concentration of organic substrates (Williams and Grey, 1970). The concentrations of amino acids in estuarine water (Hobbie *et al.*, 1968) are not notably higher than those in coastal and oceanic waters, probably as a consequence of close coupling between microbial activity and the concentration of the substrates.

12.3.2.5. *Vitamins*

The presence and importance of growth regulating organic compounds in sea water has been recognized for some time (Lucas, 1947, 1961). Although the action of such compounds may be recognized in principle and seen in culture work, little is known of their identity and action in the environment. The only group of compounds that has been studied in any detail is the vitamins of the B group (biotin, thiamine and vitamin B_{12}).

TABLE 12.9

Typical concentrations of vitamins in sea water (Partially from Carlucci, 1970).

	Depth (m)	Vitamin B_{12} ($ng\,l^{-1}$)	Thiamine ($ng\,l^{-1}$)	Biotin ($ng\,l^{-1}$)
Oceanic water	0–100	0·1	8	1·8
	200–500	4·0		
	1000+	1·0		
Coastal water		2	15	3·8

Microbiological assay procedures have been developed that permit the concentrations of these compounds to be determined directly in sea water. Vitamin B_{12} has been studied in greater detail than have the other vitamins, and a pattern of distribution is now emerging. The concentration of vitamin B_{12} in surface oceanic water usually falls in the range undetectable to $1\,ng\,l^{-1}$; the average figure obtained by Daisley and Fisher (1958) for the photic zone in the Bay of Biscay was $0·57\,ng\,l^{-1}$, Natarajan (1971) found similar concentrations in samples taken from the subarctic Pacific Ocean and Carlucci (1970) has given a figure of $0·1\,ng\,l^{-1}$ as representative of the central Pacific Ocean. Vitamin B_{12} usually shows a very characteristic intermediate maximum in vertical profiles (see Fig. 12.1c) which is unusual for dissolved organic compounds in the ocean. This is seen in the work of Daisley and Fisher (1958), Menzel and Spaeth (1962), Holm-Hansen et al. (1966), Carlucci and Silbernagel (1966). Below the first hundred metres where the concentration is typically less than 1 ng l^{-1}, there is an increase to 3–5 ng l^{-1} at depths in the region 200–500 m and finally at depths greater than 1000 m there is a fall to about 1 ng l^{-1}. Vitamin B_{12}, biotin and, to a lesser extent, thiamine have been shown to lose their activity when exposed to sunlight (Carlucci et al., 1969) and this may, in part, account for the higher levels of vitamin B_{12} below the photic zone.

Thiamine is generally required by algae in relatively larger amounts than is vitamin B_{12}, and the concentration of thiamine found in the sea is correspondingly higher. Carlucci (1970) has given 8 ng l^{-1} as an average value for the central Pacific Ocean. Ohwada and Taga (1972) have found concentrations ranging from undetectable to 12 ng l^{-1} for the N.W. Pacific Ocean, and Natarajan and Dugdale (1966) and Natarajan (1971) reported values ranging from undetectable to 445 ng l^{-1} in the Pacific Ocean, the median value being about 50 ng l^{-1}. Thiamine did not show an intermediate maximum in the vertical profiles given by Natarajan and Dugdale (1966) and this may be explained by the fact that it is less susceptible to loss of activity in sunlight. Little information is available for biotin concentrations in the sea. Carlucci (1970) has given a figure of 1·3 ng l^{-1} as an average value for the central Pacific Ocean. An almost identical mean value (1·2 ng l^{-1}) for surface waters was obtained by Ohwada and Taga (1972), who found a decrease in biotin concentration with depth. Vitamins are generally found in higher concentrations in coastal water than in offshore waters (Table 12.9).

12.3.2.6. Miscellaneous compounds

Certain compounds have been studied which do not fit into any of the preceding major categories. Urea has been determined in sea water and McCarthy (1970) has recorded that urea concentrations are typically in the range 0–0·3 μg-at urea N l^{-1}, exceptionally reaching 0·8–2·5 μg-at N l^{-1}. Remsen (1971) reported generally higher levels (0·5–5·0 μg-at N l^{-1}). When the methods used by these two workers were compared on replicate samples the agreement between them was found to be good (Pineda, 1973).

Corwin (1970) has determined certain volatile organic compounds in sea water (acetone, butyraldehyde and 2-butanone). For two vertical profiles, one in the Florida Straits and the other in the Eastern Mediterranean, he recorded concentrations of 0–50 μg l^{-1} acetone, 0–48 μg l^{-1} butyraldehyde and 0–22 μg l^{-1} 2-butanone. Although these compounds could be by-products of microbial metabolism, there are certain features in their distributions that suggest otherwise. First, they do not decrease in concentration below the euphotic zone as would be expected of a microbial by-product; secondly, whereas butyraldehyde was present in samples taken from the Florida Straits it was absent from those taken in the Mediterranean, which would be surprising if it were a general by-product of microbial metabolism.

Nucleic acids are one of the more resistant fractions of the cell and they therefore may contribute to the resistant organic fraction of the sea. Pillai and Ganguly (1970) have analysed water from Bombay Harbour Bay for nucleic acids, and found concentrations of 13–80 µg l^{-1}. It would be interesting to have data from ocean waters for dissolved nucleic acids.

The general findings of the work on individual organic compounds in sea water are that the likely intermediates of regeneration such as sugars, amino acids, etc. usually fall in the range 0·2–20 µg l^{-1} (i.e. about 10^{-7}–10^{-9} M), whereas growth regulating compounds are usually in the range 0·2–20 ng l^{-1} (i.e. about 10^{-10}–10^{-13} M). The general lack of pronounced seasonal changes and differences down the water column in the concentration of the dissolved organics contrasts with that observed for the inorganic nutrients. This suggests the existence of some sort of active biological control which maintains the concentration of these compounds for much of the time at a threshold level. There is, however, at the present no compelling experimental evidence or general microbial kinetic theory to support this hypothesis.

12.3.2.7. The uncharacterized fraction

It is apparent that the conventional approaches to the detailed organic analysis of sea water account for no more than a small part of the total dissolved organic material present (see Table 12.12). This problem is not unique to sea water and is common also to the study of the chemistry of both soil and freshwater. The residual fraction in the sea has often been termed marine humus by analogy with that of the soil but this practice can be misleading since it implies that the same type of materials are present in the two environments and at present the evidence suggests otherwise. For want of a better alternative the term *uncharacterized fraction* has been used in this account to refer to this complex of compounds.

It has been recognized for some time that the organic fraction of sea water imparts to it a yellow colour; the compounds responsible for this have been termed "gelbstoff". This material (Kalle, 1966) like the humic acid of soil (Schnitzer and Khan, 1972), shows a monotonic increase in light absorption from 700 nm into the ultra violet. Early work on the "gelbstoff" fraction of sea water was mainly restricted to examination of its distribution as determined by the light extinction of sea water at the blue end of the visible spectrum. The most important observation that came from this early work was that the "gelbstoff" present in sea water did not simply consist of the yellow humic material transported from continental sources into the sea by the rivers (see e.g. Kalle, 1966). Thus, although there was a decrease in absorption with increase in salinity, the decrease was not progressive; there was a distinct change in the relationship in the salinity region 5–10‰. At salinities between 0‰ and 5‰ there was a marked decrease in the "gelbstoff" content with salinity, whereas above 10‰ the changes with salinity were small. This is consistent with the observation that a substantial part of the river-borne organics is precipitated out on contact with sea water (Skopintsev, 1947).

Recently there have been some chemical studies on the yellow-coloured substances in sea water. Sieburth and Jensen (1968) isolated such materials from coastal sea water and freshwater by adsorbing them onto nylon and subsequently eluting them with alkali. They examined the composition of the mixture by paper chromatographic techniques and found that the spectrum of compounds present in sea water differed from that of fresh water. In a subsequent paper (Sieburth and Jensen, 1969) they studied the release of organic material by seaweeds (*Ascophyllum*, *Laminaria* and *Fucus* species) and found that components of these excretion products inter-reacted to produce "gelbstoff"-like compounds. The requirement for the reaction appears to be a phenolic precursor plus proteinaceous and carbohydrate compounds.

Khaylov and Finenko (1970) extracted a yellow-brown polymeric material from coastal sea water. It was a mixture of high molecular weight compounds, a large percentage of which had a molecular weight between 100000 and 200000. Although the material was yellow-coloured, like the soil humic acids, it did not precipitate out on acidification, as a humic acid must do by defini-tion. They, like Sieburth and Jensen, concluded that the soluble macromole-cular fraction present in sea water is not identical with the soil and freshwater humic acids. The biochemical composition of the major part of the biomass in the terrestrial and marine ecosystems is so different that one might expect pronounced differences. A substantial part of terrestrial plant tissue is secondary thickened material composed of lignin and hemi-celluloses, but these compounds are absent from, or only minor constituents of, marine algae. It is not entirely clear at present what fraction of the total dissolved material the "gelbstoff" comprises. Khaylov and Finenko estimated that the yellow-coloured material they isolated from sea water accounted for 10 % or more of the total organic material.

12.3.3. TEMPORAL AND SPATIAL VARIATIONS

Although passing reference has been made already to variations in the organic content of sea water with location and season, the topic is sufficiently important to merit fuller consideration. Although Krogh (1934) was not strictly correct when he concluded that the changes in the quantity of dissolved organic material brought about by organisms must be extremely small and gradual, there is still no general agreement about the variation of dissolved organic material in the oceans and the degree to which organisms are involved in the process.

12.3.3.1. *Seasonal variations*

Published studies on the seasonal variations of dissolved organic material are few, but almost without exception they do suggest that there are increases as a consequence of phytoplankton blooms. Early studies were based on changes in the DOP content of sea water. Redfield *et al.* (1937) found an increase in the DOP in the water of the Gulf of Maine over a depth range of 0–24 m from 0·1 to 0·5 µg-at $P \, l^{-1}$ following the spring bloom. The increase in the particulate fraction, which includes the plankton, was small in comparison with that of the soluble organic fraction. Substantially similar results were obtained by Armstrong and Harvey (1950) for Station E_1 in the English Channel for which there was an average increase in DOP in the summer months of about 0·2 µg-at $P \, l^{-1}$ above a baseline value of 0·1–0·2 µg-at $P \, l^{-1}$. At this station the depth of the water column is approximately 70 m. Again the increase in the particulate fraction was small. Banoub and Williams (1973), also working at station E_1, recorded an average increase in DOC in the summer months of about 25 µg-at $C \, l^{-1}$ (300 µg $C \, l^{-1}$), a less well defined increase in DON of 1–2 µg-at $N \, l^{-1}$ and no obvious increase in DOP. They also confirmed that the changes in the particulate fraction were proportionally smaller than those of the soluble fraction. It appears that the seasonal changes in the DOP fraction at this station are very variable; in some years there is clear evidence of an increase whereas in other years no increase is seen. The reason for this is not understood.

Duursma (1961) followed the seasonal fluctuation in the dissolved organic fraction of carbon, nitrogen and phosphorus in surface samples at a station in the North Sea off the Dutch coast. In the summer months he found an accumulation of about 1000 µg $C \, l^{-1}$ of DOC, 0·5–0·6 µg-at $P \, l^{-1}$ of DOP, but no accumulation of DON. For samples from the North Atlantic, Duursma observed a difference of 500–700 µg $C \, l^{-1}$ in the upper part of the water column between the DOC content of samples taken in early spring before the bloom and samples taken in late autumn. The peak accumulation of dissolved organic phosphorus and carbon occurred about a month later than that of the phytoplankton bloom. A similar phenomenon can also be seen from the data of Banoub and Williams; Duursma considered that this indicated that phytoplankton excretions were not a major source of the accumulating dissolved organic material, a conclusion supported by the study made by Ketchum and Corwin (1965). These workers monitored a phytoplankton bloom in a body of water in the Gulf of Maine over a period of 10 days, using a drogue to mark the water. Over the 10 day period there was a fall in inorganic phosphate, an approximately equivalent increase in particulate phosphate, but only a small increase in the soluble organic fraction. Thus, during the bloom there was little evidence for accumulation of phosphorus-containing organic excretion products.

Duursma (1963) used the seasonal increase in DOC to obtain an estimate, albeit probably a minimum one, of 52 g C m^{-2} for the amount of dissolved organic material produced annually. Similar calculations (Banoub and Williams, 1973) for data from station E_1 in the English Channel gave a figure of 34 g C m^{-2}. These values are equivalent to 15–50% of the net annual production of organic material by phytoplankton in the areas considered.

Not all studies give evidence of regular and protracted seasonal increases. Holmes et al. (1967), for example, followed the DOC content of near-shore water off La Jolla, California. Apart from pronounced increases in DOC content associated with "red tide blooms"; no regular seasonal pattern was noted. A similar conclusion was reached by Strickland et al. (1970) who found little evidence of a correlation between the DOC concentration and productivity at three coastal stations in the same area.

The data on temporal changes in the DOM content of sea water suggest that the observed increases in concentrations probably result from the decomposition of the plankton bloom rather than from excretion during the bloom.

12.3.3.2. Spatial distribution

A selection of results of the DOC content of sea water by various groups of workers are summarized in Table 12.2. If the obvious systematic differences between the results of wet and dry combustion methods are ignored, the existing data provide no evidence for significant or regular differences between the various oceans or their climatic zones. The relative constancy of the concentration of dissolved organic material in sea water was illustrated by Thomas et al. (1971). They found that there was only a small statistical difference in the DON content of nitrate-rich water (average nitrate content $7\cdot3 \text{ µg-at N l}^{-1}$) and nitrate-poor water (nitrate content, $<0\cdot1 \text{ µg-at N l}^{-1}$). Banoub and Williams (1972) also commented upon the general uniformity of the amount of dissolved organic material in the sea in comparison with the inorganic nutrients and other biologically relevant factors. For example, they observed that although the concentrations of inorganic nutrients, particulate organic material and chlorophyll in the English Channel are five to ten times greater than in the surface water of the Mediterranean, there is little difference between the amounts of dissolved organic material in the waters at the two locations. The contentious point at present is to what extent there are real and significant variations with depth and location within this pattern of general uniformity. There is some measure of agreement over the general vertical profiles of organic material; concentrations in the upper 100 m are usually 30–50 % higher than those encountered in deep water. Typical values are given in Section 12.3.1 and Tables 12.2, 12.3 and 12.4; see also Fig. 12.1. It is less easy to generalize upon the finer details of distribution. From extensive studies of the Atlantic, Menzel and Ryther (Menzel and Ryther, 1968, 1970; Menzel, 1970) have concluded that in this ocean, below about 400 m all variations in DOC, as well as POC, were less than the error of their analytical methods, and there was no compelling evidence to suggest that there were differences associated with water masses of different origins. Menzel (1967) drew a similar conclusion from analyses of DOC off the west coast of Peru. Menzel and Ryther (1968) further concluded that within the present limits of analytical precision, the water masses of the Atlantic, once removed from their source, all have essentially the same concentration of organic material. Any changes in concentration as a result of biological oxidation are again within the limits of analytical precision.

However, there are reports of significant vertical and regional variations. Rochford (1963) found pronounced maxima in the vertical profiles of dissolved organic phosphorus in the S.E. Indian and S.W. Pacific Oceans. These maxima sometimes coincided with major hydrological cores, the most consistent association being with the Antarctic intermediate salinity minimum. Menzel (1964) has given convincing evidence that the distribution of DOC in the deeper water of the western Indian Ocean can be related to water masses. The Indian Ocean Central Water contains typically $1 \cdot 5$–$2 \cdot 0$ mg C l^{-1}, the Equatorial Water $0 \cdot 6$–$0 \cdot 8$ mg C l^{-1} and the Arabian Sea Water $0 \cdot 2$–$0 \cdot 4$ mg C l^{-1}. Along two density contours, the changes in DOC content during admixing of these water masses closely followed the changes in salinity. Fraga (1966), who analysed samples collected during the same cruise, also noted regional differences, and the overall pattern of distribution which he observed for DON is similar to that found by Menzel for DOC. It is not known why the Indian Ocean should show these marked regional differences whereas the Atlantic Ocean does not.

The spatial distribution of dissolved organic material, especially the variation with depth, raises not only the problem of the relationship between the inorganic and organic forms of carbon, nitrogen and phosphorus, but also their relationship to oxygen distribution. There appear to be fundamental differences of opinion over the reasons for any such relationship. Redfield

(1934), both in his classic paper and subsequently (Redfield *et al.*, 1963), has demonstrated that there is a close relationship between the nitrate, phosphate and ΣCO_2 contents of a water sample and the calculated extent of oxygen undersaturation (the apparent oxygen utilization AOU, see Vol. 1, Section 8.6.2). They concluded that the atomic ratios in which inorganic forms of carbon, nitrogen and phosphorus appeared and oxygen was utilized in the water were $P:N:C:O_2 = 1:16:106: -276$. (The exact value for oxygen is difficult to determine and has been modified over the years.) In deep water, the inorganic nutrients were present in excess of their predicted amount, and it was proposed that the excess could be attributed to "preformed" nutrients present in the water mass when it left contact with the atmosphere. There would appear to be no general disagreement with Redfield's broad hypothesis that the depletion of oxygen in deep water is a consequence of biological oxidation of organic material. There are differing views over certain details, e.g. whether the oxidation occurs soon, or a long time, after the water mass sank and, what is basically the same problem, whether or not the variations in oxygen and nutrient concentrations known to occur in deep water are generated by *in situ* biological activity.

Menzel and his collaborators (Menzel and Goering 1966; Menzel, 1967; Menzel and Ryther, 1968, 1970; Menzel, 1970) have concluded that below the first 100 m, or so, the POM and DOM are resistant to decay. They also concluded that there was little reason to believe that there would be addition of POM by sedimentation. If this is so, there would be little oxygen consumption as a result of *in situ* decomposition of DOM. They illustrated this by reference to the oxygen minimum in the S. Atlantic along a section in the S–N flow of the Antarctic Intermediate Water from 38°S to 8°N (Menzel and Ryther, 1970). At the southernmost end of the section the water at a density $\sigma_t = 27.2$ had an oxygen content of 6.32 ml l^{-1}. At the northern end of the section, at the same density, the oxygen content had fallen to 3.67 ml l^{-1}. The decrease in oxygen content of the water of 2.65 ml l^{-1} is equivalent to a consumption of 1.1 mg C l^{-1} on the basis of the ratios given by Redfield *et al.* (1963). However, the DOC content of the water totalled only 0.4 mg C l^{-1}, and along the section variations were no greater than ± 0.1 mg C l^{-1}. Addition of POM by sedimentation had already been discounted (Menzel, 1967). They were therefore led to the conclusion that there was no opportunity for *in situ* decomposition. They suggested that the changes in oxygen concentration were a result of mixing of waters that had already completed their regeneration process. They offered further evidence for this (Menzel and Ryther, 1968, 1970) by demonstrating that there is a very close correlation between the oxygen content and salinity. In a subsequent paper, Menzel (1970) has made a more general case against the *in situ* decomposition of organic material in the

S.W. and S.E. Atlantic and the tropical eastern Pacific. Again it is argued that the observed changes in DOC are incapable of explaining the changes in oxygen and nutrient concentrations, and he showed that deep water nitrate, phosphate and oxygen concentrations correlate closely with the salinity at the salinity minimum. He further demonstrated that it is possible to contour the oxygen concentration with temperature-salinity profiles. From these observations he concluded that beneath the top 200 m of the ocean, oxygen, the inorganic nutrients, and therefore, organic material behave essentially conservatively, and he therefore rejected the possibility that long term *in situ* decomposition of organic material in deep water exerts any major influence on nitrate, phosphate and oxygen concentrations. Menzel and Ryther (1970) are of the view that the oxygen deficient water which gives rise to the oxygen minimum in the Atlantic, originates in the eastern tropical regions of the ocean. A similar conclusion was arrived at independently by Bubnov (1972). He examined data on the density and oxygen content of water from the S.E. Atlantic, and concluded that the oxygen minimum originated off the western coast of South Africa and its propagation west and northwards was by water mixing. It was not necessary to invoke *in situ* decomposition of organic material to account for the observed oxygen minimum.

A different conclusion was arrived at by Craig (1971); he argued that the phenomenon of *in situ* oxygen consumption in deep water enables the depth variation of oxygen and total carbon dioxide in the Pacific Ocean to be predicted; furthermore he considered that DOC profiles do in fact give evidence of *in situ* consumption. He has shown that a good fit could be obtained between the predicted and observed profiles only if it was assumed that decomposition of DOM occurred in deep water, and he calculated that 37 % of the oxygen consumed is used in the decomposition of DOM (Redfield's ratios were used for the calculation). This conclusion appears to be confirmed by the observations by Ogura (1970) that there was an inverse relationship between DOC concentration and the AOU in the Pacific. This relationship was most pronounced between the density layers σ_t 24·5 to σ_t 27·0 (about 100 to 500 m deep); above and below this little correlation was obtained. At depths where this relationship applied the fall in DOC again accounted for about one-third of the observed AOU. Neither Craig nor Ogura have provided any satisfactory explanations for the mechanism by which the remaining oxygen is consumed. It is difficult to decide between these two apparently opposing hypotheses, as the data is drawn from very different locations and sections.

Studies of the biological consumption of oxygen in deep water raise the question of whether there are sizeable populations of planktonic organisms present at depth. Bernard (1963) examined a large number of samples of deep and surface water from the Mediterranean and the Indian Ocean and found that in many instances the number of small phytoplankton organisms and flagellates in deep water samples was as great as, if not greater than, that found in the euphotic zone. Similar findings of relatively large numbers of alga-like cells in deep water have also been reported by Fournier (1966, 1970, 1971) for the North Atlantic Ocean and the western Mediterranean and by Hamilton et al. (1968) for the East Pacific Ocean. The number of cells appears to be of the order of 10–100 ml^{-1}; Hamilton et al. have estimated the cell biomass to be about 0.05 µg C l^{-1}. To what extent these organisms are metabolically active is unclear. At least some of the material must be alive since Hamilton and his co-workers were able to culture small green autotrophs from deep water samples. If they are alive, and not in a resting stage, it is necessary to consider the nature of their food supply. Light intensities are too low to support a photosynthetic existence. Riley et al. (1965) have argued that the amount of particulate organic material in the deep waters of the ocean is insufficient to support the bathypelagic population, and that dissolved organic material (the most abundant form of organic material) is the food source. Menzel's conclusions, of course, do not support the notion that there is a sizeable deep-water population living on dissolved organic material. Attempts to demonstrate the uptake of dissolved organic compounds by deep water populations have so far been unsuccessful (Hamilton et al., 1968; Fournier, 1970; Williams, 1970), and the problem of the deep water populations and their food supply is still a puzzle.

In a consideration of the general distribution of dissolved organic material in the sea, the anoxic environment merits special attention*. Oxygen is required for the decomposition of organic material although nitrate, nitrite, sulphate and even, apparently, carbon dioxide may substitute. The sea contains comparatively small amounts of nitrate and nitrite, but sulphate and carbon dioxide are present in much greater amounts. Although for the provision of energy, sulphate and the other alternative proton acceptors are less suitable than oxygen, there is no a priori reason to assume that organic material would decompose to a lesser extent when they substitute for oxygen. However, when Otsuki and Hanya (1972, a, b) followed the decomposition of algal cells under aerobic and anaerobic conditions they found less complete decomposition in the absence of oxygen; this was common to both the particulate and the dissolved fraction. Richards (1970), in reviewing the organic chemistry of anoxic waters, came to the conclusion that there was no evidence that the preservation of dissolved organic material is enhanced by anoxic conditions. Skopintsev et al. (1967) concluded that the concentration of DOC in the Black Sea (weighted mean 2.2 mg C l^{-1}) is higher than that in ocean

* See Vol. 3, Chapter 16.

water (weighted mean $1 \cdot 54$ mg C l^{-1}). Although the vertical profile of DOC for the Black Sea shows the typical decrease below the first two or three hundred metres (Table 12.2), the decrease was less than that in normal ocean waters. It is not clear if this is a consequence of the anoxic conditions or not.

Deuser (1971) came to a somewhat different conclusion; he reported an increase in dissolved organic carbon with depth in the Black Sea, increasing from about 2 mg C l^{-1} in the aerated zone to 6 mg C l^{-1} in the deep water. These results are particularly interesting as they were obtained by a wet combustion method and it is the only instance in which this method has given higher values than the dry combustion procedure used by Soviet workers. Deuser suggested that this could be explained by loss of volatile organics during the evaporation stage prior to the dry combustion. If this is the explanation it would mean that in the anoxic waters, most of the organic material is volatile. Volatile organic products are common end products of bacterial fermentation; thus methane, for example, shows a pronounced increase with depth in the Black Sea (see Table 12.6).

12.4. PROCESSES OF REMOVAL

Williams (1971) estimated the total input of organic material into the sea to be $3 \cdot 6 \times 10^{16}$ g C yr^{-1}; a very similar estimate ($3 \cdot 85 \times 10^{16}$ g C yr^{-1}) has been produced by Skopintsev (1971). If it is presumed that for the organic material in the ocean a steady state concentration prevails, then the annual loss must be equal to this.

The dominant source of organic material is phytoplankton which, as discussed earlier, will give rise to DOM both by excretion of photosynthetic products, and by a variety of routes involving decay etc. Although it is difficult, if not naive, to place an exact figure on the amount of DOM arising from these processes, it should fall in the range 10–50% of phytoplankton production. If the average rate of photosynthetic fixation in the euphotic zone is assumed to be 10 µg C l^{-1} day^{-1} the input of DOM from phytoplankton would be equal to 1–5 µg C l^{-1} day^{-1}. A figure for the flux of DOM may also be derived from the estimates made by Duursma (1963) and Banoub and Williams (1973). Banoub and Williams, for example, estimated that the minimum mean annual flux of DOM at station E_1 in the English Channel is 34 g C m^{-2}. The water depth at this station is 70 m, and if it is assumed that production and removal occurred over a 6-month period, then the average minimum mean daily flux over that period would be about 3 µg C l^{-1} day^{-1}. Thus the two types of calculation give estimates of the same magnitude.

Losses of dissolved organic material can occur by three principal mechanisms: (1) physicochemical removal arising from change of physical state, (2) chemical removal or modification, and (3) biological removal. There is little doubt that of these the last mode of removal predominates; nevertheless the other means of removal raise interesting chemical considerations.

12.4.1. PHYSICOCHEMICAL REMOVAL

Loss of dissolved organic material can result by transfer either to the solid or gaseous phase. Volatile organic compounds may be formed during the production and during the chemical or biological decomposition of organic material and may escape into the atmosphere. Armstrong and Boalch (1960) demonstrated the presence of volatile organic compounds in culture media of the algae *Ectocarpus*, *Enteromorpha*, *Phaeocystis* and *Phaeodactylum*. They were also able to separate from sea water 20–50 µg C l^{-1} of volatile organic material containing 1–4 µg s l^{-1} of organic sulphur. Wilson *et al.* (1970) demonstrated the photochemical production of ethylene and propylene from organic material in sea water. Carbon monoxide was also produced, but neither methane nor the higher saturated gaseous hydrocarbons were evident. Methane, however, is present in sea water (see Section 12.3.2.2 and Vol. 1, Section 8.7.2), and it is slightly supersaturated in surface water relative to the marine atmosphere (Lamontagne *et al.*, 1971), but at the present it is not clear if there is a net flux of methane into the atmosphere. Lovelock *et al.* (1972) have demonstrated that dimethyl sulphide, which is a natural product of algae (Sieburth, 1968), is present in the surface waters of the Atlantic Ocean (average concentration 12 ng l^{-1}). They considered that dimethyl sulphide emission from the sea into the atmosphere is likely to play an important part in maintaining the sulphur balance of the sea.

Losses may occur as a result of conversion of DOM into particulate material. This could arise if the solubility is exceeded, but there is no well-established instance at present of control of an organic compound by solubility alone. Even the highly insoluble DDT group of pesticides is apparently normally present in sea water at concentrations well below their solubility (Cox, 1971). Other less-well defined precipitation phenomena occur. It is well known that filtered sea water will produce aggregates of a partially organic nature. The organic material is presumed to originate from the dissolved fraction. The mechanism leading to the formation of these aggregates is undoubtedly complex; the rate of formation is known to be accelerated by bubbling, but it is not clear whether bacteria or seed particles are involved (Riley, 1963; Sheldon *et al.*, 1967; Batoosingh *et al.*, 1969). This process will be considered in detail in Section 13.4 and will not be further discussed here (see also Chapter 10, Section 10.5.2).

Sorption onto existing surfaces is often regarded as an important step in the turnover of organic material in the sea. It is a complex problem which has, at present, been insufficiently investigated. Khaylov and Finenko (1968, 1970) have carried out an interesting study of the kinetics of the uptake of polymeric material onto natural detrital particles. They found that proteins and poly-saccharides are very strongly adsorbed onto natural detrital surfaces and their calculations suggest that at equilibrium some 5% of the polymeric material will be adsorbed onto particles in the sea.

12.4.2. CHEMICAL MODIFICATION OR REMOVAL

In common with the above processes, little is known of the importance and magnitude of purely chemical reactions in the sea. Although most of the decomposition, modification and uptake of organic material is regarded as being biological, there are certain limitations to biological processes. Firstly, some organic structures are resistant to biological attack; secondly, it is probable that there is a minimum limit to the rate of microbiological processes, and also that in deep water heterotrophic processes may come to a halt insignificant before all the available substrate is exhausted.

The recognition of the biological resistance of certain synthetic chemicals, e.g. DDT, has stimulated interest in the possibility that they may be decomposed photochemically. A study of the photolysis of DDT made by Miller and Narang (1970) has shown that the process is complex. An inducer having a low ionization potential is necessary to initiate the reaction, and the extent to which such species are present in sea water is not clear. The photochemical production of PCB compounds from DDT in the vapour phase has been mentioned in Section 12.2.1.

There is some evidence that photochemical reactions occur in sea water. Carlucci et al. (1969) showed that sterile sea water solutions of vitamin B_{12} and biotin lost most of their activity when exposed to sunlight for two weeks. Curiously, the activity of thiamine decreased to 50% and then showed no further decline. It was suggested that this could have arisen as a result of transformation to an analogue. Wilson et al. (1970) have demonstrated the photochemical production of carbon monoxide, ethylene and propylene in sea water; the amount produced appears to be dependent upon the concentration of organic material; the amounts produced are small (propylene: ca. 10^{-6} ml l^{-1}day^{-1}; carbon monoxide 10^{-4} ml l^{-1} day^{-1}) but they are significant in relation to the normal concentration of these compounds in sea water (see Table 12.6). It is worth noting that the production of carbon monoxide in the sea by the above process plus, perhaps, a contribution by photosynthetic organisms, results in a diurnal supersaturation of carbon monoxide at the sea surface with respect to the atmosphere and a probable efflux of carbon monoxide into the atmosphere (Lamontagne *et al.*, 1971). It has been suggested (Swinnerton *et al.*, 1970) that the ocean may be the principal natural source of carbon monoxide (see also Vol. 1, Section 8.7.3).

The chemical decomposition of biochemical compounds is usually very slow. Degens (1965, pp. 207–8) has suggested that the time scale of purely chemical hydrolysis of proteins is of the order of hundred or thousands of years under environmental conditions. Experimental and geological evidence implies that amino acids are stable over a period of millions of years. One approach which has been used to estimate the rates of these slow chemical processes is to determine their rates at high temperatures and then to extrapolate to the required temperature using the Arrhenius equation. In order to estimate the ammonia content of the primitive ocean, Bada and Miller (1968) used this approach to determine the rate constant of deamination of the aspartate ion; at a pH between 5 and 8 the calculated half-life of the ion is 2.8×10^7 years at $0°C$. Using a similar approach, Bada *et al.* (1970) have estimated the rate of racemization of amino acids. They used the determined rate of racemization of L-isoleucine to D-allo-isoleucine to date sediments. In principle, the same type of procedure could be used to date water samples.

The "humic fraction" of sea water is probably composed partially of chemically and biologically resistant fractions derived from organisms as well as of resistant material produced by chemical reactions from perhaps otherwise labile compounds. Duursma (1965) has discussed the possible reactions between phenols, amino acids and carbohydrates that give rise to humic or melanin-type material. Sieburth and Jensen (1969) have presented experimental evidence that the chemical interaction of such classes of compounds separated from algal exudates leads to the production of yellow-coloured "humic"-like material.

12.4.3. BIOLOGICAL REMOVAL AND DECOMPOSITION

In the upper 100 m or so of the ocean the degradation of organic material is almost certainly predominantly a biological phenomenon. Two quite distinct steps may be involved in the biological utilization of dissolved organic material. The first is the hydrolysis of polymeric non-diffusible material as well as of such esters as sugar phosphates and glycerides. The second is the uptake into the cell of diffusible material produced in this or other ways. The first step will normally occur outside the cell; it will involve extra-cellular enzymes and can provide no energy. This step will not result in remineralization of carbon, nor probably of nitrogen, but it can lead to the release of phosphate. It has been suggested that this potentially rapid release may account for the short cycling period of phosphorus in natural waters. Our knowledge of the activity of extracellular enzymes in the sea is scanty; the fact that some enzymes (e.g. urease: McCarthy, 1970) are inhibited by the salts in sea water may mean that the sea and fresh water environments could show marked differences in this respect.

It has often been argued that surface reactions may play some role in facilitating this initial degradation of DOM (see e.g. Zobell, 1946, pp. 84–85). Indeed, many microbiologists believe that aquatic bacteria normally function most effectively when sited on particles, although the physico-chemical justification for this is unclear. The work of Khaylov and Finenko (1968, 1970) on the sorption of organic material onto surfaces has been mentioned in an earlier section. They also examined the decomposition of polymeric material adsorbed onto natural non-sterile surfaces and found that both the uptake and the hydrolysis of proteinaceous material was first order, the rate of hydrolysis being 33% day^{-1}.

Biological oxidation is necessary for the complete degradation of organic material to inorganic molecules, and this can occur only under the action of highly organised enzyme systems inside the cell. There are two ecological problems here: firstly, to identify the types of organism responsible for the processes, and, secondly, to measure the rates at which the processes occur.

Traditionally, bacteria have been regarded as the predominant users and decomposers of detrital and dissolved organic material in the sea. It has always been difficult to reconcile this view with the low numbers of bacteria isolated from ocean water. However, it is almost certain that the plate count procedure for assessing bacterial numbers gives an underestimate. Much higher estimates are obtained by direct counting under the microscope, but at present this technique relies on a subjective assessment of whether or not an object is a bacterium. Hobbie et al. (1972) using the latter method recorded high counts of bacteria (10^7–10^8 cells l^{-1}) in the waters of the western North Atlantic. In contrast, Pomeroy and Johannes (1968) found low numbers of bacteria, but large numbers of colourless flagellates, about 5 μm in size, in samples also taken from the western North Atlantic. Williams (1970) carried out size fractionation of the organisms taking up ^{14}C labelled glucose and amino acids in the Mediterranean and other regions and found that most of the organic substrate was taken up by organisms less than 3 μm in size. He presumed that these organisns were probably bacteria.

It is now well-established that marine organisms other than bacteria may utilize organic material. Certain marine algae may grow partially or wholly on dissolved organic material. Lewin (1963) has discussed the heterotrophic growth of marine diatoms; Hellebust (1970) has demonstrated that the marine diatom Melosira nummuloides has a well-developed active transport system for amino acids. However, he found that other algae, e.g. Coccolithus *huxleyi, Isochrysis galbana, Dunaliella tertiolecta* and *Skeletonema costatum* were unable to take up a range of sugars and amino acids. There are reports of comparatively large populations of algae or organisms related to algae occurring well below the euphotic zone (Bernard, 1963; Fournier, 1966, 1970, 1971; Hamilton *et al.* 1968). It is usually presumed that these are growing heterotrophically on dissolved organic material, although there is no proof of this. The importance of heterotrophic algal growth in the sea is, at present, uncertain.

There is some evidence that invertebrate animals may utilize dissolved organic material. The subject has been reviewed by Stephens (1967), who has given examples of animals from four phyla which are capable of taking up dissolved organic material, and he has commented that any soft bodied invertebrate, regardless of its primary mode of feeding, will have some ability to take up dissolved organic compounds such as glucose or amino acids. The problem is how important these processes are in relation to other means of gaining food and in comparison with the turnover of dissolved organic material by the rest of the heterotrophic population. Pütter (1909) argued that dissolved organic material was the major food supply of planktonic animals. The benthic annelid *Clymenella* was studied in detail by Stephens who concluded that uptake of dissolved amino acids could satisfy the daily oxygen requirement of this organism. In common with the algae, the quantitative role of heterotrophic growth of marine animals on dissolved organic material is still an unsettled question, and at present it is only possible to guess at the relative importance of bacteria, small flagellates, algae and soft bodied invertebrates as users of dissolved organic material in the sea.

The quantitative aspects of the rates of utilization and decomposition of dissolved organic material are equally poorly understood. It is very difficult to arrive at satisfactory estimates of these rates, the rate of oxygen consumption can provide only a maximum limit because the relative contributions from the oxidation of POM and DOM are difficult to assess.

It is possible to study the uptake and respiration of dissolved organic material using ^{14}C tracer techniques. This approach was introduced by Parsons and Strickland (1962) in order to obtain a relative estimate of heterotrophic activity. They added 250 µg C l^{-1} of uniformly labelled glucose or acetate to sea water samples, incubated them for periods of up to 4 h and then filtered off the cells and measured the radioactivity taken up. They found that the relationship between the substrate concentration and the rate of uptake was satisfactorily described by the Michaelis and Menton equation. Wright and Hobbie (1966), working with samples from a freshwater lake, extended the method and showed that it was possible to derive a series of parameters from a kinetic analysis of the data. The most useful of these gave the fractions of the substrate taken up and remaining in the cells after a given period. Vaccaro and Jannasch (1966) and subsequently Vaccaro *et al.* (1968) have made kinetic analyses of the glucose uptake and have also determined the glucose concentration in the original water sample. This permitted the mass of glucose taken up per unit time to be calculated (this is generally a more useful value from the ecological point of view than the rate of turnover). A portion of the material taken up by the cells is used to provide energy, the carbon part of the molecule being respired as carbon dioxide. Williams and Askew (1968) using a related procedure, measured the rate of carbon dioxide production from the added isotope. Andrews and Williams (1971) have used this technique to measure the rates of oxidation of both glucose and an amino acid mixture; they also made measurements of the original substrate concentration in order to determine the mass of the substrate respired. Hobbie and Crawford (1969) and Williams (1970) examined the relationship between the amount taken up and that respired. In general they found that about 25 % of the material taken up was respired, the rest remaining inside the cell. The results of various studies of heterotrophic processes are summarized in Table 12.10.

When the results of these studies are considered as a whole, a pattern does emerge. Both the rate of turnover and the mass of the substrate metabolized in a given time decrease seawards. In estuaries, the rates of turnover are of the order of a day, in coastal water 1–10 days and in oceanic regions 10–100 days. In ocean water, there is a decrease in rate with depth, e.g. in the Mediterranean in the upper 100 m the rate of turnover of glucose and amino acids was 10–50 % day^{-1}, at 400–600 m it was detectable but less than 1 % day^{-1}, and it was undetectable at 2000 m (Williams, 1970). Present evidence suggests that the turnover of sugars and amino acids is comparatively rapid in areas of low productivity in which the inorganic nutrient concentration is low (e.g. the Mediterranean and the Gulf Stream).

For the English Channel, Andrews and Williams (1971) found a seasonal fluctuation in both the rate of turnover and the amount of substrate metabolised, the greatest rates being found in the summer months. Similar observations were made by Hobbie (1967) for a Swedish lake, and it seems likely that such seasonal fluctuations are a general feature of locations where there are seasonal changes in planktonic activity.

The total rate of uptake of individual compounds in the upper part of the ocean ranges from undetectable to about 5 $\mu g \, l^{-1} \, day^{-1}$, and is typically between 0·1 and 1·0 $\mu g \, l^{-1} \, day^{-1}$ (Table 12.10). The estimate of the total flux of dissolved organic material derived at the beginning of Section 12.4 lay in the range 1–10 $\mu g \, l^{-1} \, day^{-1}$. It is difficult to compare the two estimates because data have been obtained with the ^{14}C method only for a limited number of

TABLE 12.10

Average or typical measurements of heterotrophic activity in the sea

Location	Depth (m)	Substrate	Rate of metabolism (% day⁻¹)			Substrate concentration (μg l⁻¹)	Rate of metabolism (μg l⁻¹ day⁻¹)			Reference
			Net uptake	Respiration	Total uptake		Net uptake	Respiration	Total uptake	
Tropical S. Atlantic	10 and 100	Glucose	12			40	4·8			Vaccaro and Jannasch (1966)
Tropical and Temperate N. Atlantic	10–100	Glucose				100–<5	0·36–<0·06			Vaccaro et al. (1968)
York River Estuary	—	Total amino acids	4–0·3			39	24			Hobbie et al. (1968)
		Individual amino acids				17–0·55	10–0·024			
English Channel (summer months)	10 and 50	Glucose		50–20		6–<1		1·1–0·2		Andrews and Williams (1971)
English Channel (summer months)	10 and 50	Total amino acids		10–4		80–10		3–1		Andrews and Williams (1971)
Pamlico River Estuary (summer months)	—	Total amino acids				200–80			15–6	Crawford (1971)
		Individual amino acids							12–0·06	
W. Mediterranean	10–100	Glucose			100–30					Williams (1970)
	400 and 600	Glucose			1					
W. Mediterranean	10–100	Total amino acids			30–25					Williams (1970)
	400 and 600	Total amino acids			1					
Florida Straits	10–100	Glucose			25					Williams and Yentsch (unpublished)
	10–100	Total amino acids			100–50					
Coastal	25	Total amino acids			16–2	9–5			3·7–0·5	Williams (unpublished)
E. Pacific	25	Individual amino acids			40–1	2–0·0			1·2–0·0	

individual compounds; no data are available for organic acids or for sugars other than glucose. Nevertheless, the two sets of estimates are of comparable magnitude and thus lend support to one another.

12.5. MISCELLANEOUS TOPICS

12.5.1. ORGANIC CHEMISTRY OF THE SEA SURFACE (see also Chapter 10)

The continuous mixing of the upper layers of the sea will result in the selective accumulation of certain types of organic molecules at the air–sea interface. At this interface, semi-polar compounds (for example fatty acids) will tend to displace more polar ones such as proteins and polysaccharides. Convergences of water, produced by Langmuir circulation cells or other mechanisms, will also produce a concentration of such material. In areas of convergence the accumulated material may modify the surface capillary wave pattern and produce areas of smooth water, known as slicks. The presence of organic films on the sea surface may also affect other physical processes. Thus, in principle the exchange rate of gases and water vapour might be reduced by surface films, although present evidence suggests that any such effects are small and localized. In order for a monomolecular film to dampen out capillary waves at the sea surface it must generate film pressures of 0.2–1.0×10^{-5} N cm^{-1} (Garrett, 1967a). From studies on the formation of surface active films on natural sea water samples, Jarvis (1967) concluded that film pressures of the above magnitude will occur only when the molecules normally present in sea surface films are horizontally compressed by light winds or convergences. Similarly, it would appear that both gas and water vapour transfer will be markedly affected only under conditions of much closer packing of the film molecules than is widely encountered in the sea (Hawke and Alexander, 1962).

The few studies on the chemistry of the sea surface which have been made suggest that it is a unique zone. Goering and Menzel (1965) and Williams (1967) found higher levels of organic material in the surface film than in the underlying water. Goering and Menzel sampled from the top 3 mm of the sea and analysed 66 paired samples for dissolved organic carbon, ammonia, nitrite and silicate. They found that the surface water had a mean DOC content which was 0.44 mg C l^{-1} greater than that of the underlying water. The ammonia concentration was higher in the surface by 0.31 μg-at N l^{-1} on average. Williams sampled from the top 150 μm of the surface and also found that it was appreciably enriched with respect to ammonia relative to water at 15–20 cm and 20–35 m below the surface. The ranges of DOC values in these three regions were 1.68–4.43, 0.96–1.24 and 0.68–1.08 mg C l^{-1} respectively. The distribution of DON was similar. The particulate fraction showed an even greater degree of concentration at the surface, the ranges being surface,

$0.17–2.5$ mg C l^{-1}; $25–74$ m, $0.03–0.26$ mg C l^{-1}. Ammonia, nitrate and inorganic phosphate all occur at higher concentrations at the surface. Since these compounds are not surface active, some special mechanism to account for their relatively high surface concentrations must be sought. It is unlikely that rain alone can give rise to the high values at the surface (Williams, 1967). The interaction of dissolved organic material and bubbles could account for the high particulate organic content of the surface water. Bacteria are also known to be concentrated by bubbles (see e.g. Bezdek and Carlucci, 1972), and this may result in rapid mineralization and accumulation of ammonia and phosphate. In theory, this system could be continually refuelled by the migration of semi-polar material to the surface.

In common with Williams, Garrett (1967b) used a fine meshed metal sieve to remove the top 150 μm of the surface. Ferric iron was added to the sea water samples to produce a hydroxide precipitate, this was collected and subsequently extracted with chloroform. The extract was subsequently analysed by gas–liquid chromatography for fatty acids and long chain alcohols. The fatty acid composition of the surface layer was at least superficially similar to that of the DOM of sea water (see Table 12.5 and Williams, 1965). Unfortunately, no quantitative data were given and it is not possible to know if there are substantial differences in absolute amounts. Garrett has found some evidence that the surface samples contained molecules which were of longer chain length and less soluble than those present in subsurface waters.

It is likely that the sea surface may play an important role in the ecology of chlorinated hydrocarbon pesticides and PCBs in the sea. Seba and Corcoran (1969) reported DDT levels of 0.017 to 3.5 μg l^{-1} in surface slicks, the concentrations in sub-surface water being below 0.001 μg l^{-1}. Duce et al. (1972) analysed two samples from Narragansett Bay and reported PCB concentrations of 4.2 and 0.45 μg l^{-1}, whereas in sub-surface samples they occurred at concentrations of 0.15 μg l^{-1} or less (see also Harvey et al., 1973).

12.5.2. CARBON ISOTOPE RATIOS

The different isotopes of carbon undergo both chemical and biological reactions at slightly different rates resulting in an enrichment or depletion of the isotopes of different mass; the effects show a progression from ^{12}C through ^{13}C to ^{14}C. The analysis of the isotope ratios (usually ^{12}C and ^{13}C) in a compound or an organism can sometimes give an indication of the events leading up to its formation. The radioactive isotope of carbon (^{14}C) can be used to obtain additional information since it provides a measure of the time scale of processes (see e.g. Chapters 9 and 18).

12.5.2.1 *Stable carbon isotope ratios*

The relative abundance of the ^{13}C isotope of carbon in a sample is referred to a standard (e.g. Chicago belemnite, PDB-1) as follows:

$$\delta^{13}C = \frac{(^{13}C/^{12}C)\text{sample} - (^{13}C/^{12}C)\text{standard}}{(^{13}C/^{12}C)\text{ standard}} \times 1000$$

These carbon isotopes ^{12}C and ^{13}C have different abundances in the atmosphere and ocean. The principal equilibrium responsible for the difference is:

$$^{13}CO_{2(gas)} + H^{12}CO_{3(water)}^{-} \rightleftharpoons {}^{12}CO_{2(gas)} + H^{13}CO_{3(water)}^{-}$$

This results in an enrichment of the bicarbonate in the sea in carbon-13 by about 7‰ with respect to atmospheric carbon dioxide (Deuser and Degens,

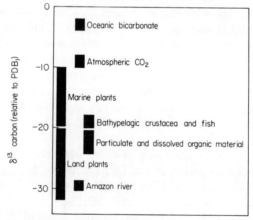

FIG. 12.2. The range of $\delta^{13}C$ for various organic and inorganic pools (Partially from Williams, 1971; and Parker and Calder, 1970).

1967). As carbon dioxide diffuses through the leaf of a higher terrestrial plant there is a depletion of ^{13}C, equal again to about 7‰, but see Fig. 12.2. The result is that the carbon dioxide reaching the photosynthetic system in marine algae is about 14‰ richer in the ^{13}C isotope than is the organic carbon of the higher terrestrial plant. Photosynthetic fixation of carbon dioxide (and to some extent the subsequent metabolism of the cell), results in a depletion of the heavier isotope; Parker and Calder (1970) in their review suggest an average figure of 17‰ for both terrestrial and aquatic plants, although the extent of fractionation during photosynthesis will depend upon the environmental conditions (see below). Ignoring this latter complication for the time being it will be seen that the above generalization does provide an adequate

explanation for the results summarized in Figure 12.2 which has been compiled from data by Williams (1971) and Parker and Calder (1970).

Degens *et al.* (1968b) concluded from a study of the effect of temperature,
aeration, pH and various carbon dioxide–bicarbonate regimes on the $\delta^{13}C$
values of cultures of *Skeletonema* and *Cyclotella*, that their observations were
compatible with the view that dissolved carbon dioxide is the source of
carbon for photosynthesis. The ecological consequences of this are worth
considering. In the sea, dissolved carbon dioxide normally represents only a
small fraction of the total inorganic carbon. At the surface it will be in isotopic
equilibrium with atmospheric carbon dioxide and some $7\%_0$ lower in ^{13}C
than the bicarbonate, the major form of inorganic carbon in the sea. Thus, if a
phytoplankton population grew under conditions in which the dissolved
carbon dioxide utilized during photosynthesis was replenished by carbon
dioxide from the atmosphere, they would use a source of carbon with a
$\delta^{13}C$ value similar to that of the atmosphere. In contrast, if the rate of removal
of dissolved carbon dioxide is sufficiently rapid, or if the population is isolated
from the atmosphere, then the condition of equilibrium with the atmosphere
is lost. Under these conditions the carbon dioxide utilized will be replenished
from the bicarbonate fraction and the $\delta^{13}C$ value of the algae will be related to
that of the bicarbonate in the sea. These latter circumstances would be normally expected to prevail in the sea, except for slow growing or small populations near the surface. The temperature of the water will also affect the
system, as low temperatures will increase the amount of dissolved carbon
dioxide in solution and will also alter the equilibrium so as to enrich the sea
with the lighter isotope.

Degens *et al.* (1968a) analysed mixed plankton from oblique hauls taken in
the Humboldt Current off Peru. The $\delta^{13}C$ value of the whole plankton ranged
from $-18\cdot0$ to $-21\cdot2\%_0$. They fractionated the cellular components and
found distinct and consistent differences between the various fractions:
e.g. the chloroform-soluble lipid fraction of the cells had $\delta^{13}C = -23\cdot0$ to
$-37\cdot4$, that of the "lignin" fraction was $-20\cdot2$ to $-25\cdot2$, that of cellulose was
$-20\cdot2$ to $26\cdot7$ and that of the sugars and amino acids was $-15\cdot9$ to $21\cdot2$. Thus
the overall isotopic ratio of the cell depends on the relative proportions of the
various cellular fractions which, in turn, are controlled to some extent by the
growth conditions.

Williams (1968) and Williams and Gordon (1970) measured the $^{13}C/^{12}C$
isotope ratios in dissolved organic material at various depths in the Pacific
Ocean (see Table 12.11). The general uniformity in the $\delta^{13}C$ values of the DOC
fraction, and the similar uniformity of that in the carbonate fraction (Deuser
and Hunt, 1969) is regarded as a sign that below the surface the dissolved
organic material is not decomposing. It is not immediately apparent how

TABLE 12.11

Isotope ratios of carbon in sea water

Fraction	Location	Depth	$\delta^{13}C$ (‰)	$\delta^{14}C$ (‰)	"Age" (years B.P.)	Reference
$^{13}C/^{12}C$ *ratios*						
Inorganic carbon	Atlantic	0–3000	+2 to −1			Deuser and Hunt (1969)
Dissolved organic carbon	N.E. Pacific	0–3000	−21·2 to −24·4			Williams and Gordon (1970)
Particulate organic carbon	N.E. Pacific	0–3000	−22·0 to −24·3			Williams and Gordon (1970)
Plankton, mixed	S.E. Pacific	0–200	−18·0 to −21·2			Degens et al. (1968a)
Bathypelagic fish and crustaceae	N.E. Pacific	400–2100	−18·0 to −20·2			Williams and Gordon (1970)
$^{14}C/^{12}C$ *ratios*						
Inorganic carbon (1958–1960)	N.E. Pacific	surface		−1 to −3		Bien et al. (1965)
Inorganic carbon (1964–1967)	N.E. Pacific	surface		+13·5 ± 1		Bien et al. (1965)
Inorganic carbon (1959)	N.E. Pacific	2000			2194 ± 70	Bien et al. (1965)
Inorganic carbon (1966)	N.E. Pacific	2000			1480 ± 80	Williams et al. (1969)
Dissolved organic carbon (1968–1969)	N.E. Pacific	1880		−351	3470 ± 330	Williams et al. (1969)
Dissolved organic carbon (1968–1969)	N.E. Pacific	1920		−341	3350 ± 300	Williams et al. (1969)
Zooplankton (1967)	Inshore Pacific	surface		+12·7 to +14·0		Williams et al. (1970)
Various bathypelagic crustaceae and fish (1966–1968)	N.E. Pacific	400–2100		+18·0 to +6·0		Williams et al. (1970)
Lepidopharres blackii (1968)	N.E. Pacific	2100		−1·4		Williams et al. (1970)

readily any such decomposition would be reflected in changes in carbon isotope ratios.

12.5.2.2. *Radioactive carbon analyses*

Williams *et al.* (1969) have used radiocarbon dating techniques to determine the "apparent age" of the DOC in two samples taken from 1880 and 1920 m in the N.E. Pacific Ocean. The δ^{14}C values (see Chapters 9 and 18) of the two samples were $-351\%_0$ and $-341\%_0$ respectively (relative to an oxalic acid standard), equivalent to an "apparent age" of 3470 ± 330 and 3350 ± 300 years before present. Inorganic carbonate samples from a similar depth in the Pacific Ocean gave "ages" of 2194 ± 70 and 1480 ± 80 years, i.e. significantly "younger". Two processes may contribute to the difference between the organic and inorganic fractions in the deep waters of the ocean. It is probable that an appreciable fraction of the DOC is not degraded and will persist, thus the organic fraction, unlike the inorganic, will not re-establish equilibrium with the atmosphere. Thus the DOC in ocean water will be radiochemically "old" before it leaves the surface. Determinations of the δ^{14}C of the DOC surface water are necessary before the importance of this effect can be ascertained. A second process that will create a difference between the "age" of the inorganic and organic carbon of deep water after it has left the surface is the "rejuvenation" of the inorganic fraction by the mineralization of recently produced organic material that has sedimented rapidly from the surface. This, of course, returns to the problem discussed earlier (see Section 12.3.3.3) of the extent of biological oxidation of organic material in deep water.

Williams *et al.* (1969) have used the mean determined "age" for DOC of 3400 years to calculate the average rate of export of the photosynthetically fixed carbon to the deep water. They calculated that this age implies an input of 1.85×10^{14} g C yr^{-1} or 0.51 g C m^{-2} yr^{-1} into deep water. This loss is equal to about 0.5% of the average primary production of the plankton.

In quite a different way, carbon-14 analyses may provide information on the time scale of processes in the ocean. Particularly during the period 1961 to 1962 there was extensive testing of nuclear weapons and this resulted in a marked increase in the ^{14}C content of the atmosphere and the surface water of the oceans. Prior to the tests, the δ^{14}C of the inorganic carbon of samples from the N.E. Pacific Ocean was -1 to $-3\%_0$, over the period 1964–1967 the δ^{14}C was $13.5\%_0$, the increase resulting from the bomb tests. This abrupt increase can provide information on the time scale of downward flux of food to the deep living organisms, about which very little is known. Williams *et al.* (1970) found that the δ^{14}C of surface plankton, and the majority of bathypelagic animals which they analysed, lay between $+6.0$ to $+18.0\%_0$, comparable to that of the inorganic carbon of the post 1962 era. Thus the time lapse between the original formation of their ultimate food supply and the time of sampling must have been less than about 8 years. These findings also rule out dissolved organic material as a starting point of the food chain for the bathypelagic forms because in deep water this has a $\delta^{14}C$ of about $-350\permil$. It seems likely that there must be a downward transport of food, either via an inter-linked system of food chains relying upon vertical migration or by a passive sinking of plankton and larger organisms. One organism did, however, have a $\delta^{14}C$ comparable with that of the inorganic carbon of the "pre-bomb" era; this was a bathypelagic fish *Lepidopharres blackii*, it is an interesting problem why the $\delta^{14}C$ of this organism should differ from that of its neighbours.

12.5.3. REGULATION OF PLANKTON GROWTH AND BEHAVIOUR BY EXTERNAL ORGANIC COMPOUNDS

As the study of biological oceanography progressed it became apparent that, whereas certain waters were conducive to growth, others were not and the notion of "good" and "bad" waters evolved. Often the effect can be related to the abundance of a mineral constituent, e.g. a micronutrient, such as nitrate or phosphate, or a trace metal such as iron. There are, however, well-established cases of difference that cannot be readily explained in this manner. Thus Provasoli (1963), in his review, cites the example of the waters off the coast of California and the British Isles which, although they have similar concentrations of inorganic nutrients, have very different productivities. It has been recognized for some time that the suitability of sea water as a medium for the growth of phytoplankton can be improved by the addition of organic extracts, and it is now becoming widely accepted that trace amounts of organic compounds can modify the growth of plankton. Khaylov (1965) reasoned that this stemmed from the fact that as an organism becomes smaller and its surface to volume ratio increases, its contact with (and therefore control by) the external environment will increase. The active and dominant population in the sea is composed of small organisms and thus external compounds, inorganic and organic, may be expected to have a more important role in the aquatic ecosystem than in the terrestrial one. That the growth and behaviour of marine organisms are controlled by such substances has been ably argued by Lucas (1947, 1961) and by Provasoli (1963).

The extent to which organic compounds participate in this activity will be considered under two major headings; the heterotrophic growth of algae will not be considered here.

12.5.3.1. *Organic algal growth promoters*

Trace additions of organic material appear to promote the growth of marine phytoplankton in at least two ways: firstly, vitamins are needed by a great proportion of the phytoplankton species and secondly, many natural organic compounds are effective chelators and modify the speciation of the heavy metal in the organism's environment.

Provasoli (1963) has assembled data on the vitamin requirements of 180 species of marine and freshwater algae. Only 56 of them had no vitamin requirement, the remainder requiring one or often more of the three vitamins: thiamine, biotin and vitamin B_{12}. The kinetics of vitamin utilization of marine algae have been studied in detail (Droop, 1968; Carlucci and Silbernagel, 1969), and it seems that the concentrations of vitamins in sea water are such that they may, on occasions, regulate phytoplankton growth and the potential size of the crop. The matter is not straightforward, as different approaches to the determination of vitamin requirements can give very different results. Furthermore, Droop (1968) has demonstrated that the organisms themselves can release material that diminishes the effective concentration of vitamin B_{12}. There have been a few environmental studies of the interrelationship between the concentrations of vitamins and phytoplankton activity. For the Sargasso Sea, Menzel and Spaeth (1962) found that an increase in the numbers of small diatoms was associated with the vitamin B_{12} concentration; when the dissolved vitamin B_{12} concentrations were low the population was dominated by *Coccolithus huxleyi*. The diatoms, when isolated in culture, were found to require vitamin B_{12}, whereas the coccolith had no such requirement. However, they observed that enrichment of the sea water with vitamin B_{12} did not increase the rate of carbon dioxide fixation. Carlucci (1970), has carried out a detailed study of the interrelationship between the concentrations of three vitamins (biotin, thiamine and vitamin B_{12}) and phytoplankton species composition and activity, and obtained no clear evidence of control by vitamins. There were instances where his results did suggest some form of correlation; for example, there were positive correlations between thiamine and the concentration of the plankton, and also between the net productivity and the vitamin B_{12} and thiamine concentrations. This type of approach gives little indication which is cause and which is effect. The general conclusion of those who have studied the interrelationships between plankton and vitamins seems to be that vitamins are not normally important in controlling and limiting production (Vishniac and Riley, 1961; Menzel and Spaeth, 1962; Carlucci, 1970). However, they may affect the floristic composition of a body of water and could be an important factor in initiating the blooming of certain species and perhaps in controlling the succession of species. Carlucci and Bowes (1970a) have made the important observation that planktonic algae can release, as well as take up, vitamins; prior to this, heterotrophic bacteria had been regarded as the probable source of vitamins in the sea (Burkholder, 1963). Carlucci and Bowes (1970b) showed that one algal species could use the vitamins excreted into the medium by a second species. This means that two vitamin-requiring organisms could, in theory, grow in a vitamin deficient medium by mutual exchange of vitamins. However, in practice it was not easy to set up experiments to illustrate such possibilities. They found that *Coccolithus huxleyi*, a thiamine-requiring species, released vitamin B_{12}, and that *Skeletonema costatum*, which required vitamin B_{12}, released thiamine. When *C. huxleyi* was inoculated into a thiamine deficient medium containing an actively growing culture of *S. costatum*, it grew using the thiamine produced by *S. costatum*. The converse was not demonstrated. When *S. costatum* was inoculated into a medium containing *C. huxleyi* its growth was inhibited rather than stimulated. Another B_{12}-producing coccolith, *C. fusiformis* caused the rapid death and lysis of *S. costatum*. The vitamin B_{12}-requiring algae seem particularly sensitive to inhibition by other organisms. Whether or not the release of vitamins by algae is important under natural conditions is not known at the present; such a process could give rise to such natural phenomena as species succession, and it would complicate the environmental relationship between vitamin concentration and the presence or activity of a species.

Vitamins are the most widely studied of the organic regulators of phytoplankton growth. However, they are not the only class of compounds that can regulate growth; Johnston (1963) studied how additions of various organic and inorganic compounds to sea water influenced its suitability as a medium for algal growth. Thiamine and vitamine B_{12} were often effective in promoting growth, but the chelating agent ethylenediamine tetraacetic acid (EDTA) was more frequently effective. The beneficial effect of additions such as soil extract probably arises through the chelating properties of some of its constituents. Barber and Ryther (1969) studied the productivity of an area of upwelling off Peru, and their results illustrate the probable importance of chelates in regulating productivity. The upwelling water was rich in organic nutrients but low in dissolved organic material. The productivity and the productivity index (productivity divided by chlorophyll *a* content) of the upwelling water were both low, and the addition of inorganic nutrients, trace metals, vitamins or amino acids, either singly or in combinations, did not improve the quality of the water for the natural population, whereas either EDTA or the filtrate of a zooplankton homogenate did. It was reasoned that, although the newly upwelled water contained an adequate supply of inorganic nutrients, it was deficient in natural organic chelators. These workers concluded that when productivity got under way, sufficient organic chelators resulted from natural processes of plankton production and decay and the environmental conditions improved. They found that, away from the immediate proximity of the upwelling area, the soluble organic carbon and the productivity index increased, although the nutrient concentrations in some instances decreased somewhat.

The manner in which chelates act in such instances is uncertain. They may chelate toxic heavy metals, or they could act by chelating and so facilitating the uptake of essential trace metals. Steemann-Nielsen and Wium-Anderson (1970) have suggested that in the instance reported by Barber and Ryther (1969) copper could be the toxic metal; this ion has been shown to be toxic towards algae at its concentration in sea water and it is probable that chelators will modify its activity.

12.5.3.2. Dissolved organic toxins

It has long been recognized that sea water exhibits antibiotic properties. This may result from the activity of an inorganic constituent. For example, Jones (1967) has found evidence that part of the bactericidal activity of sea water toward the bacterium *Escherichia coli* is attributable to copper. He has suggested that denaturation of organic compounds only partially explains the loss of toxicity which occurs on autoclaving and that part of this loss is due to the formation of precipitates containing copper. However, purely organic factors may, in some instances, exert an inhibiting effect on marine organisms Thus, it is well known that planktonic algae can release organic toxins. For example, those produced by blooms of dinoflagellates ("red tides") can lead to extensive mortalities of fish. Although the antibiotic properties of sea water are well documented, the details of the chemistry and the biology are poorly understood. Both marine bacteria (Kriss *et al.*, 1967) and marine algae (Aubert and Aubert, 1969; Sieburth, 1968; Wood, 1965) may produce organic compounds with a general or group toxicity.Certain aspects of algal toxins have been reviewed by Sieburth (1968), who drew attention to two classes of toxic materials known to be produced by plants in general, viz. allyl sulphides and the polyphenols. He suggested that they may exert their toxicity in a microzone adjacent to the algal cell surface. The importance of these compounds in the sea is not at all clear. The antibacterial activity of sea water has been studied by numerous workers (see for example Vaccaro *et al.*, 1950; Sieburth, 1971; Moebus, 1972). Vaccaro and his co-workers have studied the toxic effect of sea water on *Escherichia coli*, and Moebus extended this work to the pathogen *Staphylococcus aureus*, the marine bacterium *Serratia marinorubra* and the indigenous bacterial population. The death rates of *E. coli* and the other strains were found to vary seasonally, the organisms persisting longer in winter months.

It is recognized that trace amounts of organic compounds present in sea water not only influence the growth of micro-organisms, but also the behaviour of marine animals. Such processes as the homing of migratory fish, the settlement of larvae, mating and the sensing of predators appear, in part, to be controlled by organic compounds, almost invariably of unknown nature. It has been suggested that a subtle effect of pollutants on the ecosystem is the blocking or modification of the action of these compounds. The subject of chemical communicants in the sea is too extensive and diffuse to be considered here; the articles by Whittle and Blumer (1970) and Katona (1973) are useful sources of reference material.

12.5.4. INTERACTION BETWEEN INORGANIC AND ORGANIC COMPOUNDS (see also Section 12.3)

Two aspects of this subject will be considered, (1) the complexing of inorganic cations by organic compounds in the sea, (2) the effect of organic material on calcium carbonate precipitation.

12.5.4.1. *Organometallic complexes* (see also Vol. 1, Section 3.5)

There is now direct analytical evidence, as well as the indirect biological indications discussed in the previous section, that some of the metals present in sea water may be partially associated with organic matter. Thus, it has been shown that degradation of the organic material in sea water results in an increase in the amount of free copper (Corcoran and Alexander, 1964; Williams, 1969; Foster and Morris, 1971b) and mercury (Fitzgerald and Lyons, 1973) that may be determined in sea water.

In a few instances the metal will occur as covalently bonded organometallic compounds, which may be produced microbiologically (e.g. Hg as CH_3Hg^+ and $(CH_3)_2Hg$). More frequently the metals (e.g. copper) occur in co-ordination compounds, these complexes being either released from the cell or being formed by the combination of the metal with organic compounds released from the cell. A very full discussion of metal-organic interactions in sea water has been given by Siegel (1971). Only metal-organic co-ordination compounds will be considered here. Our present knowledge of these complexes in natural waters is very limited. Slowey *et al.* (1967) have demonstrated that 8·4–56 % of the copper in sea water is in a form that can be extracted with chloroform; preliminary evidence suggests that copper is associated with the lipid fraction. Barsdate (1970) dialysed lakewater containing the radioisotopes ^{65}Zn, ^{60}Co and ^{54}Mn, against distilled water. He demonstrated that more than half of the metal was associated with large, presumably organic, molecules incapable of passing through a dialysis membrane. He could detect little reversible complexing of copper, cobalt, maganese, lead and zinc with samples of inshore sea water. Duursma and Sevenhuysen (1966) who used quite a different approach, also found little evidence of organic complexing of

Fe^{2+}. Zn^{2+}. Ni^{2+}. Co^{2+} and Cu^{2+} in natural sea water. However. some complex formation took place if EDTA or leucine was added. A more detailed study of the kinetics of chelation has been reported by Duursma (1970) who examined the complexing of Co^{2+} and Zn^{2+} in sea water by leucine and quinoline-2-carboxylic acid (the latter being intended to simulate "humic acid"). He measured the stability constants of the complexes and calculated that leucine at a concentration of 1.8×10^{-8} M ($2.4 \mu g\,l^{-1}$) would complex only 5 % of the Co^{2+} and 1 % of Zn^{2+}. Quinoline-2-carboxylic acid at concentrations of 0.2 and 2.0 mg C l^{-1} complexed an appreciable fraction of the Co^{2+} (70 and 99 %, respectively), but a lesser amount of Zn^{2+} (8 and 51 %, respectively). From the ecological standpoint, the most important conclusion which can be drawn from this work is that, whereas the complexation of metals by compounds such as amino acids is of little significance in natural waters (see also Section 12.3), these compounds themselves may be almost completely associated with organic cations, thus, leucine is largely associated with H^+, Mg^{2+} and Ca^{2+}. The existence of such complexes raises interesting and fundamental biological questions. Firstly, to what extent are these associated metals biologically available? Secondly, at what rates are these complexes formed and dissociated. It should be noted that if ^{14}C labelled compounds are used to study heterotrophic activity, it is tacitly assumed that the isotope rapidly distributes itself uniformly through the analytically available species of the organic compound: if this does not occur then errors will result.

12.5.4.2. *Calcium carbonate and organic matter* (see also Chapter 9)

The calcium carbonate in the oceans is interesting because of the many reactions which it undergoes and because of the fact that its distribution is best described as being in a stationary state rather than in a state of equilibrium. Chave (1970) has discussed the possible role of adsorbed organic compounds in leading to the anomalous distribution of calcium carbonate. It was shown that calcium carbonate precipitates have a limited capacity to act as sites for precipitation and this capacity decreases as the organic content of the sea water increased. When examined microscopically both natural and prepared calcium carbonate precipitates appear to be coated with a layer of organic material. The layer is thought to have been formed by adsorption of dissolved organic material from the water, but the mechanism of its formation and the organic compounds involved are not known. Chave concluded that this organic layer isolates the precipitate from the aqueous phase and so removes the necessary sites for the initiation by "seeding" of precipitation from supersaturated waters and also hinders dissolution in the undersaturated deep waters.

12.6. CONCLUSIONS

The present knowledge of the organic species present in sea water and their concentrations is clearly very patchy. The broad features are known, but the details are often lacking. The gross organic content of sea water lies in the range 0.5–2.0 mg Cl^{-1}, intermediates of decomposition, such as sugars and amino acids, usually occur at concentrations of 1–10 $\mu g\,l^{-1}$ (10^{-7} to 10^{-8} M), vitamins at concentration of 1–100 ng l^{-1} (10^{-9} to 10^{-12} M). Table 12.12 is an attempt to compile a typical organic composition of sea water from existing analytical data. It is immediately evident that, unless such groups as "total carbohydrates" are included, it is possible to account for no more than 10% of the total organic material known to be present.

The factors controlling the concentration of DOM in the sea are poorly understood. This results, to some extent, from a lack of knowledge of the detailed organic composition of sea water, as well as from the lack of good methods for measuring the activity of hetrotrophic micro-organisms in the sea.

TABLE 12.12

A synopsis of the presently known average dissolved organic composition of sea water

Component	Concentration in sea water (as $\mu gC\,l^{-1}$)
Vitamin B_{12}	0·0005
Thiamine	0·005
Biotin	0·001
Total fatty acids	5
Urea	5
Total free sugars	10
Total free amino acids	10
Total carbohydrates	200
Combined amino acids	50

The precise role of DOM in the ecology of the sea is known only in general terms. Dissolved organic material constitutes a shunt of food for it is released by planktonic organisms and is subsequently taken up by the heterotrophic micro-organisms thereby re-entering the food chain. The DOM, or more correctly part of it, appears to play a role in controlling plankton growth; although this is now well-established, the exact importance of these types of compounds in the overall ecology of the oceans remains unclear, even in the well-studied instance of the vitamins.

The unsatisfactory state of the subject to some extent results from the intractable nature of the complex chemical situation presented by natural waters, but another important factor has been the paucity of the research effort into the study of the organic chemistry of sea water. Considerable progress could be made if the need arose, and major research programmes were invoked in a serious attempt to update the analytical approaches. At present this motivation does not exist and as a consequence the study is fragmentary. A more subtle obstacle to progress in this field may be the philosophy of the present approaches to the problem; a major part of the work consists of identifying and quantifying individual compounds. This traditional type of approach is limited in its scope and cannot cope with the possibility that much of the dissolved organic material in the sea may consist of compounds presently unknown and possibility not discrete molecular species but complex associations of organic and inorganic material; Degens (1970) has discussed one possible type of association, which he has termed heteropolycondensates. The nature of such associations would depend upon the physicochemical environment. Thus, analytical methods invoking different separation procedures would give varying results. In order to study this type of organic material, some radical change in thinking is needed. It may well be that, currently, the wrong type of questions are being asked and this has resulted in a tendency to retain traditional approaches.

It is probable that future research on the dissolved organic material present in sea water will fall under two broad categories: the first is likely to be a continuation of the present study of the chemistry and biology of naturally produced material and the second may well be concerned with the distribution, dynamics and ecology of man-made organic contaminants in the sea. Results so obtained will be of interest not only from the pollution point of view but in some instances they may yield information about the dynamics of oceans and their populations which has hitherto not been available.

REFERENCES

Anderson, G. C. and Zeutschel, R. P. (1970). *Limnol. Oceanogr.* **15**, 492.
Andrews, P. and Williams, P. J. le B. (1971). *J. mar. biol. Ass. U.K.* **51**, 111.
Armstrong, F. A. J. and Boalch, G. T. (1960). *Nature, Lond.* **185**, 761.
Armstrong, F. A. J. and Harvey, H. W. (1950). *J. mar. biol. Ass. U.K.* **29**, 145.
Armstrong, F. A. J., Williams, P. M. and Strickland, J. D. H. (1966). *Nature*, **211**, 481.
Arthur, C. R. and Rigler, F. H. (1967). *Limnol. Oceanogr.* **12**, 121.
Aubert, M. and Aubert, J. (1969) "Océanographie Médicale", 298 pp. Gauthier-Villars, Paris.
Bada, J. L. and Miller, S. L. (1968). *Science, N.Y.* **159**, 423.
Bada, J. L., Luyendyk, B. P. and Maynard, J. B. (1970). *Science, N.Y.* **170**, 730.
Banoub, M. W. and Williams, P. J. le B. (1972). *Deep-Sea Res.* **19**, 433.

Banoub, M. W. and Williams, P. J. le B. (1973). *J. mar. biol. Ass. U.K.* **53**, 695.
Barber, R. T. and Ryther, J. H. (1969). *J. exp. mar. Biol. Ecol.* **3**, 191.
Barbier, M., Joly, D., Saliot, A. and Tourres, D. (1973). *Deep-Sea Res.* **20**, 305.
Barsdate, R. J. (1970). *In* "Symposium on Organic Matter in Natural Waters" (D. W. Hood, ed.) pp. 485–493. University of Alaska.
Batoosingh, E., Riley, G. A. and Keshwar, B. (1969). *Deep-Sea Res.* **16**, 213.
Bernal, J. D. (1961). *In* "Oceanography" (M. Sears, ed.) pp. 95–128. American Assoc. Adv. Science, Publ. 67.
Bernard, F. (1963). *In* "Symposium on Marine Microbiology" (C. H. Oppenheimer, ed.) pp. 215–228. C. C. Thomas, Springfield, Ill.
Bezdek, H. F. and Carlucci, A. F. (1972). *Limnol. Oceanogr.* **17**, 566.
Bien, G. S., Rakestraw, N. W. and Suess, H. E. (1965). *Limnol. Oceanogr.* **10**, (suppl.) R25.
Blumer, M. (1972). *Science, N.Y.* **176**, 1257.
Bohling, H. (1970). *Mar. Biol.* **6**, 213.
Bohling, H. (1972). *Mar. Biol.* **16**, 281.
Bowman, M. C., Acree, F. and Corbett, M. K. (1960). *J. agric. food Chem.* **8**, 406.
Bubnov, V. A. (1972). *Oceanology,* **12**, 193.
Burkholder, P. R. (1963). *In* "Symposium on Marine Microbiology" (C. H. Oppenheimer, ed.) pp. 133–150. C. C. Thomas, Springfield, Ill.
Butler, E. I., Corner, E. D. S. and Marshall, S. M. (1970). *J. mar. biol. Ass. U.K.* **50**, 525.
Carlucci, A. F. (1970). *Bull. Scripps Inst. Ocean.* **17**, 23.
Carlucci, A. F. and Bowes, P. M. (1970a). *J. Phycol.* **6**, 351.
Carlucci, A. F. and Bowes, P. M. (1970b). *J. Phycol.* **6**, 393.
Carlucci, A. F. and Silbernagel, S. B. (1966). *Limnol. Oceanogr.* **11**, 642.
Carlucci, A. F. and Silbernagel, S. B. (1969). *J. Phycol.* **5**, 64.
Carlucci, A. F., Silbernagel, S. B. and McNally, P. M. (1969). *J. Phycol.* **5**, 302.
Chave, K. E. (1970). *In* "Symposium on Organic Matter in Natural Waters" (D. W. Hood, ed.), pp. 373–386. University of Alaska.
Corcoran, E. F. and Alexander, J. E. (1964). *Bull. mar. Sci. Gulf Carib.* **14**, 594.
Corner, E. D. S. and Newell, B. S. (1967). *J. mar. biol. Ass. U.K.* **47**, 113.
Corner, E. D. S., Head, R. N. and Kilvington, C. C. (1972). *J. mar. biol. Ass. U.K.* **52**, 847.
Corwin, J. F. (1970). *In* "Symposium on Organic Matter in Natural Waters" (D. W. Hood, ed.), pp. 169–180. University of Alaska.
Cox, J. L. (1971). *Fish Bull.* **69**, 443.
Craig, H. (1971). *J. Geophys. Res.* **76**, 5078.
Crawford, C. C. (1971) "The Utilization of Free Amino Acids by Estuarine Microorganisms", Ph.D. Thesis, North Carolina State University.
Daisley, K. W. and Fisher, L. R. (1958). *J. mar. biol. Ass. U.K.* **37**, 683.
Dawson, R. (1974). Ph.D. Thesis, University of Liverpool.
Degens, E. T. (1965). "Geochemistry of sediments: a brief survey" pp. 342. Prentice Hall, N.J.
Degens, E. T. (1970). *In* "Symposium on Organic Matter in Natural Waters" (D. W. Hood, ed.), pp. 77–106. University of Alaska.
Degens. E. T., Reuter, J. H. and Shaw, N. F. (1964). *Geochim. Cosmochim. Acta,* **28**, 45.
Degens, E. T., Behrendt, M., Gotthardt, B. and Reppmann, E. (1968a). *Deep-Sea Res.* **15**, 11.

Degens, E. T., Guillard, R. R. L., Sackett, W. M. and Hellebust, J. A. (1968b). *Deep-Sea Res.* **15**, 1.
Deuser, W. G. (1971). *Deep-Sea Res.* **18**, 995.
Deuser, W. G. and Degens, E. T. (1967). *Nature, Lond.* **215**, 1033.
Deuser, W. G. and Hunt, J. M. (1969). *Deep-Sea Res.* **16**, 221.
Droop, M. R. (1968). *J. mar. biol. Ass. U.K.* **48**, 689.
Duce, R. A., Quinn, J. G., Olney, C. E., Piotrowicz, R., Ray, R. J. and Wade, J. L. (1972). *Science,* **176**, 161.
Duursma, E. K. (1961). *Neth. J. Sea Res.* **1**, 1.
Duursma, E. K. (1963). *Neth. J. Sea Res.* **2**, 85.
Duursma, E. K. (1965). *In* "Chemical Oceanography" (J. P. Riley and G. Skirrow, eds.) Vol. I, pp. 433–475. Academic Press, London.
Duursma, E. K. (1970). *In* "Symposium on Organic Matter in Natural Waters" (D. W. Hood, ed.), pp. 387–397. University of Alaska.
Duursma, E. K. and Sevenhuysen, W. (1966). *Neth. J. Sea. Res.* **3**, 96.
Faust, S. D. and Hunter, J. V. (1971). *In* "Organic Compounds in Aquatic Environments" 638 pp. Marcel Dekker, New York.
Fitzgerald, W. F. and Lyons, W. B. (1973). *Nature, Lond.* **242**, 452.
Fogg, G. E. (1971). *Ergebn. Limnol.* **5**, 1.
Foster, P. and Morris, A. W. (1971a). *Water Res.* **5**, 19.
Foster, P. and Morris, A. W. (1971b). *Deep-Sea Res.* **18**, 231.
Fournier, R. O. (1966). *Science, N.Y.* **153**, 1250.
Fournier, R. O. (1970). *Limnol. Oceanogr.* **15**, 675.
Fournier, R. O. (1971). *Limnol. Oceanogr.* **16**, 952.
Fraga, F. (1966). *Deep-Sea Res.* **13**, 413.
Frankenberg, S., Coles, S. L. and Johannes, R. E. (1967). *Limnol. Oceanogr.* **12**, 113.
Fredericks, A. D. and Sackett, W. M. (1970). *J. Geophys. Res.* **75**, 2199.
Garrett, W. D. (1967a). *J. Mar. Res.* **25**, 279.
Garrett, W. D. (1967b). *Deep-Sea Res.* **14**, 221.
Goering, J. J. and Menzel, D. W. (1965). *Deep-Sea Res.* **12**, 839.
Goldberg, E. D. (ed) (1971). "North Sea Science" 500 pp. The M.I.T. Press, Cambridge, Mass.
Goldberg, E. D. (ed.) (1974). "The Sea" Vol. 5. Wiley and Sons, New York.
Gordon, D. C. and Sutcliffe, W. H. (1973) *Mar. Chem.* **1**, 231.
Hamilton, R. D., Holm Hansen, O. and Strickland, J. D. S. (1968). *Deep-Sea Res.* **15**, 651.
Handa, N. (1966). *J. Ocean. Soc. Japan,* **22**, 50.
Handa, N. (1967). *J. Ocean. Soc. Japan,* **23**, 115.
Harvey, G. R., Steinhaver, W. G. and Teal, J. M. (1973). *Science, N.Y.* **180**, 643.
Harvey, G. R., Steinhaver, W. G. and Milkas, H. P. (1974). *Nature, Lond.* **252**, 387.
Harvey, H. W. (1950). *J. mar. biol. Ass. U.K.* **29**, 97.
Hawke, J. G. and Alexander, A. E. (1962). *In* "Retardation of Evaporation by Surface Monolayers: Transport Processes" (V. K. La Mer, ed.) pp. 67–73. Academic Press, New York.
Hellebust, J. A. (1965). *Limnol. Oceanogr.* **10**, 192.
Hellebust, J. A. (1970). *In* "Symposium on Organic Matter in Natural Waters" (D. W. Hood, ed.), pp. 225–256. University of Alaska.
Hobbie, J. E. (1967). *In* "Chemical Environment in the Aquatic Habitat" (H. L. Golterman and R. S. Clymo, eds.) pp. 245–251. N.V. Noord-Hollandsche Vitgevers Maatschappij, Amsterdam.

Hobbie. J. E. and Wright. R. T. (1965). *Limnol. Oceanogr.* **10**, 471.
Hobbie, J. E. and Crawford, C. C. (1969). *Limnol. Oceanogr.* **14**, 528.
Hobbie, J. E., Crawford, C. C. and Webb, K. L. (1968). *Science, N.Y.* **159**, 1463.
Hobbie, J. E., Holm-Hansen, O., Packard, T. T., Pomeroy, L. R., Sheldon, R. W., Thomas, J. P. and Wiebe, W. J. (1972). *Limnol. Oceanogr.* **17**, 544.
Holden, A. V. (1970). *Nature, Lond.* **228**, 1220.
Holm-Hansen, O., Strickland, J. D. H. and Williams, P. M. (1966). *Limnol. Oceanogr.* **11**, 548.
Holmes, R. W., Williams, P. M. and Eppley, R. W. (1967). *Limnol. Oceanogr.* **12**, 503.
Hood, D. W. (1963). *In* "Oceanography and Marine Biology" (H. Barnes, ed.) Vol. I, pp. 129–155. Allen and Unwin., London.
Hood, D. W. (ed.) (1970). "Symposium on Organic Matter in Natural Waters" 625 pp. University of Alaska.
Jarvis, N. L. (1967). *Limnol. Oceanogr.* **12**, 213.
Jeffrey, L. M. (1970). *In* "Symposium on Organic Matter in Natural Waters" (D. W. Hood, ed.) pp. 55–71. University of Alaska.
Johannes, R. E. (1966). *Limnol. Oceanogr.* **9**, 235.
Johannes, R. E. and Webb, K. L. (1965). *Science, N.Y.* **150**, 76.
Johnston, R. (1963). *J. mar. biol. Ass. U.K.* **43**, 427.
Jones, G. E. (1967). *Limnol. Oceanogr.* **12**, 167.
Jones, K. and Stewart, W. D. P. (1969). *J. mar. biol. Ass. U.K.* **49**, 475.
Kalle, K. (1966). *Oceanogr. Mar. Biol. Ann. Rev.* **4**, 91.
Katona, S. K. (1973). *Limnol. Oceanogr.* **18**, 574.
Ketchum, B. H. and Corwin, N. (1965). *Limnol. Oceanogr.* **10**, R148.
Khaylov, K. M. (1965). *Oceanology,* **5**, 1.
Khaylov, K. M. and Burlakova, Z. P. (1969). *Limnol. Oceanogr.* **14**, 521.
Khaylov, K. M. and Finenko, Z. Z. (1968). *Oceanology,* **8**, 776.
Khaylov, K. M. and Finenko, Z. Z. (1970). *In* "Marine Food Chains" (J. H. Steele, ed.) pp. 6–18. Oliver and Boyd, Edinburgh.
Kriss, A. E., Mischustina, I. E., Mitskevich. I. N. and Zemtsova, E. V. (1967). "Microbial Population of Oceans and Seas" 287 pp. Edward Arnold, London.
Krogh, A. (1934). *Ecol. Monogr.* **4**, 430.
Lamontagne, R. A., Swinnerton, J. W. and Linnenbom, V. J. (1971). *J. geophys. Res.* **76**, 5117.
Lewin, J. C. (1963). *In* "Symposium on Marine Microbiology" (C. Oppenheimer, ed.) pp. 229–235. C. C. Thomas, Springfield, Ill.
Lewis, G. J. and Rakestraw, N. W. (1955). *J. mar. Res.* **14**, 253.
Linnenbom, V. J. and Swinnerton, J. W. (1970). *In* "Symposium on Organic Matter in Natural Waters" (D. W. Hood, ed.) pp. 455–467. University of Alaska.
Lloyd-Jones, C. P. (1971). *Nature, Lond.* **229**, 65.
Lovelock, J. E., Maggs, R. J. and Rasmussen, R. A. (1972). *Nature,* **237**, 452.
Lucas, C. E. (1947). *Biol. Rev.* **22**, 270.
Lucas, C. E. (1961). *In* "Oceanography" (M. Sears, ed.), pp. 499–517. Amer. Ass. Adv. Sci. Publ. 67.
McCarthy, J. J. (1970). *Limnol. Oceanogr.* **15**, 309.
Maugh, T. H. (1973). *Science, N.Y.* **180**, 578.
Menzel, D. W. (1964). *Deep-Sea Res.* **11**, 757.
Menzel, D. W. (1967). *Deep-Sea Res.* **14**, 229.

Menzel, D. W. (1970). *Deep-Sea Res.* **17**, 751.
Menzel, D. W. and Goering, J. J. (1966). *Limnol. Oceanogr.* **11**, 333.
Menzel, D. W. and Ryther, J. H. (1968). *Deep-Sea Res.* **15**, 327.
Menzel, D. W. and Ryther, J. H. (1970). *In* "Symposium on Organic Matter in Natural Waters" (D. W. Hood, ed.) pp. 31–54. University of Alaska.
Menzel, D. W. and Spaeth, J. P. (1962) *Limnol. Oceanogr.* **7**, 151.
Menzel, D. W. and Vaccaro, R. F. (1964). *Limnol. Oceanogr.* **9**, 138.
Miller, L. L. and Narang, R. S. (1970). *Science, N.Y.* **169**, 368.
Moebus, K. (1972). *Mar. Biol.* **13**, 1.
Natarajan, K. V. (1971). *Limnol. Oceanogr.* **18**, 655.
Natarajan, K. V. and Dugdale, R. C. (1966). *Limnol. Oceanogr.* **11**, 621.
National Academy of Sciences (1971) "Chlorinated Hydrocarbons in the Marine Environment". Washington, D.C.
Ogura, N. (1970). *Deep-Sea Res.* **17**, 221.
Ogura, N. and Hanya, T. (1967). *Int. J. Oceanol. Limnol.* **1**, 91.
Ohwada, K. and Taga, N. (1972). *Mar. Chem.* **1**, 61.
Otsuki, A. and Hanya, T. (1968). *Limnol. Oceanogr.* **13**, 183.
Otsuki, A. and Hanya, T. (1972a). *Limnol. Oceanogr.* **17**, 248.
Otsuki, A. and Hanya, T. (1972b). *Limnol. Oceanogr.* **17**, 258.
Parker, P. L. and Calder, J. A. (1970). *In* "Symposium on Organic Matter in Natural Waters" (D. W. Hood, ed.) pp. 107–127. University of Alaska.
Parker, P. L., Winters, J. K. and Morgan, J. (1972). *In* "Baseline studies of Pollutants in the Marine Environment and Research Recommendations". Working Papers pp. 555–582. The IDOE Baseline Conference, May 24–26, New York.
Parsons, T. R. and Strickland, J. D. H. (1962). *Deep-Sea Res.* **8**, 211.
Pillai, T. N. V., and Ganguly, A. K. (1970). *Current Science,* **22**, 501.
Pineda, J. P. (1973). "Aspects of the Organic Chemistry of Seawater". M.Phil. Thesis, University of Southampton.
Pocklington, R. (1971). *Nature, Lond.* **230**, 374.
Pomeroy, L. R. and Johannes, R. E. (1968). *Deep-Sea Res.* **15**, 381.
Provasoli, L. (1963). *In* "The Sea" (M. N. Hill, ed.) Vol. II, pp. 165–219, Interscience, New York.
Pütter, A. (1909). "Die Ernährung der Wassertiere und der Stoffhaushalt der Gewasser". 168 pp. Fisher, Jena.
Quinn, J. G. and Wade, T. L. (1972). *In* "Baseline Studies of Pollutants in the Marine Environment and Research Recommendations". Working Papers pp. 633–664. The IDOE Baseline Conference, May 24–26. New York.
Redfield, A. C. (1934). *In* "On the proportions of organic derivatives in sea water and their relation to the composition of plankton". pp. 177–192. James Johnstone Memorial Volume, Liverpool.
Redfield, A. C., Smith, H. P. and Ketchum, B. (1937). *Biol. Bull.* **73**, 421.
Redfield, A. C., Ketchum, B. H. and Richards, F. A. (1963). *In* "The Sea" (M. Hill, ed.) pp. 26–77. Interscience, New York.
Remsen, C. C. (1971). *Limnol. Oceanogr.* **16**, 732.
Richards, F. A. (1970). *In* "Symposium on Organic Matter in Natural Waters" (D. W. Hood, ed.) pp. 399–411. University of Alaska.
Riley, G. A. (1963). *Limnol. Oceanogr.* **8**, 372.
Riley, G. A., Van Hemert, D. and Wangersky, P. J. (1965). *Limnol. Oceanogr.* **10**, 354.

Riley, J. P. (1965). *In* "Chemical Oceanography" (J. P. Riley and G. Skirrow, eds.) Vol. II, pp. 295–424. Academic Press, London.

Riley, J. P. and Segar, D. A. (1970). *J. mar. biol. Ass. U.K.* **50**, 713.

Risebrough, R. W., Huggett, R. J., Griffin, J. J. and Goldberg, E. D. (1968). *Science, N.Y.* **159**, 1233.

Rochford, D. J. (1963). *Aust. J. Mar. Freshwater Res.* **14**, 119.

Ryther, J. H. (1963). *In* "The Sea" (M. N. Hill, ed.) pp. 347. Interscience, New York.

Sackett, W. M., Brooks, J. and Swinnerton, J. (1972). *In* "Baseline Studies of Pollutants in the Marine Environment and Research Recommendations". Working Papers, pp. 725–731. The IDOE Baseline Conference, May 24–26, New York.

Schnitzer, M. and Khan, S. U. (1972). "Humic Substances in the Environment", 327 pp. Dekker, New York.

Seba, D. B. and Corcoran, E. F. (1969). *Pesticide Monitoring J.* **3**, 190.

Sharp, J. H. (1973). *Mar. Chem.* **1**, 211.

Sheldon, R. W., Evelyn, T. P. T. and Parsons, T. R. (1967). *Limnol. Oceanogr.* **12**, 367.

Sieburth, J. McN. (1968). *In* "Advances in Microbiology of the Sea", (M. R. Droop and E. J. F. Wood, eds.) Vol. I, pp. 63–94. Academic Press, London.

Sieburth, J. McN. (1969). *J. exp. mar. biol. Ecol.* **3**, 290.

Sieburth, J. McN. (1971). *Mar. Biol.* **11**, 98.

Sieburth, J. McN. and Jensen, A. (1968). *J. exp. mar. biol. Ecol.* **2**, 179.

Sieburth, J. McN. and Jensen, A. (1969). *J. exp. mar. biol. Exp.* **3**, 275.

Siegel, A. (1971). *In* "Organic Compounds and the Aquatic Environment" (S. P. Faust and J. V. Hunter, eds.), pp. 265–295. Dekker, New York.

Skopintsev, B. A. (1947). *Akad. Nauk. SSSR, Investiga,* **11**, 22.

Skopintsev, B. A. (1960). *Trudy morsk. gidrofiz. Inst.* **19**, 3, translated in *Soviet Oceanogr,* 1963, Series No. 3, p. 1–14.

Skopintsev, B. A. (1971). *Oceanology,* **6**, 775.

Skopintsev, B. A., Timofeyeva, S. N. and Vershinina, O. A. (1966). *Oceanology,* **6**, 201.

Skopintsev, B. A., Timofeyeva, S. N., Danilenko, A. F. and Sokolova, M. V. (1967). *Oceanology,* **7**, 353.

Slowey, J. F., Jeffrey, L. M. and Hood, D. W. (1967). *Nature, Lond.* **214**, 377.

Starikova, N. D. (1970). *Oceanology,* **10**, 796.

Starikova, N. D. and Korzhikova, R. I. (1969). *Oceanology,* **9**, 509.

Stauffer, T. B. and MacIntyre, W. G. (1970). *Chesapeake Sci.* **11**, 216.

Steemann Nielsen, E. and Wium-Andersen, S. (1970). *Mar. Biol.* **6**, 93.

Stephens, G. C. (1967. *In* "Estuaries" (G. E. Lauff, ed.) pp. 367–373 Am Ass. Adv. Sci, Publ. 102 Washington.

Strickland, J. D. H., Solórzano L. and Eppley, R. W. (1970). *Bull. Scripps Inst. Ocean.* **17**, 1.

Swinnerton, J. W., Linnenbom, V. J. and Lamontagne, R. A. (1970). *Science, N.Y.* **167**, 984.

Thomas, J. P. (1971). *Marine Biol.* **11**, 311.

Thomas, W. H., Renger, E. H. and Dodson, A. N. (1971). *Deep-Sea Res.* **18**, 65.

Tolbert, N. E. and Zill, L. P. (1956). *In* "Research in Photosynthesis" (H. Gaffron, ed.) pp. 228–231. Interscience, New York.

Vaccaro, R. F. and Jannasch, H. W. (1966). *Limnol. Oceanogr.* **11**, 596.

Vaccaro, R. F., Briggs, M. P., Carey, C. L. and Ketchum, B. H. (1950). *Am. J. publ. Hlth.* **40**, 1257.
Vaccaro, R. F., Hicks, S. E., Jannasch, H. W. and Carey, F. G. (1968). *Limnol. Oceanogr.* **13**, 356.
Vallentyne, J. R. (1957). *J. Fish. Res. Bd. Canada*, **14**, 33.
Vishniac, H. S. and Riley, G. A. (1961). *Limnol. Oceanogr.* **6**, 36.
Walsh, G. E. (1965). *Limnol. Oceanogr.* **10**, 577.
Walsh, G. E. and Douglas, J. (1966). *Limnol. Oceanogr.* **11**, 406.
Watt, W. D. (1966). *Proc. R. Soc. B.* **164**, 521.
Webb, K. L. and Johannes, R. E. (1969). *Comp. Biochem. Physiol.* **29**, 875.
Whittle, K. J. and Blumer, M. (1970). *In* "Symposium on Organic Matter in Natural Waters" (D. W. Hood, ed.) pp. 495–507.
Williams, P. J. le B. (1970). *J. mar. biol. Ass. U.K.* **50**, 859.
Williams, P. J. le B. and Askew, C. (1968). *Deep-Sea Res.* **15**, 365.
Williams, P. J. le B. and Grey, R. W. (1970). *J. mar. biol. Ass. U.K.* **50**, 859.
Williams, P. M. (1965). *J. Fish. Res. Bd. Canada*, **22**, 1107.
Williams, P. M. (1967). *Deep-Sea Res.* **14**, 791.
Williams, P. M. (1968). *Nature, Lond.* **219**, 152.
Williams, P. M. (1969). *Limnol. Oceanogr.* **14**, 156.
Williams, P. M. (1971). *In* "Organic Compounds and the Aquatic Environment" (S. D. Faust and J. V. Hunter, eds), pp. 145–163. Dekker, New York.
Williams, P. M. and Gordon, L. I. (1970). *Deep-Sea Res.* **17**, 19.
Williams, P. M., Oeschger, H. and Kinney, P. (1969). *Nature, Lond.* **224**, 256.
Williams, P. M., McGowan, J. A. and Stuiver, M. (1970). *Nature, Lond.* **227**, 375.
Wilson, D. F., Swinnerton, J. W. and Lamontagne, R. A. (1970). *Science, N.Y.* **168**, 1577.
Wood, E. J. F. (ed.) (1965). "Marine Microbial Ecology", 243 pp. Chapman and Hall, London.
Wright, R. T. and Hobbie, J. E. (1966). *Ecology*, **46**, 447.
ZoBell, C. E. (ed.) (1946). "Maritime Microbiology", 240 pp. Chronica Botanica Co., Waltham, U.S.A.

Chapter 13

Particulate Organic Carbon in the Sea

T. R. PARSONS

Institute of Oceanography, University of British Columbia,
Vancouver 8, British Columbia, Canada

13.1. INTRODUCTION

Particulate organic carbon in the sea has been the subject of a number of recent reviews (e.g. Nishizawa, 1969; Riley, 1970). The term particulate organic carbon, or POC, includes both living and non-living suspended particles in the sea and generally refers to particles larger than 0·5 to 1·0 μm diameter. This size group of particles is usually decided by the smallest pore size of a membrane filter which can be readily used to filter large volumes of sea water and so concentrate the particles for analysis. The size of 0·5 to 1·0 μm is also the approximate lower limit for detailed inspection of settled material under a light microscope and is at the lower end of the range of particle sizes which will sediment and not exhibit any marked Brownian movement. Thus, the arbitrary definition of POC as organic particulate carbon larger than 0·5 to 1·0 μm diameter has some technical justification even though one must consider, in fact, that the sea contains a continuous range of microscopic particle sizes from the smallest colloids up to the largest organic aggregates and the plankton (Reiswig, 1972).

The proportion of living material in the POC has been traditionally determined by assuming that the largest fraction of living material is phyto-

365

N

plankton, and the living organic carbon has been estimated as chlorophyll *a* times a factor; this product when substracted from the total POC gives the non-living, or detrital, POC.

$$POC - Chlorophyll \ a \times f = Detrital \ POC$$
$$(Total) \quad (Phytoplankton \ POC)$$

where *f* may vary with season from ca. 25 to 250 (see Section 13.3).

However, it has become apparent that POC may contain an appreciable amount of living materials other than phytoplankton (e.g. bacterial aggregates) and that these will not be accounted for by the chlorophyll *a* determination. A better way of estimating the living fraction of POC has been given by Holm-Hansen and Booth (1966) who determined the amount of adenosine triphosphate (ATP) in the POC. Since it can be shown that ATP is rapidly destroyed following the death of an organism, it is assumed that ATP is proportional to the living organic carbon in POC. On the basis of a large number of analyses of different organisms, the average proportionality constant assumed for this conversion is 250.

The following discussion will be confined mainly to the POC, excluding the phytoplankton since these are discussed in Chapter 14. Further, it is possible to generalize on certain points regarding POC in the world's oceans and then to examine in more detail particular descriptions and processes which lead to the formation, distribution and fate of POC in the marine environment.

As a generalization, it can be stated that the quantity of detrital POC in the sea is a large fraction of the total POC, often exceeding the phytoplankton carbon by a factor of 10 or more except during large phytoplankton blooms in near surface waters. In contrast, POC is generally only a small fraction of the dissolved organic carbon (DOC). The latter may be present in open ocean water at concentrations of several $mg \ C \ l^{-1}$ (see Chapter 12) whereas the POC generally occurs in the range of 20 to $200 \ \mu g \ C \ l^{-1}$. Further, it appears that POC can be formed from DOC, either through bacterial activity or through the adsorption of DOC in organic particulate surfaces; this may be considered a reversible process so that POC can itself contribute to the level of DOC through decomposition and solution processes.

13.2. METHODS FOR THE MEASUREMENT OF POC

Particulate organic carbon is generally collected by filtration of a volume of sea water onto a fibreglass filter which has been pre-combusted to remove any background carbon. Some differences in the results obtained by different authors may be attributable to the pore size of the filters used. In one of the earliest studies on POC, Fox *et al.* (1952) collected their material on a filter bed of silica and magnesium oxide. Because of the adsorptive properties of these compounds it is probable that a great deal of colloidal and dissolved material was collected together with the POC. More recent results have generally been obtained with GF/C fibre glass filters having pore sizes of ca. 0·5 to 2 μm. However, it has been found (Quinn and Meyers, 1971) that even with these filters, certain dissolved substances such as organic acids, may be adsorbed on the filter surface and increase the apparent POC. In addition, the stated pore size of filters has been found to be a poor measure of the actual size of particles being retained. In experiments conducted by Sheldon and Sutcliffe (1969), all filters tested retained particles smaller than the stated pore size. However, the recent availability of membrane filters having pores produced by bombardment with charged particles has assisted in giving better definition to the retention size of filters (Sheldon, 1972).

Some of the earliest values for POC in sea water were obtained from the difference in weight of filtered particulate material, before and after ashing in a muffle furnace. This procedure requires a relatively large amount of material, and a more sensitive method, which can be used to measure microgram quantities, is to digest the organic carbon in sulphuric acid/dichromate and determine the amount of material oxidized either by titration or colorimetrically (e.g. as described by Strickland and Parsons, 1972). Although this method is easily applied, it suffers from a lack of accuracy because highly reduced compounds give higher values than other compounds and the results are relative to some standard organic carbon compound (e.g. glucose). More accurate and sensitive methods have been described by Dal Pont and Newell (1963) and Menzel and Vaccaro (1964). In both of these methods samples are submitted to dry combustion; in the former CO_2 is measured by conductivity and in the latter by infrared analysis. A partial review of some current techniques for measuring POC has been given by Szekielda (1966a). Because of the change in methodology from gravimetric techniques to the more precise carbon dioxide analysers since 1965, it has been considered expedient in the following text to discuss only those results obtained after that date.

Apart from the chemical determination of POC, other methods involving the counting and sizing of particles have been used in field and laboratory studies. In particular, the Coulter Counter® has been widely used for the study of particulate material in the sea (see e.g. Sheldon and Parsons, 1967). Although this instrument measures all particles, it has been shown that there is a reasonable correlation between this data and POC measured chemically (Zeitzschel, 1970).

13.3. THE ORIGIN OF POC IN THE SEA

Particulate material which has been allowed to settle out of a sea water sample can be examined with an inverted biological microscope. From such observations it is quite apparent that, apart from the readily identifiable planktonic organisms which are considered in Section 13.4, there are many fragments of detrital material which are difficult to identify. Microscopic examination of the detrital POC suggests that it may originate from either:

(i) materials of terrestrial origin, transported into the sea either by rivers, or via the atmosphere (allochthonous materials).

(ii) substances autochthonous to the marine environment. These materials may be of two types: (a) recognizable fragments of plants or animals derived from the marine food chain, and (b) POC formed *in situ* through a complex equilibrium DOC $\rightleftharpoons$ POC.

The airborne transfer of materials from the land to the sea has been most dramatically documented for certain pollutants, an example being the presence of DDT in the marine food chain in the Antarctic (George and Frear, 1966). Particles carried by rivers influence most near shore environments; in some cases this effect is apparent for a considerable distance out to sea. Thus, the Amazon river, which represents a discharge amounting to 20% of the world's river discharge, contains sufficient organic material (ca. 2 to 10 mg l^{-1}) to influence the concentration of POC in the marine environment over an area of ca. 10^7 km^2 during the time of maximum runoff (Williams, 1968). In some estuaries the contribution of allochthonous detritus is the principal form of organic production. For example, Qasim and Sankaranarayanan (1972) have found that primary production by phytoplankton represented less than 1% of the total input of detritus to a tropical estuary in southern India. Materials found in sea water which are characteristic of river runoff include wood fibres, leaf fragments, pollen grains, inorganic particles such as clays, and various specific substances depending on the proximity of domestic and industrial outfalls.

The input of terrestrial POC is largely confined to coastal environments and it may be assumed that the POC in oceanic waters is autochthonous to the marine environment. The evidence for this can be derived both indirectly from approximate calculations of the productivity of the world's oceans (e.g. Skopintsev, 1972) and directly from microscopic observations. Further evidence may be adduced from ^{13}C/^{12}C ratios, since, as Williams and Gordon (1970) have shown, the ^{13}C/^{12}C ratio of terrestrial particulate organic matter is quite different from that of the POC in the oceans.

Since the bulk of the particulate organic carbon in the sea is derived ultimately from the phytoplankton it would be anticipated that there should be a general relationship between the standing stock of phytoplankton and POC. Thus from studies of the levels of POC in the surface layer of the various oceans Hobson *et al.* (1973) concluded that high concentrations of POC were generally associated with areas of high primary production. This is also illustrated in Fig. 13.1 which shows an average seasonal cycle for the standing stock of phytoplankton (measured as chlorophyll *a* and the levels

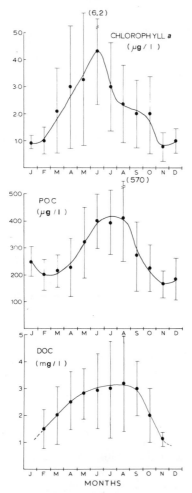

FIG. 13.1. Seasonal changes in chlorophyll *a*, particulate organic carbon (POC) and dissolved organic carbon (DOC) in the Strait of Georgia from Parsons *et al.* (1970). Monthly average values are shown for the period 1965–1968. The degree of patchiness in the data is reflected in the standard deviation shown for each mean.

of POC and DOC in the Strait of Georgia, British Columbia. Although this environment is influenced to some extent by land drainage it is apparent that a seasonal cycle exists in the levels of both chlorophyll a and POC. However, if average monthly concentrations are correlated for the whole year a poor relationship is obtained; this is in part apparent from the shape of the chlorophyll a and POC curves. Chlorophyll a increases rapidly in the spring, March to June, and then declines except for a small phytoplankton bloom indicated by the inflection during September and October. POC, on the other hand, increases slowly in the spring, reaches a maximum during July and August, and then declines. The relationship between chlorophyll a and POC has been discussed by a number of workers, and a good presentation of original data is given by Steele and Baird (1962). In this article the authors correlated chlorophyll a and POC values for different months of the year using data from Loch Nevis and Fladen Ground in the North Sea. Their results show that the ratio of organic carbon to chlorophyll a in the phytoplankton increased from 23:1 during a period of favourable growth in March/April, to ca. 250:1 in August when a lack of nutrients restricted plant growth. Although the monthly change in this ratio reflects changes in the composition of the phytoplankton, the POC value for zero chlorophyll (obtained by extrapolation) is a measure of the "detrital" POC. This value for the North Sea was found to be about 60 mg C m^{-3} during April. Another seasonal cycle in the amount of detrital POC in the sea has been described by Riley (1959) who showed that in Long Island Sound detrital POC constituted about 30% of the total POC during the spring phytoplankton bloom, but that during most of the rest of the year the detritus/phytoplankton POC ratio was between approximately 3 and 10:1. Similarly studies carried out by Nakajima and Nishizawa (1968) for the Tsugara Strait (between Honshu and Hokkaido) showed that phytoplankton POC ranged from 0·4 to 34% throughout the year with an annual average of 16% of the total POC. However, as in the example given in Fig. 13.1, the Tsugara Strait data also did not show any direct correlation between chlorophyll a and POC. A better relationship can often be observed between dissolved organic carbon (DOC) and POC, and this is indicated in Fig. 13.1. It is also apparent from the seasonal data collected by Nakajima and Nishizawa (1968) that the formation of POC was at least in part dependent on the seasonal presence of dissolved organic compounds. Since the production of dissolved organic carbon has been found to be generally out of phase with the maximum in phytoplankton production (Barber, 1967; Menzel and Ryther, 1970; Morris and Foster, 1971) it may be assumed that the most important route for POC formation in the sea is not directly from the phytoplankton; it is more likely that it is closely associated with DOC production and arises via a POC

⇌ DOC equilibrium. In contrast to this conclusion, Duursma (1963) found a close association between phytoplankton production and DOC. However, in spite of this exception, there is more direct evidence for an equilibrium between DOC and POC in the sea. From a consideration of the $^{13}C/^{12}C$ ratios of POC samples collected from deep water it is apparent that since deep water organic carbon has an accumulative residence time of several 1000 yrs (Skopintsev, 1972), any difference between the $^{13}C/^{12}C$ ratios for deep water POC and DOC would indicate a different origin for these two fractions; Williams and Gordon (1970) found, however, that the ratios were very similar. From other experimental data (see, e.g. Riley *et al.*, 1965; Sheldon *et al.*, 1967) it has also been shown that if the POC in natural sea water samples is removed by filtration, POC will reform in the filtrate in the presence of DOC.

In conclusion, it appears that the bulk of this particulate organic carbon in the sea in the immediate vicinity of some coastal areas or within the influence of large rivers is of terrestrial origin. In contrast, the POC in oceanic waters is derived mainly from the marine food chain and in particular from phytoplankton. However, the relationship between POC and phytoplankton is not direct, and this suggests that the POC is not merely fragments of the food chain. For various reasons it appears, in fact, that most of the POC in the sea is in some kind of equilibrium with the dissolved organic carbon (DOC).

13.4. THE QUANTITY AND NATURE OF POC IN THE SEA

Table 13.1 shows some general levels of particulate organic carbon in the world oceans. In oceanic areas values are highest ($>100\,\mu g\,C\,l^{-1}$) for productive waters such as those of the Peru current and of the North Atlantic during the summer. Low surface concentrations are encountered in the Sargasso Sea, the Central Pacific waters (Hawaii) and the Arctic Ocean. Deep water (below 200 m) particulate carbon values are generally reported to be less than 50 µg C l^{-1}. Menzel and Goering (1966) and Menzel (1967) have commented on the constancy of the concentration of deep water particulate carbon both with depth (below ca. 200 m and down to 5000 m), and with season. However, seasonal differences in the concentration of deep water particulate carbon have been reported both by Hobson (1967) for the North Pacific and Gordon (1970) for the North Atlantic. The latter area showed a seasonal range from a maximum of 47.3 µg C l^{-1} in January to a minimum of 8.8 µg C l^{-1} in April of the following year. These reported changes in the concentration of deep water POC may have a variety of

TABLE 13.1

Some representative values for the concentration of particulate organic carbon in the oceans. (Values reported since 1965)

Area	Particulate carbon ($\mu g\,l^{-1}$)	Method	Reference
Oceanic distribution			
North Atlantic			
10 to 100 m	80–130	Dry combustion	Szekielda (1968)
100 to 1000 m	60–120	and conductometric	
1000 to 3000 m	60–120	analysis	
Subtropical and tropical			
200 to 5000 m	10–20	Dry combustion and I.R. analysis	Menzel and Goering (1966)
Sargasso Sea			
0 to 50 m	20–60	Dry combustion	Menzel (1967)
100 to 4000 m	< 10	and I.R. analysis	
Northeast Pacific			
250 to 400 m	30–110	Wet combustion and volumetric analysis	Hobson (1967)
Central Pacific (*Hawaii*)			
0 to 100 m	5–35	Dry and combustion	Gordon (1971)
1000 to 4000 m	5–10	and chromatographic analysis	
Northwest Pacific			
Bering Sea all ⎱	50–200	Dry combustion	Nishizawa and
Oyashio current 0 to ⎰	40–300	and chromato-	Nakajima (1971)
(Philippine Sea) 300 m	40–75	graphic analysis	
Oyashio current all ⎱	60–160	Dry combustion	Handa and Yanagi
Kuroshio current 0 to ⎰	60–80	and chromato-	(1969)
Bering Sea 50 m	60–80	graphic analysis	
Eastern Tropical Pacific (Peru current)			
0 to 20 m	250–1,250	Dry combustion	Menzel (1967)
20 to 100 m	40–500	and I.R. analysis	
200 to 3000 m	< 15		
Gulf of Aden			
0 to 100 m	80–180	Dry combustion and conductometric analysis	Szekielda (1966b)

Table 13.1 (continued)

Area	Particulate carbon ($\mu g\,l^{-1}$)	Method	Reference
	Oceanic distribution		
Arctic Ocean (Amerasian Basin) 0 to 20 m	7–13	Wet and dry combustion and I.R. analysis	Kinney *et al.* (1971)
	Coastal distribution ($mg\,l^{-1}$)		
Atlantic N. American coast Charleston Harbour Hatteras Inlet	1.70 0·57	Dry combustion and and gravimetric estimation of organic matter by difference	Manheim *et al.* (1970)
Chesapeake Bay	1.0–3.0	Dry combustion and I.R. analysis	Biggs and Flemer (1972)
Pacific Gulf of California	0·06–0·40	Dry combustion and I.R. analysis	Zeitzschel (1970)
North Pacific (Aleutian Ridge) 0–50 m 100–1500 m	0·10–0·40 0·02–0·07	Dry combustion and I.R. analysis	Nakajima (1969)

causes, some of which are not truly seasonal and may reflect analytical errors arising from the difficulty of measuring very small amounts of carbon in very large volumes of water. However, in addition to the possibility of analytical error, explanations which have been given to account for changes in deep water POC include biological factors (such as the vertical migration of animals), the accumulation of particles at discontinuity layers, and differences in the POC content of different water masses. On the basis of depth distributions given by a number of authors, Hobson (1968) claims that the last of these explanations is the most acceptable.

The concentration of particulate organic carbon in coastal waters is generally one or two orders of magnitude higher than in oceanic waters. This can be attributed to the higher natural productivity of coastal waters, the addition of allochthonous organic material from the land and, occasionally, to the direct effect of particulate organic pollutants. Thus, in coastal areas such as Chesapeake Bay and Charleston harbour, the load of particulate organic material may be greater than 1 mg Cl^{-1}. However in coastal areas away from the immediate influence of urban outfalls or river discharge, appreciably lower values are found (see e.g. Gulf of California and Aleutian Ridge values in Table 13.1).

Several authors have attempted to describe the shapes and sizes of the many amorphous particles which can be seen by microscopic examination of samples of sea water. Bursa (1968) specifically identified starch grains as occurring in several oceans and particularly in the Arctic. He considered that they originated from aquatic plants (including both phytoplankton and macrophytes) following a large bloom. Kane (1967) differentiated between two particle types which he found in the Ligurian Sea. These were (i) "typical" aggregates which were composed of a gelatinous substrate to which other recognizable particles (e.g. plankton) adhered and (ii) particles which were described as a "granular" aggregates that apparently contained small inorganic particles as part of a large matrix. Gordon (1970) recognized four morphological categories of detrital organic particles in samples taken from the North Atlantic. These were aggregates, flakes, fragments and unclassifiable particles. Judging from their reaction to histochemical stains, the aggregates appeared to be chiefly carbohydrate, flakes were chiefly protein and fragments were entirely carbohydrate. Nemoto and Ishikawa (1969) also used biological stains to characterize particulate detrital material. It was found that the "typical" aggregates in samples of POC from the East China Sea described by Kane (1967) were stained with acid fuchsin which is a characteristic stain for cytoplasm. Acid fuchsin was also found to stain film-like particles which were either detrimental fragments or similar to the flakes described by Gordon (1970).

The visual appearance of most of the organic particles in the sea suggests that they are not part of any recognizable organism but may have been formed *in situ*. The processes by which such particles may be formed are difficult to study experimentally, but at least two mechanisms may be involved. The first of these is the adsorption of organic material on inorganic particles and includes the possible continual growth of inorganic-organic complexes; the second almost certainly involves bacteria and the secretion of bacterial slimes which tend to trap other particles.

The adsorption of organic material on inorganic particles has been demonstrated with respect to the adsorption of sugars, amino acids and peptone (Wilson, 1955; Bader *et al.*, 1960). Chave (1965) also showed that carbonate particles in tropical and subtropical seas have a surface layer of adsorbed organic material. The principal materials that have been identified as coatings on calcite particles are fatty acids (Seuss, 1970; Meyers and Quinn, 1971). The growth of bacteria on peptone-coated silica particles has given a further mechanism whereby such particles may form larger aggregates (Wilson, 1955).

The increase in the quantity of POC produced by the passage of bubbles through sea water (Sutcliffe *et al.*, 1963) has been the subject of much investigation.* Barber (1966) considered that bacteria were involved in this process and this is in agreement with the observations of Carlucci and Williams (1965) that bacteria were concentrated in the foam produced in this way. According to Menzel (1966), the increased amounts of POC originated from organic material in the air supply used for bubbling (this was precombusted in his experiments). However, Batoosingh *et al.* (1969) found that particles 0·22 to 1·2 μm diameter were important nuclei for the formation of POC during bubbling and that different results obtained by different authors might be in part explained by (i) the degree and duration of bubbling, and (ii) the pore size of the filters used to remove the POC. Thus, although the literature on the effect of bubbling on POC formation is confused, there appears to be a general tendency for POC to increase when sea water is aerated. This is borne out by some additional evidence. Firstly, the surface layer of the sea is much richer in POC than is the immediate subsurface water; for example, Nishizawa (1971) showed that the surface skin of sea water in the equatorial Pacific contained 418 μg C l^{-1} of POC compared with values of 39·8 μg C l^{-1} at 5 cm and 43·4 μg C l^{-1} at 10 cm. Secondly, inorganic particles which are formed by bursting bubbles at the air/water interface (see review by Blanchard, 1963) may be expected to adsorb organic material when they return to the sea surface.

Apart from mechanisms involving air/water and water/inorganic particle interfaces, bacteria are clearly involved in processes involving the formation of organic aggregates. If particulate material is removed from sea water by filtration through a membrane filter, particulate matter can be observed to reform in the filtrate (Riley *et al.*, 1965; Sheldon *et al.*, 1967). Observations made by Sheldon *et al.* (1967) and Parsons and Seki (1970) showed that the process of membrane filtration did not eliminate bacteria from the filtrate and that the resulting particle formation, including the aggregation of particles, was the result of bacterial growth, even though the concentration of organic material present in the filtrate was very low. The particle aggregation which they sometimes observed, was attributed to the presence of two or more bacterial species. This interpretation is consistent with the findings of Busch and Stumm (1968) that the aggregation of microorganisms

* See also Chapter 10, Section 10.5.2.

under their experimental conditions resulted from the interaction of polymers excreted by one or more of the microbial cells. The optimum conditions for flocculation occurred when a certain fraction of the adsorption sites on the bacteria were bridged with polymer; too little or too much polymer prevented aggregation. In other experiments, Stanley and Rose (1967) found that the tendency to clump was exhibited only by certain bacteria and that this activity could be enhanced by temperature changes and bubbling. Another mechanism for the formation of aggregates has been suggested by Sieburth (1965). Local formation of ammonia in the interstitial spaces between two or more bacterial cells, may cause a sufficient increase in the pH to bring about precipitation of carbonate. This hypothesis provides a possible explanation for the very high content of inorganic material present in the particulate matter.

13.5. CHEMICAL COMPOSITION OF POC

Particulate organic carbon in the oceans is generally associated with a large amount of inorganic material. Fox *et al.* (1953) found between 40 and 70% inorganic material in Californian coastal particulate material. Calculations based on results reported by Hobson (1967) give an approximate mean value of 80% for the northeast Pacific and oceanic average values suggested by Lisitzen (1959) and Wangersky (1965) range from 40 to 60% and 70%, respectively.

The nature of the inorganic material associated with POC in the sea has been studied by Armstrong and Atkins (1950) who showed that in oceanic waters near the coast of Britain the predominant components following ignition were silica, ferric oxide, alumina and calcium carbonate. In coastal environments where there is a heavy runoff, the inorganic material consists largely of local terrestrial minerals. For example, Manheim *et al.* (1972) found that the POC in sea water from the northern Gulf of Mexico was associated with montmorillonite, kaolinite and carbonate. In tropical and sub-tropical waters, the precipitation of carbonate causes a continual input of inorganic material which is directly associated with POC (Chave, 1965).

Very few chemical analyses have been made of the organic constituents of POC. In surface waters it is practically impossible to separate POC of detrital origin from the plankton, and in deep water the concentration of POC is so low that collection of sufficient material for detailed organic analysis is difficult. However, one important chemical property of POC which has been studied extensively is the C/N ratio. It appears from recent data that this ratio in deep water POC samples lies in the range 10 to 20 (Holm-Hansen *et al.*, 1966; Newell and Kerr, 1968; Nakajima, 1969; Gordon, 1971). Since healthy phytoplankton have a C/N ratio of ca. 6, it is apparent that there is a loss of nitrogen in POC from below the euphotic zone. This is consistent with sediment C/N ratios which are generally > 10. However, some earlier figures indicated C/N ratios of deep water detritus of less than 5 (see e.g. Parsons and Strickland, 1962; Menzel and Ryther, 1964). Although analytical error may be the explanation of these apparently anomalous results, natural processes may also be involved. For example, the C/N ratio of bacteria can be < 5; furthermore, the C/N ratio of dissolved organic material from decomposing phytoplankton has been observed to decrease from ca. 10 to 3 over a period of 200 days (Otsuki and Hanya, 1968). Thus if the DOC from decomposing phytoplankton is in equilibrium with POC through bacterial action, it is possible to obtain low C/N ratios. This would imply, however, that a large fraction of the deep water POC was composed of living bacteria. From analyses performed by Holm-Hansen and Booth (1966) it has been concluded on the basis of the ATP content that 3% or less of deep water POC was alive. It is probably true that deep water POC has a higher C/N ratio (i.e. > 10) than POC from the euphotic zone.

The carbohydrate fraction of POC has been studied by Handa and Yanagi (1969) and Handa and Tominaga (1969). Their results showed that the water soluble fraction of phytoplankton carbohydrate disappeared with depth so that between 300 and 1000 m only water insoluble carbohydrate remained. Hydrolysed samples of POC from the latter depths contained 50% less glucose than did phytoplankton, but correspondingly higher proportions of galactose, mannose, xylose and glucuronic acid. Similar results were obtained by Parsons and Strickland (1962) who showed that 70% of the carbohydrate in deep water POC samples was insoluble in weak acids and alkalis. Glucosamine and glucuronic acids were not found in appreciable quantities by Parsons and Strickland (1962). However, Wheeler (1967) has shown that copepod carcasses (containing chitin) were present at 2000–4000 m in the North Atlantic, but they would only account for between 0·5 and 5% of the total POC in the water and would not normally be detectable in terms of their glucosamine content.

The amino acid spectrum of hydrolysed samples of deep water POC has been examined by Degens (1970) who showed that below 200 m there was an apparent increase in the proportions of serine, glycine, lycine and arginine, and a decrease in alanine with depth down to 2,500 m. Cowey and Corner (1963) studied the amino acid composition of nine samples of POC taken throughout the year from the euphotic zone and concluded that shallow water POC had an amino acid composition very similar to that of phytoplankton.

Studies on the pigment composition of POC have been made by a number of authors; Yentsch (1965) showed that the chlorophyll a/phaeophytin ratio decreased with depth and attributed this to the effect of decreasing light. However, Currie (1962) showed that chlorophyll *a* was converted almost entirely to phaeo-pigments (probably phaeophorbides) during the digestion of phytoplankton by zooplankton. Saijo (1969) measured the concentration of chlorophyll *a* and phaeo-pigments down to 400 m in the northwest Pacific. He found that below 400 m the concentration of chlorophyll *a* was < 0.001 to 1.003 µg $\times$ l^{-1} but that phaeo-pigments were often present at 10 times the concentration of chlorophyll *a*. Carotenoids are less readily decomposed than are chlorophylls; for example the digestion of phyto-plankton by mussels results in the destruction of chlorophyll, but carotenoids are still present in the faeces of these animals (Fox *et al.*, 1944).

The question of whether POC is generally bio-degradable has been studied by Menzel and Goering (1966) who found that in POC samples taken from 1 m in the North Atlantic, between 16 and 52% of the detritus was bio-degradable. However, the authors could not detect any oxidation of POC in samples from below the euphotic zone (200 to 1000 m). Gordon (1970), using proteases, showed that proteases hydrolysed approximately 20 to 25% of deep water POC. Since only a small fraction of deep water POC is living (3% or less according to Holm-Hansen and Booth, 1966), the hydro-lysable material found by Gordon (1970) must represent refractory protein which is only being slowly converted to living material by heterotrophic microorganisms. However, the POC from terrestrial environments may be readily hydrolysed in near surface coastal regions as illustrated by Biggs and Flemer (1972), who found a 65% conversion of POC to DOC in an estuary. Furthermore, Seki *et al.* (1968) have found that the hydrolysates of POC are good substrates for bacteria, since sedimented organic material could be converted into bacterial biomass which could be utilized as food for the brine shrimp, *Artemia*. In some environments, POC aggregates may form an essential part of the food chain; this appears to be particularly true in coral reef communities where organic material released by the coral may serve as particulate food for zooplankton (Johannes, 1967). The transport of organic materials over very great distances may also provide a feeding mechanism for animals in ultra-oligotrophic environments. This has been postulated by Sorokin (1971) who has suggested that organic material produced by excessive primary production during the summer in temperate waters may eventually be transported into ultra-oligotrophic tropical waters. He claimed that the higher temperatures of tropical waters allow bacteria to utilize the DOC substrate and form bacterial aggregates which, in turn, are utilized as food for secondary producers.

The sinking rate of POC is an important factor in maintaining a food supply of organic carbon for the benthic community. As an approximation it is generally assumed that sinking particles will obey Stoke's Law; however, with organic aggregates there are two departures from Stoke's Law which must be taken into consideration (Riley, 1970; Kajihara, 1971). Firstly, many of the POC detrital aggregates are not spherical in shape and may be better represented either as long cylinders, or flakes. Secondly, the density of the particulate aggregates decreases as their size increases. Both these properties cause the sinking rates of large particles to be less than those predicted from Stoke's Law. In contrast, the sinking rate of debris from the food chain, such as faecal pellets from zooplankton, can be determined experimentally and values have been found to range from ca. 30 to 800 m day^{-1}, depending on the species of zooplankton and their diet (Smayda, 1969; Fowler and Small, 1972).

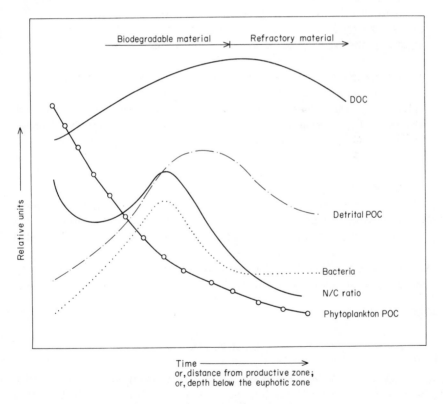

FIG. 13.2. A generalized relationship between POC and the environment.

13.6. SUMMARY

An attempt has been made in Fig. 13.2 to show the general relationship between POC and other environmental parameters. The relative change in concentration is given as a function of time following a maximum in the standing stock of phytoplankton. Since advective effects and sinking will alter the actual location of further biological and chemical activity, it is apparent that the time function may also be represented as either a lateral distance or a depth. The principal features of the summary presented in Fig. 13.2 are that detrital POC increases out of phase with phytoplankton POC but more or less in phase with the amount of DOC. Bacteria are shown to play an important (but not exclusive) role in increasing the detrital POC and in maintaining a high N/C ratio until the more biodegradable fraction of the phytoplankton has been converted to bacterial biomass. As the refractory compounds become predominant in the POC both the bacterial biomass and the N/C ratio decreases. At some time or distance from the phytoplankton bloom, the proportion of organic carbon is found to be greatest in the DOC followed by decreasing amounts of detrital POC > bacterial > phytoplanktonic carbon.

Another general summary of the distribution of POC with depth (or time) has been given by Nakajima and Nishizawa (1972). In their model they considered total POC (i.e. phytoplankton and detritus) and showed that, within a depth interval of ca. 0 to 100 m, there was an exponential decrease in total POC with depth such that

$$C = C_0\, e^{-k(Z-Z_0)}$$

where C is the concentration of carbon at depth Z and C_0 is the maximum concentration at depth Z_0. The depth of Z_0 varied from 0 to 30 m; values of k were from 0·019 to 0·039 with the highest values of k being associated with the highest concentration of C_0. The decrease in POC with depth was attributed to zooplankton grazing (ca. 30%) and to degradation processes, including bacterial decomposition.

ACKNOWLEDGEMENTS

The author is grateful to Mr. D. Blackbourn, of the Institute of Oceanography, University of British Columbia, for discussions leading to the preparation of Fig. 13.2.

REFERENCES

Armstrong, F. A. J. and Atkins, W. R. G. (1950). *J. mar. biol. Ass. U.K.* **29**, 139.
Bader, R. G., Hood, D. W. and Smith, J. B. (1960). *Geochim. Cosmochim. Acta,* **19**, 236.
Barber, R. T. (1966). *Nature, Lond.* **211**, 257.
Barber, R. T. (1967). Ph.D. Thesis, Thesis, Stanford Univ., Palo Alto. 132 pp.
Batoosingh, E., Riley, G. A. and Keshwar, B. (1969). *Deep-Sea Res.* **16**, 213.
Biggs, R. B. and Flemer, D. A. (1972). *Mar. Biol.* **12**, 11.
Blanchard, D. C. (1963). *Prog. Oceanogr.* **1**, 71.
Bursa, A. S. (1968). *J. Fish. Res. Bd. Can.* **25**, 1269.
Busch, P. L. and Stumm, W. (1968). *Environ. Sci. Technol.* **2**, 49.
Chave, K. E. (1965). *Science, N.Y.* **148**, 1723.
Carlucci, A. F. and Williams, P. M. (1965). *J. Cons. Int. Explor. Mer,* **30**, 28.
Cowey, C. B. and Corner, E. D. S. (1963). *J. mar. biol. Assoc. U.K.* **43**, 495.
Currie, R. I. (1962). *Nature, Lond.* **193**, 956.
Dal Pont, G. and Newell, B. (1963). *Aust. J. Mar. Freshwat. Res.* **14**, 155.
Degens, E. T. (1970). *In* "Organic Matter in Natural Waters" (D. W. Hood, ed.). Univ. Alaska, p. 77.
Duursma, E. K. (1963). *Netherl. J. Sea Res.* **2**, 85.
Fowler, S. W. and Small, L. F. (1972). *Limnol. Oceanogr.* **17**, 293.
Fox, D. L., Updegraff, D. M. and Novelli, G. D. (1944). *Arch. Biochem.* **5**, 1.
Fox, D. L., Isaacs, J. D. and Corocoran, E. F. (1952). *J. Mar. Res.* **11**, 29.
Fox, D. L., Oppenheimer, C. H. and Kittredge, J. S. (1953). *J. Mar. Res.* **12**, 233.
George, J. L. and Frear, D. E. H. (1966). *J. appl. Ecol.* **3** (suppl), 155.
Gordon, D. C. (1970). *Deep-Sea Res.* **17**, 233.
Gordon, D. C. (1971). *Deep-Sea Res.* **18**, 1127.
Handa, N. and Yanagi, K. (1969). *Mar. Biol.* **4**, 197.
Handa, N. and Tominaga, H. (1969). *Mar. Biol.* **2**, 228.
Hobson, L. A. (1967). *Limnol. Oceanogr.* **12**, 642.
Hobson, L. A. (1968). *Science, N.Y.* **162**, 587.
Hobson, L. A., Menzel, D. W. and Barber, R. T. (1973). *Mar. Biol.* **19**, 298.
Holm-Hansen, O. and Booth, C. R. (1966). *Limnol. Oceanogr.* **11**, 510.
Holm-Hansen, O., Strickland, J. D. H. and Williams, P. M. (1966). *Limnol. Oceanogr.* **11**, 548.
Johannes, R. E. (1967). *Limnol. Oceanogr.* **12**, 189.
Kajihara, M. (1971). *J. Oceanogr. Soc. Jap.* **27**, 158.
Kane, J. W. (1967). *Limnol. Oceanogr.* **12**, 287.
Kinney, P. J., Loder, T. C. and Groves, J. (1971). *Limnol. Oceanogr.* **16**, 132.
Lisitzin, A. P. (1959). *Preprint, Int. Oceanogr. Congress, New York,* pp. 470–471.
Manheim, F. T., Meade, R. H. and Bond, G. C. (1970). *Science, N.Y.* **167**, 371.
Manheim, F. T., Hathaway, J. C. and Uchupi, E. (1972). *Limnol. Oceanogr.* **17**, 17.
Menzel, D. W. (1966). *Deep-Sea Res.* **13**, 963.
Menzel, D. W. (1967). *Deep-Sea Res.* **14**, 229.
Menzel, D. W. and Vaccaro, R. F. (1964). *Limnol. Oceanogr.* **9**, 138.
Menzel, D. W. and Ryther, J. H. (1964). *Limnol. Oceanogr.* **9**, 179.
Menzel, D. W. and Goering, J. J. (1966). *Limnol. Oceanogr.* **11**, 333.
Menzel, D. W. and Ryther, J. H. (1970). *In* "Organic Matter in Natural Waters" (D. W. Hood, ed.). Univ. Alaska, pp. 31–54.

Meyers, P. A. and Quinn, J. G. (1971). *Limnol. Oceanogr.* **16**, 992.

Morris, A. W. and Foster, P. (1971). *Limnol. Oceanogr.* **16**, 987.

Nakajima, K. (1969). *J. Oceanogr. Soc. Jap.* **25**, 239.

Nakajima, K. and Nishizawa, S. (1968). *Rec. Oceanogr. Wks. Jap.* **9**, 219.

Nakajima, K. and Nishizawa, S. (1972). *In* "Biological Oceanography of the Northern Pacific Ocean" (A. Y. Takenouti, ed.), Idemitsu Shoten Publishing Co., Tokyo.

Nemoto, T. and Ishikawa, K. (1969). *J. Oceanogr. Soc. Jap.* **25**, 281.

Newell, B. S. and Kerr, J. D. (1968). *Aust. J. Mar. freshw. Res.* **19**, 129.

Nishizawa, S. (1969). *Bull. Plankton Soc. Jap.* **16**, 1.

Nishizawa, S. (1971). *Bull. Plankton Soc. Jap.* **18**, 42.

Nishizawa, S. and Nakajima, K. (1971). *Bull. Plankton Soc. Jap.* **18**, 12.

Otsuki, A. and Hanya, T. (1968). *Limnol. Oceanogr.* **13**, 183.

Parsons, T. R. and Strickland, J. D. H. (1962). *Science, N.Y.* **136**, 313.

Parsons, T. R. and Seki, H. (1970). *In* "Organic Matter in Natural Waters" (D. W. Hood, ed.). Univ. Alaska, p. 1.

Parsons, T. R., Le Brasseur, R. J. and Barraclough, W. E. (1970). *J. Fish. Res. Bd. Can.* **27**, 1251.

Qasim, S. Z. and Sankaranarayanan, V. N. (1972). *Mar. Biol.* **15**, 193.

Quinn, J. G. and Meyers, P. A. (1971). *Limnol. Oceanogr.* **16**, 129.

Reiswig, H. M. (1972). *Limnol. Oceanogr.* **17**, 341.

Riley, G. A. (1959). *Bull. Bingham Oceanogr. Coll.* **17**, 83.

Riley, G. A. (1970). *Adv. mar. Biol.* **8**, 1.

Riley, G. A., Van Hemert, D. and Wangersky, P. J. (1965). *Limnol. Oceanogr.* **10**, 354.

Saijo, Y. (1969). *Bull. Jap. Soc. Fish. Oceanogr., Special Number, Prof. Uda's Commemorative Papers,* 179.

Seki, H., Skelding, J. and Parsons, T. R. (1968). *Limnol. Oceanogr.* **13**, 440.

Seuss, E. (1970). *Geochim. Cosmochim. Acta,* **34**, 157.

Sheldon, R. W. (1972). *Limnol. Oceanogr.* **17**, 494.

Sheldon, R. W. and Parsons, T. R. (1967). *J. Fish. Res. Bd. Can.* **24**, 909.

Sheldon, R. W. and Sutcliffe, W. H. (1969). *Limnol. Oceanogr.* **14**, 441.

Sheldon, R. W., Evelyn, T. P. T. and Parsons, T. R. (1967). *Limnol. Oceanogr.* **12**, 367.

Sieburth, J. McN. (1965). *J. gen. Microbiol.* **41**, XX.

Skopintsev, B. (1972). *In* "Changing Chemistry of the Oceans" (D. Dyrssen and D. Jagner, eds.). Wiley Interscience, New York, pp. 205–207.

Smayda, T. J. (1969). *Limnol. Oceanogr.* **14**, 621.

Sorokin, J. I. (1971). *Int. Rev. ges. Hydrobiol.* **56**, 1.

Stanley, S. O. and Rose, A. H. (1967). *J. gen. Microbiol.* **48**, 9.

Steele, J. H. and Baird, I. E. (1962). *Limnol. Oceanogr.* **7**, 42.

Strickland, J. D. H. and Parsons, T. R. (1972). *Bull. Fish. Res. Bd. Can.* **167**, 207.

Sutcliffe, W. H., Baylor, E. R. and Menzel, D. W. (1963). *Deep-Sea Res.* **10**, 233.

Szekielda, K. H. (1966a). *In* "IBP Symposium" (A. M. Verkeggen, ed.). Koninklijke Nederlandse Akademie Van Wetenschappen, Amsterdam, pp. 314–322.

Szekielda, K. H. (1966b). *In* "IBP Symposium" (A. M. Verkeggen, ed.). Koninklijke Nederlandse Akademie Van Wetenschappen, Amsterdam, pp. 150–157.

Szekielda, K. H. (1968). *Mar. Biol.* **2**, 71.

Wangersky, P. J. (1965) *Am. Scient.* **53**, 358.

Wheeler, E. H. (1967). *Limnol. Oceanogr.* **12**, 697.

Williams, P. M. (1968). *Nature, Lond.* **218**, 937.

Williams, P. M. and Gordon, L. I. (1970). *Deep-Sea Res.* **17**, 19.

Wilson, D. P. (1955). *J. mar. biol. Ass. U.K.* **34**, 531.
Yentsch, C. S. (1965). *Deep-Sea Res.* **12**, 653.
Zeitzschel, B. (1970). *Mar. Biol.* **7**, 305.

Chapter 14

Primary Productivity

G. E. FOGG

Marine Science Laboratories, University College of North Wales
Menai Bridge, Anglesey, Wales

14.1. INTRODUCTION

During primary production there is an increase of potential chemical energy in an ecosystem; this increase derives ultimately from external energy sources which may be either radiant or chemical. Primary production is usually considered to be photosynthesis carried out by plants containing chlorophyll a, as in the equation:

$$CO_2 + H_2O \xrightarrow{\quad h\nu \quad} (CH_2O) + O_2 \qquad (14.1)$$

which represents the conversion of the radiant energy of sunlight into the potential chemical energy of the organic compounds of living organisms, for which (CH_2O) is an approximate empirical formula. Chemosynthesis, in which the potential chemical energy of inorganic compounds is converted into that of organic compounds, and bacterial photosynthesis, which depends on the availability of particular inorganic or organic hydrogen donors and in which oxygen is not evolved, may be locally important in inshore waters but make little contribution to primary production in the ocean. Under conditions in which production of new cell material takes place at the expense of allochthonous organic matter, it is difficult to define primary productivity; these conditions are most often encountered in polluted inshore waters rather than in the open sea. It is an oversimplification to regard the accumulated potential chemical energy as being only in the form of organic compounds. Both the reduction of nitrate to ammonia and the formation of polyphosphate from orthophosphate are processes resulting in the accumulation of potential chemical energy. Accurate assessment of primary production should allow for such processes, but for most purposes they are ignored.

Photosynthesis in the sea is carried out by microscopic unicellular algae—the phytoplankton—freely suspended in the water and also by attached plants, which may be microscopic algae or large seaweeds. As the attached plants are confined to waters having a depth not greater than 100 m they make only a small contribution to the total primary production of the oceans. For this reason and the less creditable one that study and measurement of their photosynthesis is much more difficult than that of phytoplankton, these plants will not be considered in this discussion.

Photosynthesis is the major biogeochemical process both on land and in the sea. Not only does it have a paramount role in determining the distribution of the various compounds of carbon and the concentration of free oxygen in the sea, but its rate indirectly determines those of nearly all other biogeochemical processes. Since all other organisms are ultimately dependent on the primary producers for food, measurement of primary production is of importance to the biologist in providing the basic data from which he can follow the flow of energy through the different trophic levels in the ecosystem. For the fisheries biologist primary production measurements give an indication of the potential yields of a fishing ground.

The rate of phytoplankton photosynthesis is primarily determined by the amount and activity of photosynthetic cell material present, by the light intensity, by the carbon dioxide concentration and by the temperature. The amount and activity of the photosynthetic cell material are, however, determined by a complex of factors including nutrient supply, which in turn depends on rates of remineralization and water movements, and zooplankton grazing, which itself is a complex function of many factors including the past abundance of phytoplankton. Such is the complication of the interactions that a comprehensive treatment of primary productivity would be too long for inclusion in this book. This chapter is, therefore, not intended to be exhaustive. However, since at least a limited understanding of algal physiology and biochemistry is essential, some account of these will be included. Because the basic features of phytoplankton photosynthesis are much the same whether the water be fresh or saline, reference will be made to freshwater as well as to marine studies where appropriate.

Accounts of phytoplankton productivity in its wider aspects have been given in a number of works. Those of Harvey (1928, 1945, 1955) and Cushing (1959a) are still well worth reading although inevitably out of date in some respects. The otherwise comprehensive treatise of Raymont (1963) does not go deeply into dynamic aspects and that by Fogg (1965) deals with the growth of phytoplankton rather than its photosynthesis. Strickland gave excellent and concise accounts of both growth and photosynthesis in the first edition of this book (Strickland, 1965) and in a posthumously published review (Strickland, 1972). Volumes containing contributions dealing with various aspects of the subject have been edited by Steemann Nielsen and Cushing (1958), Doty (1961), Riley (1963a), Hill (1963), Goldman (1965) and Costlow (1971). Manuals of methods have been compiled by Strickland and Parsons (1968) and Vollenweider (1974). Steemann Nielsen (1960, 1963, 1965), Yentsch (1963) and Platt and Subba Rao (1973) have reviewed the results of estimation of primary productivity in the oceans and Talling (1961, 1970) and Vollenweider (1970) have considered the general aspects of phytoplankton as a photosynthetic system.

No description of the organisms responsible for primary production will be given here. Table 14.1 gives the characteristics of the principal groups of marine phytoplankton algae and lists some of the more familiar genera, but further information must be sought in books such as those by Fritsch (1935, 1945), Fott (1971), Round (1965) and Morris (1967).

TABLE 14.1

Characteristics of the main classes of marine phytoplanktonic algae

Class	Distribution	Principal photosynthetic pigments	Cell character	Common or well known genera
Chlorophyceae (green algae)	Mainly freshwater and brackish	chlorophyll a chlorophyll b β-carotene lutein, violaxanthin neoxanthin	Similar to green plants; starch stored; often motile with 2 or 4 equal flagella	*Chlorella* *Scenedesmus* *Dunaliella* *Brachiomonas* *Ankistrodesmus* *Nannochloris*
Prasinophyceae	Coastal	chlorophyll a chlorophyll b β-carotene micronone, violaxanthin neoxanthin	Cell wall absent; scales present; starch-like storage product; often motile with 2 or 4 scaly flagella	*Tetraselmis* *Pterosperma* *Halosphaera*
Chrysophyceae	Coastal and oceanic	chlorophyll a chlorophyll c β-carotene fucoxanthin	Cell wall absent; siliceous scales, cysts or skeletons; chrysolaminarin stored; motile, 2 unequal flagella	*Monochrysis* *Dictyocha*
Haptophyceae	Coastal and oceanic, abundant in temperate and tropical waters	chlorophyll a chlorophyll c β-carotene fucoxanthin diatoxanthin diadinoxanthin	Cell wall absent; calcareous or organic scales; chrysolaminarin stored; motile 2 equal flagella and a haptonema	*Dicrateria* *Isochrysis* *Coccolithus* *Phaeocystis*

Bacillariophyceae (diatoms)	Coastal and oceanic, abundant from tropics to polar waters	chlorophyll *a* chlorophyll *c* β-carotene fucoxanthin diatoxanthin diadinoxanthin	Silica wall in form of box and lid; chrysolaminarin stored; normally without flagella	*Coscinodiscus* *Ditylum* *Skeletonema* *Phaeodactylum* *Thalassiosira*	*Asterionella* *Chaetoceros* *Cyclotella* *Nitzschia* *Fragilaria*
Dinophyceae	Coastal and oceanic, abundant in tropics, less so in polar waters	chlorophyll *a* chlorophyll *c* β-carotene diadinoxanthin dinoxanthin peridinin	Often with cellulose wall; starch stored; motile with 1 longitudinal, 1 encircling flagellum	*Dinophysis* *Peridinium* *Amphidinium* *Ceratium*	
Cryptophyceae	Coastal and estuarine	chlorophyll *a* chlorophyll *c* β-carotene zeaxanthin phycobiliproteins	Cell wall absent, starch stored; usually motile 2 equal flagella	*Hemiselmis* *Cryptomonas*	
Cyanophyceae (blue-green algae)	Coastal and oceanic in tropics, rare elsewhere	chlorophyll *a* β-carotene echininone myxoxanthophyll phycobiliproteins	Membrane-bound nucleus and plastids absent (prokaryotic); glycogen stored; flagella always absent	*Trichodesmium* *Synechococcus*	

14.2. The Physiology and Biochemistry of Phytoplankton Growth

14. 2.1. REQUIREMENTS FOR PHOTOSYNTHESIS AND GROWTH

14.2.1.1. *Light*

Phytoplankton utilize visible radiation of wavelength between 400 and 720 nm in photosynthesis. Near ultra-violet radiation of wavelength down to 350 nm and perhaps less may also be used (McLeod and Kanwisher, 1962; Halldal, 1964), but probably causes damage if the exposure is prolonged. The general form of the relationship of rate of photosynthesis to intensity is shown in Fig. 14.1. At limiting light intensities the quantum yield ($\Delta O_2/h\nu$ absorbed) has been found to be the same, about 0·10, for a variety of different algae under optimum laboratory conditions (Rabinowitch, 1951). At higher intensities, light becomes saturating and the curve shows a plateau of greater or lesser extent, but at intensities approaching that of full sunlight photosynthesis is inhibited. Qasim *et al.* (1972) have determined the relationship of the rate of photosynthesis to light intensity for 11 species of tropical marine phytoplankton in culture. Values of I_k (which defines the onset of light saturation, see Fig. 14.1) ranged from 11·5 klux (0·20 ly min^{-1}) for *Coscinodiscus*

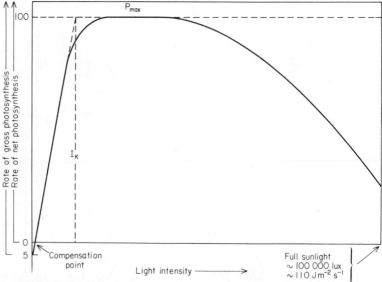

FIG. 14.1. A generalized diagram showing the relationship of rate of photosynthesis by phytoplankton to light intensity. It should be noted that the intensities at which the inflexions of the curve occur vary considerably according to species and the physiological condition of the cells. For explanation of the symbols see text.

radiatus to 34·5 klux (0·60 ly min⁻¹) for *Dinophysis miles*. However, the efficiency of light absorption and photosynthesis under sub-optimal conditions and the intensities at which saturation and inhibition occur vary greatly, not only according to the species of algae, but also with physiological condition, as will be explained later.

The distribution of the available radiant energy within the spectrum is of some importance. All algae have chlorophyll *a* as their principal photosynthetic pigment, but the accessory pigments are different for the various algal classes of phytoplankton (Table 14.1). Absorption spectra and photosynthetic action spectra of phytoplankton species do not, therefore, conform to a single pattern. Examples typical of the different algal classes have been given by Rabinowitch (1951) and Haxo (1960), and it will suffice to say here that the predominating algal classes in the phytoplankton (the Bacillariophyceae, Dinophyceae and Haptophyceae) have a predominance of carotenoid accessory pigments which gives them a maximum absorption in the blue and green in addition to a maximum in the red which is due to chlorophyll *a*. The pigmentation of phytoplankton is thus, generally speaking, such as to ensure maximum absorption of the wavelengths to which clear sea water is most transparent. Qasim *et al.* (1972) found that various tropical marine phytoplankton species photosynthesized with equal efficiency when illuminated with different parts of the spectrum, and concluded that chromatic adaptation had occurred in their cultures. It is possible that the change in spectral composition which occurs as sunlight penetrates through water may have appreciable qualitative effects on the photosynthesis and growth of phytoplankton. Complete photosynthesis is now known to depend on two linked photosystems, one of which, photosystem I, contains a pigment complex having maximum absorption at longer wavelengths than that of the other, photosystem II (see Rabinowitch and Govindjee, 1969 or Gregory, 1971 for details). Since the functions of the two photoreactions which they mediate differ, it may be that imbalance of the distribution of light energy between them may affect the pattern of metabolism and growth of an alga. It has been observed that wavelength has a marked effect on the distribution among various classes of compounds of photosynthetically fixed carbon in *Chlorella*, although it seems that this does not result from imbalance between the photoreactions (Ogasawara and Miyachi, 1970). Perhaps some such effects are responsible for the inability of certain phytoplankton species to grow in light from a fluorescent lamp having a discontinuous emission spectrum (Kain and Fogg, 1960). Nothing appears to be known of such effects *in situ* in the sea but, as Dring (1971) has emphasized, variations in the spectral composition of light in marine environments are so great that detectable biological responses to them are to be expected. It is difficult to match the spectral quality of light in laboratory incubators with that to which phytoplankton is subjected naturally (Strickland in Riley, 1963a).

Continuous illumination is normally used in growing cultures of marine phytoplankton, and there appears to be no report of growth of any species being absolutely dependent on regular alternation of light and dark periods. However, longer periods of illumination are progressively less effective in increasing photosynthetic yield and growth rate. With *Ditylum brightwellii*, Paasche (1968) found no increase in growth rate with more than 16 h illumination out of 24, and continuous illumination was actually inhibitory as compared with a cycle including 8 h of darkness. Additionally, as will be described below (p. 398), photoperiodicity has effects on the synchrony of phytoplankton growth which may be of importance in studies of primary production.

Growth of phytoplankton in the dark using organic substrates does not appear to be a common occurrence. However, it has been reported for *Nannochloris* sp. by Thomas (1966).

14.2.1.2. *Mineral elements and vitamins*

The requirements of algae for mineral nutrients are generally similar to those of other plants, and it may be stated without discussion that nitrogen, phosphorus, potassium, calcium, magnesium, sulphur, iron, copper, zinc, manganese, molybdenum and chlorine are essential for phytoplankton growth (O'Kelley, 1974). Sodium has been shown to be necessary for the growth of blue-green algae (Allen and Arnon, 1955), but although its concentration often has effects on the growth of algae belonging to other groups, its essentiality for these is not proven. Numerous algae have been shown to require vitamin B_{12} (cobalamin or its analogues), which contains cobalt, and since, among those which do not require an exogenous source of the vitamin, a requirement for inorganic cobalt has been demonstrated in a blue-green alga (Holm-Hansen *et al.*, 1954), cobalt in some form may be supposed to be essential for all algae. There still appears to be some doubt about the necessity of boron for all algae, but it has been shown to be essential for diatoms (Lewin, 1966). The requirement of diatoms for silica and the non-essentiality of this element for other algal groups such as the green algae (Chlorophyceae) are well established, but it is not clear whether members of the Chrysophyceae, which are characterized by their production of siliceous cysts or skeletons, have an absolute need for silicon. Vanadium cannot be replaced in the growth medium for the freshwater green alga *Scenedesmus obliquus* by molybdenum or any of a variety of other elements (Arnon and Wessel, 1953), but does not appear to have been shown to be necessary for any marine form. Iodine appears to be essential for the growth of the red seaweed *Polysiphonia urceolata* (Fries, 1966), but it is not known whether it is necessary for any planktonic marine algae.

Besides nitrogen and phosphorus, shortages of which have been generally regarded as most usually limiting plankton growth, it appears that certain trace elements may also often be in short supply in the sea (see p. 432). The form in which these elements are present is important, their incorporation into organic complexes evidently promoting their uptake by the cell. Strickland (1972) considered the concentrations of potential chelating agents in sea water to be normally insufficient to bind trace metals and he suggested that complexes of these released by decay of organisms might be important for the nutrition of phytoplankton. Johnston (1963) found that addition of the chelating agent ethylenediamine tetra-acetic acid to a variety of different kinds of sea water almost invariably improved them as growth media for the diatom *Skeletonema costatum* and the dinoflagellate *Peridinium trochoideum*. Prakash (1971), who demonstrated that humic substances introduced into the sea through land drainage favour high primary productivity, attributed this, at least in part, to their ability to keep these elements in solution (and therefore available) by chelating them (see also p. 52).

The majority of phytoplankton species have an absolute requirement for one or more organic growth factors or vitamins. This has been the subject of several excellent reviews (Lewin, 1961; Droop, 1962b; Provasoli, 1963; Hutner and Provasoli, 1964; Provasoli and Carlucci, 1974) and need not be discussed at length here. The commonest requirements are, in order of frequency, for vitamins B_{12} (cobalamin), B_1 (thiamine) and H (biotin or co-enzyme R). Among the algal groups represented in the marine plankton the Chrysophyceae, Haptophyceae, Cryptophyceae and Dinophyceae have the largest proportions of vitamin-requiring (*auxotrophic*) species and the Cyanophyceae, Chlorophyceae and Bacillariophyceae the least. Various compounds having B_{12}-like activity exist naturally, and specificity towards them is different in the various algal classes; most members of the Chlorophyceae, Chrysophyceae and Dinophyceae respond to B_{12} factor III as well as to cobalamin, whereas most Bacillariophyceae and Cryptophyceae have a wider specificity and will utilize pseudo-B_{12} and factors A and B as well as the compounds just named (Provasoli, 1963). It is probable that other growth factors remain to be identified. Thus, Johnston (1963) found that, after supplementing natural sea water samples with mineral nutrients, chelating agents, vitamin B_{12} and thiamine, differences in their ability to support growth of *Skeletonema costatum* and *Peridinium trochoideum* persisted. He concluded that these differences were primarily attributable to the presence in the water of labile growth-promoting substances, and were not a consequence of the presence of inhibitors, and that the former substances have different effects on different species. A requirement for an unrecognized growth factor may be a reason, although not the only one, why many common planktonic algae have defeated attempts to grow them in artificial culture.

Media for the culture of phytoplankton may be based on natural sea water, as for example the well known "Erd–Schreiber" solution, which is natural sea water fortified by addition of nitrate, phosphate and soil extract. Artificial media made up from materials of known composition are preferable for physiological and biochemical investigations. Some chemical considerations in the design of synthetic culture media for marine algae have been given by Droop (1961a) and recipes by Provasoli *et al.* (1957), Provasoli (1968) and Droop (1969). It should be noted that some media give better results than others with particular species. The reason for this is not usually apparent, but among other factors ion-ratios may be critical. A discussion by Droop (1962a) of difficulties in the culture of *Skeletonema costatum* indicates some of the physico-chemical problems which are encountered in devising culture media. It is sensible to test several when embarking on the cultivation of a species with which one has not worked before.

For many purposes it is sufficient to rely on diffusion from the atmosphere to supply carbon dioxide to the medium. Cultures may be bubbled with air enriched with low concentrations of carbon dioxide (say 0.5% v/v) but care should be taken that the pH of the medium is not altered too much as a result. Agitation of the culture may be necessary to maintain cells in suspension, and is often beneficial to growth. However, some species may suffer mechanical damage from too violent movement. A culture flask of the type designed by Walsby (1967) is very suitable for growing phytoplankton under optimum conditions of aeration.

14.2.2. GENERAL FEATURES OF GROWTH AND CULTURE METHODS

A general account of the techniques of algal culture has been given by Droop (1969) and Stein (1973). Lewin (1959), Droop (1967) and Paasche (1971) have described techniques for obtaining cultures free from all contaminating organisms (*axenic* cultures). This is usually a matter of great difficulty with truly planktonic marine algae, but is necessary for many kinds of biochemical and nutritional studies if unambiguous results are to be obtained. It is debatable whether it is necessary for much of the physiological work which is carried out with the intention of elucidating behaviour under natural conditions. It may be argued that since algae are normally associated with bacteria in nature (but see Droop and Elson 1966) experiments with bacterized cultures may give more realistic results, although artificial conditions may well shift the balance between bacteria and algae and invalidate this argument. Where possible, it is desirable to do parallel experiments on axenic and bacterized material.

Given suitable conditions of light and temperature and an appropriate medium, algae will photosynthesize and grow. The products of phytoplankton photosynthesis are used to synthesize fresh cell material which is itself capable of photosynthesis. Given steady state conditions—which, as will be seen, are only achieved statistically in large populations in specially controlled cultures—the amount of cell material increases exponentially according to the expression:

$$W_t = W_0 e^{kt} \qquad (14.2)$$

where W_0 is the amount of cell material present in unit volume of the phytoplankton suspension at zero time, W_t is the amount at time, t, e is the base of natural logarithms, and k (the symbol μ is used by some workers) is the relative growth constant, which is a measure of the efficiency of growth. This relationship is more conveniently expressed in the form

$$k = \frac{\log_e W_t - \log_e W_0}{t} \qquad (14.3)$$

or

$$k' = \frac{\log_{10} W_t - \log_{10} W_0}{t} \qquad (14.4)$$

Any attribute of the population may be used as a measure of growth and, provided that the system is in a steady state, the relative growth constant (in this case k') will be the same. If, for example, we take the number of cells per unit volume then

$$k' = \frac{\log_{10} N_t - \log_{10} N_0}{t} \qquad (14.5)$$

Eppley and Strickland (1968) have summarized information on the relative growth rates of marine phytoplankton.

In the course of growth, cell numbers increase and the amount of light incident on an individual cell decreases because of shading by others. At the same time, nutrients are taken up and so in the absence of a continued supply, their concentration in the medium decreases. Algal cells may liberate products of metabolism into the medium, and some of these extracellular products may be growth promoting or have chelating properties, whereas others may inhibit particular cell processes. Consequently, in a batch culture grown in a limited amount of medium, conditions do not remain constant. Nevertheless, exponential growth may be maintained in such cultures for a period of a week or so—corresponding perhaps to 10 generations—if the inoculum is a small one. This is possible if conditions are such that the factors affecting growth are all at saturating values, i.e. in the plateau region of a curve such as that in Fig. 14.1, the general form of which is characteristic of most biological responses such as that of relative growth rate to nutrient concentration. Light intensity and nutrient concentrations may then decrease considerably before they become limiting. *Mutatis mutandis* a similar relationship holds for the action of inhibiting substances, of which a threshold concentration must be built up before an effect on growth is perceptible. Thus, under good culture conditions the relative growth rate observed is that intrinsic to the alga at the particular temperature obtaining. The changes in light intensity and concentrations which inevitably occur may, however, affect processes in the cells differentially so that k' values based on different measures of growth differ slightly and as exponential growth continues, both the mean size and the composition of cells change (Fogg, 1959). Sooner or later in a batch culture one or more factors become limiting and the population passes abruptly or gradually into a phase in which no further cell division takes place. Although the population then remains stationary in numbers, photosynthesis continues, albeit at a reduced rate, with the result that products such as starch or fat accumulate in the cells (Fogg, 1959, 1965). The final yield in a batch culture is often taken as a measure of growth, but it is important to realise that this has an entirely different significance from the relative growth constant. The latter is an index of the intrinsic ability of the alga to reproduce itself and thus of its ability to compete with other species under a particular set of conditions. The final yield is merely a measure of the level of the limiting factors in a particular culture. This is particularly useful in bioassay, for example, vitamin B_{12} is commonly estimated from the population produced in a culture in which it is the limiting factor (Droop, 1961b), but gives no direct measure of the organism's ability to perform under natural conditions. In the stationary phase of growth, concentrations of enzymes and metabolites alter so that active exponential growth cannot be resumed without a preliminary period of reconstitution. This is one cause of the lag phase which is usually observed when fresh medium is inoculated with cells from a mature culture. Another is that a majority of the cells in the inoculum are not viable so that some time must elapse before the progeny of those cells capable of multiplication become comparable in numbers to the total inoculated.

Strickland *et al.* (1969b) have described the use of a 70 000 litre capacity tank for growing large batch cultures of marine phytoplankton algae under near natural conditions.

Conditions affecting growth may be kept constant in a culture by continous addition of fresh medium. The cell population and the extracellular products which it liberates are in this way continually diluted back to their initial levels while nutrient concentrations are maintained. There are two distinct ways of doing this. In the *chemostat*, medium in which one nutrient is at limiting concentration is fed in at constant rate. The organism then grows at a rate determined by the rate of addition of the limiting nutrient. Apparatus for the culture of algae in this way has been described by Droop (1966), by Fuhs (1969), by Caperon and Meyer (1972a, b) and by Fay and Kulasooriya (1973). In the other type of continuous culture system the *turbidostat*, the addition of fresh medium is controlled by a photoelectric device which operates to bring back the culture to a predetermined optical density. Here the alga grows at a rate determined by the light intensity and temperature, the rate of inflow of the medium being regulated automatically so as to supply the requisite amount of nutrients. The turbidostat was originally described by Myers and Clark (1944) for use with dense suspensions of *Chlorella* but a modification for use with more dilute suspensions, approximating to those found in nature, has been devised by Maddux and Jones (1964). Fuhs (1969) has also described a turbidostat unit. In both the chemostat and the turbidostat, relative growth rate may be determined from culture volumes by means of an expression analogous to (14.4) above

$$k' = \frac{\log_{10} V_t - \log_{10} V_0}{t} \qquad (14.6)$$

Culture volumes are conveniently determined from the volume of the overflow when a constant-volume device is employed. Both the chemostat and turbidostat yield cells of standard composition and physiological activity once steady-state conditions have been attained (Myers and Clark, 1944). However, sometimes, as when cells with stored polyphosphate are grown in a medium having a different phosphate concentration from that in which they were originally growing, it may take as long as two weeks for these conditions to become established (Soeder *et al.*, 1971).

A type of culture intermediate between batch and continuous culture is that in which the population is grown in a vessel with a fine sintered glass filter, or in dialysis tubing, on the other side of which the medium is continuously renewed (Jensen *et al.*, 1972). In this case the concentrations of extracellular products and nutrients may be held constant, but the effective light intensity decreases as the culture grows.

Results obtained with any of the kinds of culture so far mentioned are means for large populations and, therefore, give no indication of the changes which occur in an individual cell during its life cycle. For most planktonic algae this cycle is of the simplest type; as a result of photosynthesis the cell grows in size then, when a certain limiting size is reached, divides to give daughter individuals, usually two (but sometimes more) in number. Sexual reproduction is known to occur in some marine phytoplankton species but seems to be rare. A sample from a growing culture contains cells at all stages of the division cycle, but by various kinds of treatment divisions may be brought into step to give a synchronous culture. This may be done by physical separation of cells of different sizes, so that a new culture may be started with a fraction containing cells in the same phase. Alternatively, it may be done by subjecting the suspension to some kind of treatment, such as darkness or low temperature, which retards some part of the cycle so that individual cells are brought into phase. Many species have an endogenous rhythm, tending to have a generation time of about 24 h with division occurring in a particular part of the light/dark cycle. If appropriate photoperiodic treatment is given to such a species the division of individual cells will become entrained after a few light/dark cycles so that a synchronous culture results. Determinations made on samples from synchronized cultures, although still mean values representing large populations, give an idea of the biochemical and physiological state of individual cells at a particular stage in their life cycle. As will be seen below, the changes which occur as a cell matures and divides may be considerable. However, it is often difficult to decide whether a given change is to be attributed to a true endogenous cycle or whether it is induced by the treatment necessary to obtain synchronous cultures (see Pirson and Lorenzen, 1966). Further details of synchronous cultures of algae are given by Tamiya (1964, 1966), Hoogenhout (1963), Pirson and Lorenzen (1966) and Paasche (1967, 1968). Most photoperiodic studies with algae have been carried out with abrupt transition between light and dark. An apparatus giving continuously varying intensity similar to that found under natural conditions has been described by Quraishi and Spencer (1971). With it, they have made the interesting observation that among the five phytoplankton species which they studied, *Phaeodactylum tricornutum* and *Chlorella ovalis* divide during light periods, whereas *Brachiomonas submarina, Dunaliella primolecta* and *Monochrysis lutheri* divide preferentially in the dark.

Continuous cultures may be synchronized by imposition of appropriate light/dark cycles. Researches involving this have been described by Gimmler *et al.,* (1969) and by Eppley *et al.* (1971).

14.2.3. UPTAKE OF NUTRIENTS

The relationship at a given temperature between the uptake of a nutrient by a micro-organism and the concentration of that nutrient is usually described with reasonable accuracy in terms of Michaelis–Menton kinetics. If it is assumed for the moment that growth is proportional to uptake, then

$$k = \frac{k_\infty C}{C + K_s} \qquad (14.7)$$

in which K_s is a constant having the dimensions of a concentration and is numerically equal to the concentration giving half the maximum growth rate, k_∞. The nutrient is limiting when its concentration, C, is of the same order as K_s, but when it is larger than K_s, saturation is approached.

TABLE 14.2

Concentrations (K_s) of nutrients giving half maximum growth rate of various marine phytoplankton organisms

Nutrient	Organism	Temp. °C	K_s $\mu\,mol\,l^{-1}$	Method	Reference
Nitrate	*Asterionella japonica*	26	1·35	Cell numbers Batch culture	Eppley and Thomas (1969)
Nitrate	*Chaetoceros gracilis*	26	0·2	Batch culture	Eppley and Thomas (1969)
Nitrate	*Isochrysis galbana*	25	0·32	Chemostat	Caperon (1968)
Phosphate	*Chaetoceros gracilis*	27	0·12	Cell numbers Batch culture	Thomas and Dodson (1968)
			$\mu g\,l^{-1}$		
Vitamin B_{12}	*Cyclotella nana*	22	0·003 ⎤		
				C_{14} uptake In batch cultures	Carlucci and Silbernagel (1969)
Thiamine	*Monochrysis lutheri*	22	0·125		
Biotin	*Amphidinium carterae*	22	0·004 ⎦	Cell numbers	Carlucci and
Vitamin B_{12}	*Cyclotella nana*	22	0·0002 to 0·0004	Batch cultures	Silbernagel (1969)
Vitamin B_{12}	*Monochrysis lutheri*	19	0·002 to 0·006	Cell numbers Chemostat	Droop (1966)

In determining values of K_s for phytoplankton, the chemostat technique is the most useful approach since the concentrations at which most nutrients become limiting are so low that they can only with difficulty be maintained for long enough to make a growth measurement in batch culture. Table 14.2 presents some K_s values reported in the literature. K_s is a valuable index for understanding the spatial and temporal distribution of species, since, other things being equal, a species with a lower K_s value will be more successful than one having a higher K_s if the nutrient in question is in short supply (Strickland, 1972; Platt and Subba Rao, 1973). K_s for nitrate uptake is positively correlated with cell size, with generation time and with K_s for ammonium uptake; oceanic species have lower K_s values than neritic species (Eppley *et al.*, 1969).

As Table 14.2 shows, K_s values, although varying according to species and nutrient, are all low. Nevertheless, they are of the same order as the concentrations of these substances commonly found in sea water, for example,

5 μmol l^{-1} for nitrate and 0·001 μg l^{-1} for vitamin B$_{12}$. Hence, if the cell is motionless in unstirred sea water the rate limiting process is likely to be diffusion of the nutrient up to the cell. Munk and Riley (1952) pointed out that under these conditions motion of the cell relative to the medium is of crucial importance in steepening diffusion gradients and so increasing the rate of uptake of nutrients. Such "forced convection" is provided when cells sink, and these authors showed that division rates in diatoms are related to their cal-culated sinking rates. Forced convection can also be achieved by swimming, and it is perhaps significant that the majority of the phytoplankton species requiring vitamin B$_{12}$ possess flagella. The ability of a plankton species to absorb nutrients thus depends on, among other things, precise regulation of buoyancy in relation to turbulence in the sea. The hydromechanics of phyto-plankton have been discussed by Hutchinson (1967) and by Smayda (1970).

Various circumstances may obscure the simple relationship represented by equation (14.7). Such complications can be illustrated by reference to phos-phorus, as discussed in detail by Fogg (1973). Although phosphorus appears to be taken into algal cells only in the form of orthophosphate, phosphorus from the various organic phosphates (which are often present in sea water at concentrations equalling, or exceeding, that of orthophosphate) may also be available as a result of the activity of phosphatases which develop at the sur-faces of algae especially when phosphate is in short supply. Kuenzler (1965) has demonstrated the ability of various marine phytoplankton organisms to utilise glucose-6-phosphate. The concentration of orthophosphate in the water may thus be a misleading index of what is available for phytoplankton growth. Uptake is an active process for which energy may be supplied either by photosynthesis or respiration. Usually it is stimulated by light. *Phaeodac-tylum tricornutum* is capable of decreasing the phosphate in the medium to as little as $7·2 \times 10^{-10}$ M, a concentration much below the limits detectable by the usual analytical method (Kuenzler and Ketchum, 1962). Given a supply of phosphate, most algae appear able to accumulate an excess which is stored within the cells in the form of polyphosphate granules. In the absence of any further supply, this form of phosphorus may then be used to support consider-able growth of the population. Thus, for example, the marine diatom *Phaeodac-tylum tricornutum* was found by Kuenzler and Ketchum (1962) to accumulate an excess sufficient to provide for 5 successive doublings. In contrast, Soeder (1970) found that the freshwater green alga *Chlorella fusca* excreted consider-able amounts of organic phosphate at a particular stage of its division cycle, and there are indications that such a release may occur under natural condi-tions. Large amounts of organic phosphate may be excreted by marine phyto-plankton (Kuenzler, 1970), but it is not established that this is linked to a particular phase of the division cycle. From an investigation of the growth of two brackish water diatoms, *Cyclotella nana* and *Thalassiosira fluviatilis* in relation to phosphate, Fuhs (1969) concluded that the relative growth constant was best related to the amount of bound phosphorus per cell. Soeder *et al.* (1971), working with a freshwater diatom, *Nitzschia actinastroides*, found a relationship of the type represented by equation (14.7) except at very low concentrations. Under the latter conditions the external concentration of phosphate was found to actually increase, and at the same time there was an increase in the proportion of dead cells in the suspension. A possible explanation for this may be that the cells excreted phosphate at certain stages of their life cycle, as does *Chlorella*. This would enable some cells to continue growth at the expense of the excreted phosphate while others would pass below the critical level of cell phosphorus and die. As we have already noted, these same workers found that growth rates in continuous culture took some considerable time to stabilize after a change of phosphate concentration.

Caperon and Meyer (1972a, b) have obtained evidence of storage of excess nitrate but no reservoir of ammonium nitrogen in the flagellate *Isochrysis galbana*.

Droop (1966) likewise found that the relation of the growth of the marine flagellate *Monochrysis lutheri* to vitamin B_{12} concentration in a chemostat could not adequately be described by equation (14.7). Subsequently (1968), he established that the relative growth rate depended on the concentration of the vitamin within the cells. A non-dialysable heat labile factor was excreted from the cells and combined with the vitamin thereby rendering it inactive. Similar inhibitory factors were found to be produced by other species, some of which were not vitamin B_{12}-requiring. Finally, it should be remembered that the uptake characteristics for a particular nutrient by a given species are not constant, but depend on the level of other factors such as light and temperature (Eppley *et al.*,1969; MacIsaac and Dugdale, 1972).

14.2.4. LIBERATION OF EXTRACELLULAR PRODUCTS (see also pp. 301–306)

The idea that phytoplankton cells retain the major portion of the products of their photosynthesis so long as they are healthy has had to be reconsidered in recent years. There is now evidence that many different kinds of algae liberate organic products of various sorts into the medium while growing actively in laboratory culture. As a first approximation these extracellular products may be divided into two main classes. (a) Type I, metabolic intermediates, usually of low molecular weight, for which there is quasi-equilibrium between the intra- and extracellular concentrations. As a result of this, the amount of organic product liberated depends on the metabolic activity, the availability of a sink in the environment, and other factors. (b) Type II, consists of the end products of metabolism which are often of high molecular weight, and the rate of liberation of which does not depend on an equilibrium but is more or less proportional to the amount of growth (Fogg, 1966, 1971). Type II extra-cellular products usually seem to be quantitatively unimportant and need not concern us further, but those of Type I are of great significance in productivity studies.

The best known extracellular product of Type I is glycollic acid $CH_2OH . COOH$. The biochemistry of this compound has been the subject of a great deal of research since the discovery that it is an early product of photosynthesis which the green alga *Chlorella* may liberate into the medium in relatively large amounts (Tolbert and Zill, 1956). Reviews of this biochemistry have been given by Tolbert (1974) and Fogg (in press). Glycollic acid appears to be formed by the action of an oxygenase on ribulose-1,5-diphosphate, the carbon dioxide acceptor in the Calvin-Benson carbon dioxide fixation cycle. Its production rate is consequently lowest at high carbon dioxide concentra-tions and highest at high oxygen concentrations. It also depends directly on the relative growth rate of the alga, presumably because the concentration of ribulose-1,5-diphosphate is highest when the photosynthetic activity is greatest. In synchronized cultures excretion of glycollate is similarly greatest during the photosynthetic phase and minimal during the cell division stage. Glycollate is utilized as the substrate for photorespiration (Section 14.2.5) and in synthesis of cell constituents via oxidation to glyoxylate, then through glycine, serine and glycerate to phosphoenolpyruvate and the tricarboxylic acid cycle. Extracellular glycollate may be re-assimilated by a photosynthetic mechanism under suitable conditions. The extent of utilization by these pathways probably depends on the availability of nutrients. Ignatiades and Fogg (1973) have shown that liberation of total extracellular products, of which glycollate is thought to be a major component, is greater in nutrient deficient cultures of the marine diatom *Skeletonema costatum* than when ample supplies of these nutrients are available. These facts are consistent with the idea that the cell membranes of algae are permeable to glycollic acid, the amount excreted being determined by an equilibrium between the intra- and extracellular concentrations, the former depending on the relative rates of production and consumption in the cell. On this basis one would expect that the amount of glycollate liberated would be constant irrespective of the con-centration of cells in a culture, but although a tendency towards this had been found (Nalewajko *et al.*, 1963; Ignatiades and Fogg 1973), agreement with the simple mathematical model is poor (Watt and Fogg, 1966).

Most of the above evidence relates to freshwater green algae but the same seems to hold for marine phytoplankton. Hellebust (1965) detected glycollate in filtrates from cultures of all the 22 species, representing 5 algal classes, of marine phytoplankton which he investigated. In only four of these instances was it the major extracellular product, but experimental conditions were not such as to favour its excretion in large quantities. Hellebust (1965) found a variety of other extracellular products, but none of these could be related with certainty to photosynthesis. Watt (1969) concluded, from a review of published evidence and his own investigations with freshwater algae, that there are two types of secretion during photosynthesis; one in which glycollate is the major product and another in which polysaccharides are excreted predominantly. The same distinction may exist among marine species. Guillard and Hellebust (1971) found that *Phaeocystis poucheti* liberated large amounts of acrylic acid (but apparently not glycollic acid) and polysaccharides during photosynthesis. Huntsman (1972) studied the liberation of extracellular products of photosynthesis at various stages in the growth of batch cultures of *Dunaliella tertiolecta* and detected glycollate among them, but only in small amounts, glycerol being the major component. Maksimova and Pimenova (1969) found that contaminating bacteria had an important influence on the composition of the extracellular fraction in cultures of *Chlorella*.

14.2.5. INTERRELATIONS OF PHOTOSYNTHESIS, RESPIRATION AND PHOTORESPIRATION

Respiration is essentially the converse of the process of photosynthesis as represented by equation (14.1), except, of course, that the energy ultimately appears as heat rather than visible light. If the two processes occur concurrently then the rate of photosynthesis as actually observed (*net photosynthesis*) will be less than the total rate (*gross photosynthesis*) by an amount equal to the rate of consumption of the products of photosynthesis in respiration. At certain low values of light intensity or carbon dioxide concentration the two processes balance so that there is no net gas exchange and the photosynthetic organism is then said to be at its *compensation point*. It has usually been assumed, without real justification, that the rate of respiration remains the same in light and dark so that it is sufficient to correct observed rates of photosynthetic evolution of oxygen, say, by adding an amount equal to the uptake of oxygen in otherwise similar samples in the dark, so as to obtain a value representing gross photosynthesis.

This simple view has been complicated by the realization that there is also a process of light-stimulated *photorespiration*. This may be defined as a light dependent oxygen uptake and carbon dioxide release occurring in photosynthetic tissues. In order to measure it, it is necessary to use isotopic tracers, by, for example, supplying $^{18}O_2$ and following mass spectrometrically its uptake and the simultaneous evolution of $^{16}O_2$ during photosynthesis.

Unlike dark respiration, which is saturated at a P_{O_2} value of about 0·05 atm, the rate of photorespiration increases linearly with P_{O_2} up to 1·0 atm or above. Photorespiration is inhibited by DCMU, a specific inhibitor for the photosystem II process of photosynthesis, but cyanide has little effect. It is highly dependent on the carbon dioxide concentration, increasing to a rate approaching that of true photosynthesis when the carbon dioxide concentration is lowered to the compensation point (Lex et al., 1972). Photorespiration increases with light intensity. Low intensities of blue light have been found to increase respiration, but the available evidence suggests that this enhancement is different in nature from the photorespiration as described above (Kowallik, 1971; Tolbert, 1974). As in higher plants, the substrate for photorespiration is glycollate, which as has already been seen (Section 14.2.4), is under some circumstances a major product of algal photosynthesis. The oxidation of glycollate does not take place in mitochondria, the organelles responsible for dark respiration, and does not appear to result in the accumulation of potential chemical energy in the form of ATP (Tolbert, 1973). It may thus appear to be a biologically useless process, although it may be of value as an overflow mechanism. In bright light under nutrient deficient conditions, an alga may produce photosynthetic products more rapidly than they can be converted to cell materials or reserve product. In these circumstances, it may be advantageous to dispose of the surplus either by excretion or photorespiration.

The relative rates of photosynthesis and photorespiration vary widely according to species and the conditions to which the cells are exposed. A study of their interrelationship was made by Bunt et al. (1966) for an Antarctic sea-ice diatom Fragilaria sublinearis in axenic culture, uptake and evolution of oxygen being followed by mass spectrometry. The irradiation was carried out at two wavelengths, 678 nm and 525 nm, and three temperatures, 3°, 10° and 24°C. Increased oxygen uptake with increasing illumination was observed at both wavelengths and at all three temperatures. The light stimulated uptake reached a plateau at intensities below those required to saturate photosynthesis. At 10°C and 525 nm, light stimulated oxygen uptake declined with increasing light intensity after reaching a maximum. The reason for this aberrant behaviour is not known. Photosynthetic activity reached a maximum at about 7°C, and dark respiration had a maximum somewhere between 10° and 24°C. As a result of impaired photosynthesis and increased respiration and photorespiration, net evolution of oxygen was scarcely possible above 10°C. This appears to explain the restriction of Fragilaria sublinearis to Antarctic waters.

The significance of such interactions for the estimation of primary productivity will be discussed in Section 14.4.3.

14.2.6. VARIATION IN METABOLIC RATES AND PATTERNS DURING GROWTH

Micro-organisms in general show a much greater degree of variation in metabolism and chemical composition than do multicellular organisms and this is certainly true of phytoplankton. The variations may occur in response to changes in environmental factors, such as those which occur in the population as a whole as a batch culture grows. These variations can be eliminated by growing the organisms in continuous culture. There are also those which are related to endogenous changes during the cell cycle. These become manifest in large populations only when the organism is grown in synchronous culture. These two types of variation are, however, interrelated and not always separable.

When algae are growing exponentially with ample nutrients, the products of photosynthesis are largely used for the synthesis of new protoplasm, i.e. the main end product is protein. For *Chlorella* in this state a typical elementary analysis of the dry organic matter is $53\% C$, $7.5\% H$, $28.5\% O$ and $10.8 N$, corresponding to a protein content of about 60% on a total dry weight basis, and the overall equation for growth may be written:

$$1.0\,NO_3^+ + 5.7\,CO_2 + 5.4\,H_2O \longrightarrow C_{5.7}H_{9.8}O_{2.3}N + 8.25\,O_2 + 1.0\,OH^- \qquad (14.8)$$

or, if the nitrogen source is ammonia:

$$1.0\,NH_4^+ + 5.7\,CO_2 + 3.4\,H_2O \longrightarrow C_{5.7}H_{9.8}O_{2.3}N + 6.25\,O_2 + 1.0\,H^+ \qquad (14.9)$$

(Myers, 1962). The photosynthetic quotient $(\Delta O_2/\Delta CO_2)$ is 1.45 in the former case, 1.10 in the latter. If the cells are transferred to a medium lacking a nitrogen source there is no immediate effect on photosynthetic capacity, but intermediates previously used in protein synthesis are built up into carbohydrates. The equation for photosynthesis then is the classically accepted one (14.1) and the photosynthetic quotient becomes 1.0. To some extent redistribution of nitrogen is possible, essential components such as nucleic acids being formed at the expense of less essential ones such as chlorophyll, so that cell division can proceed. Usually, the lower limit for division is about 3% nitrogen so that one or two successive divisions may occur subsequent to deprivation of the nitrogen source. After these have taken place, only an increase in cell dry weight is possible, and as this occurs the rate of photosynthesis declines to a low value. This decline is evident under both light-limited and light-saturated conditions and does not appear to be due primarily to the associated decrease in chlorophyll (Fogg, 1959; Ebata and Fujita, 1971).

Much significance is attached to the relationship between rate of photosynthesis and chlorophyll content in primary productivity studies and since contradictory statements have been made about the effect of nutrient deficiency on this relationship, some discussion is called for at this point. The relationship is expressed in terms of what plant physiologists call the *assimilation number* or *ratio* (mg C fixed per hour per mg chlorophyll *a* at light saturation = P(c) of Strickland (1965)). Fogg (1965) has stated that this increases for nitrogen-deficient algae on the basis of measurements made by Spoehr and Milner (1949) with *Chlorella* and his own observation (1959) with *Monodus*, a Xanthophycean alga, in which the assimilation number rose from 1·6 in actively growing cells to 6·1 in nitrogen-deficient ones. Strickland (1965), on the other hand, has asserted that nitrate depletion reduces the assimilation number to one third or less of its value in the presence of excess nutrients, but based this conclusion on observations made with natural populations rather than with laboratory cultures under controlled conditions. However, falls in the assimilation number have been reported for nitrogen-deficient laboratory cultures of *Scenedesmus* (Bongers, 1956), *Dunaliella* and *Skeletonema* (McAllister *et al.*, 1964). The tropical marine diatom *Chaetoceros gracilis* grown in nitrogen-limited chemostat culture has been found to show an increase in assimilation number with increasing cell nitrogen content and increasing growth rate (Thomas and Dodson, 1972). This conforms with the finding reported by several workers (references in Platt and Subba Rao, 1973) that addition of nutrients to nutrient-deficient populations increases the assimilation number.

The causes of these contradictions, which, as will be seen (p. 424), are also apparent in results obtained from *in situ* experiments, are probably numerous. The light-saturated rate of photosynthesis depends on the activities of enzymes which bear no simple relationship to the chlorophyll concentration in a cell and which are affected by different factors. The cells studied by Spoehr and Milner (1949) and by Fogg (1959) were obtained by nitrogen starvation for extended periods and would be in a quite different physiological state from those used by Bongers (1956) and McAllister *et al.* (1964), which were subjected to much shorter periods of deficiency, and from those in Thomas and Dodson's (1972) continuous cultures, which were actively growing. As noted in Section 14.5.4, phytoplankton cells may not show physiological symptoms of nitrogen deficiency even though shortage of nitrogen is evidently limiting population size. There is no reason to suppose that cells of one species, let alone different species, in these different physiological states should, when subject to nutrient deficiency, show similar alterations in photosynthetic enzymes in relation to chlorophyll. The conditions under which the assimilation number is determined may also affect its value; exposure to high light intensities may result in reduced photosynthetic yield if the oxygen concentration is high and photo-respiration occurs, or chlorophyll may be photo-oxidized. The only possible conclusion pending a far more detailed knowledge of algal metabolism than we have at present is that assimilation numbers, to which, incidentally, plant physiologists do not attach great importance, should be interpreted with great caution.

It may also be noted here that fixation of carbon per unit amount of chlorophyll at limiting light intensity (productivity index), which, of course, varies with light intensity, must not be confused with assimilation number. For *Skeletonema costatum* at a given limiting intensity, the observed value of the productivity index rose slightly on nitrogen starvation in the experiments of McAllister *et al.* (1964) but Jørgensen (1970), working with the same species, observed a marked fall in carbon fixed per unit amount of chlorophyll at 1 klux for both phosphorus- and nitrogen-deficient cells as compared with those grown with an adequate supply.

Respiration as well as photosynthesis decreases on nitrogen starvation until a critical level of cell nitrogen content is reached when it rises again to a climacteric. This climacteric, which may be postponed by illumination, is accompanied by an increase in the dark fixation of carbon dioxide, and is evidently a symptom of the breakdown of cellular organization (Fogg, 1956).

Cells from cultures in which the nitrogen supply becomes exhausted during the course of growth behave rather differently from cells abruptly deprived of a nitrogen supply, the former often accumulating fats rather than carbo-hydrates. *Chlorella* in old batch cultures may contain as much as 85 per cent of fat on a dry weight basis (Spoehr and Milner, 1949) and the diatom *Nitzschia palea*, 42 per cent (von Denffer, 1948). This may be because staling products, accumulated in the medium, tend to block cell division at a stage characterized by production of highly reduced products (see below), or because conditions in aging cultures result in differential destruction of the enzyme systems (Fogg, 1959).

Unfortunately, few studies have been made of the effects of deficiencies of other nutrients on patterns of metabolism. A rapid decrease in the photo-synthetic activity of cultures of *Phaeodactylum tricornutum*, *Nitzschia closterium* and *Chaetoceros* sp. growing in artificial sea water medium has been attributed primarily to phosphorus deficiency (Ebata and Fujita, 1971). Ketchum *et al.* (1958) observed that in *Dunaliella euchlora* the ratio of net to gross photosynthesis fell sharply in phosphorus-deficient cells. This seems to imply a similar climacteric rise in respiration with breakdown of all organiza-tion as is observed with nitrogen deficiency. McAllister *et al.* (1964) recorded a decrease in assimilation number from 6·20 with high nutrient levels to 3·50 in phosphate-deficient cultures of *Skeletonema* but, conversely, a rise in fixation of carbon per unit amount of chlorophyll from 1·78 to 1·93 at limiting light intensities. These changes are similar to those caused by nitrogen deficiency (see above).

As a result of changes in the pattern of metabolism, the chemical composition of an alga varies within wide limits. However, the responses to particular changes in the environment seem to be the same in the different algal groups, as well as in related species, so that algae tend to have similar organic compositions when grown under similar physical and chemical conditions regardless of the size of the organism or its taxonomic position. This was demonstration by Parsons *et al.* (1961) in an investigation in which eleven species of marine phytoplankton belonging to the Chlorophyceae, Haptophyceae, Chrysophyceae, Bacillariophyceae, Dinophyceae and Cyanophyceae were grown under similar conditions and analysed during the exponential phase of growth. An exception to this generalization is that the fat accumulation under conditions of nitrogen deficiency, which is typical of most of the algal groups, does not occur in the Rhodophyceae or Cyanophyceae (Collyer and Fogg, 1955).

It is not possible to describe here the large amount of information about changes in the pattern of metabolism during the cell division cycle in algae, but reference may be made to reviews by Tamiya (1964, 1966) and Pirson and Lorenzen (1966). Cells of *Chlorella* recently produced by cell division show a rise in rate of photosynthesis to a maximum when they are a few hours old. These actively photosynthesizing cells have a low rate of respiration and high contents of cell nitrogen and chlorophyll. As they enlarge and become ripe for division their light-saturated photosynthesis rate drops to about 1/6 of what it was initially; the cell nitrogen and chlorophyll contents also fall, but the respiration rate increases somewhat. The division phase is oxygen-dependent, but can take place in the dark (Tamiya *et al.*, 1953). The phase of photosynthesis and cell enlargement is chiefly one of protein synthesis and the photosynthetic quotient is a little over unity. As the cells approach division, the photosynthetic quotient rises to between 3 and 4 and the cell material becomes correspondingly more reduced (Nihei *et al.*, 1954). The sequence of changes in rates of photosynthesis seem typical of all algae studied in synchronous culture, e.g. for *Skeletonema costatum* (Jørgensen, 1966), but the changes in photosynthetic quotient have not always been observed (Senger, 1970). It is of interest that Sorokin (1965) separated two size groups of cells from non-synchronized cultures of *Chlorella* and found that the smaller, younger, cells had a higher rate of photosynthesis than the larger, older, ones thus confirming that the cyclic change observed in synchronized cultures is not an artefact. Gimmler *et al.* (1969) found that the liberation of extracellular glycollic acid by *Ankistrodesmus braunii* changed during its life cycle. In synchronous culture, liberation was high during the phase of high photosynthetic activity, reaching a maximum 6 hours after the beginning of the light period. Periodic release of organic phosphate in synchronous cultures of *Chlorella* has been observed by Soeder (1970).

Few studies of marine phytoplankton species in synchronous culture have been made, but results reported for *Dunaliella tertiolecta* (Eppley and Coatsworth, 1966) and *Coccolithus huxleyi* (Paasche, 1967) are in general conformity with those just described for freshwater species. Eppley *et al.* (1971) studied assimilation by *Skeletonema costatum* in nitrogen limited chemostat cultures partially synchronized by light/dark cycles and found that both nitrate and ammonium nitrogen uptake occurred mainly during the day. Under similar conditions *Coccolithus huxleyi* assimilated these forms of nitrogen at a rate sufficient to keep nutrient concentrations at low levels during both day and night. The activity of nitrogen-assimilating enzymes was, nevertheless, higher in *C. huxleyi* during the day than at night indicating a similarly to *S. costatum*. Both of these species showed cell division in the dark. *Brachiomonas, Dunaliella* and *Monochrysis* were found by Quraishi and Spencer (1971) to behave similarly. *Phaeodactylum tricornutum* and the marine *Chlorella ovalis*, however showed a different pattern with division confined to the period of illumination. These laboratory studies thus suggest that the metabolic cycles of different species of planktonic algae may differ in their response to photoperiod.

14.2.7. ADAPTATION

Under this heading will be considered some of those alterations in metabolic pattern which occur in response to environmental conditions and which appear to offer a distinct biological advantage. These alterations are basically of the same nature as those discussed in the previous section. Selection of mutants capable of better performance under particular circumstances is possible with algae. Kumar (1964), for example, obtained strains of a unicellular blue-green alga resistant to various antibiotics, but studies of genetic adaptation have scarcely been conducted as yet with marine phytoplankton.

It has frequently been observed that algal cells grown at limiting light intensities contain relatively more photosynthetic pigment than do those grown at higher intensities. This was investigated with *Chlorella vulgaris* by Steemann Nielsen *et al.* (1962). Cells grown at 3 klux ("shade" cells) showed higher rates of photosynthesis per cell than those grown at 30 klux ("sun" cells) at limiting light intensities, but when these rates were expressed as carbon fixed per unit of chlorophyll they became identical. The better performance of "shade" cells at low light intensities could, therefore, be accounted for in terms of a greater amount of chlorophyll per cell giving greater light absorption per cell. The maximum rate of photosynthesis per cell under saturating intensities was somewhat greater in *C. vulgaris* grown at 30 klux than in that grown at 3 klux, indicating that the former had greater amounts of the enzymes concerned in the dark reactions of photosynthesis. On transfer from one intensity to another cells took from 24 to 36 h—a little longer than the generation time—to change to the appropriate chlorophyll concentration. Jørgensen (1969), who carried out similar experiments with other algae, found that they fell into two groups. One, which included various species of Chlorophyceae, *Monodus* and the blue-green alga *Synechococcus elongatus* behaved like *Chlorella vulgaris*. The other group, which comprised *Skeletonema costatum*, *Cyclotella meneghiniana*, *Nitzschia palea*, *N. closterium* and *Scenedesmus quadricauda*, differed in that the amount of chlorophyll per cell remained unchanged by growth at different light intensities, the only observed adaptation in photosynthesis being one involving the light-saturated rate which seems to depend on production of more of the enzymes involved in the dark reactions of photosynthesis at high light intensities. It is probably without significance that the only two marine species, *S. costatum* and *N. closterium*, investigated by Jørgensen fell into this latter group. As will be seen (p. 428), adaptation to the prevailing light intensity may occur in natural populations.

Adaptation to low temperature also occurs in algae. Steemann Nielsen and Jørgensen (1968) and Jørgensen (1968) showed that, under similar conditions except for temperature, rates of photosynthesis are nearly the same in cells of *Skeletonema costatum* grown at 7° and 20°C. The relative growth rate of the cells at the low temperature is, however, less than half what it is at the higher temperature. In accordance with this, the low temperature cells are larger and contain more protein than the others. This supports the suggestion that the retarding effect on photosynthesis of low temperature is offset in adapted cells by higher concentrations of photosynthetic enzymes. The marine flagellate *Dunaliella tertiolecta* was found by Morris and Farrell (1971) to photosynthesize more rapidly at 12°C than at 20°C, and the growth rate was almost the same at the two temperatures. Cells grown at the lower temperature contained higher concentrations of soluble protein, had higher activities of the carbon fixation cycle enzyme ribulose diphosphate carboxylase, and showed an enhanced rate of protein synthesis from photosynthetically fixed carbon dioxide. With the marine diatom *Phaeodactylum tricornutum*, cells grown at 5°C photosynthesized more rapidly than those grown at 10°C, but no similar effect was found when 10°C-grown cells were compared with ones grown at 20°C. These results appear to support the hypothesis of Steemann Nielsen and Jørgensen (1968) but Morris and Glover (1974) have pointed out that in batch cultures such as have been used in this work there is a peak of photosynthetic activity after a period which varies with temperature, and in comparing material harvested after a fixed time from low and high temperature cultures, one is comparing cells in different physiological conditions. When account is taken of this there is no evidence that adaptation to lower temperatures is a consequence of the increased levels of the enzymes required for carbon dioxide fixation. Thus, although temperature adaptation seems a real phenomenon, its explanation remains obscure. As noted elsewhere (Sections 14.2.5 and 14.6.4), the relative rates of photosynthesis and respiration are important, algae adapted to low temperatures requiring very low light intensities for compensation.

14.3. STANDING CROP: GROWTH AND INCREASE OF PHYTOPLANKTON *IN SITU*

The amount of phytoplankton present in a given water mass at a particular instant of time is termed the *standing crop* or *biomass*. Quantitative determination of this amount presents many difficulties, which it is not necessary to discuss in detail here since accounts of methodology have been given by Strickland and Parsons (1968), Strickland (1972) and Vollenweider (1974). Although it is concerned only with freshwater, an article by Lund and Talling (1957) is also useful in this context. It may, however, be useful to indicate the nature of the problems which arise.

The distribution of phytoplankton in the sea is rarely uniform and frequently extremely patchy. Horizontally, patches are usually elliptical and vary in size from a few metres to hundreds of kilometres across. Long narrow bands or streaks, a few metres in width are common (evidently resulting from wind-driven Langmuir circulation) and may form a pattern superimposed on that of the patches (Bainbridge, 1957). Steele (1961) has briefly discussed the development of patches in relation to growth rates and lateral eddy diffusion. Information about exceptionally dense populations has been summarized by Hart (1966). Species which swim or float to the surface may be concentrated against a lee shore or in regions of downwelling, giving rise to "red tides" (see Strickland, 1972). Under conditions of strong mixing, vertical distribution of phytoplankton may be uniform but, if the water column becomes stabilized, non-motile forms denser than water will tend to sink and become concentrated lower down while forms lighter than water rise to the surface and motile forms may congregate at particular intermediate depths. This happens particularly with relatively large fast-swimming species of dinoflagellates, which may become concentrated in layers only a few cm thick. Problems related to microstratification have been discussed by Cassie (1963) and a sampler for thin horizontal layers has been described by Parker et al. (1968).

Plankton nets are quite unsuitable for quantitative work in connection with primary productivity. Holmes and Anderson (1963) found that more than half the photosynthesis in sea water samples from the N.E. Pacific was carried out by algae which passed through a net with 35 μm apertures. Saijo and Takesue (1965) obtained similar results in the Indian Ocean, and Platt and Subba Rao (1973) have emphasized the importance of the contribution of the nanno-plankton (which they define as being between 5 and 65 μm in diameter) to primary production. Reynolds (1973), who used a fluorimetric method for the determination of ultraplankton (defined as being less than 15 μm in diameter), concluded that in the Barents Sea and some other northern waters they may sometimes contribute over 90% of the total chlorophyll a in the water. Samples taken with a non-metallic sampling bottle may be representative of the total phytoplankton, but it should be remembered that the usual oceano-graphic practice of taking samples at successive depths can miss highly localized concentrations of phytoplankton and may sometimes fail to give an accurate estimate of abundance. Statistical studies by Hasle (1954) and Barnes and Hasle (1957) give an idea of the variation which may be encountered. Pumps may be used to obtain large integrated samples from which representa-tive sub-samples may be withdrawn and although they have certain advan-tages they have not often been used.

The phytoplankton in a water sample may be concentrated by sedi-mentation, centrifugation or filtration and estimated by counting cells (or determining their volume) or by estimating the amount of some chemical component. Identification and counting the numbers of individual species is laborious and unnecessary for many purposes, but undoubtedly provides information which cannot be obtained by any other means. When samples are preserved for subsequent examination it should be remembered that no single preservative is suitable for all types of algae (see Strickland, 1972). Electronic dimensional particle counters (Coulter Counters) are capable of giving meaningful results with axenic laboratory cultures, but must be used with great caution on natural populations which contain much detritus as well as other organisms of different sizes and shapes. If the mean dimensions of cells are determined, counts may be converted to volumes, cell surface areas or other estimate of biomass (Reid *et al.*, 1970). Paasche (1960) has shown, with reference to samples from the Norwegian Sea, that numbers, volume and surface area give strikingly different impressions of the importance of different species, and that of these three measures the last is most closely correlated with photosynthetic activity. Chlorophyll a or some other photosynthetic pigment is frequently used as a measure of biomass (Strickland and Parsons, 1968; UNESCO, 1966), but it must be remembered that like other chemical components, the amounts per cell may fluctuate widely and that photosynthetic activity does not bear any constant relationship to pigment content (see Section 14.2.6). Similarly, expressions by which cell carbon is estimated from volume can only be regarded as approximate (Reid *et al.*, 1970; Strickland, 1972). Hobson *et al.* (1973) using such a method estimated phytoplankton carbon to vary from less than 10 µg l^{-1} in the Gulf of Mexico to 750 µg l^{-1} in nutrient-rich water off the coast of Peru. Among chemical measures of biomass, one of the best seems to be the adenosine triphosphate (ATP) content. This can be determined by the luminescence which it induces in the very sensitive luciferin-luciferase ("firefly") reaction (Holm-Hansen, 1969; Strickland, 1972).

The growth rate of the phytoplankton depends in the first place on the net rate of photosynthesis, i.e. on the excess of photosynthetic products over those consumed in respiration. Since the temperature coefficient of respiration is generally higher than that of photosynthesis, this excess is usually less the higher the temperature. Mention has already been made of an extreme example of this in *Fragilaria sublinearis* (p. 404), and the same relationship seems to hold *mutatis mutandis* with phytoplankton in general. The rate of increase in biomass in a given water mass should not be confused with the rate of growth since a natural population is subject to continual depletion by sinking and grazing. An empirical equation which represents this sort of situation is that given by Riley (1963b):

$$\frac{dP}{dt} = P(P_h - R - G) \qquad (14.10)$$

where P is the total phytoplankton population per unit of sea surface, P_h a photosynthetic quotient, R a coefficient of phytoplankton respiration, and G a grazing coefficient. P_h is estimated empirically from measurements of incident radiation, transparency of the water, depth of the mixed layer, and concentration of phosphate (taken as a measure of the general level of nutrients). Phytoplankton respiration was assessed from experimental data and was assumed to increase exponentially with rise in temperature. The grazing coefficient was assumed to be proportional to the observed herbivore population. This model gives an annual periodicity of phytoplankton abundance similar to that which actually occurs, and when data are examined using equation (14.10) by approximate integration over successive short periods of time during which environmental conditions are assumed to remain constant, curves are obtained which show reasonable agreement with those based on observed values.

From a comprehensive series of observations including measurements of chlorophyll photosynthesis and respiration during the development of a mixed phytoplankton bloom, Platt and Subba Rao (1970) concluded that the bloom passed through exponential, stationary and senescent phases analogous to those in a batch culture. No tendency to accumulate fat in the senescent phase was detected however.

14.4. The Determination of Primary Productivity

14.4.1. GENERAL REMARKS

According to the definition given on p. 386, primary productivity should be determined in terms of changes in free energy. This is not practicable for routine purposes, and instead it is usually measured as change in amount of one of the substances involved (equation 14.1). It is scarcely possible to determine changes in the amount of water and difficult to estimate the dry weight of organic matter produced. Disappearance of carbon dioxide from the water, incorporation of carbon in organic matter and production of oxygen, however, may all be determined with relative ease. Additionally, since growth is generally correlated with photosynthesis, changes in amount of nutrients utilized in growth, e.g. phosphate and silica, have sometimes been used for estimating primary productivity. Since photosynthesis is dependent on chlorophyll, the amount of this in the water is often used as a basis for predicting rates of primary productivity.

Nearly always, estimations are made in relatively small samples of water enclosed in containers of glass, plastic or, occasionally, silica. Collection of a sample and enclosure in a container immediately introduces changes which may affect the phytoplankton to a serious extent. Apparatus has been devised which enables samples to be taken and held for the period of observation at the depth from which they came, without bringing them to the surface (Dyson et al., 1965; Watt, 1965). Otherwise, care must be taken to ensure that the sample is not exposed to high light intensity when it is brought to the surface and transferred to the experimental bottle as even a few minutes exposure may result in marked inhibition of photosynthetic activity lasting several hours (Goldman et al., 1963; Doty et al., 1965). From this point of view, determinations should preferably be set up and terminated at dawn or dusk. The changes in pressure which occur when a sample is brought to the surface might also affect subsequent photosynthetic activity (Vidaver, 1972), although this possibility does not appear to have been investigated. The bottle used for the determination may alter the lighting conditions or the chemical composition of the sea water, either by solution of material from it or by adsorption of solutes. That such effects may sometimes be considerable is suggested by comparisons between determinations done in glass and plastic bottles.

Comparing primary production measurements made in Jena glass and acrylic plastic bottles, Ilmavirta and Hakala (1972) found that the latter gave lower values near the water surface, apparently because of their greater transparency towards ultra-violet light, but higher values at greater depth because they also transmitted more visible light. Some particular types of plastic may liberate toxic materials and it is a wise precaution to test any new material with which samples come into contact before beginning a series of determinations. Placing a sample in a bottle, by enormously increasing the solid surface available, promotes bacterial growth and it is to be expected that in a sample incubated for 24 h, or perhaps even for a shorter period at high temperatures, the balance between algae and bacteria will have become seriously disturbed, with corresponding effects on photosynthesis and respiration. Turbulence patterns, which may be of crucial importance for the exchange of materials between cell and water, are quite different in a bottle and in unconfined water. Finally, a sample bottle is held in a fixed position in a deck or laboratory incubator or suspended at a fixed depth by a line whereas a cell in a natural population moves actively or passively through a greater or lesser vertical distance. The difference in photosynthetic activity between a cell in a sample bottle held for a period of hours in an inhibitory intensity at the water surface and one which is circulated rapidly by vigorous natural turbulence between the surface and limiting light intensities at some depth would be expected to be considerable.

Such sources of uncertainty might be eliminated by following changes with time of carbon dioxide or oxygen in the water column itself. However, interchange with the atmosphere and (if the water is shallow) with the bottom deposit, and lateral transport give rise to uncertainty and no estimates obtained by this method can be said to have much significance. The problems which arise have been discussed by Strickland (1960). Comparisons of results obtained by determination of oxygen changes in bottles and in the water column in the much simpler situation of shallow static fresh-water, showed them to be of the same order of magnitude, but close agreement was obtained for only one out of four sets of observations (Talling, 1957a).

It is therefore customary to base estimates of primary productivity on determinations made in samples held in bottles. These may be exposed *in situ* in the sea or in an apparatus which, to a greater or lesser extent, simulates conditions in the sea. In the *in situ* method samples are collected from appropriate depths and allowed to photosynthesize at the depths from which they came. The depths should be chosen so as to give maximum information about the form of the photosynthesis versus depth curve. Thus, depths to which 100, 50, 25, 12·5, 6·25 and 1·56% of incident light penetrated might be used after a preliminary determination of the extinction coefficient of the water. The simplification of using a single plankton sample and exposing aliquots at

different depths is not to be recommended since it takes no account of possible
uneven distribution of phytoplankton or of adaptation of cells to different
light conditions. During exposure the bottles should be suspended, so as to be
shaded as little as possible by the ship or buoy, for example from a boom with
a buoy at either end.

Since the *in situ* method involves keeping a vessel hove to for several hours,
other methods are usually preferred although giving a poorer approximation
to natural conditions. Steemann Nielsen and Aabye Jensen (1957–9) described
a water bath, adjusted to sea temperature, in which bottles could be exposed to
artificial light on a rotating wheel, thus ensuring stirring and even illumination.
The determination of the effective intensity incident on the bottles is a matter
of some difficulty, and these workers used the photosynthetic response of the
plankton itself as a means of calibration. The intensity reaching individual
bottles was altered by means of wire mesh filters placed around them.
Determinations on samples taken (a) from the surface, (b) from a depth to
which 10% of incident light penetrated and (c) from a depth to which 1%
of incident light penetrated were made at corresponding intensities in the
apparatus. Production (P) in $mg\,C\,m^{-2}\,day^{-1}$ was obtained using the
empirical expression (modified by Strickland, 1960, 1965):

$$P = \tfrac{1}{3}(2P_{100} + 2P_{10} + P_1)\frac{D}{2}\,NK \qquad (14.11)$$

where P_{100} is the photosynthesis in $mg\,C\,m^{-3}$ at 1800 lux in surface water, P_{10}
that in sample b, P_1 that in sample C, D the depth in m at which the light
intensity is 1% of the total blue and green light incident at the surface and N
the number of hours from sunrise to sunset. K is a constant which depends on
the incubator and sea location; in the tropics it had a value of about 1·0 with
an incubation intensity of 18000 lux. Where the phytoplankton had a
reasonably uniform vertical distribution this method gave values agreeing
with *in situ* results to ±20%.

Less empirical approaches make use of the observed relationship between
rate of photosynthesis and light intensity (Fig. 14.1). Ryther (1956) assumed
an idealized curve for a given type of phytoplankton from which, together
with the daily radiation value a factor R could be calculated which related the
total production in the euphotic zone in $mg\,C\,m^{-2}\,day^{-1}$ to the maximum
rate of photosynthesis (P_{max}) in $mg\,C\,m^{-3}\,day^{-1}$ measured in an artificial
light incubator.

$$P = \frac{R}{k}\,P_{max} \qquad (14.12)$$

k being the vertical coefficient for light absorption. This approach has been elaborated by Sorokin to take into account uneven distribution of phytoplankton with depth. References and a summary account in English of his rather complex procedure have been given by Strickland (1965). Rodhe (1965), from an analysis based on methods developed by Talling (1957b) of photosynthesis versus depth curves obtained from a variety of lake types, has put forward an expression which gives good approximate estimates of photosynthesis in instances where phytoplankton is uniformly distributed. This is discussed in more detail in Section 14.5.1.

In designing incubators for primary productivity determinations, special attention should be paid to the light source since, as we have seen (p. 391), this may have important effects on photosynthesis. The spectral composition of daylight is so unlike that of light which has penetrated through sea water that it seems advisable, if natural light is to be used, to remove wavelengths below 400 and above 600 nm by filters. The difficulty of mimicing submarine light with artificial sources has been emphasized by Strickland (in Riley, 1963). "Daylight" and "cool-white" fluorescent lamps give light of markedly different spectral compositions neither of which resembles that of submarine light. The latter is best imitated with mixed light from green and blue fluorescent lamps or with light from a tungsten lamp with a blue filter (Strickland, 1965). Intercalibration of five different methods using incubators showed a mean coefficient of variation between them of 26%, and the values obtained were all less, sometimes only 50%, of those of parallel *in situ* determinations. Considering that primary productivity per unit area varies by as much as 20-fold from place to place and time to time, these discrepancies are not important, but it seems advisable always to check incubator methods against *in situ* determinations, (Doty *et al.*, 1965).

14.4.2. THE OXYGEN METHOD

This method, which was introduced by Gaarder and Gran (1927), consists essentially of measuring the changes in oxygen concentration in subsamples of water exposed in clear and dark bottles. Oxygen is usually measured by the Winkler technique or electrochemically (Vollenweider, 1974). The rise in the amount of oxygen in the clear bottle is a measure of net photosynthesis and, assuming that respiration is unaffected by light, the addition to this of a correction equal to the amount of oxygen consumed in the dark bottle gives an estimate of gross photosynthesis.

It has already been seen that the assumption that phytoplankton respiration is unaffected by light is an oversimplification (p. 403), and there is the additional complication that bacterial respiration may be different in the light and dark bottles. Steemann Nielsen (1955, 1958) found evidence that light in some way reduced bacterial activity in bottles containing oligotrophic sea water. He (1955) supposed that this might be due to antibiotic production by the algae in the light and, since glycollic acid excretion by photosynthesizing algae seems general (p. 402) and Sieburth (1968) has suggested that acid microzones may discourage bacterial attachment, this supposition seems reasonable.

Even if the available analytical techniques are pushed to their limits it is difficult to detect oxygen production at a rate less than $2 \text{ mg O}_2 \text{ m}^{-3} \text{h}^{-1}$ with a 12 h exposure. It is not advisable to attempt to detect lower rates by longer exposures because of bacterial growth and the uncertainties it introduces. The method is, therefore, not suitable for use with phytoplankton densities of less than 1 mg chlorophyll $a \text{ m}^{-3}$ (Vollenweider, 1974).

Although in practice it is not usually important, it should be remembered that the oxygen method cannot measure bacterial photosynthesis or photophosphorylation and processes depending on it, all of which may contribute to the accumulation of potential chemical energy in certain habitats.

14.4.3. THE RADIOCARBON METHOD

In this method, introduced by Steemann Nielsen (1952), the incorporation of ^{14}C, supplied as bicarbonate, into cell material in the light is taken as a measure of photosynthesis. Details of this technique are described and discussed in several places (Steemann Nielsen and Aabye Jensen, 1957–59; Doty and Oguri, 1958; Steemann Nielsen, 1958; Strickland, 1960; Saijo and Ichimura, 1961; Jitts, 1961; Steemann Nielsen, 1964a; Dyson et al., 1965; Strickland and Parsons, 1968; Vollenweider, 1974), but certain points should be mentioned here. It is, of course, important in the first place that the tracer should be free from toxic impurities. It should not be assumed that ^{14}C-bicarbonate solutions supplied commercially are non-toxic, and if there is doubt it is preferable to prepare the bicarbonate solution using barium ^{14}C-carbonate as the starting point (Vollenweider, 1974). Some commercial ^{14}C sodium bicarbonate solutions contain a significant amount of radioactive material which is retained by membrane filters. This extraneous material, which seems to be organic, can be removed by high intensity u.v. radiation (Morris et al., 1971a; Williams et al., 1972). Standardization of the tracer solution has given rise to difficulty; until recently most workers used a windowless gas-flow counter for measuring the soft β-radiation emitted by ^{14}C and determined the activity of their solutions after precipitation of carbon dioxide as barium carbonate. The barium carbonate gives rise to some self absorption of the β-radiation and the extrapolation of the activity to zero thickness constitutes a source of error. To avoid this difficulty Steemann Nielsen 1965 (see Vollweider, 1969) proposed the complete absorption of the carbon dioxide in an aliquot of the solution by *Chlorella*, with subsequent assay of the radioactivity under the same conditions as used for the natural phytoplankton. Measurement by scintillation counting is perhaps the most satisfactory method (Jitts, in Vollenweider, 1974). To avoid loss of soluble ^{14}C-labelled substances from damaged cells, which may sometimes be considerable (Arthur and Rigler, 1967), it is important that filtration should be carried out under low vacuum. Precipitation of ^{14}C in the form of calcium carbonate which is retained on the filter along with the algal cells is another source of error. It may be avoided by exposing the desiccated filter to hydrochloric acid fumes for a few minutes. Washing with dilute hydrochloric acid is less satisfactory (McAllister, 1961a; Steemann Nielsen, 1964a; Wetzel, 1965). This precaution should always be taken with waters containing Coccolithophoridaceae, which form calcareous scales, and perhaps should be carried out routinely with all samples since Wetzel (1965) found it necessary with a variety of freshwaters. Goldman and Mason (1962), however, found that with some freshwaters carbonate may be co-precipitated with iron in a form which is not easily decomposed by hydrochloric acid. Retention, presumably by adsorption, of ^{14}C-bicarbonate and ^{14}C-extracellular substances by membrane filters may constitute an appreciable source of error under some circumstances (Nalewajko and Lean, 1972). When scintillation counting is employed self absorption by diatoms must be taken into account (Pugh, 1970). Wallen and Geen (1968) found that there was a loss of radioactivity amounting to about 30% when ^{14}C-labelled phytoplankton on membrane filters was stored in a desiccator for 24 hrs. This loss did not increase on further storage for up to 8 weeks. This appears to be an important source of error in the ^{14}C-method which has been ignored by nearly all workers. Lastly, discrimination between the isotopes ^{12}C and ^{14}C may be surprisingly high, and Steemann Nielsen (1963) concluded that ^{14}CO$_2$ is assimilated about 5% more slowly than ^{12}CO$_2$. Correction for such discrimination is often ignored (e.g. Vollenweider, 1974), but recent work on metabolic fractionation of the stable isotopes ^{12}C and ^{13}C by marine phytoplankton (Degens *et al.*, 1968) supports the idea that it may not be negligible. When making measurements in oceanic waters it is sufficient to use tables such as those of Buch to determine the total inorganic carbon content of the water. For brackish tropical waters it is wise to use directly determined values (Subba Rao, 1965).

Intercalibration of methods by four groups using the same phytoplankton sample gave results with a coefficient of variation of only 20% as opposed to 34% when the techniques were compared against *in situ* measurements, suggesting that incubation conditions and sampling depths are major sources of variation in the ^{14}C technique (Doty *et al.*, 1965).

Even if these various sources of error are avoided, there will remain serious difficulties of interpretation. Under certain circumstances it is possible that photoassimilation of organic compounds may take place alongside or instead of photosynthetic assimilation of carbon dioxide (Wiessner, 1970). This results in less gain, per carbon atom assimilated, in potential chemical energy than with normal photosynthesis, but since ^{14}C added in inorganic form does not equilibrate with organically bound carbon, it will be missed altogether by the radiocarbon method. The organic combination in which ^{14}C is fixed may vary from carboxyl groups, which represent scarcely any gain in potential chemical energy, to hydrocarbon chains, which represent a considerable gain in potential chemical energy. Fogg (1963) suggested that the former type of fixation predominates in moribund cells. As has been seen (p. 402), products of photosynthesis escape from normal healthy cells and so will be missed by the conventional radiocarbon technique in which the radioactivity of particulate matter only is determined. Procedures for determining fixation in extracellular products have been described by Watt (1966), Fogg (in Vollenweider, 1974) and Thomas (1971). The presence in some commerical $NaH^{14}CO_3$ solutions of small amounts of ^{14}C-labelled organic matter is a potential source of error which can be avoided by photo-oxidation of the contaminating material with ultra-violet radiation (Williams et al., 1972). In experiments in which fixation in organic solutes in filtrates has been determined as well as that in particulate matter, this leakage has been found to be appreciable. Observations which will be described later (p. 423) by Antia et al. (1963) suggest that as much as 40% of carbon fixed by phytoplankton in inshore Pacific waters might be liberated in extracellular form. Fogg et al. (1965) found a value of 7% for North Sea samples; Jitts (1967) a mean value of 18% for Southern Indian Ocean samples; Horne et al. (1969) 1–2% for inshore Antarctic waters; Samuel et al. (1971) between 1 and 20% for inshore waters off Cochin, S. India; Anderson and Zeutschel (1970) between 1% and 49% for oligotrophic waters in the northern-eastern Pacific Ocean, and Thomas (1971) between 7% for estuaries in south-eastern U.S.A. and 44% for the Sargasso Sea. A rough guide to the proportion to be expected in a given sea area is provided by the observation that the relative extent of excretion increases with increasing oligotrophy (Section 14.5.5.).

Considerable uncertainty exists as to whether fixation of ^{14}C observed in light bottles represents net or gross photosynthesis or something in between. Freshly fixed carbon may be preferentially respired or preferentially used in synthesis. Since carbon dioxide is fixed by heterotrophic processes as well as by photosynthesis, and fixation of ^{14}C is observed in dark as well as light bottles, the question arises of whether light values should be corrected for this or not. These two questions are separate but interrelated. Ryther (1956) from a comparison of ^{14}C results, uncorrected for dark uptake, with parallel determinations by the oxygen method concluded that the ^{14}C method gives an estimate of net photosynthesis. Steeman Nielsen and Hansen (1959) believed that it represents something between net and gross photosynthesis and suggested factors of 0·96 and 1·06 for correction of observed values to give net and gross photosynthesis respectively. Fogg (1963) obtained results with a blue-green alga suggesting that the ^{14}C method gave a measure of gross photosynthesis. From the discussion in Section 14.2.5. it is evident that the relationships between photosynthesis, and respiration may vary according to species, the physiological condition of the cells and environmental factors such as light intensity and the concentrations of carbon dioxide and oxygen. Morris et al. (1971a) have confirmed both with cultures and natural populations that the ratio of ^{14}C incorporation in the light to that in the dark increases with increasing cell concentration—as found with fresh water phytoplankton by Fogg (1958). Without elaborate physiological investigations it seems impossible to say in any given instance what ^{14}C results represent in terms of net or gross photosynthesis. What does seem clear is that, providing the proportion of heterotrophic organisms is not unduly high, fixation of ^{14}C in the light should not be corrected by subtraction of values for fixation in dark bottles (see also Morris et al., 1971a). The Krebs tricarboxylic acid cycle, the reversible decarboxylations of which are largely responsible for the incorporation of ^{14}C from CO_2 in the dark, evidently continues in algae at the same rate in the light as in the dark (Marsh et al., 1965). Nevertheless, all the work on the path of carbon in photosynthesis goes to show that labelling of intermediates of this cycle from $^{14}CO_2$ is suppressed in the light and a different set of intermediates becomes labelled instead (Rabinowitch, 1956). Moreover, it is common experience that the incorporation of ^{14}C into particulate matter in the dark in samples at the bottom of the photic zone often equals or exceeds that in corresponding light samples. The significance of this dark fixation is obscure. Sorokin (1965, and in Vollenweider, 1974) has argued that it gives a measure of heterotrophic assimilation by bacteria.

For some hours, sometimes up to eight (Fogg et al., 1965), rates of photosynthesis determined by the ^{14}C method may remain constant. However, with longer exposures there may be deviation from linearity. On the basis of experiments with freshwater phytoplankton, in which they compared the sum of values obtained from consecutive short exposures with single long term exposures, Vollenweider and Nauwerck (1961) concluded that, in order to avoid ^{14}C losses by various processes, in situ experiments should not last for more than 4 to 6 hours.

Variations between replicate determinations by the radiocarbon method mainly arise from sampling, the difficulties of which have been outlined in

Section 14.3. Dyson *et al.* (1965) made replicate estimations with the same samples of surface water on three different occasions and obtained a mean coefficient of variation of 17%. The mean coefficients of variation between samples taken from the same depth in quick succession, and between samples taken from a water column in successive casts were respectively 21% and 11%. They concluded that two measurements might be considered significantly different if one were 80% greater than the other.

Watt (1971) has adapted the radiocarbon technique to enable primary production rates to be estimated for individual species in a mixed phytoplankton population. This is done by combining liquid scintillation counting with species identification, cell counts, and silver grain counts from radioautographs. Data from the north-east Atlantic show primary production by nannoplankton species to be usually greater than their biomass would suggest. Several species were found to contribute significantly to the biomass but not to primary production.

14.4.4. COMPARISONS OF THE OXYGEN AND RADIOCARBON METHODS

Since these two methods are subject to different errors they should not be expected always to be in agreement, but with active, relatively concentrated, algal suspensions under moderate conditions they usually yield concordant results with a photosynthetic quotient of a little more than unity. This has been found for laboratory cultures (Ryther, 1956; Steemann Nielsen, 1965; Fogg, 1963) and for natural phytoplankton in a eutrophic tropical estuary for the majority of sets of results obtained over a year (Qasim *et al.*, 1969). For another tropical situation, a bay of Nosy-Be (Madagascar), Sournia (1968) found a good agreement between the radiocarbon and oxygen methods but with an apparent value for the photosynthetic quotient of 2·76. However, McAllister (1961b) found considerable discrepancies when shade adapted cultures of *Syracosphaera carterae* were exposed to high light intensities. In this example oxygen was actually taken up, presumably as a result of photorespiration and photo-oxidation exceeding photosynthesis while ^{14}C fixation remained relatively high probably because of incorporation by carboxylation reactions without subsequent reduction. The results of Takahashi *et al.* (1971) with *Phaeodactylum tricornutum* grown at low light intensity differed considerably from those of McAllister. However, their results showed a marked time dependence. When this was allowed for, the two methods gave concordant results. McAllister (1961b) exposed oxygen and ^{14}C bottles for the same time so that the effect he observed was evidently real. Strickland and Terhune (1961) made determinations of carbon dioxide removed from the water and carbon fixed in cell material as well as oxygen and radiocarbon measurements in *in situ* experiments in the sea using a large plastic bag. Good agreement was found between carbon dioxide uptake and oxygen output, with a photosynthetic quotient of 1·1, but both the absolute carbon determination and the radiocarbon technique consistently showed that only about 60% of the carbon dioxide disappearing from the water appeared as carbon within the phytoplankton cells (McAllister *et al.*, 1961; Antia *et al.*, 1963). As mentioned above, this seems to be a clear indication of liberation of extracellular products of photosynthesis, since the oxygen method measures total photosynthesis whereas the unmodified radiocarbon technique measures only fixation in the cells. McAllister *et al.* (1964) concluded from laboratory experiments with *Skeletonema costatum, Dunaliella tertiolecta* and other algae that there was a similar lack of agreement, this varying in extent with species, physiological condition and light intensity.

To sum up, it should be realized that both methods have serious limitations. Provided that primary production is enough for accurate measurements to be made in 24 h or less, the oxygen method seems to give more meaningful results. The radiocarbon method is, however, the more sensitive, and the only one which can be satisfactorily used in oligotrophic waters, but allowance should then be made for liberation of extracellular products of photosynthesis.

14.4.5. THE CHLOROPHYLL METHOD

Photosynthesis is dependent on chlorophyll so that it may be supposed that the chlorophyll content of a water sample may be taken as an index of its photosynthetic potential. Since the chlorophyll content of sea water is readily determined, it is tempting to use this assumption as a basis for the estimation of primary productivity in spite of its questionable theoretical validity. With a process so complex as photosynthesis, no constant relationship is to be expected between the photosynthetic rate (even under standard conditions of light intensity) and any single participant factor in the process. Nevertheless, Ryther and Yentsch (1957) have suggested the use of the following expression for the determination of the rate of photosynthesis, P, in g C m^{-2} day^{-1}

$$P = \frac{R}{k} \times C \times 3·7 \qquad (14.13)$$

R being the relative photosynthesis, determined empirically, for the appropriate value of surface radiation, k the extinction coefficient per metre of the water, C the g of chlorophyll m^{-3}, and the factor 3·7 a mean value for the assimilation number. Values estimated by this method and those determined by the oxygen technique agreed reasonably well for a number of sea areas in which primary productivity varied by an order of magnitude. This agreement

may be somewhat misleading since some of the *in situ* oxygen determinations
were made over rather long periods. However, Platt and Subba Rao (1973)
found for a spring phytoplankton bloom in an inlet on the coast of Nova
Scotia that 64% of the variance in their photosynthetic index k_b^z (see p. 430)
could be accounted for by changes in the concentration of chlorophyll *a* as
compared with 9% by depth, 7% by cell number and 4% by nutrients.

For a particular situation the chlorophyll method may yield satisfactory
results, but it should not be used without full consideration of the sources of
error. As indicated in Section 14.2.6. the assimilation number varies considerably in laboratory grown algae and there is confusion as to the effect on
it of conditions such as nutrient deficiency. The problem is similar for natural
populations. Platt and Subba Rao (1973) have tabulated values from the
literature. Although some workers, e.g. Steele and Baird (1961), have obtained
fairly consistent values, others, e.g. El-Sayed (1967), have recorded a variation
of over a hundred-fold which cannot, at present, be related to particular
factors. Strickland (1965) concluded that nutrient deficiency in the sea
reduced the assimilation number of the phytoplankton, but other observations suggest the contrary. Thus, Steele and Baird (1961) recorded seasonal
trends in assimilation number, with rises from about 1 in the spring to
about 2 in the summer and a decline in the autumn, in two areas of the North
Sea, consistent with the idea that nutrient deficiency causes an increase.
Taguchi (1970) found a similar trend in Akkeshi Bay, Hokkaido, the assimilation number being 0·83 in February, when nutrients were at a high level, but
6·7 in May, when the spring maximum was beginning to decline. Steemann
Nielsen and Hansen (1959) and Yentsch (1965) found a relationship between
the assimilation number and the light intensity to which the plankton had
been adapted, cells from the bottom of the photic zone showing low assimilation numbers. Statistical analysis of the extensive data obtained by El-Sayed
(1967) indicated that the observed variations could not be accounted for by
the presence of inactive chlorophyll or chlorophyll derivatives in the samples.
No difference in assimilation number for Argentine shelf stations was found
between summer and winter. In the Drake Passage, however, there was a
statistically significant difference between the summer (9·98) and winter
(0·83) values.

It is always necessary, of course, to distinguish between photosynthetically
active chlorophyll and inactive chlorophyll or its derivatives in moribund
material. Ryther and Yentsch (1957) measured chlorophyll spectrophoto-
metrically at the chlorophyll *a* absorption peak in the red; this method does
not entirely exclude interference from degradation products. Yentsch (1965)
determined chlorophyll and its decomposition product, phaeophytin,
separately by a fluorimetric method and found that the presence of increasing

amounts of inactive phaeophytins as the samples were taken from deeper down was one factor, among others, contributing to the decrease in assimilation number with depth. However, as Anderson (1964) found in N.E. Pacific waters, this source of error is usually negligible. It should also be remembered that light absorbed by the accessory pigments—chlorophyll c, carotenoids and phycobilins—may also be effective in photosynthesis. Currie (1958), in fact, found a better correlation of photosynthesis with total pigments than with chlorophyll a alone.

14.4.6. OTHER METHODS

As has been seen (p. 415) accurate estimation of photosynthesis by determination of carbon dioxide uptake from unconfined water *in situ* is usually not possible. Under favourable circumstances carbon dioxide uptake may be determined from pH changes or by direct analysis, but because of the comparatively high carbonate content, the accuracy is not high, and the method is impracticable if the production rate is less than $250 \, \text{mg} \, \text{C} \, \text{m}^{-3}$ day^{-1} (Strickland, 1965). Antia *et al.* (1963) used the pH method successfully in a large volume of sea water enclosed *in situ* in a plastic bag. Indirect estimates of primary production on the basis of uptake of nutrients such as nitrate and phosphate have been made by Cooper (1958), Steele (1958) and Dugdale and Goering (1967), for example, but in view of the rather flexible relationships between photosynthesis and nutrient uptake, which were discussed in Section 14.2, and the rapid recycling of elements such as phosphorus which is now known to occur (Watt and Hayes, 1963) it does not appear that this approach can do more than give a lower limit of net primary production. Zlobin (1972) however, has computed primary productivity by assuming an autocatalytic type of phosphate uptake during the spring bloom and making allowance for recycling.

14.5. THE RELATION OF PRIMARY PRODUCTION RATES *IN SITU* TO ENVIRONMENTAL FACTORS

14.5.1. THE DEPTH PROFILE

Radiant energy decreases exponentially as it penetrates through optically uniform water (Fig. 14.2a) according to the expression

$$I_z = I_0 \exp\left(-k_e z\right) \tag{14.14}$$

where I_0 is the intensity of the light crossing the water surface, I_z the intensity at depth z and k_e the vertical extinction coefficient for the wavelength or wavelengths being considered. From this relationship and that between rate of photosynthesis and light intensity described in Section 14.2.1.1. it is possible to predict that the variation of the instantaneous rate of photosynthesis with depth may be represented by a family of curves of which one limit will be that for a dull day (*ca.* 85 J cm^{-2} day^{-1}) when the incident light is of an intensity already limiting for photosynthesis and the photosynthesis-depth curve will parallel that for for light intensity-depth. On a bright day (>859 J cm^{-2} day^{-1}) intensities near the surface will be inhibitory so that the rate of photosynthesis increases with depth to reach a maximum which may be as much as 5 m or more below the surface, below which light becomes limiting and the exponentially decreasing curve is followed (Fig. 14.2b). The form of the depth curve will, of course, vary during the day so that at a given depth near the surface under bright conditions the time course of photosynthesis will show peaks in the morning and evening, with a mid-day depression, and lower down, where light is always limiting, there will be a single maximum at mid-day. Such curves have been obtained by Takahashi *et al.* (1973) for the Fraser River estuary. Consecutive determinations of photosynthesis at intervals over a day have, however, rarely been made and the usual photosynthesis-depth curves are based on determinations made over a day or half-day.

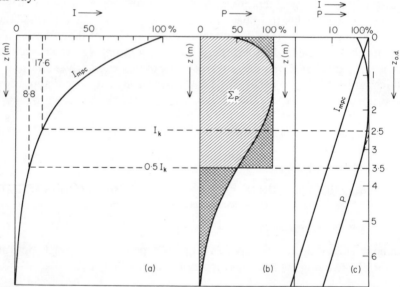

FIG. 14.2. Talling's model of the photosythesis/depth relationship as modified by Rodhe to take account of light inhibition. 0.5 I_K is assumed to be at I_{mpc} (intensity of the most penetrating component) $= 8.8\%$ of full surface illumination, which corresponds to $Z_{o.d.} = 3.5$. (a) Light penetration, (b) photosynthesis and (c) light penetration and photosynthesis plotted logarithmically against optical depth. For further explanation see text (Redrawn after Rodhe, 1965).

Usually, the object of primary productivity determinations is to obtain a value for the photosynthesis accomplished in unit time in the water column under unit area of water surface. This may be achieved by planimetric integration of photosynthesis-depth curves, but attempts have been made to construct mathematical models with the aid of which integrated values may be obtained. Talling (1957b, 1961) found that if surface inhibition and saturation effects are neglected, the total photosynthesis in a homogenous water column may be represented by the expression:

$$\Sigma P = \frac{nP_{max}}{k_e} \ln \frac{I_0}{0.5\, I_k} \tag{14.15}$$

in which n is the density of the phytoplankton population and the other symbols are as defined above (p. 425) and in Fig. 14.1. This implies that the integrated value for photosynthesis is numerically equal to the area of a rectangle one side of which is proportional to the maximum photosynthetic rate shown by the population and the other to the depth, $z_{0.5\, I_k}$, at which light saturation of photosynthesis ceases. Rodhe (1965) used this as a basis for comparison of the primary productivity of lakes. When observed rates of photosynthesis in these were plotted as a percentage of the maximum rate against optical depth $z_{o.d.}$ (each unit of which corresponds to a layer in metres which causes a halving of the intensity of wavelengths of maximum penetration) the curves, even for the most diverse lake types, were more or less coincident, showing the same response to light. Values obtained by Steemann Nielsen and Aabye Jensen (1957–59) for sea areas of widely different productivity, when plotted in the same way, gave an almost identical picture. The model, which is represented in general terms in Fig. 14.2, gave estimates for twelve lakes, varying almost 40-fold in primary productivity, agreeing satisfactorily with observed values, individual deviations not exceeding $\pm 7\%$. Often, however, it is not possible to determine $0.5I_k$, but since it is found that this value is usually attained at an optical depth between 3.0 and 3.5, where the light intensity corresponds to between 12.5 and 8.8% of the incident value, use can be made of the approximation that $z_{0.5\, I_k}$ equals the depth in metres to which 10% of the light penetrates, i.e.

$$\Sigma P = P_{max} \cdot z_{0.1\, I\, mpc} \tag{14.16}$$

Where $z_{0.1\, I\, mpc}$ is the depth at which the intensity of the most penetrating component, usually green, is reduced to 1/10 of its surface value. This gives a less accurate but still acceptable estimate of primary productivity per unit area.

Vollenweider (1965) stressed the importance of distinguishing between *instantaneous rate integrals* and *day rate integrals*. As has been seen (Section

14.4.3.) the radio-carbon method may be subject to error if experiments are continued for longer than 4 to 6 h. However, if determinations are made over short periods only, there remains the problem of converting the results to day rates. Since the specific shape of the photosynthesis-depth curve is not of great importance for estimation of the day rate in comparison with the diurnal variation in light intensity, the most straightforward way of doing this is to obtain an irradiance curve for the day in question and to determine plani-metrically the area under this curve and that fraction of it during which the exposure was made. The ratio between these two areas can then be used as a conversion factor to estimate the day rate from the exposure rate (Vollen-weider, 1974). Such conversions would be facilitated if the durations of exposures, instead of being a fixed time, were chosen to be a standard propor-tion of the light day, e.g. by dividing the period between sunrise and sunset into five equal portions and making exposures over periods 2 and 3. Calcula-tion has shown that about 55–60% of the total daily production takes place over these two periods regardless of the magnitude of the rate decrease during the day (Section 14.5.2.).

Such models, although adequate under many circumstances, are over-simplified. The factors concerned in the inhibition of photosynthesis near the water surface are not fully understood. Often inhibition may be more apparent than real, the decrease in rate of photosynthesis towards the water surface being merely the result of decrease in cell numbers as a result of settling out. Ultra-violet and visible radiation are both concerned in the inhibition in freshwater (Amah Belay, unpublished). Culture experiments show that intensities of visible light of about $\frac{1}{3}$ full sunlight begin to be inhibitory (Qasim et al., 1972). Takahashi et al. (1971) have shown with both cultures of Phaeodactylum tricornutum and natural populations that photoinhibition is due to an actual decrease in photosynthetic rate rather than to any increase in excretion of products. Inhibition in P. tricornutum only developed in material grown at low light intensity and exposed to high light intensity for 1 or 2 hours. With waves acting as cylindrical lenses, cells may be exposed to intensities several times that of full sunlight for periods approaching one second, but from laboratory experiments it does not appear that such short exposure to intense light has any inhibitory effect (Rabinowitch, 1945). Conditions of oxygen saturation have been found to depress photosynthesis in freshwater algae in culture (Stewart and Pearson, 1970) and may well have similar effects in situ. Nutrient deficiency appears to predispose phytoplank-ton to photoinhibition (Takahashi et al., 1973). Susceptibility to these factors varies at different stages in the cell cycle (Tamiya, 1964) and the inter-actions may be complex. Adaptation, in situ, of marine phytoplankton to light intensity similar to that found in laboratory cultures (Section 14.2.7.) has been demonstrated by Ryther and Menzel (1959) and by Steemann Nielsen and Hanson (1959). This adaptation occurs only when the water column is stable and samples taken from different depths under conditions of mixing all show the same response of photosynthesis to light intensity. Adapted cells from near the surface show little inhibition in full sunlight whereas those taken from a depth to which only 1 % of the incident light penetrates are considerably inhibited by intensities of only one quarter full sunlight, although at low intensities they are photosynthetically more efficient than the surface cells. The effect of this adaptation is to increase photosynthetic rates near the surface and in the lower part of the photic zone under bright conditions, but to decrease rates near the surface under dull conditions. These deviations are manifest on comparing observed photosynthesis-depth curves with the standard curve provided by Rodhe's model (Fig. 14.2).

Attempts have been made to construct models to take account of inhibition and adaptation. Fee (1969) has put forward a numerical solution of Vollen-weider's model, providing for varying types and degrees of inhibition of photosynthesis by light. A digital computer is used to estimate the inhibition statistics and to integrate the model numerically over time and depth. The data required are the photosynthesis-depth curve, P_{max}, k_e, and the variation with time of irradiance. Vollenweider (1970) has proposed more refined models which take into account adaptation of phytoplankton to changing light conditions. A model used by Takahashi et al. (1972) takes the major factors, light intensity, temperature and nitrate into account on the assumption that only one is limiting at any given time and place. This model gives reasonable agreement with observed values except for surface samples, for which the model seems to over-emphasize photoinhibition.

In equation (14.15) the terms n and k_e are, to some extent, inversely related, since phytoplankton itself absorbs some of the penetrating light. Hence there is a tendency for low rates of photosynthesis per unit volume to be offset by photosynthesis taking place over a greater depth. Although the range of chlorophyll concentrations encountered in the sea is more than a thousand fold, that of primary productivity per unit area is less than fifty-fold. In water free from particles and coloured organic solutes, the depth of the photic zone (to which 1 % of the incident light penetrates) would be about 140 m. In one of the least productive areas of the oceans, the Sargasso Sea, the depth of the photic zone is about 120 m. Steemann Nielsen (1963) gave a curve relating depth of the photosynthetic layer and the maximum rate of photosynthesis per unit area. Takahashi and Parsons (1972) found that the standing stock of chlorophyll a per unit area decreases logarithmically with the depth of the euphotic zone and have given a table of calculated daily net photosynthesis for various amounts of chlorophyll a under different levels of irradiance.

P

According to this, primary productivity reaches a maximum of $1.8 \, \text{g m}^{-2}$ day^{-1} at the maximum chlorophyll value of $260 \, \text{mg m}^{-2}$ chlorophyll a, assuming an assimilation number of 1.0 in a water column of less than 1 m depth.

Platt (1969) has defined an index, k_b^z, (dimensions m^{-1}) which represents that contribution to the absorption coefficient of light energy at any depth which is due to photosynthesis. It should equal the absorption of light by photosynthetic pigments multiplied by a factor representing the efficiency of light utilization in photosynthesis, and is approximately equal to the ratio of photosynthesis to light energy at a given depth when each is measured in calories. This may be useful as an index for comparison, but there are obvious difficulties in its determination.

14.5.2. DIURNAL VARIATION IN PHOTOSYNTHETIC RATES

As the light intensity varies during the day so the photosynthetic rate at any given depth will follow the characteristic course already described (Section 14.5.1) provided that the average cell in the population maintains a constant metabolic pattern and activity. Such constancy is, however, unlikely to occur; photosynthesis and growth will produce changes in the external and internal environment which may affect the rate of photosynthesis, and endogenous rhythms in metabolism may manifest themselves if the population becomes synchronized to any extent in response to the natural alternation of light and dark. It is well established that diurnal rhythms in photosynthetic activity are shown by phytoplankton. Doty and Oguri (1957), using the radiocarbon technique, found that a sample of tropical phytoplankton taken from the surface had a photosynthetic activity at 8 a.m. nearly 6 times as great as that which it showed under the same conditions at 7 p.m. Others have obtained similar results, e.g. Yentsch and Ryther (1957) reported a two-fold variation during the day in samples taken off Woods Hole. Similar rhythms have been recorded for freshwaters (e.g. Ohle, 1961; Vollenweider and Nauwerck, 1961). The amplitude of the variation appears to vary with latitude. Thus, variations of as much as ten-fold were observed at the equator whereas no variation appears to occur at 75° (Doty, 1959), but data tabulated by Platt and Subba Rao (1973) show only a rather irregular trend of this kind. The magnitude of these variations indicate that they must be associated with the photosynthesis itself rather than with respiration and they are accompanied by corresponding changes in chlorophyll (Yentsch and Ryther, 1957), although variation is still evident when rates are reported per unit amount of chlorophyll (Sournia, 1967). Various explanations of the mechanism have been put forward. It might, in part, be due to nutrient consumption during intense assimilation. Steemann Nielsen and Jørgensen (1962) considered that the rate of chlorophyll synthesis is reduced at high light intensities and that, because of grazing, a greater proportion of the pigment is in an inactive form in the later part of the day. Ohle (1961) found no evidence of endogenous rhythms in the freshwater populations which he studied, and he postulated that accumulation of waste products within the cells was the cause of the decline. Nevertheless, it seems likely that endogenous rhythms, which do occur in marine phytoplankton (Sweeny and Hastings, 1962; Eppley *et al.*, 1971), are the principal factor involved. If division of cells takes place during the hours of darkness, then from laboratory studies (Section 14.2.6.) it would be expected that the young cells predominating in the morning would be more active in photosynthesis than the mature cells produced later in the day. A 12 hours light: 12 hours dark cycle such as is found in the tropics would be expected to produce a greater degree of synchrony than the almost continuous daylight of high latitudes in the summer.

Vollenweider (1965) has proposed a mathematical model which takes into account the asymmetry of the daily course of photosynthesis.

14.5.3. CARBON DIOXIDE

Free carbon dioxide and undissociated carbonic acid are taken up directly by photosynthesizing algal cells, but there is still uncertainty as to whether bicarbonate ion is directly utilized by the generality of species. Degens *et al.* (1968) concluded from experiments on the effect of temperature on the relative rates of uptake of ^{12}C and ^{13}C that *Skeletonema costatum* and *Cyclotella nana* make direct use of molecular carbon dioxide rather than other forms. If only undissociated forms are taken up, these may sometimes be rate limiting in sea water, as Paasche (1964) found for *Coccolithus huxleyi* at high light intensities. On the other hand, the bicarbonate and carbonate ions form a reservoir of inorganic carbon which is rarely reduced appreciably by photosynthesis in the open sea. Talling (1960) found no reduction in the rate of photosynthesis of *Chaetoceros affinis*, at cell concentrations of 19 to 46 mm^{-3} in natural sea water, after incubation for periods of 1–3 h *in situ*. Deuser (1970), however, found a measurable reduction in available carbon during periods of intense phytoplankton growth in the Black Sea.

14.5.4. NUTRIENT CONCENTRATIONS

Although there can be no doubt that there is a general proportionality between primary productivity and the availability of nutrients at any given moment there may be little relationship between the rate of photosynthesis by the phytoplankton and the concentration of particular nutrients. This is to be expected in view of the capacity of algae to store nutrients, such as phosphate and nitrogen, surplus to requirements and to accumulate photosynthetic products, not taken up by growth, as carbohydrate or fat. Platt and Subba Rao (1973) have pointed out that using replicate *in situ* determinations of primary productivity, statistically significant differences can be detected between results from stations within a limited sea area. Such differences may be due to nutrient supply, but refined experimental design would be necessary to establish this.

Stimulation of photosynthesis (as measured by the ^{14}C technique) by added nutrients provides a sensitive means of assay for nutrient limitation in natural waters. In water samples from the Sargasso Sea, Ryther and Guillard (1959) found in 24 h experiments of this type that, surprisingly, nitrate or phosphate had little effect, but stimulation occurred on addition of silicate or of an iron-trace element mixture. Thomas (1969) however, found that nitrogen was the most likely limiting element in the eastern equatorial Pacific Ocean (see p. 439) and it seems likely that this is true for most sea areas (Strickland, 1972). McLaughlin (in Oppenheimer, 1966) has reported the use of a similar technique for assaying for vitamin B_{12} deficiency. Addition of this vitamin to B_{12}-depleted cells caused increases in $^{14}CO_2$ assimilation after 24 hrs. The effects of increased availability of nutrients *in situ* is most clearly exemplified by results obtained for the Sargasso Sea. Here a pronounced peak in primary production in March and April has been related to upwelling (Menzel and Ryther, 1960) bringing in nutrients among which vitamin B_{12} appears to be of special importance (Menzel and Spaeth, 1962). However, in other sea areas vitamin concentrations do not seem to limit primary productivity (Strickland, 1972).

Morris *et al.* (1971b) have used enhancement of dark $^{14}CO_2$ fixation by ammonium nitrogen as a measure of the degree of nitrogen deficiency in natural phytoplankton. They concluded that although nitrogen supply may limit population size, in much the same way as it might in a chemostat, the populations which they studied were not physiologically nitrogen deficient.

14.5.5. LIBERATION OF EXTRACELLULAR PRODUCTS OF PHOTOSYNTHESIS

It has already been seen that this is a normal occurrence in the sea. From observations made in freshwater as well as in marine environments it seems possible to make three valid generalizations concerning the extent of this liberation:

(1) It is not dependent on the presence of particular species or classes of algae. It has been found, for example, in cultures of *Tetraselmis* (Prasino-phyceae) and *Dicrateria* (Haptophyceae) as well as in natural populations dominated by diatoms and by Dinophyceae (Samuel *et al.*, 1971).

(2) Although the relative extent tends to be constant throughout the depth of the photic zone it rises markedly towards the surface when photosynthesis is light-inhibited. This is shown most clearly by results obtained for freshwater (Watt, 1966), but has been reported for marine phytoplankton by Horne *et al.* (1969) and by Thomas (1971). Occasionally as much as 90% of the carbon fixed may be excreted. A similar increase is observed in laboratory experiments at high light intensity and the phenomenon seems best interpreted as an overflow resulting from light inhibition of the carbon fixation cycle being less relative to that of the uptake of its products in growth.

(3) Although the absolute amount of extracellular products decreases as the phytoplankton population density decreases, the relative amount rises. Anderson and Zeutschel (1970) found the percentage of the total carbon fixed which was released in this form was greatest in oligotrophic waters in coastal and offshore areas of the north-eastern Pacific, e.g. as much as 49% compared with 10% for most eutrophic waters. They gave two equations relating x, the dissolved organic matter released in $mg\,C\,m^{-3}\,(\frac{1}{2}\,day)^{-1}$, and y, the particulate fixation in $mg\,C\,m^{-3}\,(\frac{1}{2}\,day)^{-1}$. For individual measurements off the Washington and Oregon coasts between 15 and 31 July 1969 they found:

$$\log x = 0.68 \log y - 0.46; r = 0.865 \tag{14.17}$$

and for the euphotic zone:

$$\log x = 0.66 \log y + 0.03; r = 0.932 \tag{14.18}$$

r being the correlation coefficient. Thomas (1971) found a similar trend ranging from 7% in estuaries of the Georgia coast to up to 21% in coastal waters off south-eastern U.S.A. and even to 44% in the Sargasso Sea. Horne *et al.* (1969) found only 1–2% in highly productive Antarctic waters with between 10 and 50 mg chlorophyll *a* m^{-3} whereas Samuel *et al.* (1971) recorded between 5% for samples with 117 mg chlorophyll *a* m^{-3} and 18.9% for those with 0.54 mg chlorophyll *a* m^{-3} for inshore tropical waters near Cochin. Again these findings parallel those of laboratory experiments. Possibly the phenomenon is partly the result of a quasi-equilibrium between intracellular and extra-cellular concentrations (Section 14.2.4), but again it appears to be largely an overflow effect resulting when the uptake of photosynthetic products during growth is limited by nutrient deficiency (Ignatiades, 1973).

14.5.6. SEASONAL PERIODICITY

Seasonal variations in phytoplankton biomass tend to conform to patterns which vary according to latitude (Cushing 1959b). In polar seas there is a single peak of phytoplankton in early summer in which the population density may be as much as a thousand times as great as that in winter. In temperate regions a similar peak develops in spring to be followed by a decline in summer and, usually, by a secondary peak in autumn. In tropical seas the usual state is one in which only minor oscillations, of five-fold or so, of phytoplankton biomass occur. The factors determining these seasonal patterns are many, and the interactions between them are complex (see Fogg, 1965, for a discussion), but to a first approximation they may be explained in terms of Riley's equation (14.10) in which the major determinants are a photosynthetic factor, respiration and grazing, the latter two varying according to temperature. Since photosynthesis is assumed to be proportional to biomass in this equation, it should be expected that the seasonal variation in primary productivity would show similar patterns according to latitude. Although series of measurements continuing over a growing season in a given sea area have rarely been made, the information available largely bears out this expectation. Off Signy Island, South Orkney group (60°43'S), Horne et al. (1969) found that primary production paralleled algal biomass during a 60-day period in the austral summer, showing a single peak in early February. Winter values were not determined, but were presumably negligible and the range observed was between 75 and 250 mg C m^{-2} h^{-1}. Anderson (1964) found that the primary productivity in oceanic waters off the Washington and Oregon coasts (40°–48°N) showed a winter minimum, followed by a large increase in spring, then a summer minimum and a smaller autumn pulse. However, there was little seasonal change in the standing stock, and the range of primary production values observed was only 5 to 25 mg C m^{-3} day^{-1}. Ryther and Yentsch (1958) concluded that variation in primary productivity of between 5 and 10-fold was typical for waters off New York (39° to 41°N). In a tropical estuary on the S.W. coast of India (9°58'N) there was no distinct pattern of variation in primary productivity and a range from 121 to 586 mg C m^{-2} day^{-1} (Qasim et al., 1969). A gradation between little seasonal difference in primary productivity at 4°S to a well marked periodicity with maxima in spring and autumn and minima in winter and summer at 12°S was observed by Guillén et al. (1971) for coastal Peruvian waters. However, in view of the complications introduced by upwelling in this region this apparent perfect fit with the theoretical picture must be regarded as probably fortuitous.

Apart from the amount of photosynthetic biomass, the other factor which is clearly important in determining the seasonal periodicity of primary production is the amount of solar radiation. Ryther (1956) calculated the daily relative photosynthesis in the euphotic zone for days ranging between the darkest and brightest days of the year, and obtained a curve of gradually decreasing slope which did not, however, reach saturation. Nevertheless,

Qasim *et al.* (1969) concluded that in a tropical estuary at latitude 9°58′N, monthly variations in total solar radiation were not sufficiently large to produce seasonal changes in production. Another important factor is loss of phytoplankton cells by sinking or grazing. McAllister *et al.* (1960) who determined productivity over a six-week period in July and August, at a latitude 15° 50′ N in the N.E. Pacific Ocean found a fairly uniform rate of primary production. Nutrients, temperature and light did not limit phytoplankton growth, and it appeared that primary production was being kept at a relative low level by grazing and sinking. Menzel and Ryther (1960) have cited an example of a pronounced peak (in an otherwise uniformly low rate of primary production maintained for the greater part of the year) resulting from the introduction of nutrients by upwelling.

An attempt to correlate the seasonal variation in primary productivity with simultaneously determined values of all the major factors that may be envisaged as affecting it, has been made in the freshwater Lake Maggiore in Italy by Goldman *et al.* (1968). Stepwise multiple regression analysis was carried out with data obtained by measuring 28 variables at 9 depths on 14 different days between May and December. A close relationship was found between primary productivity and the solar radiation incident on the water surface during each ^{14}C incubation period, but the correlation was not absolute. In an attempt to determine the factors causing the divergencies, the values were transformed to eliminate the effects of unequal vertical distribution of phytoplankton and differences in the absolute value of photosynthesis. The resulting values of relative photosynthesis, when plotted against light intensity for each occasion, revealed no simple consistent relationship and there was no indication in the correlation matrix that any of the 27 other variables could adequately account for the unexplained variability in primary productivity. It was concluded that undetermined variables such as the physiological state of the algae must have had important effects. As has already been emphasized (p. 405), the physiological state of algae may bear little relationship to the nutrient concentrations in the water surrounding them at a given moment, but possibly it might be fruitful to investigate the correlation between relative photosynthesis and nutrient concentrations at some previous time.

Similar studies in the sea have been less extensive. Platt and Subba Rao (1970) made observations during the spring diatom bloom in St. Margaret's Bay, Nova Scotia, and found that 64% of the variation in the photosynthetic index k_b (see p. 430) could be accounted for by variation in chlorophyll concentration and, as in Lake Maggiore, primary productivity was not correlated significantly with nutrient concentrations. Perhaps the most interesting outcome was confirmation that there is a negative correlation between primary productivity per unit of biomass (measured by k_b) and species diversity.

14.6. REGIONAL STUDIES OF PRIMARY PRODUCTIVITY

14.6.1. GENERAL REMARKS ON SPATIAL AND TEMPORAL VARIATIONS

It has been seen that the estimation of the primary productivity which takes place below unit area of sea surface in a 24-h period is fraught with complications and great effort must be expended even to get a moderately accurate result. Because of patchiness in the distribution of phytoplankton and variations in weather, estimates made on consecutive days may differ considerably and give an impression of apparent randomness (Steele, 1961). It is necessary to make regular determinations at frequent intervals for two or more seasons if a reliable assessment of the primary production of a given sea area is to be obtained.

Fortunately, even a few determinations can be used to give an approximate picture of the spatial and temporal variation of primary productivity in a given sea area if ancillary information is available and used intelligently. Day by day variations caused by differing light conditions are small during summer in temperate regions, but large in winter (Steemann Nielsen and Hansen, 1961). Berge (1958), who made an extensive series of determinations of primary productivity in the Norwegian Sea in late May and June, showed that production rates were characteristic of water-masses. A productive area extending southwards of Spitzbergen towards Iceland was found to be coextensive with water of salinity between 35·00 and 35·15‰ and also to coincide with a herring feeding area. Thus, position is not crucial in making determinations near the centre of a water mass and it suffices to make sufficient measurements to establish the seasonal fluctuation. Primary production at a station where two different water masses meet will, however, show great variation within short spaces of time, according to which predominates. Knowledge of the hydrography of the area will assist in selecting stations to give most information. Thus, a study in the Baie des Chaleurs, Canada (48°N 65°W) showed that the horizontal pattern of primary productivity fits that to be expected theoretically from the existence of a cyclonic gyre system. High productivity is associated with cold nutrient-rich water brought close to the surface by the gyre. East of the gyre, in the Gulf of St. Lawrence, primary production is less (Legendre and Watt, 1970). The conditions off a coast where upwelling takes place may be more complicated. Strickland et al. (1969a) encountered two distinct kinds of conditions in coastal waters off northern

Chile and Peru, a region of intense upwelling. In one, the water contained abundant nutrients and an actively growing standing stock of phytoplankton the biomass of which was, however, relatively low (blue water). The other had lower nutrient concentrations and an abundant standing stock (brown water). The brown water had the higher productivity per unit area. The reason for these differences could not be established with certainty, but it appears probable that the phytoplankton in the blue water is kept in check by grazing, the effect of which is perhaps augmented by dilution of the phytoplankton population with upwelling phytoplankton-free water. As pointed out by Goldberg (1971) the introduction of river water may affect primary productivity in various ways:

(a) By bringing in, diluting, or removing by sedimentation plant nutrients.

(b) By altering light penetration by bringing in suspended matter or dissolved coloured substances.

(c) By affecting the stability of the water column; a low density surface layer increases stability and may increase production by reducing the tendency of cells to be carried below the critical depth for photosynthesis. One of the major effects of the Columbia River discharge is to alter the timing of events in the seasonal cycle by affecting the depth of the mixed layer. However, the total annual production was similar inside and outside the plume. The nutrients introduced by the river water appeared to be used in high production areas immediately off the mouth of the river (Anderson, 1964).

The possibility of long-term changes in the primary productivity of a given sea should not be overlooked. Russell et al. (1971) have reported changes in the occurrence of nutrient salts, plankton and young fish which have taken place off Plymouth since 1924. This area supported a herring fishery in the 1920's but this declined in the 1930's. In recent years there has been an increase in nutrient salt concentrations and a corresponding increase in primary production from which it was correctly predicted that herring would return. This seems to be one facet of the general changes in the North Atlantic which Robinson (1970) has demonstrated by Principal Component Analysis of results obtained with the Continuous Plankton Recorder during the period 1958 to 1969. Although there has been an increase of phytoplankton in the English Channel, over the North Atlantic as a whole there has been a marked decline, and there has been a progressive delay in the timing of the spring outbreak in both the North Sea and the North Atlantic during the period 1948 to 1969. Possibly these changes are related to variations in overspill of Arctic water into the Atlantic (Cooper, 1955).

Although it is scarcely possible to review all the work done on the primary productivity of different sea areas, it seems desirable to give a few examples and key references. Unless otherwise stated the values given are radiocarbon values uncorrected for respiration.

14.6.2. TEMPERATE REGIONS

The most extensive studies on the primary production of the North Sea were made prior to the general use of the radiocarbon method. Steele, (1956, 1958) developed a mathematical model of the plant production cycle by means of which he made theoretical estimates of production and compared them with estimates based on consumption of phosphate. For the northern North Sea (58 to 61°N, 2°W to 3°E) the theoretical estimates ranged between 60 and $120 \text{ g C m}^{-2} \text{ yr}^{-1}$, whereas the observed values varied between 45 and $110 \text{ g C m}^{-2} \text{ yr}^{-1}$ in a broadly similar pattern. Because recycling of phosphate was considered negligible, these latter estimates are probably somewhat low. In the same sea area a comparison between the phosphate and radiocarbon methods showed reasonably good agreement (Steele, 1957) but, in view of probable errors in the standardization of the tracer solution used (see Steemann Nielsen, 1965) and neglect of the liberation of extracellular products, estimates made by the latter method are also likely to have been low. From radiocarbon measurements made from a Danish light-ship in the Kattegat (58° 50′N 11° 40′E) over a period of 7 years, Steemann Nielsen (1964b) estimated the gross primary production to be $67 \text{ g C m}^{-2} \text{ yr}^{-1}$ on average. An interesting feature found at this station was the high rate of production occurring during the winter as a result of both stratification of the water column and ineffective grazing by zooplankton. In the Great Belt (55° 20′N 11° 0′E) another series of observations made over 5 years gave an average of $59 \text{ g C m}^{-2} \text{ yr}^{-1}$. Here vertical mixing was greater in the winter and the corresponding production was less at this season. For continental shelf waters off New York (39 to 41°N 71 to 74°W) Ryther and Yentsch (1958) using the radiocarbon and chlorophyll methods found daily production to range from roughly 0.20 to $1.00 \text{ g C m}^{-2} \text{ day}^{-1}$. Estimated annual production decreased in a seaward direction from $160 \text{ g C m}^{-2} \text{ yr}^{-1}$ in the shallow coastal waters to $100 \text{ g C m}^{-2} \text{ yr}^{-1}$ on the continental slope.

In the north-east Pacific Ocean (50°N 145°W) the photosynthetic productivity was found to average about $0.20 \text{ g C m}^{-2} \text{ day}^{-1}$ over a six week period in the summer (McAllister *et al.*, 1960). Practically all the photosynthesis was carried out by organisms passing through the finest plankton net. Anderson (1964) found that off the Washington and Oregon coasts (43 to 48°N 124 to 128°W) the mean daily rates were between 0.17 and $0.42 \text{ g C m}^{-2} \text{ day}^{-1}$ corresponding to an annual production rate of around $60 \text{ g C m}^{-2} \text{ yr}^{-1}$ both inside and outside the plume of the Columbia River, and $152 \text{ g C m}^{-2} \text{ yr}^{-1}$

in a region of upwelling. During the summer off the south coast of Japan (30 to 35°N, 136 to 141°E) daily production rates were found to range between 0·07 and 0·15 g C m^{-2} day^{-1} in oceanic waters, and from 0·30 to 0·50 g C m^{-2} day^{-1} in inshore waters (Ichimura and Saijo, 1959).

14.6.3. TROPICAL REGIONS

Steemann Nielsen (1959) has pointed out that, in open tropical seas, productivity is not correlated with temperature but, as elsewhere, replenishment of nutrients is decisive. Thus low primary production is found in lower latitudes in the central part of the anticyclonic eddies, and high primary production occurs where currents diverge and upwelling takes place, as for example at the boundaries between the South Equatorial Current and the Counter-current in the eastern parts of all three oceans. Correspondingly, there is an inverse relationship between the magnitude of primary production and the depth of the thermocline.

Primary production in one area of anticyclonic circulation, the Sargasso Sea, has been a subject for dispute since Steemann Nielsen (1952) found, using the radiocarbon method, a value, 0·05 g C m^{-2} day^{-1}, much lower than that of 0·46 reported by Riley et al. (1949) who used the oxygen method. In addition to the errors which arise because of the insensitivity of the oxygen method when applied to oligotrophic waters, seasonal variation contributes to this discrepancy. As already noted (p.432) productivity in this area rises to a high value at times of upwelling. Menzel and Ryther (1960) found values similar to those found by Steemann Nielsen over much of the year, but noted that net production reached a peak of 0·83 g C m^{-2} day^{-1} in the spring, giving an annual production of 72 g C m^{-2} yr^{-1}, at 32°N, 65°W. Iron appears to be the nutrient which limits primary production in this area (Ryther and Guillard, 1959). Thomas (1969) showed from nutrient enrichment experiments in the eastern equatorial Pacific Ocean, that nitrogen was the most likely limiting nutrient in water of low productivity both north and south of the equator, whereas in equatorial upwelling water no nutrient appeared to limit phytoplankton growth. Hobson et al. (1973) have reported a maximum rate of 9 g C m^{-2} day^{-1} in nutrient-rich water off Peru; this appears to be the highest daily rate yet recorded. The effect of the monsoons in the region of the Indian sub-continent is considerable. Off the west coast of India increased productivity, generally in July, is associated with the south-west monsoon which causes bottom and surface waters to mix (Subrahmanyan, 1960). In the Indian ocean, nannoplankton of size less than 110 μm is responsible for the major part of the photosynthetic activity (Saijo, 1964; Saijo and Takesue, 1965). Inshore waters in the tropics generally show high primary productivity which may be related to high rates of nutrient regeneration in sediments consequent upon the high temperature (Steemann Nielsen, 1959). In protected inshore waters off Puerto Rico (18°N 67°W), 0·5 to 5 m deep, rich blooms of phytoflagellates were found to give primary productivity values between 0·8 and 8·3 g C m^{-2} day^{-1} (Burkholder et al., 1967). In a turbid and polluted tropical estuary, the Cochin Backwater (9° 58'N, 76° 15'E), light, not nutrients, limited production (Qasim et al., 1969). Net primary production was estimated by three independent methods (radiocarbon, oxygen and chlorophyll) which gave concordant results with a mean of 124 g C m^{-2} yr^{-1}.

14.6.4. ARCTIC AND ANTARCTIC REGIONS

Although it appears that the phytoplankton of high latitudes in both north and south is dominated by diatoms, this conclusion may be erroneous since most sampling has been done with nets (Allen, 1971). Zernova (1970) found that diatoms constituted over 99% of the total number of cells in Nansen bottle samples taken in the Southern Ocean. However, Balech (1970) states that the silicoflagellate Dictyocha speculum sometimes outnumbers any species of diatom. Hasle (1969), who examined Nansen bottle samples from the Pacific Southern Ocean, concluded that diatoms were dominant in the Antarctic waters although nannoplanktonic flagellates were usually numerically more abundant. She found dinoflagellates to be scarce. Phaeocystis is abundant at times in both the Arctic and Antarctic (see Allen, 1971, for references). The physiology of cold water phytoplankton requires further study. The problem of adaptation to low temperatures is unresolved (see Section 14.2.7), and indications that assimilation of dissolved organic substances plays some part in survival during long periods of darkness need investigation (Fogg and Horne, 1970; Allen, 1971).

Hydrographically, the Arctic and Antarctic Oceans are quite different. The former is surrounded by land, and water penetrating through the few openings from the Atlantic and Pacific Oceans produces complex zones of mixing. The Southern Ocean surrounds the continent of Antarctica. Prevailing westerly winds and the absence of any major land masses combine to create a relatively uniform circumpolar environment with the Antarctic Convergence as a distinct boundary to the north. Water movements in the Southern Ocean bring nutrient-rich bottom water to the surface whereas there is little upwelling in the Arctic region. Consequently, there are marked differences in the concentrations of nutrients such as phosphate, nitrate and silicate, average values quoted by Allen (1971) being from 3 to 40 times higher for the Antarctic than for the Arctic. In view of this and the fact that the typical phytoplankton species in the two oceans are different, it is not altogether surprising that there have been suggestions that there are also differences in the physiology of the algae (Allen, 1971).

The Arctic region, defined as the water mass which, at least down to 100–200 m, is derived solely from the polar sea, is generally considered to be unproductive. Observations made in the central North Polar Sea from Drift Station Alpha confirm this (English, 1959). Sokolova and Solov'yeva (1971), who made determinations from April to October for inshore waters on the Murmansk Coast, found the highest values of primary productivity to be about 30 mg C m^{-3} day^{-1}. Their results suggest an annual production of 20 g C m^{-2}. Subarctic waters are well known to be highly productive although determinations of primary productivity in them seem to be few.

Data on primary productivity in the Southern Ocean have been reviewed by El-Sayed (1966, 1967, 1970a). Relatively low photosynthetic activity (< 15 mg C m^{-2} h^{-1}) has generally been found in the Drake Passage, the eastern Weddell Sea, the Bellingshausen sea, the Pacific sector between 75° and 150°W (El-Sayed, 1967) and the Antarctic and Indian Ocean sector between 40°W and 100°E (Saijo and Kawashima, 1964). High values (> 15 mg C m^{-2} h^{-1}) have been found southeast of Tierra del Fuego, in the vicinity of the South Orkney Islands, in the south-western Weddell Sea, in the Ross Sea and in the waters west of the Antarctic Peninsula (El-Sayed, 1967). El-Sayed (1967) recorded 3·2 g C m^{-2} day^{-1} in February 1965 in the Gerlache Strait (64° 30′ S, 62° 30′ W) and Mandelli and Burkholder (1966) reported a value of 3·62 g C m^{-2} day^{-1} near Deception Island (63° 0′ S, 60° 30′ W). Horne et al. (1969) who made observations for inshore waters off Signy Island (60° 43′ S, 45° 38′ W) found a peak in primary productivity of 250 mg C m^{-2} h^{-1} in early February 1967 and estimated the total for the season to be 130 g C m^{-2}. A dense phytoplankton bloom in a part of the Weddell Sea (74° 59′ S, 60° 57′ W) only accessible by ice-breaker was found by El-Sayed (1971) to have a primary production of 65 mg C m^{-2} h^{-1}, equivalent on a daily basis to 1·56 g C m^{-2}. These values (which are nearly as high as any recorded for the oceans), show that low temperature is not limiting and they are consistent with what would be expected with long days and ample nutrient supply. The problem is to explain the low values which prevail over most of the Southern Ocean. It seems likely, as Saijo and Kawashima (1964) surmised, that the depth of the mixed layer is usually greater than that of the euphotic layer so that the average amount of light received per cell is low. This view is supported by the finding that primary productivity is low at the Antarctic Convergence, where vertical mixing is intense (El-Sayed and Mandelli, 1965; Hasle, 1969) and high under stable conditions among slush and pancake ice (Hasle, 1969; El–Sayed, 1971).

Sea ice. which covers 12 % of the world ocean surface may harbour prolific growths of algae. It provides a remarkably complex environment (Bunt, 1963; Weeks, 1966; El-Sayed, 1970b) with a range of habitats the extremes of which are represented by the dimly lit under-layer of loosely aggregated ice-crystals and by the lenses of low salinity melt water on the surface of the dense and hard upper layer. Diatoms, chrysophyceans and green flagellates are frequently abundant, dinoflagellates less so. This ice-flora appears to be distinct from the true planktonic flora (Bunt, 1966). Studies with axenic cultures of sea-ice diatoms have shown them to be adapted to low temperature (see Section 14.2.5) and to low light intensity. Bunt (1967) found that the compensation intensity for *Fragilaria sublinearis* at 6·8°C was about 0·0005 ly min^{-1} or 70 lux, about 0·17 of the intensity usually taken as the compensation point for planktonic algae, and at lower temperatures even lower intensities would be expected to suffice. Primary productivity in sea-ice is difficult to measure, but Bunt (1966), who mentions a peak rate of 3·8 mg C m^{-3} h^{-1}, makes a tentative estimate of 0·1 metric tons hectare^{-1} yr^{-1} (10 g m^{-2} yr^{-1}) which is low compared with values for plankton in open Antarctic waters. Measurements made with ^{14}C by SCUBA diving under arctic sea-ice gave values ranging from 7·7 mg C m^{-2} h^{-1} near the peak of an algal bloom to a low of 0·3 mg C m^{-2} h^{-1} just before the thaw (Clasby *et al.*, 1973).

14.7. THE TOTAL YIELD OF PHOTOSYNTHESIS IN THE OCEANS

Riley (1944) estimated the annual gross primary production of the hydrosphere to be 14·6 × 10^{10} metric tons of carbon fixed, assuming an average production of about 1 g C m^{-2} day^{-1} as indicated by oxygen measurements. Steemann Nielsen and Aabye Jensen (1957–59), on the basis of their determinations by the radiocarbon method, considered the average daily gross production in all seas to be about 0·15 g C m^{-2} day^{-1}, giving a correspondingly smaller total. Vallentyne (1965) revised these estimates. Using a photosynthetic quotient of 1·25 and assuming net productivity to be 60% of the gross value, Riley's estimate for the marine environment is brought down to 6 × 10^{10} metric tons of carbon per year. After correction for error in the estimation of self absorption and back-scattering in standards (× 1·47) and for the liberation of extracellular products of photosynthesis (× 1·25), Steemann Nielsen's and Aabye Jensen's estimate of net productivity is revised to 2·2 to 2·8 × 10^{10} metric tons of carbon per year. From a recalculation of previous Russian estimates using radiocarbon, Bruyevich and Ivanenkov (1971) arrived at a figure of 4·4 × 10^{10} metric tons of carbon per year. Platt and Subba Rao (1973) taking new data into account, have given an estimate of 3·1 × 10^{10} metric tons of carbon a year. Neither of these last two pairs of authors appear to have taken extracellular products into account, and their estimates may therefore be too low. It may be noted that the yield from the oceans is about the same as that estimated for the land, 2·2 to 3·2 × 10^{10} metric tons of carbon. In terms of utilization of radiant energy reaching the Earth's surface, the seas have a lower efficiency in primary production, namely about 0·2% as compared with 0·4% for the land.

14.8. THE POSSIBILITY OF INCREASING THE PRIMARY PRODUCTIVITY OF MARINE PHYTOPLANKTON

Many methods of obtaining increased yields of food from the sea, e.g. shellfish culture, depend on the organisms concentrating food produced over an area much larger than that which they themselves occupy and thus do not involve any increase in primary production (Ryther et al., 1972).

In attempting to increase primary production itself, a limit is set by the nature of the photosynthetic mechanism and the penetration of light into water (Section 14.5.1). The maximum chlorophyll a content of the photic layer appears to be about 0.30 g m^{-2}, less than the 1.0 g m^{-2} found in closed terrestrial communities. in which the higher plants constitute a much more efficient light-diffusing and absorbing system (Talling, 1961; Takahashi and Parsons, 1972). It seems unlikely that this limitation inherent in the aquatic system can be overcome without light-diffusing and stirring devices which would be impracticable on a large scale.

However, it is only in especially favourable areas that this limit is reached, and normally the availability of nutrients determines the amount of primary productivity. Addition of fertilizers which, of course, is only feasible in enclosed sea areas, undoubtedly stimulates phytoplankton production although in practice it has not proved easy to ensure that this results in increased production of some desirable secondary producer (Costlow, 1969). The possibility of pumping deep water, rich in nitrates, phosphates and other nutrients to the surface where it can support increased phytoplankton productivity, has been considered (Roels et al., 1971; Da Silva, 1971). Where deep water is close inshore the cost of pumping might be offset by using the cold water for the production of freshwater, air conditioning or electrical power by the Claude process. In the open sea, the deep water might be a by-product of sea-bed mining. A pilot scheme operated on the northern shore of St. Croix (17° 40′ N, 64° 50′ W), has shown that, with deep water containing at least 35 μg atoms $NO_3^- \text{-N } l^{-1}$, yields approaching $11 \text{ g C fixed m}^{-2} \text{ day}^{-1}$ may be obtained (Roels et al., 1971). However, the economic feasibility of this scheme remains to be established.

The combined culture of algae and disposal of sewage or other organic wastes, which has been shown to be practical in freshwater and about which a great deal of knowledge exists (Oswald, 1970), may also be operated in sea water. It depends on associated growth of algae and bacteria, the latter providing carbon dioxide and ammonia for algal growth by breakdown of the organic matter, and the oxygen produced in algal photosynthesis stimulating the activity of the bacteria. Possibly direct photo-assimilation of some organic matter by the algae also occurs. The yield of algal material may be as much as 1·5 metric tons dry weight per million gallons of sewage or 150 metric tons per hectare per year (approximately $16·5 \text{ g C m}^{-2} \text{ day}^{-1}$). It might be possible to use algal material produced in this way as human or animal food, but under present conditions it is more acceptable to convert it to shell-fish or other animal meat. Highly eutrophic conditions tend to be unstable, oscillating between aerobic and anaerobic conditions according to the amount of light received and other conditions, and are unsuitable as a habitat for secondary producers useful as food. Ryther *et al.* (1972) have described pilot experiments, in what they term controlled eutrophication, in which sea water enriched with sewage effluent is used to grow phytoplankton, the water then being circulated through tanks containing oysters or other molluscs. The algal cultures achieved a yield of about $8 \text{ g dry weight m}^{-2} \text{ day}^{-1}$ and over a period of 30 days at 7–10°C the oysters removed 77% of the algae entering the system, converting 22% of the filtered material to new oyster flesh. These results give promise of an effective means of simultaneously avoiding pollution and obtaining more food from the sea.

REFERENCES

Allen, M. B. (1971). *Ann. Rev. Ecol. System.* **2**, 261.
Allen, M. B. and Arnon, D. I. (1955). *Physiol. Plant.* **8**, 653.
Anderson, G. C. (1964). *Limnol. Oceanogr.* **9**, 284.
Anderson, G. C. and Zeutschel, R. P. (1970). *Limnol. Oceanogr.* **15**, 402.
Antia, N. J., McAllister, C. D., Parsons, T. R., Stephens, K. and Strickland, J. D. H. (1963). *Limnol. Oceanogr.* **8**, 166.
Arnon, D. I. and Wessel, G. (1953). *Nature, Lond.* **172**, 1039.
Arthur, C. R. and Rigler, F. H. (1967). *Limnol. Oceanogr.* **12**, 121.
Bainbridge, R. (1957). *Biol. Rev.* **32**, 91.
Balech, E. (1970). *In* "Antarctic Ecology" (M. W. Holdgate, ed.) Vol. I, pp. 143–147. Academic Press, London.
Barnes, H. and Hasle, G. R. (1957). *Nytt Magasin for Botanikk,* **5**, 113.
Berge, G. (1958). *Rapp. Cons. Explor. Mer.* **144**, 85.
Bongers, L. H. J. (1956). *Meded. Landbouwhogesch. Wageningen, Ned.* **56**, 1.
Bruyevich, S. and Ivanenkov, V. N. (1971). *Oceanology,* **11**, 694.

Burkholder, P. R., Burkholder, L. M. and Almodovar, L. R. (1967). *Bull. Mar. Sci.* **17**, 1.
Bunt, J. S. (1963). *Nature, Lond.* **199**, 1255.
Bunt, J. S. (1966). *In* "Symposium on Antarctic Oceanography" (R. I. Currie, ed.) pp. 198 Scott Polar Research Institute, Cambridge.
Bunt, J. S. (1967). *Antarctic Research Series* **11** (Biology of the Antarctic Seas, III), pp. 1–14.
Bunt, J. S., Owens, O., Van, H. and Hoch, G. (1966), *J. Phycol.* **2**, 96.
Caperon, J. (1968). *Ecology*, **49**, 866.
Caperon, J. and Meyer, J. (1972a). *Deep-Sea Res.* **19**, 601.
Caperon, J. and Meyer, J. (1972b). *Deep-Sea Res.* **19**, 619.
Carlucci, A. F. and Silbernagel, S. B. (1969). *J. Phycol.* **5**, 64.
Cassie, R. M. (1963). *Oceanogr. Mar. Biol. Ann. Rev.* **1**, 223.
Clasby, R. C., Horner, R. and Alexander, V. (1973). *J. Fish. Res. Bd. Can.* **30**, 835.
Collyer, D. M. and Fogg. G. E. (1955). *J. Exp. Bot.* **6**, 256.
Cooper, L. H. N. (1955). *Deep-Sea Res.* **3**, (Suppl.) 212.
Cooper, L. H. N. (1958). *Rapp. Cons. Explor. Mer.* **144**, 35.
Costlow, J. D., Jr. (1969). "Marine Biology" (J. D. Costlow, Jr., ed.), Vol. V, 606 pp. Gordon and Breach, New York.
Costlow, J. D., Jr. (1971). "Fertility of the Sea" (J. D. Costlow, Jr., ed.), 2 vols., 622 pp. Gordon and Breach, New York.
Currie, R. I. (1958). *Rapp. Cons. Explor. Mer.* **144**, 96.
Cushing, D. H. (1959a). Fisheries Investigations. Series II, Vol. **22**, (6) London, H.M.S.O. 40 pp.
Cushing, D. H. (1959b). *J. Cons. Int. Explor. Mer.* **24**, 455.
Da Silva, P. de C. M. (1971). *In* "Fertility of the Sea" (J. D. Costlow, Jr., ed.), Vol. II, pp. 463–468. Gordon and Breach, New York.
Degens, E. T., Guillard, R. R. L., Sackett, W. M. and Hellebust, J. A. (1968). *Deep-Sea Res.* **15**, 1.
Denffer, D. von. (1948). *Biol. Zentr.* **67**, 7.
Deuser, W. G. (1970). *Science, N.Y.* **168**, 1575–1577.
Doty, M. S. (1959). *J. Mar. Biol. Ass. India.* **1**, 66.
Doty, M. S. (1961). Proceedings of the Conference on Primary Productivity Measurement, Marine and Freshwater. (M. S. Doty, ed.), 237 pp. U.S. Atomic Energy Commission TID-7633.
Doty, M. S. and Oguri, M. (1957). *Limnol. Oceanogr.* **2**, 37.
Doty, M. S. and Oguri, M. (1958). *Rapp. Cons. Explor. Mer.* **144**, 47.
Doty, M. S., Jitts, H. R., Koblentz-Mishke, O. J. and Saijo, Y. (1965). *Limnol. Oceanogr.* **10**, 282.
Dring, M. J. (1971). *In* "Fourth European Marine Biology Symposium" (D. J. Crisp, ed.), pp. 375–392. Cambridge University Press.
Droop, M. R. (1961a). *Bot. Mar.* **2**, 231.
Droop, M. R. (1961b). *J. Mar. Biol. Ass. U.K.* **41**, 69.
Droop, M. R. (1962a). *In* "Beiträge zur Physiologie und Morphologie der Algen", pp. 77–82. Deuts. Bot. Gesellsch. Gustav Fischer Verlag, Stuttgart.
Droop, M. R. (1962b). *In* "Physiology and Biochemistry of Algae" (R. A. Lewin, ed.), pp. 141–159. Academic Press, New York.
Droop, M. R. (1966). *J. Mar. Biol. Ass. U.K.* **46**, 659.
Droop, M. R. (1967). *Br. Phycol. Bull.* **3**, 295.

Droop, M. R. (1968). *J. Mar. Biol. Ass. U.K.* **48**, 689.

Droop, M. R. (1969). *In* "Methods in Microbiology" (J. R. Norris and D. W. Ribbons, eds.), Vol. IIIB, pp. 269–313. Academic Press, London.

Droop, M. R. and Elson, K. G. R. (1966). *Nature, Lond.* **211**, 1096.

Dugdale, R. C. and Goering, J. J. (1967). *Limnol. Oceanogr.* **12**, 196.

Dyson, N., Jitts, H. R. and Scott, B. D. (1965). C.S.I.R.O. Australia. Division of Fisheries and Oceanography Technical Paper No. 18. 12 pp.

Ebata, T. and Fujita, Y. (1971). *Plant Cell Physiol. Tokyo,* **12**, 533.

El-Sayed, S. Z. (1966). *In* "Symposium on Antarctic Oceanography" (R. I. Currie, ed.) pp. 227–239. Scott Polar Research Institute, Cambridge.

El-Sayed, S. Z. (1967). *Antarctic Research Series* **11**, (Biology of the Antarctic Seas III) pp. 15–47.

El-Sayed, S. Z. (1968). *In* "Primary Productivity and Benthic Marine Algae of the Antarctic and Subantarctic", Folio 10 Antarctic Map Folio Series, pp. 1–6. American Geographical Society.

El-Sayed, S. Z. (1970a). *In* "Antarctic Ecology" (M. W. Holdgate, ed.), Vol. I, pp. 119–134. Academic Press, London.

El-Sayed, S. Z. (1970b). In *"Symposium on Antarctic Ice and Water Masses"* Tokyo. pp. 35–54.

El-Sayed, S. Z. (1971). Antarctic Research Series, **17**, (Biology of the Antarctic Seas IV), pp. 301–312.

El-Sayed, S. Z. and Mandelli, E. F. (1965). Antarctic Research Series **5**, (Biology of the Antarctic Seas II), pp. 87–106.

English, T. S. (1959). *Proc. Int. Cong. Oceanogr., New York, 1959.* pp. 338–389.

Eppley, R. W., Rogers, J. N. and McCarthy, J. J. (1969). *Limnol. Oceanogr.* **14**, 912.

Eppley, R. W., Rogers, J. N., McCarthy, J. J. and Sournia, A. (1971). *J. Phycol.* **7**, 150.

Eppley, R. W., Carlucci, A. F., Holm-Hansen, O., Kiefer, D., McCarthy, J. J., Venrick, E. and Williams, P. M. (1971). *Limnol. Oceanogr.* **16**, 741.

Eppley, R. W. and Coatsworth, J. L. (1966). *Arch. Mikrobiol.* **55**, 66.

Eppley, R. W. and Strickland, J. D. H. (1968). *In* "Advances in microbiology of the Sea" (M. R. Droop and E. J. Ferguson Wood, eds.), pp. 23–62. Academic Press, London.

Eppley, R. W. and Thomas, W. H. (1969). *J. Phycol.* **5**, 375.

Fay, P. and Kulasooriya, S. A. (1973). *Br. Phycol. J.* **8**, 51.

Fee, E. J. (1969). *Limnol. Oceanogr.* **14**, 906.

Fogg, G. E. (1956). *Ann. Bot. N.S.* **20**, 265.

Fogg, G. E. (1958). *Rapp. Cons. Explor. Mer.* **144**, 56.

Fogg, G. E. (1959). *Symp. Soc. Exp. Biol.* **13**, 106.

Fogg, G. E. (1963). *Brit. Phyc. Bull.* **2**, 195.

Fogg, G. E. (1965). "Algal Cultures and Phytoplankton Ecology", 126 pp. University of Wisconsin Press, Madison.

Fogg, G. E. (1966). *Oceanogr. Mar. Biol. Ann. Rev.* **4**, 195.

Fogg, G. E. (1971). *Arch. Hydrobiol. Beih. Ergebn. Limnol.* **5**, 1.

Fogg, G. E. (1973). *Water Res.* **7**, 77.

Fogg, G. E. (in the press). *In* "Photosynthesis and Productivity" (J. P. Cooper, ed.). Cambridge University Press.

Fogg, G. E. (in the press). *In* "Photosynthesis and Productivity" (J. P. Cooper, ed.). Vol. II. Academic Press, London and New York, pp. 632–638.

Fogg, G. E., Nalewajko, C. and Watt, W. D. (1965). *Proc. R. Soc. B,* **162**, 517.

Fott, B. (1971). "Algenkunde" 2nd Ed. 581 pp. VEB Gustav Fischer Verlag, Jena.

Fries, L. (1966). *Physiol. Plant.* **19**, 800.

Fritsch, F. E. (1935). "Structure and Reproduction of the Algae" Vol. I, 791 pp. Cambridge University Press.

Fritsch, F. E. (1945). "Structure and Reproduction of the Algae" Vol. II, 939 pp. Cambridge University Press.

Fuhs, G. W. (1969). *J. Phycol.* **5**, 312.

Gaarder, T. and Gran, H. H. (1927). *Rapp. Cons. Explor. Mer.* **42**, 1.

Gimmler, H., Ullrich, W., Domanski-Kaden, J. and Urbach, W. (1969). *Plant Cell Physiol. Tokyo.* **10**, 103.

Goldberg, E. D. (1971). *In* "Fertility of the Sea" (J. D. Costlow, Jr., ed.), pp. 143–156. Gordon and Breach, New York.

Goldman, C. R. (1965). *In* "Primary Productivity in Aquatic Environments" (C. R. Goldman, ed.), 457 pp. *Mem. Ist. Ital. Idrobiol.* **18**, Suppl.

Goldman, C. R. and Mason, D. T. (1962). *Science, N.Y.* **136**, 1049.

Goldman, C. R., Mason, D. T. and Wood, B. J. B. (1963). *Limnol. Oceanogr.* **8**, 313.

Goldman, C. R., Gerletti, M., Javornicky, P. M. Melchiorri-Santolini, U. and Amezaga, E. de. (1968). *Mem. Ist. Ital. Idrobiol.* **23**, 49.

Gregory, R. P. F. (1971). "Biochemistry of Photosynthesis", 202 pp. Wiley-Interscience, London.

Guillard, R. R. L. and Hellebust, J. A. (1971). *J. Phycol.* **7**, 330.

Guillén, O., Mendiola, B. R. de and Rondán, R. I. de. (1971). *In* "Fertility of the Sea" (J. D. Costlow, Jr., ed.) pp. 157–185. Gordon and Breach, New York.

Halldal, P. (1964). *Physiol. Plant.* **17**, 414.

Hart, T. J. (1966). *In* "Some Contemporary Studies in Marine Science" (H. Barnes, ed.) pp. 375–393. George Allen and Unwin, London.

Harvey, H. W. (1928). "Biological Chemistry and Physics of Sea Water", 194 pp. Cambridge University Press.

Harvey, H. W. (1945). "Recent Advances in the Chemistry and Biology of Sea Water," 164 pp. Cambridge University Press.

Harvey, H. W. (1955). "The Chemistry and Fertility of Sea Waters", 224 pp. Cambridge University Press.

Hasle, G. R. (1954). *Nytt Magasin for Botanikk,* **2**, 121.

Hasle, G. R. (1969). Hvalrådets Skrifter, No **52**, 1.

Haxo, F. T. (1960). *In* "Comparative Biochemistry of Photoreactive Systems" (M. B. Allen, ed.) pp. 339–360.

Hellebust, J. A. (1965). *Limnol. Oceanogr.* **10**, 192–206.

Hill, M. N. (ed.) (1963). "The Sea", Vol. II, 554 pp. Interscience, New York.

Hobson, L. A., Menzel, D. W. and Barber, R. T. (1973). *Marine Biology,* **19**, 298.

Holmes, R. W. and Anderson, G. C. (1963). *In* "Symposium on Marine Microbiology" (C. H. Oppenheimer, ed.), pp. 241–250. Springfield, Ill.

Holm-Hansen, O., Gerloff, G. C. and Skoog, F. (1954). *Physiol. Plant.* **7**, 665.

Holm-Hansen, O. (1969). *Limnol. Oceanogr.* **14**, 740–747.

Holm-Hansen, O. (1971). *In* "Fertility of the Sea", Vol. I, (J. D. Costlow, Jr., ed.), pp. 197–207. Gordon and Breach, New York.

Hoogenhout, H. (1963). *Phycologia,* **2**, 135.

Horne, A. J., Fogg, G. E. and Eagle, D. J. (1969). *J. Mar. Biol. Ass. U.K.* **49**, 393.

Huntsman, S. A. (1972). *J. Phycol.* **8**, 59.

Hutchinson, G. E. (1967). "A Treatise on Limnology" Vol. II, 1115 pp. John Wiley and Sons, New York.
Hutner, S. H. and Provasoli, L. (1964). Ann. Rev. Plant. Physiol. 15, 37.
Ichimura, S. and Saijo, Y. (1959). Bot. Mag. Tokyo, 72, 193.
Ignatiades, L. (1973). J. Mar. Biol. Ass. U.K. 53, 923.
Ignatiades, L. and Fogg. G. E. (1973). J. Mar. Biol. Ass. U.K. 53, 937.
Ilmavirta, V. and Hakala, I. (1972). Ann. Bot. Fennici, 9, 77.
Jensen, A., Rystad, B. and Skoglund, L. (1972). J. Exp. Mar. Biol. Ecol. 8, 241.
Jitts, H. R. (1961). In "Proceedings of the Conference on Primary Productivity Measurement, Marine and Freshwater" (M. S. Doty, ed.), pp. 114–120. U.S. Atomic Energy Commission, TID-7633.
Jitts, H. R. (1967). C. S. I. R. O. Division of Fisheries and Oceanography, Ann. Rep. 1966–67, p 36.
Johnston, R. (1963). J. Mar. Biol. Ass. U.K. 43, 427.
Jørgensen, E. G. (1966). Physiol. Plant, 19, 789.
Jørgensen, E. G. (1968). Physiol. Plant, 21, 423.
Jørgensen, E. G. (1969). Physiol. Plant. 22, 1307.
Jørgensen, E. G. (1970). Physiol. Plant. 23, 11.
Kain, J. M. and Fogg. G. E. (1960). J. Mar. Biol. Ass. U.K. 39, 33.
Ketchum, B. H., Ryther, J. H., Yentsch, C. S. and Corwin, N. (1958). Rapp. Cons. Explor. Mer. 144, 132.
Kowallik, W. (1971). In "Photosynthesis and Photorespiration" (M. D. Hatch, C. B. Osmond and R. O. Slatyer, eds.), pp. 514–522. Wiley-Interscience. New York.
Kuenzler, E. J. (1965). J. Phycol. 1, 156
Kuenzler, E. J. (1970). J. Phycol. 6, 7.
Kuenzler, E. J. and Ketchum, B. H. (1962). Biol. Bull. 123, 134.
Kumar, H. D. (1964). J. Exp. Bot. 15, 232.
Legendre, L. and Watt, W. D. (1970). Mar. Biol. 7, 167.
Lewin, J. C. (1966). J. Exp. Bot. 17, 473.
Lewin, R. A. (1959). Revue. Algologique, 4, 181.
Lewin, R. A. (1961). In "Encyclopedia of Plant Physiology" (W. Ruhland, ed.), Vol. 14, pp. 401–417.
Lex, M., Silvester, W. B. and Stewart, W. D. P. (1972). Proc. R. Soc. B. 180, 88.
Lund, J. W. G. and Talling, J. F. (1957). Bot. Rev. 23, 489.
MacIsaac, J. J. and Dugdale, R. C. (1972). Deep-Sea Res. 19, 209.
Maddux, W. S. and Jones, R. F. (1964). Limnol. Oceanogr. 9, 79.
Maksimova, I. V. and Pimenova, M. N. (1969). Mikrobiologiya, 38, 609.
Mandelli, E. F. and Burkholder, P. R. (1966). J. Mar. Res. 24, 15.
Marsh, H. V. Jr., Galmiche, J. M. and Gibbs, M. (1965). In "Biochemical Dimensions of Photosynthesis" (D. W. Krogmann and W. H. Powers, eds.), pp. 95–107. Wayne State University Press, Detroit.
McAllister, C. D. (1961a). Limnol. Oceanogr. 6, 447.
McAllister, C. D. (1961b). Limnol. Oceanogr. 6, 483.
McAllister, C. D., Parsons, T. R. and Strickland, J. D. H. (1960). J. Cons. int. Explor. Mer. 25, 240.
McAllister, C. D., Parsons, T. R., Stephens, K. and Strickland, J. D. H. (1961) Limnol. Oceanogr. 6, 237.
McAllister, C. D., Shah, N. and Strickland, J. D. H. (1964). J. Fish. Res. Bd. Can. 21, 159

McCleod, G. C. and Kanwisher, J. (1962). *Physiol. Plantarum,* 15, 581.
Menzel, D. W. and Ryther, J. H. (1960). *Deep-Sea Res.* 6, 351.
Menzel, D. W. and Spaeth, J. P. (1962). *Limnol. Oceanogr.* 7, 151.
Morris, I. (1967). "An Introduction to the Algae", 189 pp. Hutchinson University Library, London.
Morris, I. and Farrell, K. (1971). *Physiol. Plant.* 25, 372.
Morris, I., Yentsch, C. M. and Yentsch, C. S. (1971a). *Limnol. Oceanogr.* 16, 854.
Morris, I., Yentsch, C. M. and Yentsch, C. S. (1971b). *Limnol. Oceanogr.* 16, 859.
Morris, I., and Glover, H. E. (1974) *Mar. Biol.* 24, 147.
Munk, W. H. and Riley, G. A. (1952). *J. Mar. Res.* 11, 215.
Myers, J. (1962). *In* "Beiträge zur Physiologie und Morphologie der Algen". Vorträge a.d. Gesamtgebiet d. Botanik, Deutsch. Bot. Ges. N.F. 1, 13–19.
Myers, J. and Clarke, L. B. (1944). *J. Gen. Physiol.* 28, 103.
Nalewajko, C., Chowdhuri, N. and Fogg, G. E. (1963). *In* "Studies on Microalgae and Photosynthetic Bacteria". Japanese Society for Plant Physiology, University of Tokyo Press.
Nalewajko, C. and Lean, D. R. S. (1972). *J. Phycol.* 8, 37.
Nihei, T., Sasa, T., Miyachi, S., Suzuki, K. and Tamiya, H. (1954). *Arch. Mikrobiol.* 21, 155.
Ogasawara, N. and Miyachi, S. (1970). *Plant. Cell Physiol., Tokyo,* 11, 1.
Ohle, W. (1961). *Verh. Int. Ver. Limnol.* 14, 113.
O'Kelley, J. C. (1974). *In* "Algal Physiology and Biochemistry" (W. D. P. Stewart, ed.) pp. 610–635. Blackwell, Oxford.
Oppenheimer, C. H. (1966). "Marine Biology II" (C. H. Oppenheimer, ed.), 369 pp. New York Academy of Science, New York.
Oswald, W. J. (1970). *In* "Prediction and Measurement of Photosynthetic Productivity", pp. 473–487. Centre for Agricultural Publishing and Documentation, Wagenigen.
Paasche, E. (1960). *J. Cons. Int. Explor. Mer.* 26, 33.
Paasche, E. (1964). *Physiol. Plantarum,* Suppl. 3, 1.
Paasche, E. (1967). *Physiol. Plant.* 20, 946.
Paasche, E. (1968). *Physiol. Plant.* 21, 66.
Paasche, E. (1971). *J. Cons. Int. Explor. Mer.* 33, 509.
Parker, B. C., Leeper, G. and Hurni, W. (1968). *Limnol. Oceanogr.* 13, 172.
Parsons, T. R., Stephens, K. and Strickland, J. D. H. (1961). *J. Fish. Res. Bd. Can.* 18, 1001.
Pirson, A. and Lorenzen, H. (1966). *Ann. Rev. Plant. Physiol.* 17, 439.
Platt, T. (1969). *Limnol. Oceanogr.* 14, 653.
Platt, T. and Subba Rao, D. V. (1970). *J. Fish. Res. Bd. Can.* 27, 887.
Platt, T. and Subba Rao, D. V. (1973). Fisheries Research Board of Canada, Technical Report, No. 370.
Prakash, A. (1971). *In* "Fertility of the Sea" (J. D. Costlow, Jr., ed.), Vol. II, pp. 351–368. Gordon and Breach, New York.
Provasoli, L. (1963). *In* "The Sea" (M. N. Hill, ed.), Vol. II. pp. 165. Interscience Publishers, New York.
Provasoli, L. (1968). *In* "Cultures and Collections of Algae" (A. Watanabe and A. Hattori, eds.), pp. 63–75. Japanese Society for Plant Physiology, University of Tokyo Press.

Provasoli, L., and Carlucci, A. J. (1974). *In* "Algal Physiology and Biochemistry", (W. D. P. Stewart, ed.) pp. 741–787. Blackwell, Oxford.

Provasoli, L., McLaughlin, J. J. A. and Droop, M. R. (1957). *Arch. Mikrobiol.* **25**, 392.

Pugh, P. R. (1970). *Limnol. Oceanogr.* **15**, 652.

Qasim, S. Z., Wellershaus, S., Bhattathiri, P. M. A. and Abidi, S. A. (1969). *Proc. Indian Acad. Sci. B,* **69**, 51.

Qasim, S. Z., Bhattathiri, P. M. A. and Devassy, U. P. (1972). *Mar. Biol.* **16**, 22.

Quraishi, F. O. and Spencer, C. P. (1971). *In* "Fourth European Marine Biology Symposium" (D. J. Crisp, ed.), pp. 393–408. Cambridge University Press.

Rabinowitch, E. I. (1945). "Photosynthesis and Related Processes. Vol. I. 599 pp. Interscience Publishers, New York.

Rabinowitch, E. I. (1951). "Photosynthesis and Related Processes" Vol. II. Part 1. pp. 603–1208. Interscience Publishers, New York.

Rabinowitch, E. I. (1956). "Photosynthesis and Related Processes" Vol. II. Part 2. pp. 1211–2088. Interscience Publishers, New York.

Rabinowitch, E. I. and Govindjee. (1969). "Photosynthesis" 273 pp. John Wiley & Sons, New York.

Raymont, J. E. G. (1963). "Plankton and Productivity in the Oceans", 660 pp. Pergamon Press, London.

Reid, F. M. H., Fuglister, E. and Jordan, J. B. (1970). *Bull. Scripps Instn. Oceanogr.* **17**, 51.

Reynolds, N. (1973). *Br. phycol. J.* **8**, 135.

Riley, G. A. (1944). *Amer. Scientist.* **32**, 129.

Riley, G. A. (1963a). *In* "Marine Biology I". (G. A. Riley, ed.) 286 pp. American Institute of Biological Sciences, Washington, D.C.

Riley, G. A. (1963b). *In* "The Sea". (M. N. Hill, ed.), Vol. II. pp. 438–463. Interscience Publishers, New York.

Riley, G. A., Stommel, H. and Bumpus, D. F. B. (1949). *Bull. Bingham Oceanogr. Coll.* **12**, 1.

Robinson, G. A. (1970). *Bull. Mar. Ecol.* **6**, 33.

Rodhe, W. (1965). *Mem. Ist. Ital. Idrobiol.* **18**, Suppl, 365.

Roels, O. A., Gerrard, R. D. and Bé, A. W. H. (1971). *In* "Fertility of the Sea" (J. D. Costlow, Jr., ed.), Vol. II. pp. 401–415.

Round, F. E. (1965). "The Biology of the Algae", 269 pp. Edward Arnold, London.

Russell, F. S., Southward, A. J., Boalch, G. T. and Butler, E. I. (1971). *Nature, Lond.* **234**, 468.

Ryther, J. H. (1956). *Limnol. Oceanogr.* **1**, 61.

Ryther, J. H. and Guillard, R. R. L. (1959). *Deep-Sea Res.* **6**, 65.

Ryther, J. H. and Menzel, D. W. (1959). *Limnol. Oceanogr.* **4**, 492.

Ryther, J. H. and Yentsch, C. S. (1957). *Limnol. Oceanogr.* **2**, 281.

Ryther, J. H. and Yentsch, C. S. (1958). *Limnol. Oceanogr.* **3**, 327.

Ryther, J. H., Dunstan, W. M., Tenore, K. R. and Huguenin, J. E. (1972). *Bioscience,* **22**, 144.

Saijo, Y. (1964). *J. Oceanogr. Soc. Jap.* **19**, 187.

Saijo, Y. and Ichimura, S. E. (1961). *In* "Proceedings of the Conference on Primary Productivity Measurement, Marine and Freshwater" (M. S. Doty, ed.), pp. 91–96. U. S. Atomic Energy Commission TID-7633.

Saijo, Y. and Kawashima, T. (1964). *J. Oceanogr. Soc. Jap.* **19**, 190.

Saijo, Y. and Takesue, K. (1965). *J. Oceanogr. Soc. Jap.* **20**, 10.
Samuel, S., Shah, N. M. and Fogg, G. E. (1971). *J. Mar. Biol. Ass. U.K.* **51**, 793.
Senger, H. (1970). *Planta (Berl.)* **90**, 243.
Sieburth, J. McN. (1968). *In* "Advances in Microbiology of the Sea" (M. R. Droop and E. J. Ferguson Wood, eds.), Vol. I, pp. 63–94. Academic Press, London.
Smayda, T. J. (1970). *Oceanogr. Mar. Biol. Ann. Rev.* **8**, 353.
Soeder, C. J. (1970). *Arch. Hydrobiol.*, Suppl **38**, 1.
Soeder, C. J., Müller, H., Payer, H. D. and Schulle, H. (1971). *Mitt. int. Ver. Limnol.* **19**, 39.
Sokolova, S. A. and Solov'yeva, A. A. (1971). Oceanology, **11**, 386.
Sorokin, C. (1965). *Biochim. Biophys. Acta,* **94**, 42.
Sorokin, Ju. I. (1965). *Mem. Ist. Ital. Idrobiol.* **18**, Suppl. 187.
Sournia, A. (1967). *C.R. Acad. Sci. Paris,* **265**, 1000.
Sournia, A. (1968). *Int. Revue ges. Hydrobiol.* **53**, 1.
Spoehr, H. A. and Milner, H. W. (1949). *Plant. Physiol.* **24**, 120.
Steele, J. H. (1956). *J. Mar. Biol. Ass. U.K.* **35**, 1.
Steele, J. H. (1957). *J. Mar. Biol. Ass. U.K.* **36**, 233.
Steele, J. H. (1958). "Plant production in the northern North Sea, 36 pp. Scottish Home Dept., Marine Research Series No. 7. H.M.S.O.
Steele, J. H. (1961). *In* "Oceanography", (M. Sears, ed.) pp. 519–538. American Association for the Advancement of Science, Publ. No 67.
Steele, J. H. and Baird, I. E. (1961). *Limnol. Oceanogr.* **6**, 68.
Steemann Nielsen, E. (1952), *J. Cons. Int. Explor. Mer.* **18**, 117.
Steemann Nielsen, E. (1955). *Deep-Sea Res.* **3**, Suppl. 281.
Steemann Nielsen, E. (1958). *Rapp. Cons. Explor. Mer.* **144**, 38.
Steeman Nielsen, E. (1959), *J. Mar. Biol. Ass. India.* **1**, 7.
Steeman Nielsen, E. (1960). *Ann. Rev. Plant Physiol.* **11**, 341.
Steemann Nielsen, E. (1963). *In* "The Sea" (M. N. Hill, ed.), Vol. II, pp. 129–164. Interscience Publishers, New York.
Steemann Nielsen, E. (1964a). *J. Ecol.* **52**, (Suppl.). 119.
Steemann Nielsen, E. (1964b). *Meddelelser fra Danmarks Fiskeri-og Havunder-søgelser. N.S.* **4**, 31.
Steemann Nielsen, E. (1965). *Limnol. Oceanogr.* **10**, (suppl.) R247.
Steemann Nielsen, E. and Aabye Jensen, E. (1957–59). *In* "Galathea Report" (A. F. Brunn, S. Greve and R. Spärck, eds.), Vol. I, pp. 49–136. Copenhagen.
Steemann Nielsen, E. and Cushing, D. H. (1958). *Rapp. Cons. Explor. Mer.* **144**, 158 pp.
Steemann Nielsen, E. and Hansen, V. K. (1959). *Deep-Sea Res.* **5**, 222.
Steemann Nielsen, E. and Hansen, V. K. (1961). *Physiol. Plant.* **14**, 595.
Steemann Nielsen, E. and Jørgensen, E. G. (1962). *Arch. Hydrobiol.* **58**, 249.
Steemann Nielsen, E. and Jørgensen, E. G. (1968). *Physiol. Plant.* **21**, 401.
Steemann Nielsen, E., Hansen, V. K. and Jørgensen, E. G. (1962). *Physiol. Plant,* **15**, 505.
Stein, J. R. (1973). "Handbook of Phycological Methods", (J. R. Stein, ed.) 448 pp. Cambridge University Press, London.
Stewart, W. D. P. and Pearson, H. W. (1970). *Proc. R. Soc. B,* **175**, 293.
Strickland, J. D. H. (1960). *Fish. Res. Bd. Can. Bull.* No. 122, 172 pp.
Strickland, J. D. H. (1965). *In* "Chemical Oceanography" (J. P. Riley and G. Skirrow, eds.), Vol. I, pp. 477–610. Academic Press, London.

Strickland, J. D. H. (1972). *Oceanogr. Mar. Biol. Ann. Rev.* **10**, 349.
Strickland, J. D. H. and Parsons, T. R. (1968). *Fish. Res. Bd. Can. Bull.* **167**. 311.
Strickland, J. D. H. and Terhune, L. D. B. (1961). *Limnol. Oceanogr.* **6**, 93
Strickland, J. D. H., Eppley, R. W. and Mendiola, B. R. de (1969a). *Bol. Inst. Mar. Peru-Callas,* **2**, 4.
Strickland, J. D. H., Holm-Hansen, O., Eppley, R. W. and Linn, R. J. (1969b) *Limnol. Oceanogr.* **14**, 23.
Subba Rao, D. V. (1965). *Aust. J. Mar. Freshw.* **16**, 273.
Subrahmanyan, R. (1960). *J. Indian Bot. Soc.* **39**, 78.
Sweeney, B. M. and Hastings, J. W. (1962). *In* "Physiology and Biochemistry of Algae". (R. A. Lewin, ed.), pp. 687–700. Academic Press, New York.
Taguchi, S. (1970). *Bull. Plankton Soc. Jap.* **17**, 65.
Takahashi, M., Fujii, K. and Parsons, T. R. (1973). *Mar. Biol.* **19**, 102.
Takahashi, M. and Parsons, T. R. (1972). *Indian J. Mar. Sci.* **1**, 61.
Takahashi, M., Shimura, S., Yamaguchi, Y. and Fujita, Y. (1971). *J. Oceanogr. Soc. Jap.* **27**, 43.
Talling, J. F. (1957a). *Proc. R. Soc. B.* **147**, 57.
Talling, J. F. (1957b). *New Phytologist,* **56**, 29.
Talling, J. F. (1960). *Limnol. Oceanogr.* **5**, 62.
Talling, J. F. (1961). *Ann. Rev. Plant Physiol.* **12**, 133.
Talling, J. F. (1970). *In* "Prediction and Measurement of Photosynthetic Productivity", pp. 431–445. Centre for Agricultural Publishing and Documentation, Wageningen.
Tamiya, H. (1964). *In* "Synchrony in Cell Division and Growth" (E. Zeuthen, ed.), pp. 247–305. Wiley, New York.
Tamiya, H. (1966). *Ann. Rev. Plant. Physiol.* **17**, 1.
Tamiya, H., Iwamura, T., Shibata, K., Hase, E. and Nihei, T. (1953). *Biochim. Biophys. Acta.* **12**, 23.
Thomas, J. P. (1971). *Mar. Biol.* **11**, 311.
Thomas, W. H. (1966). *J. Phycol.* **2**, 17.
Thomas, W. H. (1969). *J. Fish. Res. Bd. Can.* **26**, 1133.
Thomas, W. H. and Dodson, A. N. (1968). *Biol. Bull.* **134**, 199.
Thomas, W. H. and Dodson, A. N. (1972). *Limnol. Oceanogr.* **17**, 515.
Tolbert, N. E. (1974). *In* "Algal Physiology and Biochemistry" (W. D. P. Stewart, ed.) pp. 474–504. Blackwell, Oxford.
Tolbert, N. E. and Zill, L. P. (1956). *J. Biol. Chem.* **222**, 895.
UNESCO (1966) "Monographs on Oceanographic Methodology I". UNESCO Paris, 69 pp.
Vallentyne, J. R. (1965). *Mem. Ist. Ital. Idrobiol.* **18**, Suppl. 309.
Vidaver, W. (1972). *Symp. Soc. exp. Biol.* **26**, 159.
Vollenweider, R. A. (1965). *Mem. Ist. Ital. Idrobiol.* **18**, Suppl. 425.
Vollenweider, R. A. (1970). *In* "Prediction and Measurement of Photosynthetic Productivity", pp. 455–472. Centre for Agricultural Publishing and Documentation, Wageningen.
Vollenweider, R. A. (1974). "A Manual on Methods for Measuring Primary Production in Aquatic Environments." IBP Handbook No. 12, 2nd Ed. 225 pp. Blackwell, Oxford.
Vollenweider, R. A. and Nauwerck, A. (1961). *Verh. int. Ver. Limnol.* **14**, 134.
Wallen, D. G. and Geen, G. H. (1968). *J. Fish. Res. Bd. Can.* **25**, 2219.

Walsby, A. E. (1967). *Biotechnology and Bioengineering*, **9**, 443.
Watt, W. D. (1965). *Limnol. Oceanogr.* **10**, 298.
Watt, W. D. (1966). *Proc. R. Soc. B.* **164**, 521.
Watt, W. D. (1969). *Ann. Bot.* **33**, 427.
Watt, W. D. (1971). *Deep-Sea Res.* **18**, 329.
Watt, W. D. and Fogg. G. E. (1966). *J. Exp. Bot.* **16**, 117.
Watt, W. D. and Hayes. F. R. (1963). *Limnol. Oceanogr.* **8**, 276.
Weeks, W. F. (1966). *In* "Symposium on Antarctic Oceanography" (R. I. Currie ed.), pp. 173–190. Scott Polar Research Institute, Cambridge.
Wetzel, R. G. (1965). *Ecology,* **46**, 540.
Wiessner, W. (1970). *In* "Photobiology of Microorganisms" (P. Halldal, ed.), pp. 95–133.
Williams, P. J. le B., Berman, T. and Holm-Hansen, O. (1972). *Nature. New Biology,* **236**, 91.
Yentsch, C. S. (1963). *Oceanogr. Mar. Biol. Ann. Rev.* **1**, 157.
Yentsch, C. S. (1965). *Mem. Ist. Ital. Hydrobiol.* **18**, suppl., 323.
Yentsch, C. S. and Ryther, J. H. (1957). *Limnol. Oceanogr.* **2**, 140.
Zernova, V. V. (1970). *In* "Antarctic Ecology" (M. W. Holdgate, ed.), Vol. I. pp. 136–142. Academic Press, London.
Zlobin, V. S. (1972). *Oceanology,* **12**, 550.

Chapter 15

The Hydrochemistry of Landlocked Basins and Fjords

K. GRASSHOFF

Institut für Meereskunde, Kiel, Federal German Republic

15.1. INTRODUCTION

Several seas and numerous fjords are connected to the open ocean or to the adjacent sea only by narrow channels and in some instances water exchange is further restricted by a shallow sill. The hydrographical and hydrochemical regimes of such landlocked basins and fjords are usually influenced to only a small extent by either the horizontal and vertical patterns of water circulation or the hydrographic events in the adjacent ocean. The enclosed seas develop their own individual stratification, and the motions of their water masses frequently differ fundamentally from those of the neighbouring open sea. For these reasons the landlocked seas and fjords have a distinctive hydrography, hydrochemistry and biology; it is therefore appropriate to discuss them separately from the open ocean.

Usually the chemistry of semi-enclosed seas is highly dependent on the prevailing hydrographical conditions, and cannot be discussed without a thorough understanding of the latter.

The hydrographical features of the deeper water in landlocked seas are governed by two principal factors (1) the differences in the density stratification between the enclosed sea and the adjacent open sea, and (2) the inclinations of the sea surface and the internal boundary layers; these in turn are dependent on both the water exchange with the open ocean and the addition or removal of water from the enclosed part by run-off or evaporation.

If the exchange of heat (Q_Σ) and the water exchange (W_Σ) allow a deep reaching thermo-haline convection to become established in the enclosed sea, the water at the depth of the sill will attain a higher density than that at the same depth in the open sea. The result will be an overflow of deeper water into the open sea.

If there is a general horizontally decreasing density profile from the open sea into the enclosed part, and if there is also a considerable input of fresh water into the epicontinental sea, there will normally be an outflow of surface water and a penetration of ocean water into the deeper layers of the land-locked basin or fjord.

In this Chapter discussion of semi-enclosed seas and fjords will be restricted to those which have a relatively narrow and shallow connection with the open ocean. In general, these enclosed basins can be divided into two main categories:

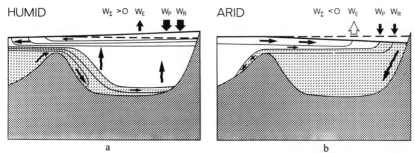

FIG. 15.1. (a) Schematic section through a semi-enclosed sea in a humid zone with positive water balance. (b) Schematic section through a semi-enclosed sea in an arid zone with negative water balance.

(i) Those which are located in a humid zone with a positive water balance ($W_\Sigma > 0$) relative to the adjacent ocean, and in which the combined rates of precipitation and input of fresh water from the surrounding drainage area exceed that of removal by evaporation and the inflow of oceanic water (Fig. 15.1a).

(ii) Those which are located in an arid zone having a negative water balance ($W_\Sigma < 0$) where the evaporation exceeds the total input of fresh water (Fig. 15.1b).

The exchange of water between all semi-enclosed basins and the oceans is very much dependent upon the topography of the transition area—usually a strait. As a rule, pronounced development of distinct water layers with marked thermoclines and haloclines can be observed. It is possible to distinguish two types of land-locked basin according to the slope of the sea surface along its longitudinal axis: (i) those in which the slope is negligible, i.e. where the sum of river input (W_R), and precipitation (W_P) just balances evaporation (W_E) i.e.

$$W_R + W_P = W_E$$

and (ii) those in which the slope has a pronounced influence on the directions and velocities of the currents in the transition area. The inclination of the sea surface will be greater if the relationship between the water budget ($W_\Sigma/\mathrm{yr}^{-1}$) within the basin and the cross section of the narrowest parts of the transition area leads to considerable current velocities in the straits. Only those seas having a pronounced positive or negative water balance will be discussed in this account.

The surface layers of those basins having a positive water balance have significantly lower salinities. This produces an outflow of low salinity water which can be observed in the surface layers of the transition area and can be traced far out into the adjacent ocean.

Because of the strong influence of river water the composition of this low salinity water usually differs appreciably from that of ocean water. Typical representatives of this category are the Baltic and the Black Sea. A considerable number of fjords, especially those in humid zones, also belong to this category.

The depth of the halocline and the associated salinity gradient, and the horizontal gradient of the salinity from the transition area to the innermost parts of the enclosed basin depend to a large extent on the sill depth. The development of haloclines and thermoclines strongly restricts the exchange of material between the water layers, and can, under certain circumstances, lead first to an accumulation of micro-nutrients and then to a total and permanent depletion of the dissolved oxygen; this may be followed subsequently by the generation of hydrogen sulphide. The development of permanently or intermittently anoxic conditions is a general feature of those semi-enclosed seas which have a positive water balance, a strong halocline and basins which are considerably deeper than the sill. The oxygen depletion causes a drastic change in the redox conditions and, therefore, in the hydrochemical regime of the deeper water.

In those enclosed seas which have a negative water balance there is, generally, not only an ingoing surface current from the open ocean to balance the water lost by evaporation, but also an outgoing current of deeper more saline water. The latter cascades into the adjacent ocean basin in which it can often be traced for long distances from its source. This current results when the increase in density caused by the surplus of evaporation counterbalances the decrease which occurs because of warming of the surface. A convection is thus established in those areas which have large evaporation rates especially during the colder seasons. Thus, the water at the sill depth and below in the enclosed sea has a greater density than that at the sill depth in the ocean outside. Examples of this category of landlocked seas are the Mediterranean, the Red Sea and the Arabian Gulf.

Since all the enclosed seas in both of these two categories have many points in common regardless of their geographical location, the hydrochemistry of those with a positive water balance will be illustrated by reference to the Baltic and Black Seas and Oslo Fjord. The Red Sea will be used to typify the category characterized by a negative water balance.

15.2. THE BALTIC

15.2.1. THE GEOLOGICAL HISTORY OF THE BALTIC

The Baltic is a relatively young postglacial sea, and its hydrochemical environment has changed several times since its formation via the melting of the Fenno–Scandian glaciers about 13000 years ago. The sea floor was alternately elevated and depressed in sympathy with changes in the ice burden of glacial and post glacial times which reached a maximum of about 3000 m. Of at least six ice ages, only the last three were of consequence to the present day Baltic. The rapid withdrawal of the glaciers at the end of the last ice age, the so-called Vistula period, led to a dramatic uplift of extensive areas in the Baltic region; (with a maximum elevation amounting to 300 m over 10 000 years in the western part of the Bothnian Bay) together with a downward displacement in the southern North Sea of about 30 m in 10 000 years. Even today the displacement of the sea floor is still continuing.

In addition to these isostatic movements of the surface of the earth in the Baltic area, eustatic changes in the level of sea water have occurred as a result of changes in the distribution of water between its liquid and crystalline phases. Thus, because of the melting of the ice caps the sea level is today increasing at a rate of $\sim 1 \cdot 1$ mm yr^{-1} (Dietrich, 1954) and according to Shephard and Curray (1967) has risen 10 m during the last 9000 years and by 100 m during the last 20 000 years.

The postglacial history of the Baltic was determined by the interaction of the isostatic and eustatic effects, and it is difficult to ascertain the changes in shape which the Baltic has undergone since its initial formation. Four principal periods in the oceanographic history of the Baltic can be recognized from the results of geological, morphological and palynological investigations, together with data from radiometric age determinations. Each of these periods was characterized by either an open or closed connection with the open ocean. Much of the information on the early history of the climate and the hydrochemical environment of the Baltic has been obtained from an examination of the fossils of its sediments.

Figures 15.2(a)–(d) show the contours of the post-glacial Baltic corresponding to these four periods. In the first of these periods it is referred to as the "Baltic Ice Sea". This fresh water inland sea which existed from 12 000 to 8000 B.C. was the drainage basin of the withdrawing Fenno-Scandian glaciers, and its sediments contain no residues of marine molluscs. The second period was contemporary with a general rise in sea level in the world ocean. The Baltic at that time (8000–7250 B.C.) occasionally became a strait extending from the North Sea to the White Sea and having an almost oceanic salinity. Its fauna was characterized by arctic species, e.g. *Yoldia arctica* (also known as *Portlandia arctica*) and for this reason the Baltic during this period is called Yoldia Sea. During the third period at the beginning of a climatic optimum, the isostatic rise balanced the eustatic rise of sea level and the connections to the North Sea and the White Sea became

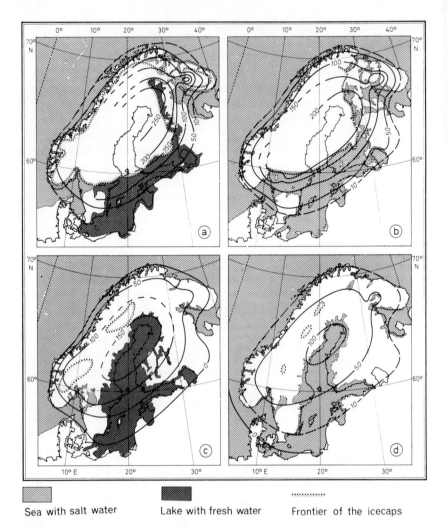

Sea with salt water   Lake with fresh water   Frontier of the icecaps

FIG. 15.2. The postglacial development of the Baltic. (a) Baltic Ice Sea; 12,000 B.C.–8,000 B.C. (b) Yoldia Sea; 8000–7250 B.C. (c) Ancylus Sea; 7250–2000 B.C. (d) Littorina Sea (after *Littorina ovata*), 2000 B.C.–500 A.D. (Dietrich, 1974).

blocked. The enclosed Baltic became warmer and its salinity decreased, finally attaining fresh water conditions. The dominant fossil during this period was the fresh water snail *Ancylus fluvialis*. Already the Ancylus Sea considerably resembled the present Baltic. Because of excessive run-off from the surrounding continent, the Ancylus Sea had a river-like connection to the open ocean, at first through the middle part of Sweden and later further to the south. This river gave rise to many of the features of the present transition area, e.g. the furrows in the Danish Belt, the Fehmarn Belt and the Kadett Furrow. During the fourth period the eustatic rise of the ocean level matched the isostatic lift. Salt water was able to penetrate through the furrows mentioned above, and the Baltic became a brackish semi-enclosed sea. The brackish water snail *Littorina littorea* gave its name to this period of the Baltic, which lasted from 5100 to 2000 B.C. After this, input of saline water decreased and as a consequence the salinity also decreased. *Littorina* was not able to adjust to the diminished salinity. During the period 2000 B.C. to 500 A.D., the water snail *Littorina ovata* appeared, and this name is applied to the Baltic of that period. Subsequently, further brackish water organisms appeared, among them the mussel *Mya arenaria*. On the basis of this the present Baltic is referred to geologically as the Mya Sea.

15.2.2. THE TOPOGRAPHY OF THE BALTIC

The present topography is important in determining the hydrography of the Baltic and especiallly its hydrochemical environment. In particular, the chemical processes in the central part of the Baltic cannot be understood without a knowledge of the topography and of the oceanographic processes in the entrances and in the associated chain of basins and sills.

The main features of the Baltic sea floor resulted from the postglacial geology. As the ice moved along the axis of the present Baltic it pushed away older gravel residues and deposited them elsewhere during quiet periods. Differences in the crystalline base rocks have led to the sequence of more or less separated basins constituting the present day Baltic; each of these has a slightly different hydrographical structure and hydrochemistry.

The topography of the Baltic is well known (Schulz, 1956; Dietrich and Ulrich, 1968; Hollan, 1973). Wattenberg (1949) recognized the following areas:

  (i) The transition area comprising the Kattegat, the Belt Sea, the Kiel Bight and the Bight of Mecklenburg.
  (ii) The main Baltic together with the Arcona Sea, the Bornholm Sea, the Gdansk Bight and the Eastern and Western Gotland Seas.
  (iii) The Bight of Riga.
  (iv) The Gulf of Bothnia, the Åland Sea and the Bay of Bothnia.
  (v) The Gulf of Finland.

It should be noted however that the Arcona Sea differs considerably from the eastern parts of the Baltic proper with respect to its hydrophysical and hydrochemical features and that it resembles the transition area more closely.

The various parts of the Baltic are linked via narrow straits and sill regions. Thus, the hydrography of the Belt Sea is governed by the shallow sills near the Kattegat, with depths of 26 m off Samsoe and only 7 m at the Drogden Sill. A sill with a depth of 18 m between the Fehmarn Belt and the Arcona Sea off Darss further restricts the entrance to the central Baltic. The Arcona Sea, which is a shallow basin with a maximum depth of 50 m, may be called the mixing chamber of the Baltic. The Bornholm Sea, with an average depth of about 80 m, has its deeper entrance with a sill depth of 45 m. to the west in the Bornholm Gatt. The shallows between the island of Bornholm and the island of Ruegen allow only the surface water to exchange with the Arcona Sea. Towards the east, the Bornholm Basin is separated by the Slupsk Furrow with sill depths of 68 m and 71 m. Shallows to the north separate the Western Gotland Basin from the Bornholm Basin. At the eastern end the eastern sill of the Slupsk Furrow slopes into the Eastern Gotland Basin and the Bay of Gdansk, with maximum depths of 247 and 118 m respectively. A ridge separates the Bay of Gdansk from the Gotland Basin, which is the largest basin of the Baltic. To the north, a series of smaller deeps, separated by sills, leads to the northern part of the Gotland Basin which levels to the east without marked sills into the Gulf of Finland. A sill which is located between the island of Gotland and the Swedish coast separates the Eastern and the Western Gotland Basins. The Baltic attains its greatest depth (456 m) in the Landsort Deep south of the archipelago of Stockholm. This unusual topography is the reason for the peculiar hydrography and chemistry of this deep. The Western Gotland Basin is shallower to the south and has at present no deep water connection to the Bornholm Sea. An archipelago with many thousands of small islands separates the Gulf of Bothnia from the Baltic proper. A narrow channel with a depth of 70 m forms the only deep water connection between the central part of the Baltic and the Åland Sea and the Gulf of Bothnia. The sill depth of this channel is of extreme importance in controlling the hydrochemistry of the Gulf of Bothnia, which is also structured into deeper basins separated by sills. The areas, and volumes of the various parts of the Baltic together with their maximum, mean and sill depths are given in Table 15.1.

15.2.3. WATER BALANCE AND EXCHANGE

Since the volume of water in the Baltic is, on average, constant a steady state water balance is maintained. In considering this it is necessary to take five factors into account. Influx of water occurs as a result of: the addition of water from the North Sea (W_I), continental run off (W_R) and precipitation

TABLE 15.1

Geographical features of the main basins of the Baltic (ICES, 1970)

Name of basin or deep	Area km²	Volume km³	Max. depth m	Mean depth m	Sill-depth m
Kattegat	23 000	580		25	
Belt Sea	21 000				
Baltic proper	202 000	13 600	459	67	17
1. Arcona basin (below 30 m).		70	55		17
2. Bornholm basin (below 60 m)		160	105		45
3. Central basin (below 60 m)		4 100	459		60
A. Eastern Gotland basin (below 100 m)		920	249		60
a. Gdansk basin (below 100 m)		10	116		88
b. Gotland Deep (below 150 m)		196	249		60
c. Fårö Deep (below 150 m)		25	205		140
B. Northern Central basin (below 100 m)		558	459		115
a. Northern basin (below 100 m)		228	219		115
b. Landsort Deep (below 100 m)		270	459		138
C. Western Gotland basin (below 100 m)		101	205		100
a. Norrköping Deep			205		100
b. Karlsö Deep			112		101
4. Bay of Riga	16 700	460	51	28	c. 20
Gulf of Finland	29 500	1 125	100	38	
Baltic proper + Gulf of Finland (0–60 m)	249 000	9 500			17
Baltic proper + Gulf of Finland (total)	249 000	15 190	459	61	17
Åland Sea	5 200	405	301	77	
Archipelago Sea	8 300	195		23	
Bothnian Sea	66 600	4 595	293	69	40 (70?)
Bothnian Bay	37 000	1 540	126	42	25
Baltic Sea (total)	366 000	21 960	459	60	17

(W_P). Water is removed from the Baltic by an outgoing current (W_O) and also by evaporation (W_E). These different factors are related by the equation

$$W_R + W_P - W_E = W_O - W_I.$$

Since the total salt content of the Baltic is constant, or only varies slowly, this relationship, in principle, allows the average water balance to be calculated provided that some of the terms are known. Thus,

$$W_O \times S_O = W_I \times S_I$$

where S_O and S_I are, respectively, the salinities of the water leaving and entering the Baltic. The total water balance was first studied in detail by Brogmus (1952) and several further investigations of it have been made recently (see Table 15.2) There are considerable difficulties in assessing the budget and

TABLE 15.2
Water Balance of the Baltic (Values in $km^3 yr^{-1}$)

Source	Fresh water run off	inflow	outflow
	(W_R)	(W_I)	(W_O)
Brogmus (1952)	471	471	942
Soskin and Rova (1957)	473	1187	1660
Lisitzin (1967)	491	1263	1754
Svansson (1972) I	472	576	1048
Svansson (1972) II	472	1246	1718
Kullenberg (1973)	471	154	942
		(upper layers)	
		317	
		(deep water)	

for this reason the various estimates differ considerably. Thus, the inflow and outflow undergo large short and long term variations which are caused by changes in the atmospheric circulation and seasonal processes affecting the density of the water in the entrances of the Baltic. Meteorological processes influence not only the rates of precipitation and evaporation, but also the water level and the wind-driven transport of water out of the Baltic. Depending on the prevailing conditions in the transition area (i.e. in the Belt Sea and Danish Sounds) and in the Western Baltic, very complicated mixing processes occur before water finally leaves or enters the Baltic. The complexity of the processes is illustrated by Fig. 15.3; however, it should be noted that there is no general agreement about the absolute values of the figures given.

Brogmus (1952) calculated the water budget using the average salinity values for the inflowing and outflowing waters at the Gjedser Rev light vessel compiled by Knudsen (1899). Svansson (1972) based his first alternative water budget on all the salinity values reported by Knudsen (1899),

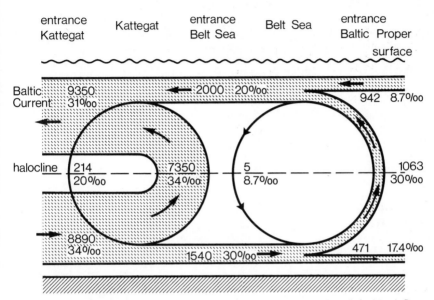

FIG. 15.3. Schematic display of the water exchange between the Baltic and the North Sea.

including those below 15‰ which had been omitted by Knudsen because he was of the opinion that the density of these waters was too low for them to have penetrated into the Baltic Proper. The second estimate by Svansson (1972) was based on a 30 year series of salinity observations at Gjedser Rev light vessel. Lisitzin (1967) used sea level variations in the Aland Sea for her calculations. Kullenberg (1973) assumed that about one third of the inflowing water is mixed into the surface water and thus never reaches the deeper layers in the central part of the Baltic. The areas and volumes of the various basins of the Baltic are shown in Table 15.1 and data on the major rivers and their average water transport are given in Table 15.3. The large fraction of the total fresh water input into the Baltic carried by the river Newa (18.2%) is noteworthy. It should be mentioned that the water transport of the rivers shows large seasonal variations, maximum values being reached in February (23 km³) and in May (66 km³). The latter maximum is caused by spates of the Newa and Scandinavian rivers.

Precipitation plays an important role in the water budget of seas in humid areas. There is an almost complete dearth of data on precipitation for the open Baltic. The estimate of the total amount of precipitation is based on data for about 200 stations along the coasts and on islands. It has been observed that precipitation data for islands are generally about 20% below the values for coastal stations. The total estimated annual precipitation for

TABLE 15.3

Major Rivers of the Baltic Drainage Area and their average discharges (Mikulski, 1970)

River or Basin	Watershed Area (km^2)	Average flow (m^3 s^{-1})	Average flow (l s^{-1} km^{-2})
	Gulf of Bothnia		
Umeälv-Skellefteälv	6·170	62	10·0
Skellefteälv	11·640	148	12·7
Skellefteälv-Piteälv	7·700	86	11·2
Piteälv	11·210	156	13·9
Piteälv-Luleälv	1·670	17	10·1
Luleälv	25·250	477	18·9
Luleälv-Kalixälv	6·360	64	10·0
Kalixälv	23·580	298	12·6
Kalixälv-Torneälv	2·000	22	11·0
Torneälv (Tornionjoki)	40·000	444	11·1
Tornionjoki-Merenkurkku	45·665	456	10·0
Kemijoki	51·400	581	11·3
Iijoki	14·385	181	12·6
Oulujoki	22·925	232	10·1
Total	269·955	3224	11·8
	Bothnian Sea		
Norrtäljeån-Dalälven	4·500	31	6·88
Dalälven	29·040	340	11·7
Dalälven-Ljusnan	5·040	43	8·53
Ljusnan	19·820	230	11·6
Ljusnan-Ljungan	5·380	50	9·29
Ljungan	12·840	131	10·2
Ljungan-Indalsälven	0·610	6	9·99
Indalsälven	26·740	420	15·7
Indalsälven-Ångermanälven	1·500	15	10·0
Ångermanälven	31·890	465	14·6
Ångermanälven-Umeälv	15·340	160	10·4
Umeälv	26·730	400	15·0
Merenkurkku-Lakalahti	13·385	119	8·88
Kokemäenjoki	27·100	224	8·26
Total	219·915	2·634	12·0
Bothnian Sea including Gulf of Bothnia	489·870	5·858	12·0

TABLE 15.3—*continued*

River or Basin	Watershed Area (km²)	Average flow (m³ s⁻¹)	Average flow per unit area (l s⁻¹ km⁻²)
	Gulf of Finland		
Hanko-Newa (Hanko-Neva)	19·460	189	9·71
Kymijoki	37·230	309	8·30
Newa (Neva)	281·100	2·600	9·25
Newa-Narwa (Neva-Narva)	4·310	43	10·0
Luga (Luga)	13·010	97	7·46
Narwa (Narva)	56·190	403	7·18
	9·690	97	10·0
Total	420·990	3·738	8·88
	Gulf of Riga		
Virtsu-Kalkasrags	34·160	256	7·50
Gauja	8·900	82	9·16
Dzwina (Daugava)	87·900	688	7·83
Total	130·960	1·026	7·89
	Central Baltic		
Öresund-Helgeån	1·890	16	8·46
Helgeån	4·780	55	11.5
Helgeån-Mörrumsån	1·460	13	8·90
Mörrumsån	3·380	27	7·99
Mörrumsån-Lyckebyån	3·140	22	7·00
Lyckebyån	0·850	6	7·06
Lyckebyån-Emån	4·730	33	6·98
Emån	4·460	31	6·95
Emån-Motalaström	6·180	37	5·99
Motalaström	15·470	97	6·27
Motalaström-Mälaren	7·450	49	6·58
Mälaren	22·600	173	7·65
Mälaren-Norrtäljeån	1·900	13	6·88
Lakalathi-Hanko	9·780	91	9·30
Gulf of Finland-Gulf of Riga	4·000	40	10·0
Gulf of Riga-Kurishes Gulf	19·870	199	10·0
Direct basin of the Kurishes Gulf and the Baltic	2·150	21	10·0
Neman	98.200	674	6.86

TABLE 15.3—*continued*

River or Basin	Watershed Area (km²)	Average flow (m³ s⁻¹)	Average flow per unit area (l s⁻¹ km⁻²)
Waters of the Kurishes Gulf	1·610	—	—
Taran-Vistula cape	23.600	117	4·95
Waters of the Vistula Gulf	1·610	—	—
Vistula	193·910	954	4·92
Vistula-Szczecin Gulf	17·240	142	8·43
Direct basin of the Szczecin Gulf and the Baltic	11.350	45	4·00
Odra	118·610	464	3·91
Waters of the Szczecin Gulf	0·910	—	—
Szczecin Gulf-Flensburg	16·960	68	4·00
Danish coast of the Baltic	10·410	51	4·90
Total	607·730	3·438	5·69
Baltic Sea			
Gulf of Bothnia	269·955	3·224	11·8
Bothnian Sea	219·915	2·634	12·0
Gulf of Finland	420·990	3·738	8·88
Gulf of Riga	130·960	1·026	7·89
Central Baltic	607·730	3·438	5·69
Baltic Sea, Total	1 649·550	14·060	8·52

the Baltic amounts to 183 km³, which corresponds to an average precipitation of 474 mm (range 405 to 580 mm depending on the area) (see Table 15.4). The precipitation varies markedly with season, reaching a minimum in March, and a maximum in August.

The estimation of the evaporation from the Baltic is a matter of considerable difficulty, and most estimates have been made by semi-empirical methods. The problem is complicated by the fact that large areas of it are covered by ice for several months in winter. The best available estimates seem to be those by Brogmus (1952), which show that the total average evaporation for the whole Baltic is ∼183 km³ yr⁻¹ (Table 15.5). The available data suggest that precipitation approximately equals evaporation.

In general, too little is known about water transport and mixing processes to enable a reliable water balance to be drawn up for the Baltic, let alone for each of its hydrographic regions. However, when the results of the many

TABLE 15.4

Precipitation (in mm yr^{-1} and km^3 yr^{-1}) in the Baltic (Brogmus, 1952).

Area	mm yr^{-1}	km^3 yr^{-1}
Bay of Bothnia	405	15·0
Gulf of Bothnia	425	28·5
Åland Archipelago	520	6·8
Gulf of Finland	560	16·8
Bay of Riga	580	9·3
Gotland Sea	470	69·6
Arcona Sea ⎱ Bornholm Sea ⎰	480	25·9
Belt Sea	515	10·8
Total Baltic	474	182·7

TABLE 15.5

Evaporation (in mm yr^{-1} and km^3 yr^{-1}) from the Baltic (Brogmus, 1952).

Area	mm yr^{-1}	km^3 yr^{-1}
Bay of Bothnia	331	12·3
Gulf of Bothnia	404	27·0
Åland Archipelago	394	5·2
Gulf of Finland	398	12·0
Bay of Riga	361	5·8
Gotland Sea	553	81·9
Arcona Sea Bornholm Sea	528	28·5
Belt Sea	505	10·6
Total Baltic	474	183·3

national and international research projects which are at present in progress become available it will be possible to deduce much more about the water balance and exchange of water and material in the Baltic.

15.2.4. THE DISTRIBUTION OF SALINITY AND TEMPERATURE

The Baltic is the largest brackish water area in the world. Its salinity ranges horizontally from almost zero in the innermost parts of the Bothnian Bay and the Gulf of Finland to nearly that of ocean water in the entrances of the Skagerak. As can be seen from Fig. 15.4 the largest salinity gradients in both

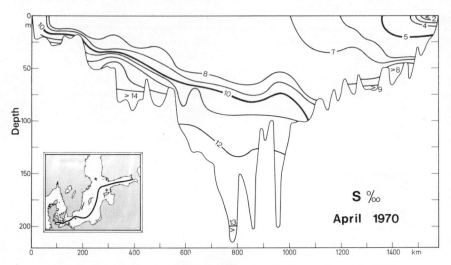

FIG. 15.4. Distribution of salinity along a longitudinal section of the Baltic.

the surface and deeper layers occur in the transition area, i.e. west of the Fehmarn Belt.

The vertical distribution of salinity is marked by a significant discontinuity layer except in the inner parts of the Gulfs of Finland and Bothnia, in which it is not very pronounced. This halocline is permanent except in those parts of the Gulfs of Finland and Bothnia which are subject to convectional overturn of the surface water during winter. The depth of the halocline in the transition area is about 10–20 m, but changes in meteorological conditions, and the outflow of Baltic surface water and inflow of more saline water from the North Sea may give rise to considerable variations. In the Arcona Sea the permanent halocline is at a depth of about 35–40 m, but here again large fluctuations are common. In the Bornholm Sea the halocline is still deeper, and usually lies at a depth of about 50 m (see Fig. 15.4). Further to the east, the boundary layer between the Baltic surface water and the deep water slopes down to a depth of 65 to 70 m. The depth of the permanent halocline is governed by the depth of the sills separating the different basins of the Baltic. It has been observed (Hela, 1965) that occasionally a secondary halocline develops in the central Baltic (Fig. 15.6).

The permanent halocline has a pronounced influence on the vertical exchange of material in the Baltic and is the cause of the stagnant conditions in the deeper layers. Although the halocline is a barrier to exchange between the deep water and the surface water, some slow mixing does occur and this gives the surface water an appreciable salinity.

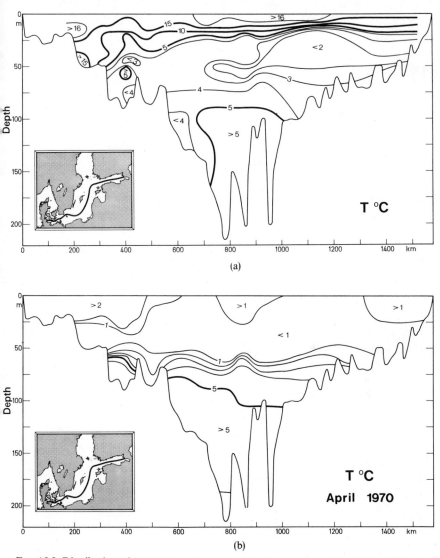

FIG. 15.5. Distribution of temperature along a longitudinal section of the Baltic. (a) August 1969, (b) April 1970.

The surface water of the Baltic shows considerable seasonal variations (see Fig. 15.5). During the summer a distinct thermocline or a series of thermoclines develops in all parts of the Baltic. The change from warm surface water

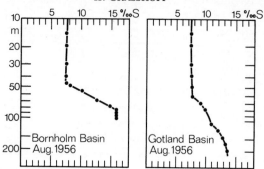

Fig. 15.6. The structure of the halocline in the Bornholm Basin and the development of a secondary halocline in the Gotland Basin (Hela, 1965).

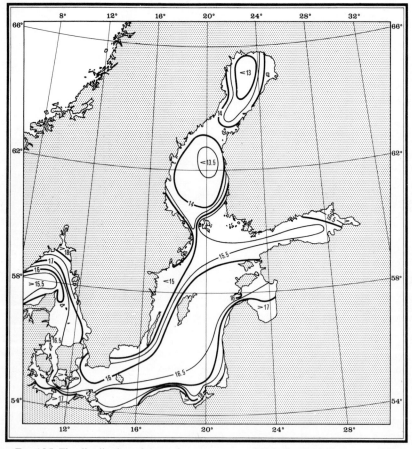

Fig. 15.7. The distribution of the surface temperature (°C) in the Baltic in August 1969.

to underlying cold water layers can be very sharp, with temperature gradients of up to $2°C\,cm^{-1}$. The formation of the upper thermocline starts in March–April in the western parts and in April–May in the central Baltic and the Gulfs of Finland and Bothnia. Whether either one or several thermoclines develop depends on the meteorological conditions in spring and early summer and the depth to which warming of the surface water penetrates. The distribution of the surface temperature in August is shown in Fig. 15.7 and the mean seasonal variation of temperature with depth is indicated in Fig. 15.8.

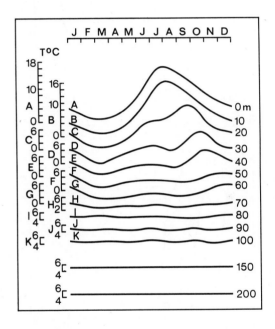

FIG. 15.8. Mean annual variation of temperature in the Gotland Deep from 1902–1970 (Matthäus, 1973).

During September and October the thermoclines are usually destroyed by storms and the general cooling of the surface water. As cooling proceeds, a convectional process is set up which usually extends down to the permanent halocline. In mild winters, however, the density of the surface water may not exceed that of the waters of recent winters, with the result that convection fails to reach the halocline; vertical mixing between deep water and surface water is therefore inhibited.

Fig. 15.9 shows some temperature profiles for the central Baltic and demonstrates the different stages in the development of a thermocline. At least in the central Baltic, the water below the permanent halocline has a fairly

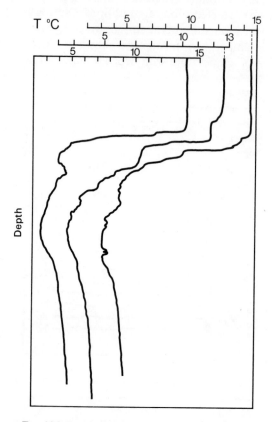

Fig. 15.9. Temperature profiles for the Central Baltic.

constant temperature as seasonal temperature variations do not penetrate into the deep water. However, in the western part of the Baltic and the transition area, the temperature is highly dependent on that of the inflowing water, i.e. on the prevailing meteorological conditions (see above).

A significant part of the northern and eastern Baltic is covered with ice in winter, and in severe winters this ice cover may extend even to the entrance of the Baltic. The long term frequency of maximal ice coverage in the Baltic is shown in Fig. 15.10. The ice cover has, of course, a significant influence on the exchange of gases between the atmosphere and the water.

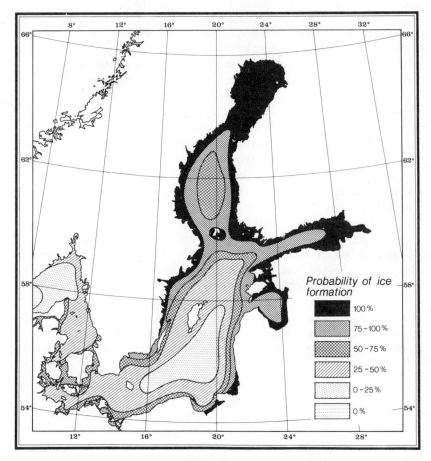

FIG. 15.10. The long term frequency of maximal ice coverage in the Baltic. (Dietrich, 1975).

15.2.5. CURRENTS IN THE BALTIC

Three factors govern the currents in the Baltic: the wind, the slope of the sea surface and the horizontal density gradients. In addition, large oscillating currents can be observed at various depths; these arise from internal waves and sea level fluctuations (seiches). Internal waves and seiches have been studied by Krauss (1963) and by Magaard (1965). These internal water motions are generated by atmospheric processes, e.g. by the passage of depressions across the Baltic area. Tidal currents are of only minor importance except in the Belt Sea. The wind generated surface drift current has an important influence on the water circulation in all parts of the

Baltic. A constant wind velocity of about 10 m s^{-1} generates a surface current of approximately 15 cm s^{-1}. Because of the deflection of the current to the right relative to the wind direction, the prevailing westerlies result in a weak counterclockwise surface current system. However, the topography and the shape of the coastline have a considerable bearing on the direction and magnitude of the currents. The action of the wind is not limited only to the production of drift currents. Wind pressure causes an updamming of the water in shallow areas; this in turn results in a slope current in the whole water column. Such slope currents can reach a considerable velocity in extensive shallow areas, in straits and also in exposed stretches of the coast. They are of great importance in the entrances to the Baltic. The outflow of the surplus water in the Baltic is the result of the sea surface slope. This current is deflected to the eastern coast of Sweden and augments the wind driven circulation of the surface water in the Baltic.

In addition to the drift and slope currents a third component adds to the final motion field of the Baltic surface waters. This is a density compensation current which results from the fact that the drift current of the surface water, which has a lower density than does the underlying water, is diverted to the right by the Coriolis force. Denser water therefore accumulates on the left side of the wind direction. The horizontal proximity of water of two different densities results in an inner field of force which leads to the development of a density compensation current. A detailed description of the relation

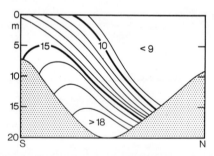

FIG. 15.11. Distribution of salinity (‰) along a section Warnemünde–Gjedser (Kadett Furrow) in May 1952 (Wolf, 1973).

between wind direction and speed and the resulting surface currents has been given by Dietrich (1951) and a treatment of the complicated water motions in the Fehmarn Belt has been published by Wyrtki (1954).

The water motions in the Fehmarn Belt, at the Darss Sill and in the Danish Belt have some common features which can be summarized as follows:

(i) the current direction is determined by the topography of the strait, i.e. only two directions are dominant;

(ii) prevailing easterlies and westerlies cause the strongest currents;

(iii) outgoing currents are caused by winds which blow from the directions NNW through E to SSW, and ingoing currents result from winds blowing from SSW through W to NNW;

(iv) an outgoing residual current can be observed if the wind speed is below approximately 4 m sec^{-1}.

Surface drift currents have a significant influence on the currents in the deeper layers and this is particularly true for the entrances to the Baltic and its shallower western parts. An outgoing surface current generates an ingoing counter-current of more saline water.

15.2.6. MECHANISMS OF INFLOW INTO THE BALTIC PROPER

The inflow of high salinity water through the Danish sounds and Belts and further into the Baltic proper is highly dependent on the meteorological conditions in the area of the Skagerak and Kattegat. The distribution and tracks of depressions and high pressure areas (usually from west to east) and persistent strong northwesterly winds favour an inflow into the Baltic. However, the extent of penetration, if any, of the incoming water also depends on the density of the water in the different basins and on the stability of the layering.

Three water layers can be distinguished in the Baltic proper during most times of the year. (i) A surface layer having a relatively low salinity and a temperature which varies seasonally, (ii) an intermediate layer of colder water having a salinity almost equal to that of the surface water (this layer is usually renewed regularly from the converging surface water during the winter season) and (iii) a bottom layer below the permanent halocline consisting of water which, although somewhat saline, is far less so than is ocean water. The temperature of this deep water is moderate and only varies slightly.

Small amounts of incoming water are trapped in the Arcona Basin in which they become partially mixed with the surface water and are subsequently transported out again. It is for this reason that the Arcona Basin is sometimes referred to as the mixing chamber of the Baltic. If sufficient saline water enters, penetration into the Baltic proper may occur. One of the following three inflow models might then be applicable.

In general, the bottom water passing through the Bornholm Gatt has a density which is too low for it to replace the deep water in the Bornholm Basin. It is, however, dense enough to intrude between the Baltic surface water and the old water below the permanent halocline. This intrusion will occur below the halocline, which may therefore be elevated slightly. This incoming water influences the hydrographic conditions in the Hanö Bight in which intermixing of old and new water may be observed. Most of the inflowing water progresses further into the Baltic. The slope of the halocline favours this eastward motion. However, the water has a density which is usually insufficiently great for it to displace the old bottom water in the basins of the Baltic proper. Eastward of the Slupsk sill there are no further sills having depths of less than 100 m except that which separates the Åland Sea from the Central Baltic. The sills and the small basins northward of the Gotland Basin act as hydrographical filters. Only water having a very narrow density range is able to penetrate any further. (This density range is subject to long term changes.) If the water finally succeeds in reaching the Landsort Deep and the Western Gotland Basin the short term changes in density will be very small indeed.

This type of inflow behaviour is by far the most frequent one and is almost certainly the cause of the density structure shown in Fig. 15.12.1; however, it has no bearing on the exchange of the deep water in the central basins.

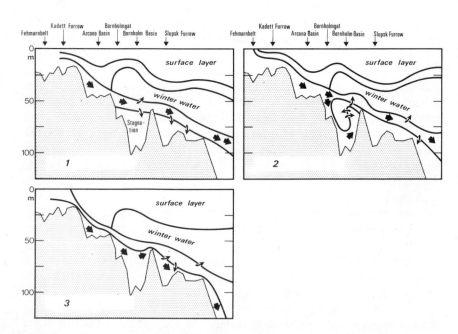

FIG. 15.12. The three different inflow conditions in the Western Baltic and in the Bornholm area.

On the contrary, the exchange of water from the deeper layers (with its accumulated dissolved material) with the productive surface layer is inhibited.

Sometimes favourable wind conditions promote the intrusion of relatively dense water through the Bornholm Gatt into the Bornholm Basin. The formation and advance of denser deep water takes place more frequently in the cold seasons. This incoming water may be dense enough to replace the present deep water and lift it upwards. Such a process is favoured by a longer period of stagnation in the basin since this always results in a decrease in density because of the slow downward mixing of Baltic surface water. Often vertical salinity and temperature profiles show step-like distributions. In most instances the rate of inflow is just great enough to replace the bottom water in the Bornholm Basin, but is insufficient to fill the basin up to the height of the sill. This type of inflow is illustrated in Fig. 15.12.2.

Ocassionally, there is enough incoming water first to replace the deep water in the Bornholm basin, to pass over the Slupsk Furrow and then to penetrate across the eastern sill of the furrow. There the boundary slopes down either to the Gdansk Deep, or more gradually down into the Gotland Deep. The old stagnant water, which is usually rich in nutrients, is lifted upwards and may eventually be displaced into regions in which the density gradient is too small for upward mixing to be inhibited. The mixing which takes place brings about a significant fertilization of the surface layer. This rather uncommon type of inflow phenomenon is illustrated in Fig. 15.12.3. It seems probable that after it has replaced the water in the Bornholm Deep the inflowing water is no longer sufficiently dense to intrude also below the older bottom water and that it moves further north as a layer just below the halocline and is therefore out of contact with the deep water.

Actual inflow patterns of water are seldom as clear cut as the above models would suggest and, in fact, a great variety of intermediate combinations may occur. It will be appreciated that the mechanisms outlined above (which may well be applicable to other basins in the Central Baltic) are complicated and that it is not possible at present to predict what will happen under any particular meteorological conditions. It seems probable that it will be possible to make forecasts about the exchange of the deep water of the Baltic proper only on the basis of long-term analysis of hydrographical and meteorological records, and it is important that efforts be made to accumulate such data. A knowledge of these exchange and renewal processes is of considerable value as it can give information about the drastic changes in the environment which occur during periods of stagnation and successive intrusions of new water.

The renewal of the deep water in the Landsort Deep, south of the archipelago of Stockholm takes place by a different mechanism. The narrow

Landsort Deep (456 m) is the deepest spot in the Baltic, but is only ~3·2 km broad and 20 km long. The sea floor slopes downwards from a depth of less than 200 m to the deep within a few kilometres. Between the Landsort Deep and the Arcona Basin there are altogether 5 sills and 5 basins; these sills and basins act as density filters. Water which succeeds in passing the last sill has a very restricted range of densities. If the incoming water is slightly denser than the old water in the deep it will tend to sink to the bottom. However, because there is a negligible density gradient in the Landsort Deep below 150 m and also because of the hole-like topography it is possible for the old water to mix thoroughly with the new water as the latter flows downwards. The amount of energy necessary for such mixing seems to be very small. For this reason the intrusive new water becomes completely mixed with deep water from 150 m to 456 m having a similar potential temperature profile. This ease of mixing is the reason why the Landsort Deep retained its aerobic conditions until as late as 1969 (see Fig. 15.13), even when anaerobic deep

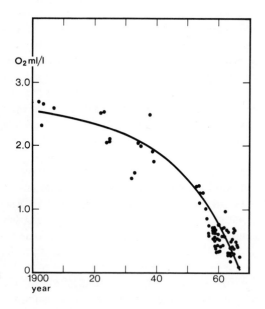

FIG. 15.13. Mean values of oxygen (ml l^{-1}) in the Landsort Deep from 1902 to 1967 (Fonselius, 1969).

water had been formed elsewhere, and also why long term trends in stagnation can be best observed in the Landsort Deep. It is for these reasons that despite its narrowness the Landsort Deep deserves detailed hydrographical examination.

15.2.7. ANOMALIES IN THE COMPOSITION OF BALTIC WATER

The dissolved components in sea water may be divided into the conservative and the non-conservative ones. Nutrients and dissolved organic compounds belong to the latter category as also do many of those metals which are indispensable to living organisms. The concentrations of conservative elements in sea water depend on a number of factors, both physical and chemical. Because of their relatively high concentrations they are not affected to a significant extent by biological processes. Thus, although potassium is essential for growth of organisms it shows no detectable variation in its concentration relative to chloride. Under certain circumstances the chemical and biological behaviours of the dissolved inorganic components are controlled by physical processes, e.g. the uptake of nutrients in response to light in photosynthetic processes, and the pressure adaptation of organisms by rearranging the Na/K/Ca/Mg ratios in their cell fluids.

Thus, the concentration of conservative elements in any body of water will be controlled macroscopically only by physical (e.g. horizontal and vertical exchange) and geochemical processes. It would therefore be expected that the relative concentrations of these elements in sea water will be approximately constant. However, it must be remembered that the attainment of geochemical equilibrium is not instantaneous, and, if the water is not well mixed, deviations from such constant relationships are to be expected. This is especially so for areas in which waters of completely different compositions are admixed (e.g. river water with sea water), and also for water bodies subject to rapid changes of temperature.

The relative proportions of the conservative elements in sea water are usually expressed relative to chlorinity, and anomalies are defined relative to the proportions of these elements in ocean water. Because of the unusual hydrographical features of the Baltic it might be expected that these elements would show significant anomalies. Discharge from more than 200 rivers adds annually about 480 km^3 of fresh water from drainage areas of a variety of different geological types (Mikulski, 1970). In addition, the water brought into the Baltic by precipitation (183 km^3 yr^{-1}; Brogmus, 1952) has a composition, which differs significantly from that of distilled water. Equilibria are further complicated by the presence of dissolved organic substances having considerable complexing properties, e.g. humic substances, and by variations in the redox conditions. When these factors are borne in mind it is surprising that the relative proportions of the major elements in Baltic water are so similar to those in ocean water.

15.2.7.1. *Physical relevance of the anomalies*

Knudsen (1903) established a relationship between salinity and chlorinity in sea water, viz.

$$S\%_0 = 1\cdot8050 \ Cl\%_0 + 0\cdot0300$$

this equation being based on the analysis of only 13 different samples, 8 of which were taken from the Baltic and the transition area. The constant term takes account of the mean anomalies of the Baltic water which derive from the dilution of ocean water with river water. More recently, salinity has been redefined on the basis of the relative conductivity of sea water (Cox *et al.*, 1962) and the relation between salinity and chlorinity has been internationally defined by the expression (Wooster *et al.*, 1969):

$$S\%_0 = 1\cdot80655 \ Cl\%_0$$

Of the more than 100 sea water samples from all parts of the world ocean which were used as the basis for this equation, only a few were from the Baltic. It has, however, been stated that deviations from the relationship of more than $0\cdot02\%_0$ might occur in waters with a salinity below $20\%_0$.

Salinity itself is of secondary interest to the physical oceanographer since it is used for calculation of the density, a parameter which is extremely difficult to determine directly. In practice, the density is therefore usually calculated from measurements of either the relative conductivity, or the chlorinity. However, if a water of anomalous composition is being examined the relationship between the measured parameter and the density (more correctly specific gravity) will be different from that of normal sea water. With such waters it will be necessary to know the magnitude of these deviations if reliable estimates of density are to be made from measurements made by these techniques. Under these circumstances it might be worthwhile for particular water bodies to derive other empirical relationships between the density and conductivity or chlorinity. This would be permissible only if it is certain that the major ions in the water do not vary significantly in their relative proportions with either time or location.

15.2.7.2. *Sampling problems*

When attempting to prove the existence of statistically significant anomalies in the composition of Baltic water, it is essential that when sampling is done due regard should be paid to the hydrographical conditions, i.e. to the layering, and the morphological division of the area. Samples from the highly saline stagnant deep water, the intermediate layers and the surface waters must be considered individually. Furthermore, attention should be paid to the possibility that seasonal and long term changes in composition may occur. It must be emphasised that only accurate and well-tested analytical methods should be used as the variations encountered are usually only small. Precise complete analyses for the major constituents of Baltic waters covering all parts of the Baltic over a period of five years have been published recently by Kremling (1969, 1970, 1972).

15.2.8. CATION ANOMALIES

15.2.8.1. *Sodium*

Kremling (1969) has found an average Na/Cl ratio of 0·5547 (based on 40 samples) for the entire Baltic. On the basis of the calculated standard deviations of the individual measurements of ($\pm$ 0·0021) this value is not significantly different from that of 0·5555 for ocean waters (Culkin, 1965). Indeed, no significant anomaly would be expected as the ratio for river water is 0·4–0·6. Therefore even in Baltic waters having such low chlorinities as 1·213‰ (Bay of Bothnia) and 0·434‰ (Newa estuary) the deviations from the oceanic Na/Cl ratio are insignificant and too small to be determined with the available methods.

15.2.8.2. *Potassium*

Using a gravimetric method employing precipitation as potassium tetraphenyl boron, Kremling (1969) has found a K/Cl ratio for Baltic water ranging from 0·19 to 0·213. The relative standard deviation of $\pm$0·48% corresponds to a scatter in the ratio of $\pm$ 0·0001. These values are to be compared with the mean values of 0·0206 (with extreme values of 0·0203 and 0·0209) for oceanic waters found by Cox and Culkin (1966). Samples from 17 different stations gave an average K/Cl ratio of 0·0206 which is similar to that found for oceanic waters. However, the standard deviation for 40 individual determinations from all parts of the Baltic was 0·00059 which is about six times the analytical error of the potassium determination. This large deviation is mainly caused by the abnormal K/Cl ratio found for the Fehmarn Belt, the Bornholm Gatt and the Bornholm Basin as well as by the high values of 0·0214 and 0·0216 in the Åland Sea and the Bay of Bothnia respectively. These anomalies are not adventitious but have been found repeatedly over several years; their cause is uncertain. Data on the K/Cl ratios of 100 Baltic samples are summarized in Table 15.6.

15.2.8.3. *Calcium*

It has been proved by numerous workers (v. Brandt, 1936; Gripenberg, 1937; Wittig, 1940; Rohde, 1966; Kremling, 1970) that of all the major element/

TABLE 15.6
Mean values of the K/Cl ratio in the Baltic from 1966–1968 (Kremling, 1970).

Sampling	Number of samples	Range of Cl‰	$\text{K/Cl} \left[\dfrac{\text{g/kg}}{\text{‰}}\right]$ mean value	Standard deviation (mean value)	Standard deviation (measurement)
Spring 1966	40	0·434–10·981	0·0206	±0·0006	±0·0001
Spring 1967	16	4·079–12·108	0·0203	±0·0005	±0·0001
Fall 1967	22	3·669–11·577	0·0204	±0·0004	±0·0001
Spring 1968	32	3·953–11·016	0·0205	±0·0005	±0·0001
Ocean (Cox and Culkin, 1966)			0·0206	±0·0002	

chlorinity ratios of Baltic water that of calcium is the one showing the greatest excess over the ratio in ocean water. This is the result of the high Ca/Cl ratio of the river waters which are discharged into the Baltic. For example a value of approximately 1·5 has been found for the Vistula over the period between 1958 and 1960 (Trzosinka, 1968); this should be compared with the value of 0·02125 for ocean water (Cox and Culkin, 1966). It is therefore not surprising that Ca/Cl ratios as high as 0·0268 occur off the Vistula estuary (Kremling, 1969). In contrast, the calcium concentrations in the intermediate and deep waters are relatively constant in relation to chlorinity (Table 15.7).

For those surface waters having chlorinities below 4·5‰ there is a relatively large scatter in both the slope and the constant term of the regression equations for Ca and Cl. The reasons for this scatter are mainly statistical. Over this small range of chlorinities (most samples lie between 3·9 and 4·3‰) the relative large error of the calcium determination (coefficient of variation ± 0·43%) causes a large scatter in the slope of the equation. However, in addition, variations in ths discharge of calcium by rivers will significantly affect those surface waters having low chlorinities. For such waters it is necessary to take both long-term and seasonal fluctuations of the river discharge into consideration.

15.2.8.4 *Magnesium*

Although Voipio (1957) has claimed that the Mg/Cl ratio of Baltic water (0·06692) is almost the same as that of ocean water (0·06680; Cox and Culkin, 1966) more recent analyses have shown that this is probably not the case. Numerous determinations for the Western and Central Baltic waters have been reported (see e.g. Nehring and Rohde, 1966). Rohde (1967) found a Mg/Cl ratio of 0·0674 with a standard deviation of ± 0·0004 and Kremling

TABLE 15.7

Ca/Cl‰ Correlation in the Western and Central Baltic

Sampling period	Area	Number of samples	Cl-range	Equation Ca (g kg^{-1})	Source
1967	Western and Central Baltic			$0·0204 \times Cl‰ + 0·0226$ $0·0204 \times Cl‰ + 0·0239$	Gripenberg (1937) Trzosinka (1968)
1967	Western and Central Baltic	12	$<4·5‰$	$0·0216 \times Cl‰ + 0·0174$	Kremling (1968)
1967	Western and Central Baltic	20	$>4·5‰$	$0·0200 \times Cl‰ + 0·0247$	Kremling (1970)
1968	Western and Central Baltic	15	$<4·5‰$	$0·0201 \times Cl‰ + 0·0234$	Kremling (1970)
1968	Western and Central Baltic	13	$>4·5‰$	$0·0204 \times Cl‰ + 0·0218$	Kremling (1970)
	Baltic (general)		$>4·5‰$	$0·0228 \times Cl‰ + 0·0127$	Trzosinka (1968)

(1969, 1970) reported values of 0.0671 ± 0.003 for 1967 and of 0.0672 ± 0.002 for 1968 for all parts of the Baltic on the basis of 38 and 32 samples respectively (Table 15.8). These recent findings point to a slightly positive magnesium anomaly for the Baltic. However, because of the relatively large error

TABLE 15.8

Mean values of the Mg/Cl ratio in the Baltic from 1966–1968 (Kremling, 1970)

Sampling period	Number of samples	Range of chlorinities	$Mg/Cl \left[\dfrac{g/kg}{\%_0}\right]$ mean value	Standard deviation (mean value)	Standard deviation (measurement)
Spring 1966	39	1·213–10·981	0·0672	± 0.0003	± 0.0004
Spring 1967	16	4·079–12·108	0·0671	± 0.0003	± 0.0004
Autumn 1967	22	3·669–11·577	0·0672	± 0.0003	± 0.0004
Spring 1968	32	3·953–11·016	0·0672	± 0.0002	± 0.0004
Ocean (Cox and Culkin, 1966)			0·06692	± 0.00004	

in the magnesium determination it is not possible to state this with certainty. The probability of there being a positive magnesium anomaly in the Baltic is strengthened by the fact that there is a positive difference between the measured excess of anions and cations; this cannot be explained without invoking a magnesium anomaly.

15.2.9. ANION ANOMALIES

15.2.9.1 *Sulphate*

Available data suggest that the more saline Baltic waters ($S > 8\%_0$) do not show a sulphate anomaly (see Table 15.9). However, the relatively large standard deviations of the results which are about four times the analytical error point to local deviations from the oceanic ratio (0·1400; Morris and Riley, 1966). A significant deficit has been observed (Kremling, 1970) during the spring and autumn of 1965 and 1968, especially in the deeper layers of the Fehmarnbelt. The SO_4/Cl ratio was as low as 0·315 in Spring 1967, but the origin of this large negative anomaly could not be traced. Average SO_4/Cl ratios for all areas of the Baltic for the years 1967 and 1968 are summarized in Table 15.9.

The formation of hydrogen sulphide in the Bornholm Basin and in the

TABLE 15.9

SO_4/Cl *ratios in the Western and Central Baltic (Kremling, 1970)*

Sampling period	Number of samples	Chlorinity range	$SO_4/Cl \left[\dfrac{g/kg}{\%_{oo}} \right]$	Standard deviation
1967	12	$< 4.5\%_{oo}$	0.1410	± 0.0009
1967	26	$> 4.5\%_{oo}$	0.1400	± 0.0009
1968	15	$< 4.5\%_{oo}$	0.1405	± 0.0009
1968	17	$> 4.5\%_{oo}$	0.1400	± 0.0007

Eastern and Western Gotland Basins should theoretically lead to a negative sulphate anomaly as most of the sulphide–sulphur originates from the reduction of sulphate (according to Deuser (1970) only about 3% of the sulphide sulphur is derived from organically bound sulphur). Some of the sulphur is lost from the cycle when the sulphide is oxidized to sulphur and thiosulphate. Sulphide is also removed from the cycle when it combines with iron or other metal ions and is lost to the sediment. Significant sulphate anomalies have been observed in waters with a very high content of hydrogen sulphide, e.g. the deeper waters of the Black Sea (Redfield *et al.*, 1963). As the maximum concentration of hydrogen sulphide which has been observed in the Central Baltic is only ~ 40 μg-at l^{-1} it would not be expected that significant sulphate anomalies would occur in the anaerobic waters. According to Bather and Riley (1954) the precision of the gravimetric sulphate determination is about $\pm 0.16\%$. This implies an analytical error of approximately ± 1.9 mg kg^{-1} in Baltic water (sulphate content 1.18 g kg^{-1}). The cause of the sulphate anomalies can not therefore be loss through sulphide formation. Earlier analyses (Thompson *et al.*, 1931) for waters of Bothnian Bay and the Eastern Gotland Sea gave SO_4/Cl ratios of 0.1412–0.1419 for the surface water and 0.1404 for the deep water. Zarin and Ozolin (1935) have reported values of 0.1410 for the Gotland Sea and 0.1411 for the Lettonian coastal water. Trzosinska (1967) found a mean ratio of 0.1436 for 26 samples for the Bornholm Sea, the Arcona Sea and the Bay of Gdansk. However, some of her individual values were considerably higher. The main cause of the positive sulphate anomaly of the surface waters is certainly dilution with river water which has a SO_4/Cl ratio of 0.3–0.6 (Rubey, 1951).

15.2.9.2 Fluoride

The F/Cl ratios of Baltic waters vary significantly from one area to another and with depth. In addition, seasonal variations have been observed. Thus, values in the autumn have been found to be significantly greater than those in the spring (see Fig. 15.14; Table 15.10). These changes have been found

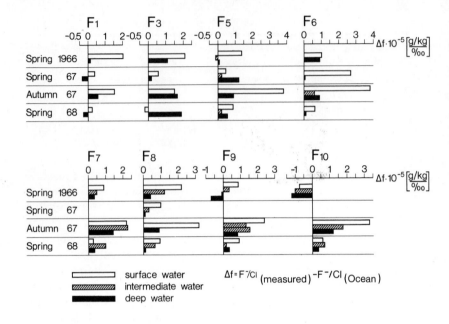

FIG. 15.14. Anomalies of the F^-/Cl ratio in the Baltic between spring 1966 and spring 1968 (Kremling, 1970).

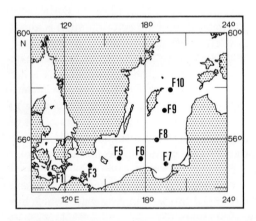

to be most pronounced in the surface waters at stations in the Central Baltic for which striking differences have been observed (Kremling, 1969, 1970). The F/Cl ratios during autumn, 1967, were larger than that of ocean water (Greenhalgh and Riley, 1963), The average decrease of the fluoride content

TABLE 15.10

F/Cl ratios of waters of the western and central Baltic from 1966–1968 (Kremling, 1970)

Sampling	Number of samples	Range of Cl‰	$F/Cl \times 10^5 \dfrac{g\,kg^{-1}}{\permil}$			Standard deviation (measurement)
			Average	Maximum	Minimum	
Spring 1966	32	3·813–10·981	7·24	8·87	5·51	±0·07
Spring 1967	16	4·079–12·108	7·33	9·37	6·37	±0·07
Autumn 1967	22	3·669–11·577	8·45	10·55	7·29	±0·07
Spring 1968	32	3·953–11·016	7·12	8·59	6·37	±0·07
Ocean			6·70			±0·07
(Greenhalgh and Riley, 1963; Riley, 1965)						

of Baltic waters between autumn and spring for water of chlorinity 5‰ amounted to about 0·07 mg kg^{-1}. There are three possible explanations of these changes.

(1) They may be produced by seasonal variations in the F/Cl ratios of the water entering the Baltic. Values of 0·605 to 0·090 with an average of 0·166 mg Fl^{-1} have been reported for Swedish rivers (Kullenberg and Sen Gupta, 1972). However, it is not possible to explain positive deep water anomalies in this way.

(2) As suggested by Sillén (1961), fluoride may be taken up by phytoplankton along with silicate, and may replace O^{2-} or OH^- groups and subsequently be released during decomposition.

(3) They may be caused by biological uptake of fluoride; in this context it is of interest to note that some mysidae, crabs and snails are able to bind considerable amounts of calcium fluoride.

A significant reduction in the F/Cl ratio has been observed for stagnant waters (Kullenberg and Sen Gupta, 1972). The fluoride loss over a period of 3 months in 1966, when conditions changed from oxic to anoxic amounted to 17% in the Bornholm deep water and to 30% in the Gotland deep water. This decrease in the fluoride concentration might be explicable by the hypothesis (Arrhenius, 1963) that accumulation of skeletal apatite occurs in those shallow areas in which the low pH of the bottom waters and high phosphate concentration causes it to be supersaturated with respect to this phase. This suggests that fluoride is lost from the water to the sediment, at least during periods of stagnation in the deeper parts of the Baltic, in association with magnesium ions or with the formation of apatite.

The average cation and anion concentrations for all parts and depths of the Baltic over the period 1966–1968 are summarized in Table 15.11.

TABLE 15.11

Range of mean relationships between ionic concentrations ($g \ kg^{-1}$) and chlorinity ($\%_0$) in the Baltic from 1966–1968 (Kremling, 1970)

Ion	Baltic	Ocean $\dfrac{g \ kg^{-1}}{\%_0}$	Excess (+) deficiency (−)
Na^+	0·5547	0·5555	0
K^+	0·0203–0·0206	0·0206	0 (but local anomalies)
Ca^{2+}	$<4·5\%_0$ Cl: $=0·0201$ Cl$\%_0$ + 0·0234 $=0·0216$ Cl$\%_0$ + 0·0174		+
	--------------------	0·02166	
$(+Sr^{2+})$	$>4·5\%_0$ Cl: $=0·0200$ Cl$\%_0$ + 0·0247 $=0·0204$ Cl$\%_0$ + 0·0218		+
Mg^{2+}	0·0671–0·0672	0·0669	(+)
Alkal.	0·303–0·342 $\left[\dfrac{\text{meq } l^{-1}}{\%_0}\right]$	0·123 $\left[\dfrac{\text{meq } l^{-1}}{\%_0}\right]$	+
SO_4^{2-}	$<4·5\%_0$ Cl: 0·1405–0·1410		+
	--------------------	0·1400	
	$>4·5\%_0$ Cl: 0·1400–0·1403		0 (but local anomalies)
Br^-	$=0·00328$ Cl$\%_0$ + 0·0005 $=0·00348$ Cl$\%_0$ + 0·0004	0·00347	−
H_3BO_3	B (mg kg^{-1}) = 0·220 Cl$\%_0$ + 15	0·230 $\left[\dfrac{\text{mg kg}^{-1}}{\%_0}\right]$	+
F^-	spring 1966: $7·24 \times 10^{-5}$ spring 1967: $7·33 \times 10^{-5}$ autumn 1967: $8·45 \times 10^{-5}$ spring 1968: $7·12 \times 10^{-5}$	$6·70 \times 10^{-5}$	+

15.2.10. DETERMINATION OF SALINITY

From the above it is evident that the constancy of the relative proportions of the major ions observed for ocean waters is not found for Baltic water. For this reason there is considerable difficulty in applying normal methods for the determination of the salinity to the latter as they are based on the concept that the major ions are present in the same relative proportions as they are in sea water. The salinity of Baltic water can be estimated most accurately by summation of the concentrations of the individual major cations and anions, correction being made for conversion of bromide to chloride and for removal of carbonate as required by the definition of salinity (Knudsen, 1903). At low salinities this method is more reliable and exact than is the direct gravimetric determination of the salt content (Morris and Riley, 1966). The summation method has been found to have an accuracy of $\pm$ 0·0075‰ S at a chlorinity of 5‰ (Kremling, 1969).

It is of importance to the physical oceanographer to know the magnitude of the difference between the salinity determined by means of a salinometer (conductivity measurement) and that estimated by the summation method, or by chlorinity titration. The latter correlation is necessary since chlorinity determinations are still carried out. Such comparisons are also necessary as they enable earlier data to be re-evaluated. In addition, it is of interest to compare salinities obtained by the summation method with those calculated using the currently accepted conversion formula S‰ = 1·80655 (Cox et al., 1962). As shown in Fig. 15.15 the application of the constant conversion factor leads to salinities which are low in comparison with the true salt content. For the Baltic, the deviations range from approximately 0·015 to 0·030 at low chlorinities to less than 0·010 at higher ones. The mean deviation of all values from 1966 to 1968 is 0·014–0·018‰. This is not surprising because the conversion of chlorinity to salinity by means of a constant factor does not take into account the variable ion/Cl ratios of the river water which enters the Baltic annually in an amount equal to about 1/40 of its total volume. It is not valid to calculate the salinity in this way if the chlorinity lies below 10‰ (see e.g. Cox et al., 1962). At higher chlorinities the composition of the Baltic water approximates to that of mean ocean water. The deep water from the Fehmarnbelt which shows appreciable negative anomalies in its K/Cl and SO_4/Cl ratios is unusual. These negative anomalies result, even at relatively large chlorinities, in a significant negative difference between the true salt content and the calculated salinity. The differences varied between $-0·059‰$ and $-0·031‰$ during 1967 and 1968.

Salinities of Baltic waters evaluated from relative conductivity measurements agree well with those assessed by the summation method. Especially in the low chlorinity range the differences are, with few exceptions, less than 0·01‰. Kremling (1970) has observed that for these exceptions the deviations, which are caused by the relative small contribution to the partial conductivity of the excess ions, are almost always positive, and it is possible to account for them using calculations based on partial molar conductivities. Large, and at

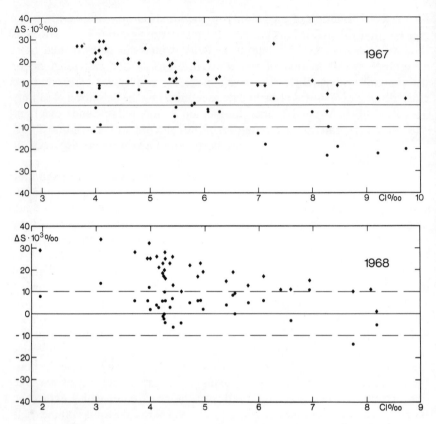

FIG. 15.15. Differences between "true" and indirectly determined salinity. (●) $\Sigma : \Sigma(C^+ + A^-)-$ S‰ (salinometer); (◆) $\Sigma : \Sigma(C^+ + A^-)-1\cdot80655\cdot Cl$ (titrated). (Kremling, 1972).

present unexplained, negative deviations are observed for the deep waters of the Baltic. These are particularly evident for the Fehmarnbelt, for which variations of between $-0\cdot096$‰ and $-0\cdot059$‰ were observed in spring 1967 and 1968. It is well known (Park, 1968) that biochemical processes change the conductivity without changing the salinity. However, as the parameters of the carbonate system and the pH do not differ significantly from those of the Standard Sea water used for calibration of the salinometer (Copenhagen Standard Sea Water) it is unlikely that primary production or processes connected with remineralisation are responsible for the anomalies. This is also true for the deep water of the Baltic proper, in which the cumulative effects of the mineralisation and nutrients accumulation can hardly affect the conductivity by more than the equivalent of $0\cdot02$‰ salinity.

It has been stated above that the relationship between the true salt content, i.e. the sum of all major cations and anions, and the chlorinity can be approximated satisfactorily by equations similar to that proposed by Knudsen (1903). It would be expected that it would be necessary to use different equations for the waters originating from the various catchment areas; these equations would probably differ regionally or according to the different degrees of admixture of fresh water with saline water. Kremling (1970) has classified 108 samples from all regions of the Baltic into 3 groups according to their chlorinities and he has proposed equations with a constant term for each range of chlorinities as shown in Table 15.12. The standard deviations

TABLE 15.12

Relationships between S‰ (cations + anions) and Cl‰ in the Baltic from 1966–1968 (Kremling, 1970)

Sampling	Number of samples	Range of Cl‰	Correlation	Standard deviation (equation)
1966	38	3·196–10·981	$S‰ = 1·802_8 \, Cl‰ + 0·041$	$S_a = \pm 0·006_5$ $S_b = \pm 0·000_6$
1967	36	3·669–9·734	$S‰ = 1·802_5 \, Cl‰ + 0·040$	$S_a = \pm 0·005_5$ $S_b = \pm 0·000_6$
1968	34	1·959–8·193	$S‰ = 1·802_5 \, Cl‰ + 0·039$	$S_a = \pm 0·006_0$ $S_b = \pm 0·000_8$

S_a is the standard deviation for the factor a and S_b that for the term b in the equation $S‰ = a \cdot Cl‰ + b$.

within these three chlorinity ranges is surprisingly low and much less than the one found in Knudsen's investigation. For the measurement of the salinity and the calculation of the density of Baltic waters, conductivity measurements are superior to the conventional (titrimetric) methods.

5.2.11. THE DISTRIBUTION OF DISSOLVED OXYGEN (see also Chapter 16)

A permanent oxygen deficiency exists in the Baltic below the halocline. There are three principal reasons for this (i) the limited exchange of water with the open ocean, (ii) the existence of a permanent halocline in almost all parts of the Baltic, (iii) the stabilization of the water during certain seasons of the year as additional thermoclines and haloclines develop. The impoverishment of the oxygen is increased by the influx of organic material brought into the Baltic by the rivers etc. It can be assumed that a more or less pronounced oxygen deficit has prevailed over extensive deep areas of the Baltic since it attained its present form. The exchange and renewal of deep water is highly

R

dependent on meteorological conditions and the frequency with which aerated water enters from the open ocean is linked with local climatic fluctuations. During unfavourable periods oxygen may be totally stripped from the semi-stagnant deep water and the generation of hydrogen sulphide may ensue. Hydrogen sulphide has also been observed in the Bornholm Basin and in the entrance of the Gulf of Finland, but geological and sedimento- logical observations suggest that periods of hydrogen sulphide development occurred in the middle ages and during the 18th century.

Figure 15.16 is a longitudinal section through the Baltic showing the distribution of the dissolved oxygen in August 1969. As can be seen, all the deeper parts of the Baltic proper are oxygen deficient and hydrogen sulphide is present in the Gotland Deep.

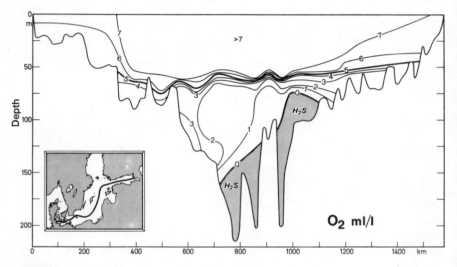

FIG. 15.16. The distribution of dissolved oxygen along a longitudinal section in the Baltic in August 1969 (IBY data).

15.2.12. VARIATION OF SALINITY, TEMPERATURE AND DISSOLVED OXYGEN

Since about 1900 the hydrography of the Baltic has been studied intensively and it is, in fact, the only enclosed sea for which there are records not only for the coastal zones, but also for the open sea. With the exception of the years during the two world wars regular cruises have been carried out since 1902 and repeated observations have been made at a number of selected stations over a long period. These observations have revealed that the mean oxygen content of the deep water has decreased steadily since 1900. This is a matter of grave concern because of its deleterious effects on the Baltic as a whole.

15.2.12.1. *Variation of salinity*

The salinity of both the deep and surface water show considerable horizontal variations. In general, these variations are more common and are larger in the western parts of the Baltic. Three kinds of salinity variations can be distinguished: seasonal fluctuations, irregular short term variations and long term fluctuations with periods of several years up to decades and possibly even centuries.

Regular seasonal changes in salinity can be observed in the surface water of the Baltic (see e.g. Fig. 15.17). They are caused mainly by the seasonal variations in the river discharge and are of minor importance to the marine environment.

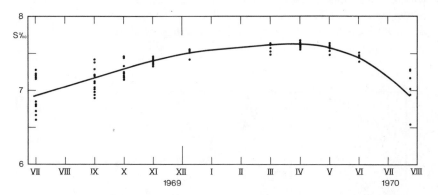

FIG. 15.17. Annual variations in the salinity of the surface water of the Gotland Sea.

Numerous investigations (Fonselius, 1962, 1967, 1969a, b; Granquist, 1952; Hela, 1966; Segerstråle, 1951; Nehring *et al.*, 1971) have shown that there has been a significant increase in the mean salinity and in the mean temperature of the Baltic since about 1900. This has been found to be the case for the whole water column of the Bornholm Sea, the Gotland Sea and even for the northern and eastern parts of the Baltic. The increase is most evident for the deep water in which seasonal changes are less pronounced. Earlier observations by Segerstråle (1951) suggested that the salinity of Baltic deep water has increased on average by 0·5‰ since the beginning of the 20th century. However, in later work Fonselius (1969a, b) found an increase of more than 1‰ in certain areas. This increase has not been continuous and the average salinity decreased for a period. This decrease was followed by a further increase during 1938/39 which probably continued during the war years. Since 1947 the salinity of the surface water has remained above 7‰ and that of the deep water above 12‰. During 1951/52 a rather large inflow of saline water further increased the salinity in the deeper layers of the central Baltic (Wyrtki, 1954), and salinities of 13‰ occurred in the Gotland Deep at depths greater than 150 m. Subsequently, the salinity decreased slightly until the autumn of 1969 when a significant increase set in and was still continuing in 1973. Fonselius (1970) has studied these intermittent alternating periods of inflow of saline water from the Kattegat and of stagnation. Each inflow coincided with an increase in the salinity of the deep water and this resulted in turn in a temporary increase in the stability of the deeper layers. In addition to the general increase of the salinity which may have serious consequences for the Baltic, and particularly for the Gulf of Bothnia, another phenomenon has been observed. The 8‰ isohaline, which, according to Hela (1966), can be chosen as the upper boundary of the permanent halocline in the Baltic proper has risen from a depth of about 80 m at the beginning of the 20th century by about 20 m. This means that the volume of the deep water having limited exchange with the surface

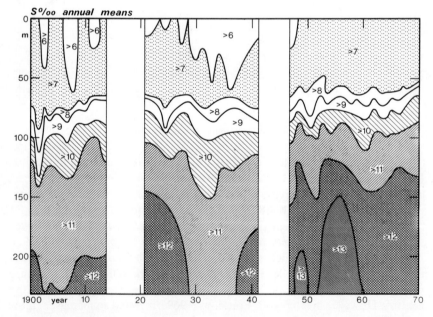

Fig. 15.18. Yearly salinity means at a station in the Northern Gotland Sea (Fonselius, 1971).

layers has increased by approximately $200 \, km^3$, i.e. by about 10% of its total volume.

The salinity isopleths in the Central Baltic from 1900 to 1970 are shown in Fig. 15.18. It is evident that not only the average salinity, but also the stability have increased (Hesselberg, 1918) in the layers between the $8\%_0$ and $10\%_0$ isohaline. Such an increase in the stability of the boundary between the relatively well mixed and aerated surface water and the deep water (with its tendency to permanent stagnation) will lead to less vertical exchange of dissolved matter, both upwards and downwards. The process of stagnation

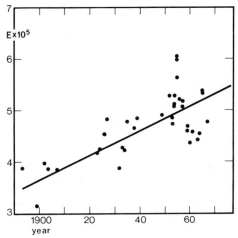

FIG. 15.19. Stability of the 50–100 m layer of the Landsort Deep; June, July and August 1893–1968 (Fonselius, 1970). ($E = 1/\rho \cdot d\rho/dz$ where ρ is the density of the water and z is depth).

is thereby accelerated. Figure 15.19 shows the average increase of the stability of the 50–100 m layer in the Landsort Deep from 1893–1968.

Even in the northern part of the Baltic irregular fluctuations with a general trend towards increasing salinity values have been observed (Hela, 1966). There is no doubt that the rise of the halocline since the beginning of the 20th century has permitted the intrusion of water from below the halocline, usually through the Åland Channel into the Gulf of Bothnia. Since 1933 a significant increase in the salinity of both the deep water and the upper water has been observed. By 1951 the salinity in the upper waters of the Bothnian Sea had increased from about $5 \cdot 3\%_0$ in 1936 to $6 \cdot 1\%_0$ (Fig. 15.20). There is a clear correlation between this increase and that in the Baltic proper (Hela, 1966).

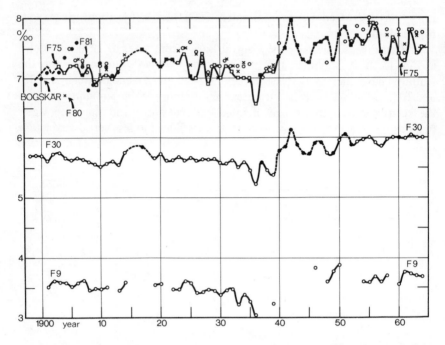

FIG. 15.20. The secular changes in the salinity of the upper waters in the northern part of the Baltic proper (uppermost curves), in the Bothnian Sea (F30) and in the Bothnian Bay (F9) (Hela, 1966).

The changes in the mean hydrographical conditions in the Baltic arose as a consequence of an alteration of the atmospheric circulation pattern which led to climatic changes in the European region (Hupfer, 1962). This also caused a decrease in the average river discharge into the Baltic (see Fig. 15.21). However, decreased precipitation is not the only reason for the increase in salinity. Dickson (1971, 1972) has pointed out that the major incursions of ocean water into the Baltic are connected with year-to-year fluctuations of the salinity in the European shelf seas. Each incidence of high-salinity conditions appears to be associated with the re-establishment of a persistent anomalous pressure pattern over the North Atlantic. In addition, the rise of the Fennoscandian shield led to a fall in the sea level in the northern parts of the Baltic; in contrast, tide gauges in the southern parts showed a rise (Nehring and Francke, 1971). Diminution of the river discharge also caused a decrease in the slope of the sea level in the Baltic from west to east; this in turn may be the reason why the level of the permanent halocline in the Baltic proper has risen. Such a rise would facilitate the ingress of water with higher

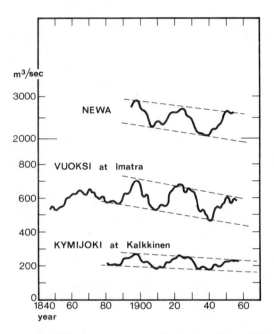

FIG. 15.21. 10 year means of the water transport of the rivers Newa, Vuoksi and Kymijoki (Fonselius, 1969a, b).

salinity through the different basins and over the various sills. This supposition is supported by the fact that the salinity and stability have increased in the Landsort Deep since the beginning of the 20th century.

15.2.12.2. *Variations in the water temperature*

As has been noted above, the temperature of the surface layer in the Baltic shows pronounced seasonal variations which at least in the Central Baltic, extend down only to the permanent halocline (see also Fig. 15.8). The temperature of the bottom water is mainly determined by that of the intruding saline water. Depending on the density structure of the water of an individual basin the inflowing water either forms the bottom layer or intrudes into intermediate layers. In general the fluctuations of the bottom water temperature are greatest in the western parts of the Baltic and least in the central basins (Fig. 15.22).

In addition to the general trend of increasing salinities in the Baltic, it has been proved statistically that there has been an overall increase in the mean temperatures since the beginning of the 20th century (Matthäus, 1973). The

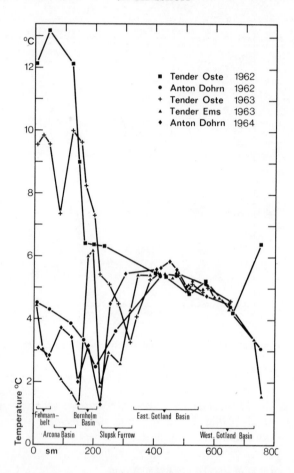

FIG. 15.22. Temperatures of the Baltic bottom waters 1962–1964.

rise has been about 2°C in the deeper layers of the Bornholm Basin, 1·4°C
in the Baltic proper and 1·2°C in the Landsort Deep. Kullenberg (1970) has
pointed out that the temperature of the water below the permanent halocline has a direct bearing on the oxygen content. The rise in temperature
affects the amount of dissolved oxygen in at least two ways. Firstly, the
oxygen content of the incoming water will decrease with increasing temperature because of the decrease in solubility of the gas at higher temperatures.
This will influence the more saline surface water in the Belt Sea, which is
likely later to penetrate into the deeper layers of the Baltic proper. In this way
the actual rise in temperature will lower the original oxygen content by

about 2·5–5%. Secondly, the oxygen content may decrease because processes causing oxygen consumption which are accelerated by about 7% per °C rise in temperature. This corresponds to an increase in the rate of consumption of about 15% in the Bornholm Basin and 7% in the Landsort Deep.

15.2.12.3. *Long term and short term fluctuations in the dissolved oxygen content of the Baltic*

The limited water exchange with the open ocean, the development of a permanent halocline in almost all parts of the Baltic and the topographical structure are the main reasons for the oxygen deficiency in the water below the halocline. The amount of oxygen which diffuses downwards from the relatively well aerated layers above the permanent halocline is not sufficient to balance that consumed by the remineralization of the abundant organic material which results from primary production or is carried into the Baltic by rivers and outfalls. Thus, the only way in which the oxygen concentration in the deeper layers can be enhanced is by means of saline water intrusions from the open sea. However, these processes occur only rarely, to variable extents and at unpredictable times. The oxygen regime in the Baltic appears likely to deteriorate further because the absolute amount of oxygen contained in the incoming water is decreasing both as a result of the temperature effect and because of the increasing load of organic material brought in by discharge of sewage outfalls and industrial wastes (Hermann and Olsen, 1970). In addition, the increasing discharge of phosphorus and nitrogen compounds into the Belt Sea stimulates primary production and yet further increases the supply of oxidisable organic material.

The hydrographical structure of the Baltic suggests that its deep water has always been undersaturated with oxygen since the development of the permanent halocline. Fonselius (1969a, b, 1970) has studied the development of the dissolved oxygen conditions in the Baltic since the first reliable data became available, i.e. since the beginning of the 20th century.

The oxygen distribution in a typical longitudinal section of the Baltic in August 1969 is shown in Fig. 15.16. A strong gradient can be observed in almost all parts of the sea. This gradient coincides with the halocline and is semipermanent in all areas east of the Bornholm Gatt except the inner parts of the Gulf of Finland and the Gulf of Bothnia. The section also demonstrates the development of anaerobic conditions in the Gotland Basin.

Stagnation leading to the formation of hydrogen sulphide in the bottom water of the Gotland Deep was first observed in 1931 (Kalle, 1943). Before 1914 the oxygen concentrations in the bottom water at 200 m in the Gotland Deep was always above 1 ml l^{-1}. During the 1920's, the oxygen values at that depth was generally below $1\,ml\,l^{-1}$, but around 1928–1929 a sudden increase in the oxygen content of the water below the halocline was observed. This was caused by an intrusion of high salinity water which seems to have remained in the basin for several years before stagnant conditions were established in the bottom water. These stagnant conditions came to an end in 1933 (Fonselius, 1969a, b).

From 1952–1961 a long lasting period of stagnation led to a severe decrease in the oxygen content of the water below the halocline in the Central Baltic. During the summer of 1961 the bottom water in the Gotland Basin was replaced with new saline water. It is probable that this decade of stagnation

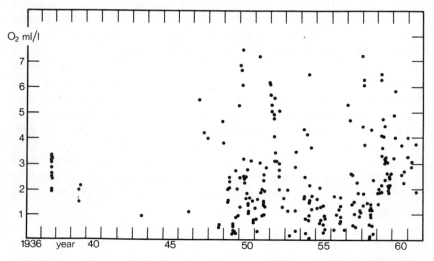

FIG. 15.23 (a). Dissolved oxygen concentrations at 80 m in the Bornholm Basin from 1937–1961.

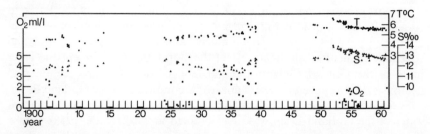

FIG. 15.23 (b). Temperature, salinity and dissolved oxygen concentration at 200 m in the Gotland Basin from 1900–1961 (Fonselius, 1969a, b).

was caused by a large intrusion of water with a comparatively high density in November and December 1951 (Wyrtki, 1954). This intrusion led to a sharp increase in the salinity of the deep water in all parts of the Baltic. In the Gotland Basin the increase at 200 m was 1·2‰ and the temperature rose by about 1°C. Data on the temperature, salinity, and dissolved oxygen values from 1900–1961 in the Gotland and Bornholm Basins are shown in Fig. 5.23.

After the intrusion of new water in 1961 another stagnation period began and eventually lead to the development of anoxic conditions during 1963. The deep water was again renewed in January 1964, and another period of reduced deep water exchange ensued. This alternation between intrusion of new bottom water and stagnation has been studied intensively by Fonselius (1962, 1967, 1969a, b) who was able to show that on average the dissolved oxygen concentration has tended to decrease in the Central Baltic and its northern and eastern parts. As can be seen from Fig. 15.24, the oxygen

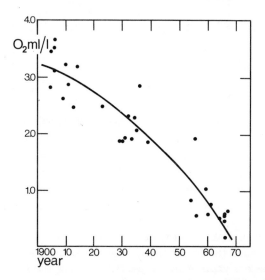

Fig. 15.24. Oxygen content at 100 m in the northern Central Basin 1904 to 1967 (Fonselius, 1968).

content of the deep water in the Northern Central Basin was about 3 ml l^{-1} at the beginning of the 20th century and has decreased to values close to zero at the present time. This general fall has been interrupted from time to time by the inflow of saline water with higher oxygen content. An increasing dissolved oxygen deficiency has also been observed in the Gulf of

Bothnia. Fig. 15.25 shows the mean oxygen values at 175–200 m in the Ulvö
Deep from 1900–1967. The oxygen content of the deeper layers of the Åland
Sea has also significantly decreased. However, this is not the case for the

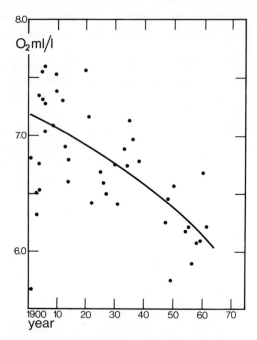

FIG. 15.25. Oxygen content at 175–200 m in the Bothnian Sea from 1900–1967 (Fonselius, 1968).

Bothnian Bay, as there winter cooling processes lead to convergence of the
surface water, the salinity being sufficiently low to inhibit permanent halocline formation.

Both medium and short term fluctuations on these secular changes of
the oxygen content of the deeper layers in the Baltic Basins are superimposed.
These variations are greatest in the western parts of the Baltic and even in
the Bornholm Basin the oxygen content of the deep water shows considerable
variations (see e.g. Ruppin, 1912; Soskin, 1963; Fonselius, 1967, 1968, 1969a, b,
1970; Matthäus, 1973). In a way the Bornholm Basin acts as a trap for less
intense inflows of new water into the Baltic, and, as has been mentioned in
Section 15.2.6, inflows of type (3) occur only rarely. The magnitudes of such
short term fluctuations can also be deduced from Fig. 15.22. Because of
the intensification of hydrographic research in the Baltic it has also been
possible to study the short term fluctuations of the oxygen content in the

deeper layers of the Gotland Basin. A compilation of the medium term variations of oxygen and hydrogen sulphide concentrations at 240 m in the Gotland Deep from 1954 to 1971 is presented in Fig. 15.26. From this it can

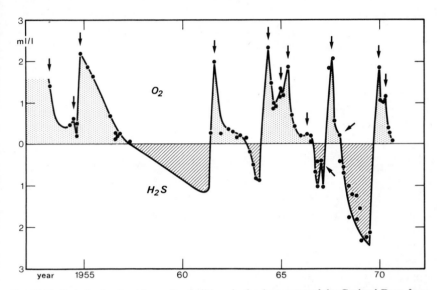

FIG. 15.26. Periods of oxic and anoxic conditions in the deep water of the Gotland Deep from 1950 to 1971 (Fonselius, 1972). Arrows indicate major and minor intrusions of new bottom water.

be seen that there have been a large number of minor intrusions of oxygenated water. It seems likely that the duration of anoxic periods in the deep water is becoming shorter. The increasing amount of hydrogen sulphide in the bottom water suggests that the intensity of the development of anoxic water is becoming greater. This self-amplification of the anoxic periods can be explained as follows:

Micronutrient elements are released into the deep water following mineralization of organic matter. Particularly during periods of stagnation, the major part of the released nutrients accumulate in the water beneath the permanent halocline. During periods of inflow of denser water from the west, and also as a result of "normal" vertical diffusion, some of the released nutrients will reach the photic layers and lead to increased primary production and to considerable plankton blooms. The decomposition of the latter will cause a "secondary" oxygen utilization, which may be between two and five times larger than the original primary oxygen utilization (Fonselius, 1972).

Oxygen and H_2S isopleths for a typical intrusion of new water into the Gotland Deep during the International Baltic Year (1969–1970) are shown in Fig. 15.27.

It will be seen that the inflow of saline water displaced the bottom water upwards and that the H_2S which it contained was oxidised. The initial oxygen content of the intruding water was reduced from more than $1\,ml\,l^{-1}$ to $0.5\,ml\,l^{-1}$, and after 7 months the oxygen content of the water had again been reduced to less than $0.5\,ml\,l^{-1}$.

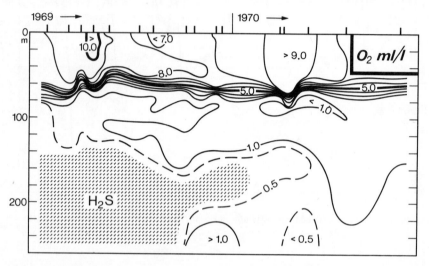

FIG. 15.27. Medium term variations of the oxygen content in the Gotland Deep from January 1969 to November 1970 (Nehring and Francke, 1971).

The short term variations of the hydrographical parameters are super-imposed on these medium term fluctuations. They are probably caused by inhomogeneity in the distribution of nonconservative parameters, oscilla-tions of the water masses and residual net transport of deep water. The inter-connection between water transport and short term fluctuation of chemical parameters has been studied by Holland (1969). Fig. 15.28 shows the short term variations of the oxygen content observed for the water below the halocline in the Gotland Deep in September 1967 and May 1968. The magnitude of these short term changes suggests that if medium and long term variations are to be identified with certainty it will be necessary to make a large number of observations extending over long periods. In some layers the magnitudes of the hourly and daily oxygen variations may even be similar to those of the long term variations. Such variations in the

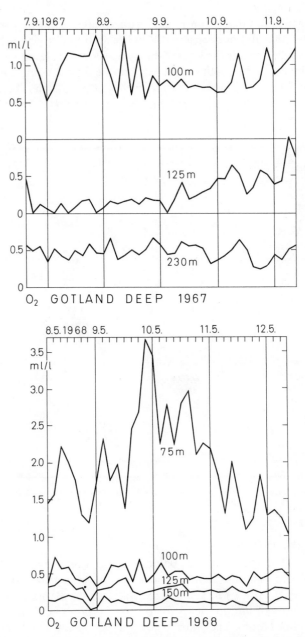

FIG. 15.28. Short term variations of the oxygen content in the Gotland Deep in September 1967 and May 1968 (Gieskes and Grasshoff, 1969).

oxygen content can often be correlated with variations in other parameters, both conservative and nonconservative (Gieskes and Grasshoff, 1969).

During anoxic periods the level of the hydrogen sulphide extends from the bottom to higher layers. During the very severe anoxic conditions of 1969 the Gotland Basin was anoxic up to a level of about 120 m. The maximum horizontal extent of anoxic conditions during 1968 and 1969 is shown in Fig. 15.29.

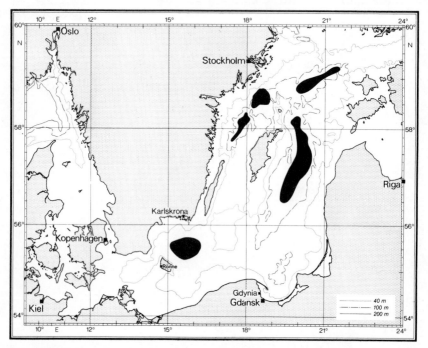

FIG. 15.29. The maximum extension of anoxic conditions in the deep water of the Baltic in 1969.

15.2.13. NUTRIENTS IN THE BALTIC

15.2.13.1. *Inorganic phosphate*

The permanent halocline in the Baltic is a diffusion barrier not only for oxygen but also for phosphate. In general, the phosphate concentration $(0–0·3 \,\mu\text{g-at } PO_4^{3-}–P\,l^{-1})$ in the surface water of the Baltic proper is very close to the growth-limiting level for primary producers, at least after the spring plankton bloom. The water below the permanent halocline acts as a nutrient trap. Especially during periods of stagnation, enormous amounts of phosphate accumulate; thus, up to $9 \,\mu\text{g-at } PO_4^{3-}–P\,l^{-1}$ have been observed in the deep water of the Gotland Basin. This accumulation of dissolved phosphate arises partially from mineralization of organic phosphorus compounds present in organic material but also to a varying extent from leaching of phosphates from the sediment (*vide infra*). The distribution of dissolved phosphate in a longitudinal section through the Baltic in August 1969 is shown in Fig. 15.30. It will be observed that the surface layer is divided from the deep layer by a steep phosphate concentration gradient. This gradient corresponds with the thermo-halocline.

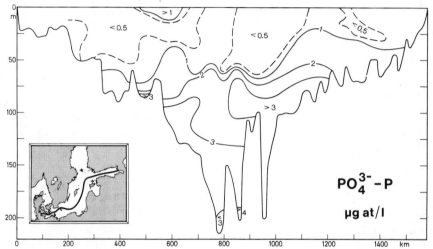

FIG. 15.30. Longitudinal section of the Baltic showing the dissolved inorganic phosphate content in 1969.

The convergent movement of the surface water resulting from cooling in autumn and winter also causes upward mixing of the phosphate-rich water from below the halocline with the surface layer. This process leads to regular fertilization of this layer. However, the amount of phosphate involved is comparatively small and does not produce dense plankton blooms. Annual peaks in the phosphate concentration of the surface water occur (Fig. 15.31), but they do not often exceed 0.3–0.4 µg-at PO_4^{3-}–$P\,l^{-1}$, and disappear in April and May after the spring plankton bloom. Sometimes values as high as 1 µg-at l^{-1} are attained. These are usually associated with significant intrusions of bottom water which interrupt periods of stagnation and phosphate accumulation in the deep water and which cause uplift of the old water which is admixed with the surface layers.

The total volume of the surface layer of the Baltic down to 50 m is about $10^4\,km^3$. During "normal" winters the phosphate content rises by about 0.3 µg-at l^{-1}. This corresponds to an addition of approximately 10^5 tons of

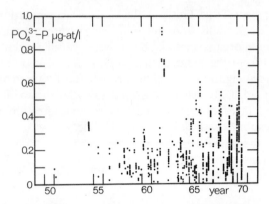

Fig. 15.31. The increase of dissolved phosphorus in the surface water (0–10 m) of the Central Baltic Basin from 1950 to 1970 (Fonselius, 1972).

phosphate. According to Fedosov and Zaitsev (1959) the total amount of phosphorus brought into the Baltic annually by rivers is 3 500 tons. Mikulski (1970) considers that the average discharge for the Baltic rivers from October to May to be 63% of the total discharge. This implies that during this period approximately $2·2 \times 10^3$ tons of phosphorus enter the Baltic through river discharge, if the average phosphorus load per unit volume of river water is about constant. This is not really so (see below), but changes in this figure will not significantly affect the conclusion that only 2–3% of the winter increase of phosphate in the surface layer is due to river discharge. During periods of overturn, as for example in the winter of 1962, about $1·1 \times 10^5$ tons of phosphorus will be brought additionally into the surface layer (Fonselius, 1966). Recent observations suggest that this extra phosphate disappears very quickly in spring and early summer. Calculations based on the rate of water transport out of the Baltic and the volume of the surface layer show that it would take only about 10 years to carry the additional phosphate out of the Baltic. It must, therefore, be assumed, that the phosphate is rapidly removed from the water *via* primary production. In periods following the fertilization zooplankton are abundant. Thus Nicolaev *et al.* (1962) found populations at such times exceeding 180% of the long period means. For instance, the average abundance of Crustacea in the 0–100 m water layer of the Baltic in August 1962 was 17,400 individuals m^{-3}, which is 187% of the long period mean value. The biomass amounted to 318 mg m^{-3} during the same time, which corresponds to 159% of the long period mean. It is thus evident that vertical transport of phosphorus from the deeper layer into the productive surface water has considerable importance for the fertility of the Baltic. Even if the transport does not reach the upper 20 m, i.e. the euphotic layer, but only penetrates the permanent halocline, convection in the following winter will normally mix the water column from the surface down to the halocline and thus bring the phosphate into the surface layer after a few months delay.

There are two reasons why the phosphate released in the deep water normally reaches the surface only during winter. The thermocline, which usually develops between April and June, is very stable and effectively isolates the deep water from the surface (Fonselius, 1966). In addition, the density gradient in the permanent halocline is weakened during the convective processes in winter, with the result that upward diffusion of phosphate, which has accumulated in the deep water, is favoured. This process is also encouraged by upward displacement of deeper water following the inflow of bottom water, which usually takes place in winter. Especially in the inner parts of the Gulf of Finland a type of upwelling may occur which brings large quantities of phosphate enriched water into the surface layers. This phosphate is then distributed in the surface waters of the Baltic proper. The halocline in the entrance to the Gulf of Finland shifted upwards by 25 m in 1962. This process is not easily studied because of the ice cover during the "critical" months, but its existence was forecast by Buch (1932).

Superimposed on these seasonal and sporadic variations in the phosphate content of the surface water there seems to be a general trend towards an increase in phosphate concentration in the surface waters of the Baltic proper (see Fig. 15.31). The increased addition of phosphate to the surface water seem to be correlated with the increase of accumulated phosphate in the deep water. As has been mentioned previously, the values from the Landsort Deep are very suitable for demonstrating long term changes in the Baltic Deep water, because of the filtering effect of the different basins and sills on short and medium term fluctuations. Figure 15.32 shows a compilation of the mean dissolved phosphate concentrations from 1938 to 1970, at depths between 100 and 400 m. The tendencies towards longterm increases or decreases can be more easily observed in these deep waters than in the surface ones because of the absence of seasonal fluctuation caused by primary production. The increased phosphorus content of the deep water and the occasional extra fertilizations of the surface layer in 1962 and 1966 are clearly shown in Fig. 15.33.

It is now necessary to consider the source of the accumulated phosphate in the deep water. It is well known that the average composition of plankton is relatively constant. In normal subsurface waters there is usually a clear correlation between the oxygen consumption and the released phosphate (Redfield et al., 1963). The classical theoretical molar relationship between released phosphorus and apparent oxygen utilization (AOU) should be

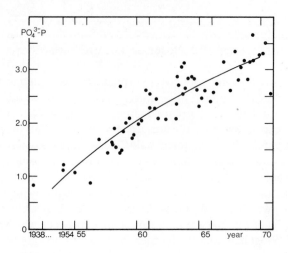

FIG. 15.32. The increase of dissolved phosphorus in the deep water of the Landsort Deep as mean values from the 100, 200, 300, 400 m levels in 1938 and 1954–1970 (Fonselius, 1972).

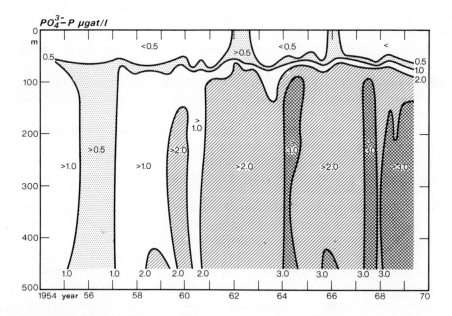

FIG. 15.33. Isopleth diagram of the dissolved phosphate concentration in Landsort Deep from 1954 to 1970 (Fonselius, 1972).

1:276 by atoms (Fleming, 1940). According to Sen Gupta (1973) the relation is 1:360 for the Baltic down to the nitrate maximum (if the surface samples which might be supersaturated with respect to oxygen are omitted). The deviation of the Baltic relationship from the classical one may be the result of the low overall phosphate content of the productive layers, leading to the production of a plankton population having a relatively low phosphorus content.

Fonselius (1967) and Gieskes and Grasshoff (1969) have pointed out that the relationship between AOU and phosphate concentration in stagnant waters is markedly different from that in ocean waters especially when anoxic conditions develop. In the latter instance AOU must be assessed from the sulphate reduced and the relationship becomes approximately 1:22. This magnitude of the ratio shows that it is necessary to invoke a different mechanism to that involved in ocean water to explain the occurrence of such high amounts of dissolved phosphate. It has been assumed that phosphatic materials dissolve from the sediment if the pH of the anoxic water drops to values close to 7·0 (Fonselius, 1967; Gieskes and Grasshoff, 1969; Koroleff, 1968). A log–log plot of hydrogen ion activity versus the concentration of dissolved phosphate (Fig. 15.34), shows that a correlation between the pH and the phosphate exists down to a pH value of 7·1. Below this pH the

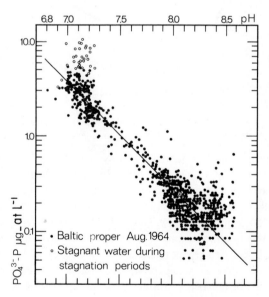

Fig. 15.34. The relationship between dissolved inorganic phosphate and the pH in the Central Baltic (Fonselius, 1967).

phosphate concentration rises sharply from about 2·5 to 10 μg-at PO_4^{3-}–P l^{-1} Recent experiments by Hallberg *et al.* (1972) have supported the belief that phosphate dissolves from the sediment under anaerobic conditions, especially if it contains high amounts of organic material. This process seems to be reversible because the phosphate content of the supernatant water drops to "normal" values when the water is reaerated. These experiments were carried out in closed systems so that the possibility of admixture with water rich in phosphate was excluded. The most probable explanation is the dissolution and reprecipitation of phosphates such as those of iron, the dissolution of which is favoured by the reduction of iron(III) to iron(II) and the subsequent formation of iron(II) sulphides. Figure 15.35 depicts the dissolution of phos-

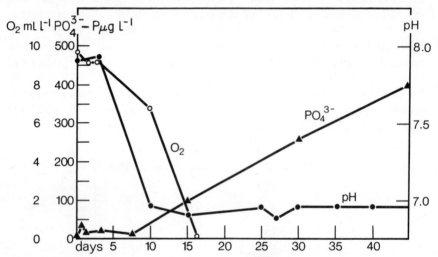

Fig. 15.35. Oxygen, dissolved phosphate and pH as a function of time in a trapped water body with sediment rich in organic substances (Hallberg *et al.*, 1972).

phates from sediments in a closed *in situ* system. It is also possible that in the system Na_3PO_4–$CaCO_3$–H_2O at pH values above 7 carbonate ions in the crystal lattice are replaced by phosphate ions so as to form carbonate apatite and *vice versa* at lower pH (Ames, 1959). It might be noted that in the Cariaco Trench the pH does not fall below 7·8 (Richards and Vaccaro, 1956) and the AOU: PO_4^{3-} ratio is almost normal (corresponding to the decay of organic matter). In the Black Sea the concentration of phosphate is higher, but the AOU values calculated allowing for the hydrogen sulphide are also higher and the pH does not fall below about 7·5. Evidently, there is no dissolution of phosphate from the sediment in these two instances. Because of the much greater depth of the Black Sea (2000 m) and the Cariaco Trench (1550 m) there seems to be little sedimentation of phosphatic material and little phosphate enrichment of the sediments. Anoxic conditions are permanent in the Black Sea and almost permanent in the Cariaco Trench.

The anaerobic conditions are caused by decaying organic material, which falls from the photic layer into the water below the permanent halocline. This organic material has a carbon-phosphorus ratio of approximately 156:1 (Sen Gupta, 1973) which is not dissimilar to that found for oceanic plankton (106:1; Fleming, 1940). Decay of the organic matter would be expected to lead to the accumulation of a predictable amount of phosphate in the water underlying the discontinuity layer, as in those other parts of the world ocean in which a stable layering prevents vertical mixing. However, in addition to this phosphate release, a much larger amount of phosphate is liberated from the sediment. If this phosphate together with the organogenic phosphate is transported into the surface layer and retransformed into organic bound phosphorus by primary production, and if the C:P ratio is still the same (and there are no signs that it is not), the result will be an increased production of organic matter which, in turn, will consume more oxygen during the remineralization and release more phosphate from the sediment. This cyle, which is self-amplifying was first suggested by Fonselius (1966). Figure 15.36 is a schematic display of this process which is of the

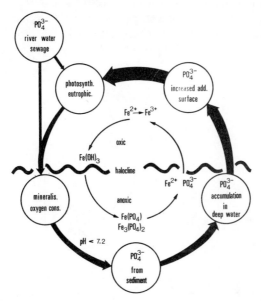

FIG. 15.36. Schematic display of the cycling of phosphorus between the aerobic and anaerobic regimes in the Baltic.

utmost importance for the Baltic. Sjöberg *et al.* (1972), who have tried to model this process in the Baltic have pointed out that the oxygen content of the water below the halocline is quite sensitive to small additions of phosphorus to the Baltic system and that the whole water column below about 70 m is in danger of becoming, if not anoxic, then depleted in dissolved oxygen to such an extent as to make this deeper water inhospitable to higher organisms, e.g. fish. Because of the long residence time of the deep water (10 to 40 years) the whole system reacts very slowly, and changes in the phosphorus input, e.g. from the continent, might begin to cause deleterious effects after a few decades. For this reason, a complete understanding of the phosphorus cycle is of the utmost importance, if measures are to be taken to protect the Baltic.

Phosphorus is brought into the Baltic both by discharge of rivers and outfalls and by precipitation of airborne phosphate. In comparison with the amount of phosphorus carried by rivers or discharged in other ways, the amount of airborne phosphate is of minor importance. Investigations of Polish rivers (Andruliewicz *et al.* 1972) and data from coastal stations (Majewski *et al.*, 1972) have shown that the phosphate content of the rivers is highly variable, possibly because of the irregular wash out of phosphates used in fertilizers, and that particularly in coastal regions the phosphate concentration is strongly affected by the discharge from the continent. The monthly variations of the phosphate concentration in the Polish rivers Wisla and Odra during during 1971 and 1972 are shown in Fig. 15.37. During that period the total annual amounts of phosphate phosphorus discharged by

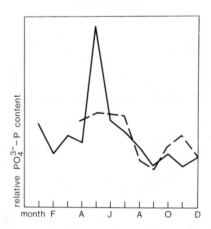

FIG. 15.37. Relative content of PO_4^{3-}–P in Wisla (—) and Odra Rivers (– – –) during 1971 and 1972 (Andruliewicz *et al.*, 1972).

these two rivers were 3800 and 2600 tons respectively. These recent observations cast doubts on earlier budgets of Baltic phosphate. Sen Gupta (1973) has attempted to establish a revised phosphate budget based on Kullenberg's (1973) estimates of the water exchange, and more recent information about the phosphorus inputs from rivers and waste outfalls (see Fig. 15.38). This

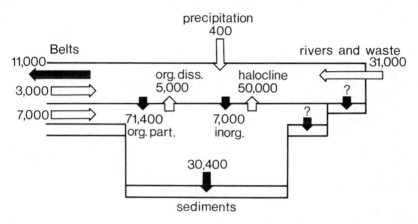

FIG. 15.38. Phosphorus balance of the Baltic proper under aerobic conditions (in tons $P\,yr^{-1}$) (Sen Gupta, 1973).

budget is only an approximate one, and it makes no attempt to estimate the amount of phosphorus which is sedimented either as phosphate mineral or as organically bound phosphorus in the shallower parts of the Baltic. When considering this model it should be borne in mind that the figures for the water exchange with the open sea and especially the vertical exchange numbers are only preliminary ones.

As has already been seen, the transition from anoxic to oxic conditions in the deep water drastically alters the dissolved phosphate concentration. Fig. 15.39 demonstrates that the re-establishment of an oxic environment causes some phosphate to disappear not only by entering the layer above the halocline by a mechanism which is not yet known with certainty, but also by precipitation.

15.2.13.2. Distribution of nitrogen compounds

Mainly because of analytical difficulties, few reliable systematic studies of nitrogen species in the Baltic were carried out before 1964. However, as a result of the International Baltic Year 1969/70, a considerable amount of data has now been amassed. The available data have been evaluated by Sen Gupta

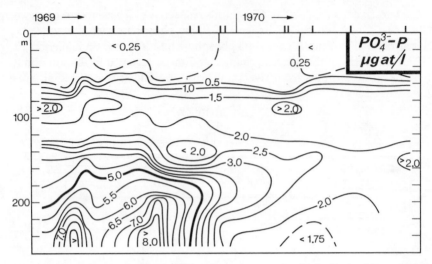

FIG. 15.39. Variations in the dissolved phosphate concentration in the Gotland Deep during an inflow of new water (Nehring and Francke, 1971).

(1973) who has attempted to establish a stoichiometric model and a balance for the nitrogen compounds in the Baltic.

Observations of the surface layer indicate that nitrate disappears almost completely (<0.1 mg $NO_3^- - N\, l^{-1}$) from the water column down to the thermocline during spring following the plankton bloom and remains low during the summer. Since detectable amounts of phosphate are still present it can be assumed that the nitrogen is the main limiting factor for primary production in the Baltic. All the nitrate which has been generated from organically bound nitrogen and from inorganic nitrogen compounds having a lower state of oxidation, and which has been brought into the photic layer during winter by thermoconvection is rapidly consumed and is totally exhausted by June–July. After that, primary production depends on the usable nitrogen compounds set free in the surface layer or on the ammonia brought into the Baltic in precipitation. It has been observed that, especially in late summer, large populations of blue–green algae appear. However, it is not yet known to what extent these algae rely on fixation of molecular nitrogen rather than the uptake of ammonia and the fixed forms of nitrogen. In autumn, when the water becomes cooler, regeneration of nitrate commences. No seasonal variations occur in the deeper layers; in these layers the nitrate concentration depends on the rate of upward diffusion of ammonia and on the subsequent rate of oxidation.

In conformity with other anoxic waters (see Chapter 16), nitrate is never present in the hydrogen sulphide containing waters of the Baltic, and in these the only stable nitrogen compound is ammonia which is liberated both from amino nitrogen and from nitrate through denitrification.

The nitrate distribution along a typical longitudinal section of the Baltic in June, 1969 is shown in Fig. 15.40. The concentration of nitrate in the

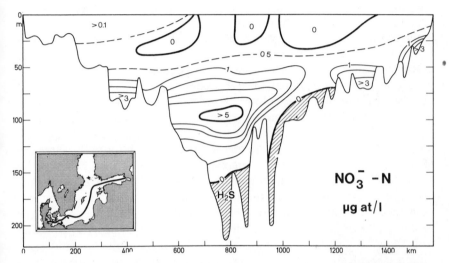

FIG. 15.40. Longitudinal section of the nitrate distribution in the Baltic in June 1969 (IBY data).

surface layers is <0.1 µg-at $NO_3^- - N l^{-1}$ everywhere even though it was relatively high in March of the same year. Even in the winter, water below the thermocline (which at that time lies at a depth of about 15 to 20 m), has only a low nitrate content. At the depth of the permanent halocline a sudden increase of the nitrate concentration to values exceeding 5 µg-at $NO_3^- - N l^{-1}$ can be observed. However, these are still low in comparison with values for deep waters from oceanic areas having a permanent pycnocline. At depths at which the content of dissolved oxygen tended to zero, the concentration of nitrate also fell owing to denitrification. Wherever hydrogen sulphide was present, no nitrate could be observed. Although it is sometimes asserted that nitrate and hydrogen sulphide can coexist there is no evidence for this in the Baltic; however, this may happen in areas in which turbulent mixing processes occur, or may result from rapid oxidation of ammonia after sampling when the sample is in contact with oxygen from the atmosphere.

The ammonia content of the surface layers is probably controlled either by direct uptake by the phytoplankton or, less probably, by its oxidation and uptake as nitrate. Because of analytical difficulties there are few reliable data on the occurrence of ammonia in the Baltic. In the open water the concentration of ammonia in the oxygen containing waters lies below 1·0 μg-at NH_3–N l^{-1}. During the spring plankton bloom this ammonia is rapidly consumed and values around 0·2–0·3 μg-at NH_3–N l^{-1} occur.

As long as the oxygen content of the deeper water is $>0·5$ ml l^{-1} the ammonia content remains low. However, when anoxic conditions develop, usually as a result of a very restricted horizontal and vertical exchange ammonia accumulates and values of ~ 5 μg-at NH_3–N l^{-1} are commonly found in the stagnant waters of the Baltic, although maximum values of as much as 9 μg-at l^{-1} have been observed occasionally.

As long as oxygen is present, the nitrite content is usually very low through-out the whole of the water column. However, intermediate nitrite maxima may occur even in the presence of oxygen. One of these maxima is produced when the first spring plankton crop grows in water having a relatively high nitrate content. This results in "luxury feeding" which leads to the excretion of nitrite (see Chapter 11). This maximum is usually found at depths of 15–20 m. The second maximum is the result of nitrification and occurs just above the sharp increase in nitrate concentration which lies below the permanent halocline. It is not very pronounced and usually does not exceed 0·5 μg-at NO_2^-–N l^{-1}. Vertical profiles of NH_3, NO_2^-, and NO_3^- together with those for oxygen, pH and dissolved phosphate at the Gotland Deep station in September 1967 are shown in Fig. 15.41. It can be seen that an intrusion of aerobic water has lifted the hydrogen sulphide containing water upward. Some of the ammonia in that bottom water is at once oxidized to nitrate, possibly passing transiently through the nitrite stage so that no significant increase in the nitrite content could be observed. This process occurs even if the oxygen level is very low, e.g. $<0·5$ ml l^{-1}.

Regular measurements of total nitrogen have not been made until very recently. It appears that between 70 and 90% of the nitrogen found in the water is organically bound (Sen Gupta, 1973). In the open Baltic there is always an accumulation of organic matter in, or just below, the permanent halocline which leads to an increase in the concentration of the organically bound nitrogen. Below this depth, the total nitrogen concentration increases with depth during prolonged anoxic conditions and seems to change little when the bottom water is reoxygenated. During the initial stage of an anoxic regime the total nitrogen content of the anoxic waters and of the layers lying immediately above them appears to decrease. This is perhaps caused by the rapid decomposition of organic matter (Sen Gupta, 1973). If the anoxic regime prevails, the rate of mineralization decreases and consequently the organic matter accumulates in the deep water. The slow decomposition rate

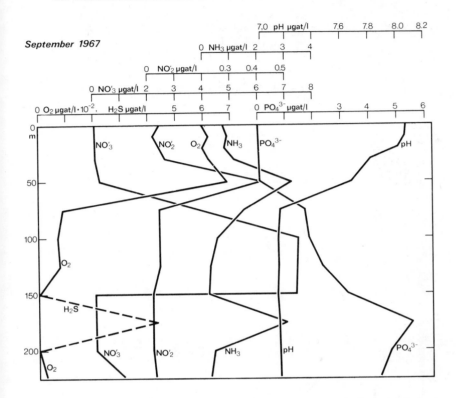

FIG. 15.41. Vertical profiles of NH_3, NO_2^-, NO_3^-, O_2, pH and dissolved PO_4^{3-} in the Gotland Deep in September 1967.

causes the organic fraction to be several times greater than the sum of the inorganic nitrogen compounds. This is in contrast to the vertical distribution of the nitrogen compounds in the open sea in which the organically bound fraction slowly decreases with depth to very low values and the nitrate fraction is the dominant species of fixed nitrogen. Table 15.13 is a compilation of mean values for the different nitrogen species for the Central Baltic during the years 1970 and 1971. Figure 15.42 shows the seasonal variations of the different nitrogen and phosphorus fractions in the Central Baltic from 1969.

According to Sen Gupta (1973) humus is the major contributor to the total nitrogen in the upper layer. This material may have two origins. Some may be "marine humus" (plankton and detritus), a major part of which becomes oxidized at depth (Skopintsev, 1959). The remainder is derived from the large amounts of terrigenous humus which are carried into

TABLE 15.13

Average concentrations of nitrogen compounds in the Baltic in 1970 and 1971 (in μg-at l⁻¹)

Depth m	1964–66 NO$_3$	NO$_3$ (1)	NO$_3$ (2)	NO$_2$ (1)	NO$_2$ (2)	NH$_3$ (1)	NH$_3$ (2)	Total—N (1)	Total—N (2)	Inorganic—N (1)	Inorganic—N (2)	Organic—N (1)	Organic—N (2)
0	0·36	0·47	0·75	0·10	0·19	0·49	0·87	16·89	17·28	1·06	1·81	15·83	15·47
5	0·43	0·45	0·80	0·08	0·13	0·84	0·66	—	—	1·37	1·59	—	—
10	0·45	0·53	0·73	0·08	0·13	0·60	0·60	—	—	1·21	1·46	—	—
15	0·48	0·51	0·66	0·07	0·14	0·67	0·52	18·05	15·79	1·25	1·32	16·80	14·47
20	0·49	0·48	0·71	0·07	0·09	0·47	0·50	—	—	1·02	1·30	—	—
30	0·51	0·47	0·71	0·12	0·14	0·41	0·51	18·19	19·64	1·00	1·36	17·19	18·28
40	0·67	0·70	0·85	0·11	0·17	0·52	0·53	—	17·38	1·33	1·55	—	15·83
50	0·93	0·92	1·26	0·12	0·18	0·35	0·53	14·56	18·58	1·39	1·97	13·17	16·61
60	1·44	1·54	2·22	0·07	0·12	0·23	0·47	13·43	16·11	1·84	2·81	11·59	13·30
70	2·50	1·60	2·76	0·05	0·06	0·25	0·57	17·08	17·87	1·90	3·39	15·18	14·48
80	3·30	1·70	2·89	0·02	0·06	0·36	0·73	—	17·50	2·08	3·68	—	13·82
90	4·50	1·88	2·54	0·01	0·09	0·57	1·00	—	20·44	2·46	3·63	—	16·81
100	4·93	1·96	2·70	0·05	0·03	0·63	0·94	17·16	20·70	2·64	3·67	14·52	17·03
125	5·24	1·97	3·02	0·01	0·05	0·83	0·94	—	21·20	2·81	4·01	—	17·19
150	5·34	2·09	2·64	0·02	0·08	0·91	0·96	—	16·79	3·02	3·68	—	13·11
175	5·33	2·72	2·28	0·02	0·10	1·09	1·72	20·66	—	3·83	4·10	16·83	—
200	5·88	2·88	2·27	0·03	0·07	0·92	1·82	20·63	—	3·83	4·16	16·80	—
225	6·41	5·06	2·89	0·05	0·05	0·58	1·84	—	17·75	5·69	4·78	—	12·59
250[3]	6·03	5·57	2·84	0·06	0·05	0·72	2·50	—	20·26	6·35	5·39	—	14·87

1. Average of values from measurements during 1970.
2. Average of values from measurements during 1971.
3. Average of values between 230 and 245 metres.

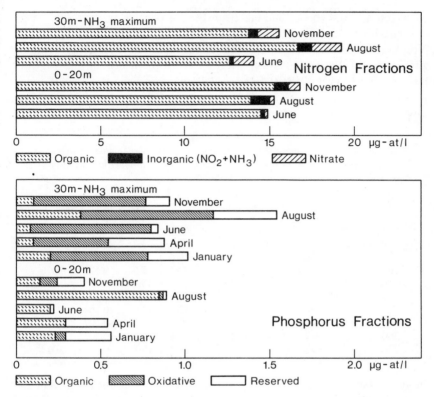

FIG. 15.42. Seasonal variations in the phosphorus and nitrogen fractions in the Central Baltic in 1969 (Sen Gupta, 1973).

the Baltic every year by rivers. This material is comparatively stable and resists mineralization. Humus lignin forms about 60% of the total carbon content in the Baltic deep sediments and is characterized by its comparatively high nitrogen content.

The variations in the concentration of organic nitrogen suggest that there are three decomposition zones for organic matter: the first, and probably the most important of these lies above the halocline is the zone in which all the remineralized nitrogen and phosphorus become available for reuse by the primary producers either immediately or after the winter thermoconvection of the surface layer. The second zone lies between the primary and secondary halocline and is characterized by its very low oxygen content; in this zone the amino nitrogen is oxidized to nitrate. The third zone lies near the bottom where anaerobic conditions exist. The C:N:P ratio of an average Baltic plankton appears to be different from that of oceanic species. Only a few analyses for the C:N:P ratio of plankton from the Baltic have been published. Voipio (1973) found an average ratio of 101:19:1 for samples consisting of plankton and detritus collected with a 150 μm mesh net during the International Baltic Year 1969/70. For the Baltic proper, Erhardt (1969) found an average N:C ratio of 1:88 for the organic material collected on a glass fibre filter of 1 μm pore size. The data in Table 15.14 suggest that the ratio tends

TABLE 15.14

Concentrations of dissolved and particulate organic carbon and values of the nitrogen/carbon ratio in the Central Baltic in May 1968. (Ehrhardt, 1969)

Depth (m)	$\mu g\,C_{part}\,l^{-1}$	$mg\,C_{diss}\,l^{-1}$	C_{diss}/C_{part}	N/C_{part}
1	259·8 $+$ 38·0	4·54 $+$ 1·14	18·0 $+$ 4·7	0·120 $+$ 0·009
10	228·8 $+$ 49·5	3·3 $+$ 0·5	14·9 $+$ 3·2	0·130 $+$ 0·011
20	183·7 $+$ 50·9	3·38 $+$ 0·7	19·8 $+$ 6·9	0·128 $+$ 0·015
30	138·4 $+$ 30·1	3·7 $+$ 0·7	27·5 $+$ 7·6	0·123 $+$ 0·010
50	132·6 $+$ 29·3	4·6 $+$ 1·4	35·9 $+$ 13·8	0·115 $+$ 0·010
75	107·5 $+$ 33·2	3·0 $+$ 0·6	29·7 $+$ 7·6	0·117 $+$ 0·016
100	83·4 $+$ 30·3	3·2 $+$ 1·1	40·9 $+$ 13·1	0·104 $+$ 0·019
150	99·2 $+$ 40·2	3·68 $+$ 0·96	43·9 $+$ 20·3	0·107 $+$ 0·018
200	91·3 $+$ 27·5	4·7 $+$ 1·8	55·6 $+$ 24·1	0·097 $+$ 0·016
230	97·7 $+$ 23·3	3·23 $+$ 1·19	34·4 $+$ 12·7	0·108 $+$ 0·013

to be higher for the particulate matter from the productive layers and those immediately below them. Koroleff (personal communication, 1973) has found an average N:P rate of 13·3:1 for both phyto- and zooplankton. (Table 15.15).

If all the phosphate and nitrate present in subsurface water had a biogenic origin, the slopes of the relationship between AOU and dissolved phosphate and nitrate should have the same C:N:P ratio as do plankton. Combination of the ΔAOU:ΔP and ΔAOU:ΔN values (Sen Gupta, 1973) gave an N:P atomic ratio of 4:1. Although this ratio is not very reliable because of the large scatter shown by the ΔAOU:ΔN values, it does strongly suggest that the ratio of the concentrations of nitrogen and phosphorus in the plankton is different from that in the water of the Central Baltic. The reason for this is the dissolution of mineralogenic phosphate from the sediment; a significant proportion of the nitrogen derives from the decomposition of terrigenous organic material, including sewage.

Fonselius (1972) has estimated the organic material transported into the Baltic by certain rivers to be 575,000 tons of carbon yr^{-1}. By extrapolation to all Baltic rivers he arrived at 1·73 million tons of C yr^{-1}. Neuman *et al.*

TABLE 15.15

Concentrations of Nitrogen and Phosphorus in Plankton (Koroleff—personal communication, 1973)

Collection period	Station No.	Composition	N P (µg-at/45 ml)		N:P (by at)
16 May 1973	24	Mainly phytoplankton	0·90	0·07	12·9:1
	24	,,	2·10	0·17	12·4:1
	24	,,	1·32	0·12	11·0:1
22 May 1973	25	,,	1·98	0·18	11·0:1
	25	,,	2·54	0·24	10·6:1
26 June 1973	LL 3	,,	7·38	0·67	11·0:1
to	24	,,	2·58	0·24	10·8:1
4 July 1973	25	,,	5·13	0·35	14·7:1
	26	,,	5·87	0·40	14·7:1
	20	,,	2·76	0·20	13·8:1
	18	,,	3·71	0·25	14·8:1
	12	,,	2·32	0·16	14·5:1
	9	,,	2·78	0·19	14·6:1
	8	,,	4·65	0·37	12·6:1
	Gdansk Deep	,,	0·76	0·08	9·5:1
	7	,,	4·54	0·35	13·0:1
	6	,,	1·91	0·12	15·9:1
	5	,,	3·32	0·26	12·8:1
	2	,,	4·03	0·28	14·4:1
	38	,,	2·32	0·15	15·5:1
	36	,,	2·09	0·14	14·9:1
	35	,,	2·39	0·17	14·1:1
	32	,,	2·34	0·15	15·6:1
	29	Mainly zooplankton	3·74	0·22	17·0:1
	20	,,	2·52	0·22	11·5:1
	18	,,	7·38	0·70	10·5:1
	12	,,	9·22	0·79	11·7:1
	9	,,	9·49	0·71	13·4:1
	8	,,	11·44	0·73	15·7:1
Average of all samples			3·91	0·30	13·3:1
Estimated from analyses of sea water					4·0:1

(1959) have estimated that organic material equivalent to 220 000 tons of carbon per year reaches the Baltic in precipitation. About 1·3 million tons of organic carbon also enter the Baltic yearly from outfalls and sewers. The total annual input of organic carbon is thus about 3·2 million tons. If this is compared with the estimated annual primary production of 26×10^6 tons carbon (Fonselius, 1972) it will be realised that the organic matter of non-marine origin is equivalent to approximately 15% of the primary production. It is therefore apparent why there is not a constant relationship between $\Delta AOU : \Delta P$ and ΔN in the Baltic.

It has been postulated that nitrate, nitrite and ammonia are not the only nitrogen compounds which are of importance in the marine ecosystem. Because of analytical difficulties, only occasional attempts have been made to detect other nitrogen compounds. Thus, hydroxylamine and urea have been found in significant amounts in the surface layer. Sen Gupta (1973) found $0.2–0.3$ µg-at $NH_2OH-N l^{-1}$ in the surface waters of the Central Baltic in November 1969, the concentration decreasing with increasing depth.

There is evidence that urea plays an important role as a potential source of nitrogen for primary producers. Using a new and highly sensitive method, Koroleff (1974) found concentrations as high as 2 µg-at urea–$N l^{-1}$; the number of samples examined was too small for any clear correlations with other parameters to be discerned. However, these values suggest that urea is likely to be an essential source of nitrogen for phytoplankton, particularly when other ones have become exhausted.

In order to establish a complete balance for nitrogen compounds in the Baltic it will be necessary to take into account metabolic and intermediate products in the nitrogen cycle and their role in the stimulation of the primary production. At present our ability to do this is limited by lack of information on the distribution of the various species and their variation in time and space. The same is also true about our knowledge of the rates of oxidation of ammonia to gaseous nitrogen, and of bacterial reduction of nitrate to molecular nitrogen even in the presence of oxygen (Sillén, 1966). Water which sinks from the surface to deeper layers during autumn and winter has insufficient time to become saturated with respect to oxygen before it leaves the surface. This is also true for nitrogen and a variable degree of saturation with respect to this gas which is caused solely by physical processes will be found in deeper waters. The amount of molecular nitrogen likely to be produced by oxidation of ammonia will be so small in comparison with that already present that it will be undetectable even with the most sophisticated modern techniques. It seems possible that the oxidation of ammonia to molecular nitrogen and thus its withdrawal from the nitrogen cycle may be of importance for the Baltic. However, nitrogen lost in this way may be counterbalanced by the fixation of nitrogen by blue green algae.

15.2.13.3. The distribution of silicate

Silicate is brought into the Baltic mainly by the Scandinavian rivers. Average contents of 57 µg-at Si l^{-1} have been reported for rivers which flow from a catchment area rich in granite (Hofman-Bang, 1904), and rivers draining quaternary deposits usually have a silicate content three times as high as this. In contrast, the rivers from the European continent have considerably lower silicate contents than do those from Scandinavia.

The water entering the Baltic through the Danish Sounds is usually a mixed water from the surface of the Kattegat, which contains only a few µg-at Si l^{-1}. Exceptionally heavy intrusions of deeper water may occur, but even these do not contain more than 20 µg-at Si l^{-1}.

In general, the silicate conception of the Baltic surface waters decrease, from the inner parts of the Gulf of Finland and the Gulf of Bothnia to the entrances of the Baltic. As a consequence of the uptake of silicate by diatoms, the concentration may vary considerably. The explosive blooming of diatoms occurs soon after the melting of the ice in the inner parts of the Baltic (Voipio, 1961) and a second, but much smaller, maximum occurs in late summer. Thus, silicon shows a strong seasonal dependence in the surface waters. These variations are influenced by the availability of nitrogen and phosphate species. The silicate is, however, never a growth limiting factor in the Central Baltic. Figure 15.43 shows the distribution of silicate along a longitudinal section through the Baltic for August 1965. The highest silicate contents in the surface water, sometimes reaching 50 µg-at Si l^{-1}, are to be found in the Bay of Bothnia (Voipio, 1961). Because of the low silicate concentration of

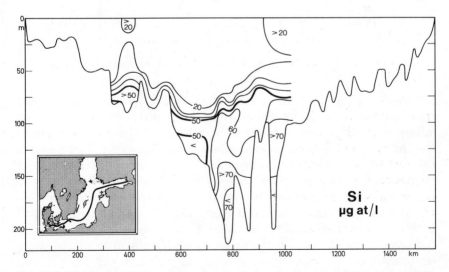

FIG. 15.43. Distribution of silicate along a longitudinal section through the Baltic in April 1970.

the rivers Newa and Narwa it would be expected that the content in the surface water would decrease from west to east in the Gulf of Finland. The distribution of silicate in an estuary with archipelagic character is shown in Fig. 15.44. This figure also shows that silicate is a useful parameter for the study of mixing processes in an estuary because of the large differences between the silicate contents of open sea and river water.

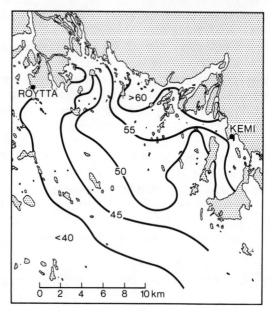

FIG. 15.44. Distribution of a silicate in the delta area of the river Kemijoki, 1959, in µg-at Si l^{-1} (Voipio, 1961).

Although the silicate concentration of the surface water is comparatively uniform down to the permanent halocline, particularly following the winter mixing, it increases sharply beneath the halocline. Depending on the area and the depth of the basin, concentrations of 60–80 µg-at Si l^{-1} are found in the deep water of the Baltic proper. Silicate accumulates during periods of stagnation. It has been observed that the fine structures of the skeletons of diatoms dissolve more rapidly under anoxic conditions, and they are completely destroyed by the time the shells reach the sea floor if reducing conditions prevail, e.g. in the Landsort Deep, possibly because of the reduction of iron(III), a coating of which prevents the skeletons of diatoms from attack during oxic conditions. The accumulation if dissolved silicate in the stagnant deep water is possible because the sea water is undersaturated with respect to

(Krauskopf, 1956; Rankama and Sahama, 1955). Biogenic processes are probably responsible for the removal of silica from the water and since these affect only the transport of silicate from surface to deep water (and to the sediments) variations of the silicate content of the deep water can be used in the evaluation of horizontal and vertical mixing processes in the Baltic deep water. On the basis of the assumption that about $1.4\,\mu$g-at Si l^{-1} is removed annually by primary production and that the silicate once biologically deposited will not dissolve in the layer above the permanent halocline, Voipio (1961) has calculated renewal times for the deep water of the Central Baltic and Gulf of Bothnia to be ~ 25 and ~ 14 years respectively.

Silicate distribution can also be used to examine local processes of stagnation and reduced water exchange. This is illustrated in Fig. 15.45 which shows the distribution of silicate along a longitudinal section through a fjord in the Western Baltic, in which the renewal of water is strongly dependent on the wind driven circulation (Fig. 15.45).

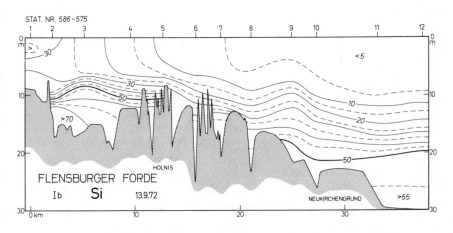

FIG. 15.45. Distribution of silicate in the Flensburg Fjord 1972 (in μg-at Si l^{-1}).

15.3. FJORDS

15.3.1. INTRODUCTION

Two principal types of fjord can be distinguished depending on whether or not water exchange with the open sea is restricted by one or more sills at the mouth. Exchange is most strongly inhibited when the sills rise close to the surface or, when relative to sill depth, there is a comparatively high river flow into the inner part of the fjord. An unusual hydrochemical regime may be produced if there is no significant discharge of fresh water into the fjord and if the connection to the open sea is restricted. The hydrographical regime of fjords without sills differs little from those of a "normal" coastal region and will not be considered here. The hydrochemistry of those fjords which possess sills is best illustrated by reference to particular examples. Oslo Fjord in Norway is a typical example of a fjord possessing a relatively shallow sill and having a large discharge of fresh water. One of its side fjords is an excellent example of a well stratified fjord, its density stratification resulting from the combination of heavy discharge of fresh water with a very narrow and shallow connection between it and the main fjord.

The Flensburg Fjord in the Western Baltic at the boundary between Germany (FRG) and Denmark is an example of a brackish water fjord having a restricted exchange of water with the open Baltic and receiving only a relatively small discharge of fresh water.

15.3.2. HYDROGRAPHY OF OSLO FJORD

15.3.2.1. *Topography and water exchange*

Oslo Fjord can be subdivided into three major parts, (i) the outer fjord from the entrance to the constriction between Horten and Moss, (ii) the middle fjord between the first constriction and the narrow entrance to the inner fjord at Dröbak, including also the open area at Breiangen and the entrance to the Dramsfjord; (iii) the inner fjord from Dröbak to the innermost end, including the appendix-like Bunnefjord. The section shown in Fig. 15.46 and the following figures follows the deepest part of the Oslo fjord from the fjord mouth to the Bunnefjord. It is apparent that the fjord has a complicated topography. Numerous islands, ridges, and bays are found nearly all along its length and give the fjord the character of an archipelago. At the innermost end the axis of the fjord bends through an angle of almost 180°.

The bottoms of the various basins are mainly covered with clay, but the ridges separating the basins are more or less rocky in character. Four major ridges can be distinguished, which are all of importance for the basins lying behind them. The most important of them is the Dröbak bar, which separates the middle and inner parts (Beyer, 1970). Water exchange through the straits at Dröbak is restricted both by the narrowness of the strait and by the shallowness of the sill, the western half of which is only 1 m deep and there is in fact only a narrow gap having depths greater than 15 m (max. 19·5 m = 12% of maximum the depth of the fjord).

Water transport in the uppermost layers of the fjord is controlled by the

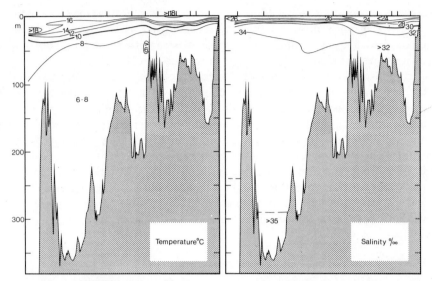

FIG. 15.46. Longitudinal section of Oslo Fjord showing the temperature distribution (ICES, 1970b).

FIG. 15.47. Longitudinal section of Oslo Fjord showing the salinity distribution (ICES, 1970b).

tide, the local wind, the meteorological conditions, the sea level in the Skagerak, and by the input of fresh water into the different parts of the fjord (Beyer, 1970). Water transport is mainly along the fjord axis, but it is influenced by the topography of the fjord, for example, northerly winds drive the surface water towards the end of the Bunnefjord, but towards the open sea in the main fjord. This water will mainly pass the constriction at Dröbak through the deeper gap in the eastern half of the sill. Winds from the southerly quarter generate a surface current which flows into the main fjord but out of the Bunnefjord. The wind driven currents are modified by the action of the tides which although they are weak may have a marked influence on the currents and therefore on the water exchange.

As the major discharge of river water passes the Dramsfjord, it mixes with sea water and is then diverted to the east by islands. As a rule it then takes a north-south direction, sometimes marked by a visible convergence.

The complicated interaction of wind, tides, sea level and bottom topography causes extraordinary current gradients especially in the sill areas. The current boundaries are also marked by strong salinity gradients—as great as $\Delta S\%_0/\Delta z = 4{\cdot}5\%_0\,\mathrm{m}^{-1}$ (Beyer, 1970) and comparable gradients probably exist for other chemical parameters. Water sampling in such areas with conventional sampling procedures (e.g. using Nansen bottles) may lead to the collection of non-representative samples. In addition, the vertical gradients are accompanied by a complicated horizontal structure caused by eddies of various size. Gradients of $3\%_0$ S m^{-1} have been regularly observed.

Discharge of fresh water into the inner part of the fjord is rather small, and for this reason a typical estuarine circulation is not very well developed there. Indeed, because large rivers discharge into the middle and outer fjord the estuarine circulation may occasionally be inverted (Gade, 1967). The amount of fresh water discharged into the inner fjord roughly compensates that lost by evaporation during the summer (Gade, 1968). This implies that, if wind stress generates a sea-going surface current, a compensation current may raise water from deeper parts of the fjord basin on the windward side and give rise to a kind of upwelling. Under favourable conditions new salt water can penetrate into the inner parts of the fjords.

15.3.2.2. Distribution of salinity and temperature

The boundary of the deep water and the surface water is marked by the 7°C isotherm (Fig. 15.46). Although the deep water is thermally uniform the temperature of the surface layer increases rapidly above the 7° isotherm because of seasonal heating of the surface. Because of the effect of the sills and shielding by surrounding mountains and hills the mixing of the surface layer is very much restricted in the inner parts of the fjord, resulting in a steep thermocline at a depth of about 10 m. This thermocline forms a strong diffusion barrier and reduces the vertical exchange during the summer season.

Because of the influence of the Baltic current which carries mixed Baltic water along the Swedish west coast towards the entrance of the Oslo Fjord, the salinity of the surface water in the entire area off the fjord mouth is relatively low (Beyer, 1970). Furthermore, the discharge of river water into the Oslo Fjord considerably reduces its surface salinity. Two major Norwegian rivers flow into the fjord—the Glomma, which discharges into the outer fjord and the Dramselv which runs into a side fjord. The water delivered from these gives rise to a marked salinity stratification in the whole fjord with a pronounced gradient of about $5\%_0$ Sm^{-1} at a depth of 10 m during the summer season. This, combined with the thermocline generates a very strong pycnocline at that depth. Because of the salinity stratification, seasonal warming of the water is confined to the upper 10 m in the middle and inner fjord. The distribution of the salinity is shown in Fig. 15.47. When Figs. 15.46 and 15.47 are compared it will be seen that the isopleths of salinity and temperature are not parallel. This is the result of the complex pattern of surface currents and eddies, which causes the local mixing processes to be very variable.

According to Beyer (1970), the distinct vertical temperature and salinity gradients are mainly caused by the fact that both the seasonal heating and the dilution originate in the surface layer. It must be borne in mind that a reduction in salinity has a much greater effect on density (and therefore on the stability of the layering) than does a change in temperature.

Most other chemical parameters do not show the same horizontal stratification as do salinity and temperature. The distributions of these are probably influenced by different mechanisms such as biological activity, biodegradation, chemical reactions under conditions of low redox potential and by the discharge of domestic and industrial sewage from the urban areas of Oslo and other towns.

15.3.3. NON-CONSERVATIVE CHEMICAL PARAMETERS

15.3.3.1. Oxygen

The sill at Dröbak forms a marked barrier against exchange of deep water from the inner fjord (see Fig. 15.48). In both the outer and middle parts of the fjord there is no discontinuity in the oxygen concentration-depth profile. This almost homogeneous distribution is a consequence of the fact that, for the same percentage of oxygen saturation, the warmer surface water contains less oxygen than does the cooler sub-surface water. This effect is partly counterbalanced by the reduced salinity of the surface water, but saturation is about 96 %, indicating that considerable oxygen is consumed in the surface water of the outer fjord through biodegradation of the organic load which enters it from the inner fjord and the rivers. In the layers below the pycnocline the absolute oxygen content is around 5·5 ml l^{-1}, corresponding to a degree of oxygen saturation of about 80–85 %. Such values are similar to those of the subsurface water in the adjacent open sea.

A marked stratification of the oxygen content can usually be observed in the inner fjord during the summer season. Below the pycnocline the oxygen is rapidly consumed by the biodegradation of organic material which may consist of both material brought from the land and dead algae from the surface layers. The rate of primary production is high in the inner fjord as a result of intense fertilization caused by sewage. Dense crops of phytoplankton make the water turbid and reduce the penetration of light, thus confining the intense primary production to the immediate vicinity of the surface. In this surface layer a biogenic oxygen supersaturation is the rule during spring and summer. Thus, the degree of saturation exceeded 125 % in the upper 5 m at some places in the inner fjord in August–September 1969.

Below the discontinuity layer the oxygen content drops to quite low values. There is sufficient water exchange to maintain aerobic conditions even in the innermost parts of the Bunnefjord. As can be seen from Fig. 15.48, an oxygen minimum layer develops in the innermost parts of the fjord at a

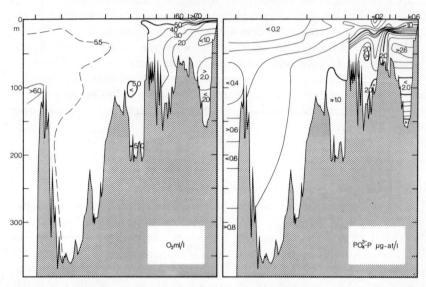

Fig. 15.48. Longitudinal section of Oslo Fjord showing the oxygen distribution (ICES, 1970b). Fig. 15.49. Longitudinal section of Oslo Fjord showing the dissolved phosphate distribution (ICES, 1970b).

depth of 30–50 metres. This indicates that some dense water regularly intrudes over the sill at Dröbak, as is also indicated by the vertical oxygen isolines just inside the Dröbak sill. There is evidence from work on benthic fauna that the oxygenation of the deeper water in the fjord is deteriorating. In contrast to the Baltic, this worsening is mainly not the result of natural processes (e.g. climatic changes) but is caused by the increasing discharge of organic material from settlements along the fjord.

15.3.3.2. *Phosphate*

There is a close inverse correlation between the distributions of phosphate and oxygen in the fjord (Fig. 15.49). In the outer fjord phosphate concentrations are low, increasing from less than 0·2 µg-at $PO_4^{3-}-P\ l^{-1}$ to 0·8 µg-at l^{-1} at a depth of about 300 m. The middle fjord is characterized by a slight accumulation of phosphate in the deeper water compared to the outer fjord. This accumulation below about 100 m increases towards the Dröbak bar. This increase is mainly the result of discharge of phosphate and phosphorus containing organic matter from the rivers and settlements. Furthermore, there is occasionally some spillage of phosphate-rich water from the inner fjord over the Dröbak sill. The distribution of phosphate in the inner fjord generally shows a more complex picture (Føyn, 1970). Various maxima and minima appear in the deeper water, but even so the oxygen varies inversely with phosphate. At depths greater than 60 m the water is more or less stagnant and this leads to a significant accumulation of dissolved phosphate. At a depth of about 100 m there is a distinct phosphate minimum in both the inner part of the fjord and the Bunnefjord. This can again be explained by the occasional intrusion over the sill at Dröbak of water which is insufficiently dense to replace the bottom water in the inner fjord. The phosphate content of the overflowing water is, however, much less than that of the deeper water in the inner fjord. As exchange of the bottom water of the Bunnefjord occurs only rarely, the phosphate content is particularly high.

There is a very steep gradient in the phosphate distribution in the surface layer. The thermo-halocline is an effective diffusion barrier for the upward transport of the phosphate from the deeper layers.

Because the considerable amounts of phosphate which are discharged into the fjord are rapidly taken up by phytoplankton, the phosphate concentration in the surface layer is low except in Bunnefjord. After death, the plankton fall through the water column, penetrating the pycnocline as they do so; their mineralisation leads to phosphate accumulation in the deep water.

15.3.3.3. *Nitrate*

The effect of the presence of the sill at Dröbak is also shown by the distribution of nitrate (Fig. 15.50). In the middle and outer fjord the nitrate content of the surface water increases slightly with depth and this is also true for the deeper water of the middle fjord near the sill, This distribution was probably brought about by the same factors that were responsible for the distribution of phosphate. In late summer, at least, nitrate does not seem to be growth limiting for primary production. The nitrate content of the surface waters varies between 2 and 4 μg-at $NO_3^- - N\ l^{-1}$. The inner fjord shows the same accumulation for nitrate as for phosphate with a distinct minimum at about 50 m in the Bunnefjord. This minimum is caused by the occasional intrusion from the middle and outer fjord of water which contains considerably less nitrate than the water normally found at the 50 m level.

15.3.3.4 *Ammonia*

The distribution of ammonia is complex. An intermediate maximum of uncertain origin is observed between 50 and 100 m in the outer fjord. There appears to be no general tendency for the ammonia to increase with depth or towards the inner fjord (Fig. 15.51). Large amounts of ammonia and of amino nitrogen are discharged into the inner fjord, but in the presence of oxygen a stationary state seems to exist in the deeper layers between the rate of input of ammonia from the settlements plus that from decaying organic material and that of bacterial nitrification. This balance maintains the ammonia level at about 1 µg-at $NH_3 -N\ l^{-1}$. This picture is likely to change drastically if anoxic conditions develop.

According to Føyn (1970) the horizontal distribution of dissolved oxygen and nutrients points to the sewage of the city of Oslo being the main source of the pollution of the fjord. The input of organic material and of nutrients from this urban area influences the conditions throughout the whole of the water system. Mixing and aeration of the deep water is strongly reduced by the existence of the thermo- and haloclines.

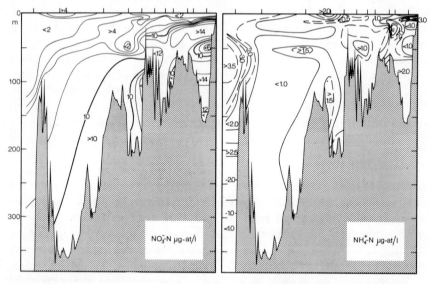

FIG. 15.50. Longitudinal section of Oslo Fjord showing the nitrate distribution (ICES, 1970b).
FIG. 15.51. Longitudinal section of Oslo Fjord showing the ammonia distribution (ICES, 1970b).

15.3.4. HYDROCHEMISTRY OF FLENSBURG FJORD

The 30 km long Flensburg Fjord, which lies on the border between Germany and Denmark, has been selected as an example of a fjord system lacking any significant input of fresh water. The hydrographical and chemical

properties of this fjord are controlled by its highly complex bottom topography by its marked thermo-haline stratification during the summer and by
water movements caused by the winds.

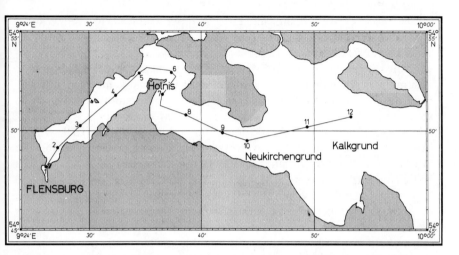

FIG. 15.52. Map of Flensburg Fjord indicating the longitudinal section of the subsequent figures.

The duration, direction and force of the winds and the interaction and
linkage of the fjord system to the sea level fluctuations in the Baltic (seiche
period $\sim$ 13·5 hours) are responsible for the water exchange, especially in
the inner part of the fjord which frequently becomes stagnant and even
anaerobic in its deeper layers during the summer months. The sewage from
Flensburg (population 96 000) and several small rural communities is
discharged into the fjord. Only sewage from Flensburg is biologically
treated.

A section showing the salinity distribution along the axis of the fjord in the
late summer 1972 is depicted in Fig. 15.53a. The outflowing Baltic surface
water, after considerable mixing, enters the Flensburg Fjord and causes the
surface water to have a low salinity. Incoming North Sea water penetrates
through the Small Belt, through the Als Sund and through a trough connecting
the Large Belt with the entrance of the Flensburg Fjord and forms the bottom
water of the fjord. A double layer salinity structure thus develops in which
the surface salinity is 16‰ and the bottom water has a salinity ranging from
26‰ in the outer fjord to 21‰ in the inner fjord. The halocline is at the same
depth as a series of sills, especially in the narrowest part of the fjord. The
effects of this halocline, which constitutes a diffusional barrier between the

surface and the deeper water, is further augmented by one (Fig. 15.53b) which forms in April and persists until its destruction following the autumn cooling of the surface water and the storm generated turbulence; this change may occur within a period of one day.

The locally derived organic and nutrient input load leads to a reduction in the concentration of oxygen in the water below the halocline in both the inner and outer fjords (Fig. 15.53c) and frequently to the formation of hydrogen sulphide in the inner fjord. Under these stagnant conditions phosphate accumulates not only from mineralization of the organic material, but also from the dissolution of phosphates from the bottom sediments. This can be seen from the N/P ratios which are sometimes as low as about 2.5:1, which contrasts strongly with the ratio found for marine plankton (10:1 to 20:1 ave. 16:1). The high surface water phosphate content can probably be attributed to the large number of sewage outlets in the fjord, these outlets being concentrated towards the inner fjord (Fig. 15.54a). Below the halocline and in the inner fjord, where the nitrate is frequently reduced during anoxic or nearly anoxic periods, the nitrate content of the water of low or zero oxygen content is very low. In the outer fjord nitrate accumulates even when thermohaline layering persists, indicating that the oxygen is not completely depleted (Fig. 15.54b). The very high nitrate values (e.g. >25 µg-at NO_3^- – Nl^{-1} in the innermost part of the fjord observed on 12th September 1972), are partly the result of nitrate discharge, but are mainly derived from the oxidation of ammonia, either brought to the surface by upwelling or introduced via sewage (Fig. 15.54c).

The upwelling was caused by a shift in the wind direction and force. South-easterly winds of force 3 to 4 prevailed for 3 days until September 11th, when a change to moderate westerly or north-westerly direction occurred. The easterly winds caused the sea level in the fjord to rise and because of the accompanying introduction of warm and low salinity water to the fjord the thermohaline structure was further stabilized. When the onshore wind pressure is reduced or the current changes to an offshore surface drift, sea level falls and warm surface water leaves the fjord area and is partially replaced by colder and more saline water from deeper layers. When this occurs the water temperature may drop from 20°C to 15°C within a few hours. At the same time, ammonia which is brought into the surface layers is rapidly oxidized to nitrate and this stimulates a local plankton bloom. These wind generated local upwelling processes associated with pronounced sea level fluctuations are typical of the shallow fjords of the Baltic during the summer months. During calm late summers the oxygen depletion may become critical and severe fish kills have often occurred both in some fjords and near coastal areas when the wind has changed from easterly to westerly and suddenly

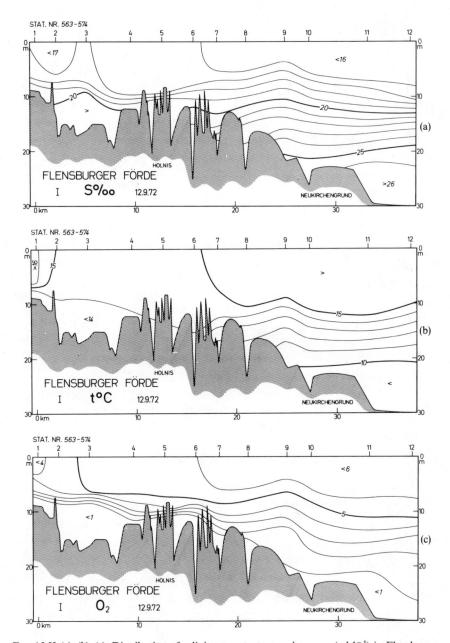

FIG. 15.53 (a), (b), (c). Distribution of salinity, temperature and oxygen (ml l^{-1}) in Flensburg Fjord in late summer (12th September 1972).

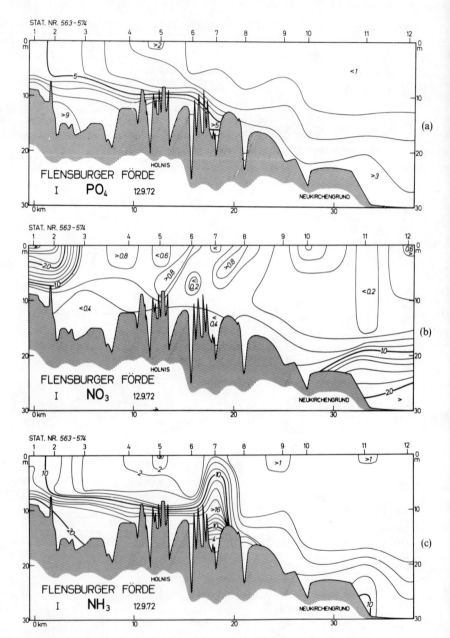

FIG. 15.54 (a), (b), (c). Distribution of dissolved phosphate, nitrate and ammonia (all as μg-at l⁻¹) in Flensburg Fjord in late summer.

brought water with a low oxygen content into otherwise aerated regions. At present, the Flensburg Fjord is polluted with domestic and rural sewage discharges and this has influenced its chemistry and has led to eutrophication. Fig. 15.55 shows the high primary production as indicated by the chlorophyll

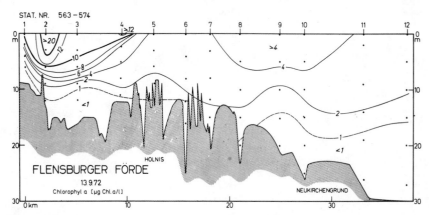

FIG. 15.55. Distribution of chlorophyll a in Flensburg Fjord after local upwelling of bottom water (13th Sept. 1972).

a content, which reached values of more than 20 µg of chlorophyll $a\,l^{-1}$ on September 13th, two days after the fertilization of the surface layer by upwelling. The combined oxygen demand from sewage and dead phytoplankton may lead to anoxic conditions below the pycnocline. As is typical for a fjord system with no significant source of fresh water the thermo-haline layering disappears almost completely during the winter period (Fig. 15.56), and the whole fjord is well aerated from surface to the bottom. At that time of year the strong winds cause complete renewal of the water and discharge of sewage does not cause difficulties. The phosphate content is 1 µg-at $PO_4^{3-}-P\,l^{-1}$ throughout the whole fjord at all depths. Only the ammonia and nitrate content indicates the sewage load of the inner fjord but the ammonia is rapidly oxidized and decreases to values below 1 µg-at $NH_3-N\,l^{-1}$ only a few kilometres from the points of discharge. The high nitrate content of the water in the inner fjord is also rapidly reduced to values below 1 µg-at $NO_3^--N\,l^{-1}$ by mixing processes which are dominant at that time of the year. Starting in late March or in April the pycnolines tend to build up again (Fig. 15.57) and the fjord again becomes a poorly oxygenated, highly eutrophic area.

The reduced horizontal and vertical water exchange, the discharge of

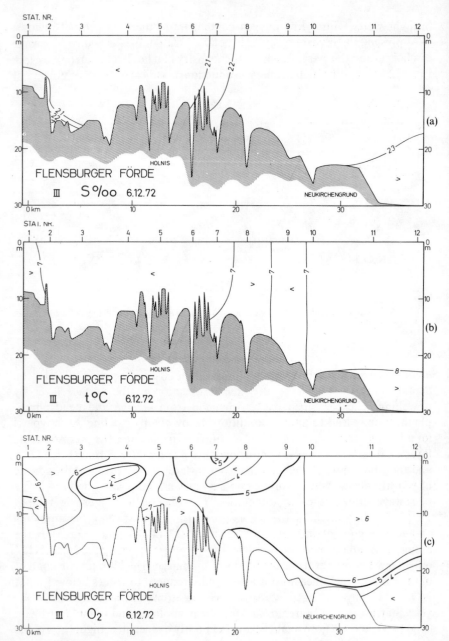

FIG. 15.56 (a), (b), (c). Distribution of salinity, temperature and oxygen (ml l^{-1}) in Flensburg Fjord in winter (6th Dec. 1972).

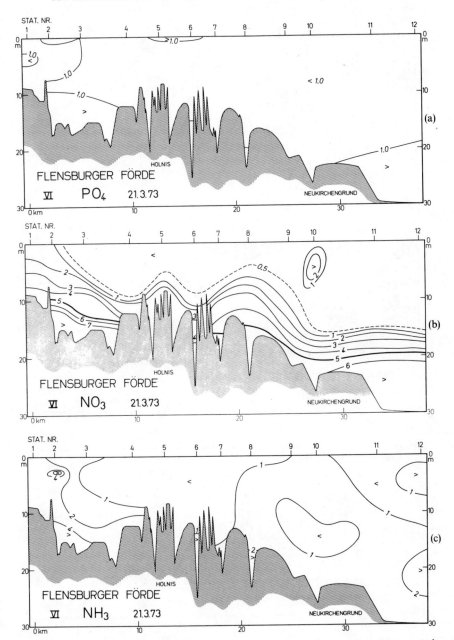

FIG. 15.57 (a), (b), (c). Distribution of dissolved phosphate, nitrate and ammonia (all as µg at l^{-1}) in Flensburg Fjord in early spring (21th Mar. 1973).

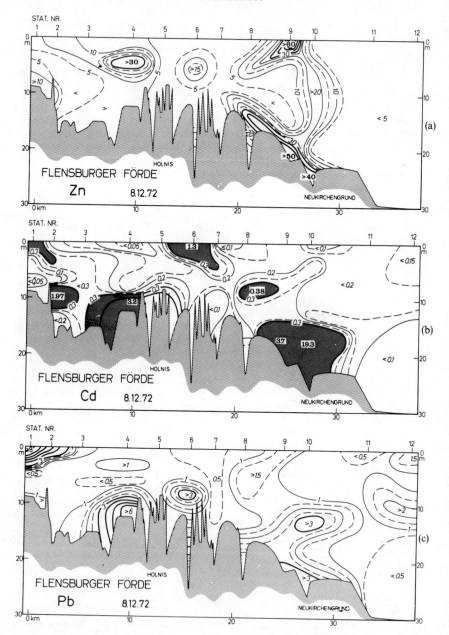

FIG. 15.58 (a), (b), (c). Distribution of dissolved (ionic) zinc, cadmium and copper ($\mu g\ l^{-1}$) in Flensburg Fjord (8th Dec. 1972) (Kremling, unpublished).

sewage and the reducing conditions in the bottom water cause the concentrations of potentially toxic trace metals in the Flensburg Fjord to be relatively high (see Figs. 15.58a, b, c). Trace metals accumulate in those coastal areas with restricted water exchange when pycnoclines impede vertical exchange of dissolved matter. Some contribution may also be derived from domestic sewage. These phenomena are observed in most Baltic fjords.

5.4. THE BLACK SEA

15.4.1. INTRODUCTION

In terms of its chemistry the Black Sea is a highly atypical water body. It is connected by a narrow strait to the Mediterranean which it itself a semi-enclosed sea with its own hydrographic properties. The Black Sea has much in common with the Baltic inasmuch as the total input of fresh water into it via runoff and precipitation greatly exceeds that lost by evaporation. Even though the Black Sea is situated in a semi-arid climatic zone, precipitation is very high on its south-eastern part with the result that, on average, evaporation only slightly exceeds precipitation. However, the principal catchment area of the Black Sea is in a humid climatic zone and it is this fact that leads to the excess input of fresh water and to the low salinity of the surface waters. As a consequence, a permanent halocline is present similar to that in the Baltic. In contrast to the Baltic, which is the largest brackish water area in the world, the Black Sea has a much greater average depth, so that its total volume is more than 35 times that of the Baltic.

In the Baltic the volume of the sub-halocline water is less than that of the surface water which participates in convective mixing during the winter. In contrast, in the Black Sea the volume of the sub-halocline water is more than eight times greater than that of the surface water. The latter, which is 100–240 m in thickness, is aerated by the seasonal convection and covers 2000 metres of permanently anoxic water. The permanent halocline of the Black Sea, like that of the Baltic, effectively inhibits exchange of dissolved matter. Over large areas of the Black Sea the permanent halocline coincides with the permanent boundary between oxic and anoxic water and has a profound influence on the chemistry of this sea.

Despite the many unique features of the Black Sea, the available oceanographic and marine chemistry texts rarely devote much space to it. However, it has been studied in detail in recent years and the literature has been reviewed by Grasshoff (1969) and Laking (1974) who have collected more than 2000 references dealing with its physical oceanography, hydrochemistry, geochemistry and biology.

15.4.2. PHYSICAL GEOGRAPHY

As with all semi-enclosed seas, a knowledge of local and bottom topographical features (especially those of the transition area between it and the adjacent sea) and of the hydrography is vital if the development of the chemical features is to be fully appreciated. It has been seen (Section 15.2) that for the Baltic the shapes of the interconnected basins and the sill regions between them in combination with fluctuations in the regional climate cause both long and short term changes in the hydrographical and also therefore in the hydrochemical properties. This is also true for the Black Sea; however, in this case the topography is the reason for the stability of the hydrographical features, and changes in the climate do not lead to major alterations in the hydrographical and hydrochemical regimes. For this reason, a short description of the geographical and morphological features of the Black Sea precedes the treatment of the chemical features.

The Black Sea, together with the Azov Sea, covers an area of 462,000 km²; the area of the latter sea being 48,500 km². Its dimension east to west is ∼ 1150 km and that from north to south is ∼ 610 km, There are few major topographical features except the Crimean Peninsula which divides the Black Sea in two, narrowing the coastal separation of the coasts to 267 km. The volume of the Black Sea is estimated to about 537,000 km³.

The northern boundary of the Mediterranean mountain chain region which extends from the Balkans to the Jaila Mountains and to the Caucasus divides the coastal structure into two regions. In the southern region, south of Varna the coast is mainly steep and rocky along the coasts of Bulgaria, Turkey and Caucasian Russia. The slope along that coast is mainly 4°–6°, but in places it increases to 12°–14°. There are no large peninsulas, no islands or large fjords. This part of the coast is similar to that of the western Mediterranean. North of this boundary, the coast is extremely flat. Three coastal structures predominate: low cliffs formed of reddish clay, flooded river mouths (Limans) and sandy beaches and dunes. Bars of sand dunes often cut off large inshore water areas (backwaters). The Crimean Peninsula extends far into the Black Sea basin and the Peninsula of Kerch and Taman, extending westward and eastward, separate the shallow Sea of Azov which is connected to the Black Sea proper by the narrow Strait of Kerch.

The northern boundary line is a chain of mountains which divides the Black Sea into two parts. North of this boundary the shallow shelf sea is predominant. The north-western part of the Black Sea, north of the line Varna–Sevastopol, is shallower than 100 m, and north of the estuary of the Danube and Mys Tarkhankut, is less than 50 m. The Sea of Azov has a maximum depth of 13 m. The rest of Black Sea, i.e. by far the largest part, is a deep down-folded basin.

The coastal strip is narrow and shelves steeply to a depth of 1500–2200 m. The maximum depth of the Black Sea is 2234 m, (150 km south of Sevastopol). Its average depth including the shelf areas in the north west is 1,200 m.

In the northern part of the Black Sea, the most abundant sediments are sand, sandy gravel and muddy gravel. Along the west side of the Crimean Peninsula there is a band of mud containing isolated large stones. At the southern boundary of the northern part of the Black Sea, very mechanically weak mud is dominant on the shelf. The narrow shelf off the southern coasts is often rocky and along the Caucasian coast a shelf is absent.

The deeper sediments of the central part of the western basin consist mainly of calcareous ooze. They are surrounded by a zone of carbonaceous mud containing sapropelic clay layers (Fig. 15.59). In the vicinity of the Bosporus there is a smaller area of grey clay extending to depths of 2000 m, and west of this clay area, a zone of transitional mud occurs along the coast of the Anatolian shelf.

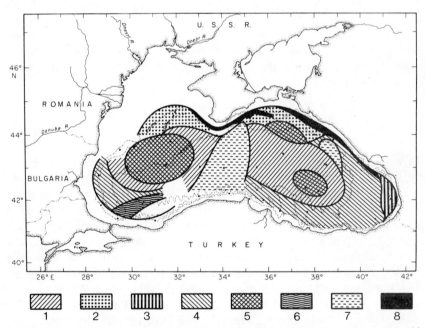

FIG. 15.59. Deep sea sediments in the Black Sea (Arkhangel'skii, 1928). (1) calcareous ooze with clay layers, (2) carbonaceous mud, (3) transitional mud, (4) grey clay, (5) transitional mud with clay and sand layers, (6) grey clay with mud layers, (7) transitional mud with clay layers, (8) no sediments.

In the corridor between the Crimean Peninsula and Anatolia there is an extensive area of grey clay interspersed with mud layers. The south east slope of the Crimean shelf is devoid of sediments. The central part of the eastern basin is covered with calcareous ooze which often contains sapropelic clay and only rarely is pure calcareous ooze encountered. The sediments of the transition shelf consists of transitional muds, similar to those of the western and eastern basins. In addition, the central area is surrounded by broad zones of transitional mud with clay layers and there are no sediments on the steep slopes along the Caucasian coast. Off Batum an area of transitional mud with very fine sand and clay layers has been observed.

The average rate of sedimentation for the whole Black Sea basin has been estimated to be 0·5 mm yr^{-1}. There are, however, considerable regional differences; thus, the rate of deposition of deep sea clay is estimated to be 0·2 mm yr^{-1}, and that for calcareous clay is even less (0·04–0·08 mm yr^{-1}) (Strakhov, 1954). The amounts of sediments brought into the Black Sea by rivers are difficult to estimate. For the river Danube it is calculated at 48×10^6 tons yr^{-1} on the basis of an average of 211 mg of solid matter per litre. The small rivers in the south-east carry very large amounts of clastic material into the Black Sea; this is deposited near to the point of influx. Generally, winter ice coverage only extends to the northern parts over the main shelf and the Azov Sea, although occasionally a small band along the western coast as far as the Bosporus may also be covered. The central Black Sea is always free of ice.

15.4.3. WATER BALANCE

As with all semi-enclosed seas the water balance influences not only the hydrographic features but also the hydrochemistry. The Black Sea water balance is of category (i) (see Section 15.1 and Fig. 15.1a). Despite the fact that evaporation exceeds precipitation, the large discharge of fresh water from the enormous catchment area maintains an essentially brackish surface layer. There are considerable variations in the precipitation between the various areas. Large deviations from the monthly means occur, especially along the mountain coasts in the south east. According to Sverdrup et al. (1942), the mean annual precipitation into the Black Sea is 240 km^3 yr^{-1}. In addition to this, run-off introduces a further 320 km^3 yr^{-1}. The major sources of run-off are the rivers Danube and Dnieper with catchment areas of 840.000 and 502.000 km^2 and average discharges of 193 and 52 km^3 yr^{-1} respectively. Additional run-off is provided during certain seasons by the rivers of the Batumi region which is subject to tropical storms and has an

annual mean rainfall of > 2500 mm. The spatial distribution of the mean annual precipitation over the Black Sea is shown in Fig. 15.60.
There are large discrepancies between the reported rates of inflow and

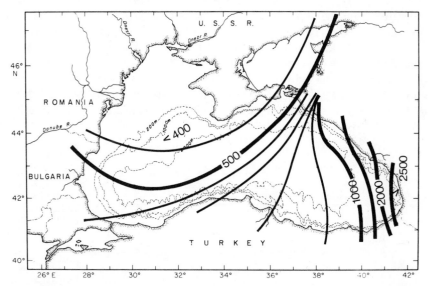

FIG. 15.60. Mean annual precipitation (mm) in the Black Sea area (Source, Morskoi Atlas, Vol. II).

outflow through the Bosporus. The earliest scientific account of the current system in the Bosporus occurs in letters by Count Marsilli to Queen Kristina of Sweden (1681). No large scale direct long term current measurements have been made. There are in fact a number of considerable difficulties in studying water movements in this region: (i) strong currents make it difficult to keep the instruments at a fixed depth; (ii) the cores of the currents meander within the Strait region; (iii) the currents and the depth of the boundary between ingoing and outgoing currents fluctuate markedly and are strongly influenced by the local and regional meteorological conditions, (iv) the straits are narrow and much used by shipping.

Indirect calculations of the water exchange through the Bosporus are imprecise, because of the lack of reliable information on runoff, precipitation and evaporation. Although some authors consider the inflow of salt water from the Sea of Marmara to be small or negligible (see e.g. Ullyot and Ilgaz, 1946), this cannot in fact be the case because a considerable amount

of salt is transported out of the Black Sea by the surface outflow and yet
there is no evidence that the salinity conditions in the Black Sea deep and
surface water have changed since reliable observations first became available,
70 years ago.

Merz (1918a, b, 1921) and Merz and Möller (1928) carried out a considerable number of individual current measurements in the Bosporus
and on the basis of the constant salinity calculated an average annual
outflow of 397 km^3 and an inflow of 190 km^3. Data published for both
directions may vary between 3,000 and 30,000 $m^3 s^{-1}$. The average ratio
of the salinity of the inflowing water to that of the outgoing surface water is
about 2.2:1 (38·5‰ :17·5‰), however these figures must still be considered
to be preliminary values. Fonselius (1973) has proposed a box model to
account for the balances of the Black Sea (Fig. 15.61).

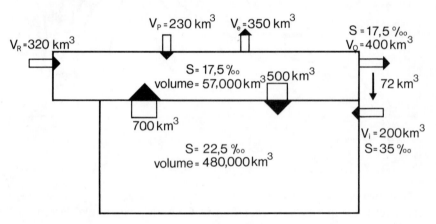

FIG. 15.61. A box model of the Black Sea (Fonselius, 1973).

15.4.4. GEOLOGICAL HISTORY

Like the Baltic, the Black Sea has frequently alternated in character between
saline and brackish or fresh water conditions during its geological development. However, the Black Sea is much older than the Baltic and even as
early as the late Tertiary period it had become increasingly isolated from the
open ocean. Caspers (1957) has summarized its geological history.

During the early and middle Miocene the basin was filled by a salt sea, a
part of the vanishing Tethys. Connection with the ocean was probably lost
by the end of the Middle Miocene, and a progressive trend towards less
saline conditions began. This process affected different parts of the area

in different ways, since the mid-Miocene Sea was separated into distinct basins. By the late Miocene, the Sarmatian Basin had been formed (Fig. 15.62a), saline water species of plankton had disappeared and the surface water of the Sarmatian Basin finally became almost fresh. However, denser saline water stagnated at greater depths and as a result hydrogen sulphide accumulated.

The progressive restriction of the connection between the western and eastern parts of the sea continued until the formation of the so-called Meotic Basin in the Miocene (Fig. 15.62b). At this period connection with the ocean was re-established, and the salinity of the water increased, at least in the western area. At the end of this period the oceanic connection narrowed. In the eastern basin, which at this stage had only a narrow connection with the western one, the deep water continued to stagnate and to accumulate hydrogen sulphide.

At the beginning of the Pliocene, the Pontic Sea developed (Fig. 15.62c); this was completely isolated from the ocean. By the end of the Pontic Stage elevation of the land severed the connection between the western and eastern basins, and the Caspian Basin became isolated (Fig. 15.62d). Later the connection between the two basins was renewed for a short period and the Caspian received a new influx of water from the Black Sea.

The Black Sea basin had essentially assumed its present day extent by the end of the Tertiary. During the Quaternary, the salinity was determined

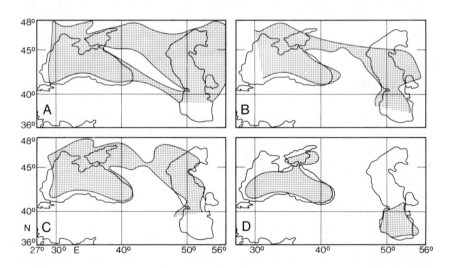

FIG. 15.62. The four main stages in the geological development of the Black Sea (Caspers, 1957).

TABLE 15.16

Correlation of the stages of development of the Black Sea with the glacial epochs;
North American equivalents in parentheses (Caspers, 1957)

Sea	Salinity state	Geological period
Modern Black Sea	Brackish, av. 18‰ S	Recent
Older Black Sea Basin	Weak increase of salinity	Post-glacial ("Climatic Optimum")
Newer Euxenian Sea	Brackish	Würm (Wisconsin) glaciation
	(Mediterranean closed at end of the stage)	
Karangat Sea	Increased salinity	Riss–Würm (Sangamon) interglacial stage
Post-Uzunlar Basin	Brackish to fresh	Riss (Illinoian) glaciation
	(Mediterranean connection)	
Uzunlar Basin	Brackish to saline	Mindel–Riss (Yarmouth) interglacial stage
Older Euxinian Sea	Brackish to fresh	Mindel (Kansan) glaciation
Claudinian Sea	Brackish to fresh	Gunz (Nebraskan) glaciation

by the alternate opening and closing of the connection with the Mediterranean. The low salinity at the beginning of the period was reduced still further, and at one stage it was virtually a fresh water lake. Reconnection then led to a minor influx of salt water, but as the seas were again separated, freshening was resumed. This failed to progress as far as it did in the earlier stage, and at the end of the glacial period salt water again entered the brackish-water sea via a new Mediterranean connection. In general, periods of glacial advance corresponded with periods of freshening, and interglacial periods coincided with intrusions of salt water. These effects are related to eustatic changes of sea level, transmitted via the Mediterranean. Table 15.16 shows the correlation between glacial and interglacial stages and the various stages in the development of the Black Sea basin.

The high level of the Black Sea in the Uzunlar stage was followed by a considerable regression. The post-Uzunlar sea again became an inland basin with a brackish and fresh-water fauna. The surface was 15 m below the present sea level, and the shore line lay well out on the continental shelf. This fall of water level was essentially a eustatic regression conditioned by the Riss glaciation period, although there were also epeirogenic movements on some stretches of coast. The ensuing interglacial period, between the Riss and the Würm glacial advances, was marked by a considerable increase in the amount of water as a result of the melting of the glacial mass. The level rose to 12–15 m above that at present, as demonstrated by the numerous terraces found today along Mediterranean and Black Sea coasts. The Russian term for this period is the Karagat stage.

The Sea of Marmara flowed through a narrow gap in the coastal mountain chain into the Manyas-Apolyont depression, extending from Iznik and probably from Brusa in Asia Minor. There was at that time no Bosporus. The Sea of Marmara was connected to the Black Sea via a prolongation of the graben-faulted gulf from Izmid to Adapazar. During this period the water level was high and the Black Sea spread out via the lowlands of Rostov-on-Don and the Manych depression to the Caspian basin. During this inter-glacial period the whole flow of the Danube was through the now broad and dry valley of the Karasu. The modern Danube delta had not yet been formed, and that area was then merely an estuary for streams entering from the Carpathian Mountains and Bessarabia.

The succeeding Würm glaciation lowered the water level to 80–100 m below the present one, and a broad coastal plain was laid bare. This was the post-Tyrrhenian regression, and the eustatic lowering of sea level isolated the Sea of Marmara, from both the Black Sea and the Mediterranean and gave the European continent a broad connection with Asia Minor.

During the Newer Euxenian phase the Bosporus was a meandering valley. Water entering from the Valley of the Golden Horn discharged into the Black Sea, which gradually became fresher. A land bridge connected the Rumanian region with the Crimea and the Sea of Azov did not yet exist. The mouth of the Danube became a deep erosional valley, and because of the eustatic lowering of sea level, all the river valleys entering the Black Sea extended across the continental shelf. The estuaries ("Limans") of the present day occupy the deeply eroded valleys of that period. In contrast with modern conditions, there must have been a deficit in the water balance in the Black Sea. As a result, the Bosporus became a river valley discharging into the Black Sea. The amount of water lost by evaporation was not replaced during Würm I and for this reason the sea level became low. It seems likely that tributary streams delivered little water, since much of the precipitation was locked up as ice. Certainly, the flow of the Ukrainian streams would have been low because of the low precipitation rates encountered in the vicinity of the ice mass (this phenomenon has been recognized for recent lakes in Greenland and Antarctica). The evaporation rate must have exceeded the combined rates of precipitation and inflow, with the result that in New Euxinian times the unstable water balance of the Black Sea became negative and the shelf was laid dry.

Following the Würm I glaciation, there was again a rise of water level, interrupted by temporary regressions in Würm II and Würm III. There were warm interstadial periods between each of these readvances. The topography shown in the modern map was gradually reached with the Post-Glacial Transgression. By the beginning of Würm II, the first salt-water plankton entered the Black Sea via the Bosporus which was now a marine strait. The level of the Bosporus at the end of the Würm II/III interstadial was still 11 m lower than today.

As the salinity increased, the abyssal depths became filled with more saline water and a stable permanent halocline developed which effectively isolated the deep water from direct contact with the atmosphere. Anoxic conditions then developed and copious amounts of hydrogen sulphide were formed.

15.4.5. CURRENT SYSTEM

15.4.5.1. Surface currents

As with the Baltic, the surface and deep currents exert a determining influence on the hydrography of the Black Sea. An understanding of the fundamental oceanographic features is essential if the complicated horizontal and, particularly, the vertical exchange processes are to be appreciated.

The average distribution of the surface currents is illustrated in Fig. 15.63. However, large deviations from this main pattern may occur in sympathy with local weather conditions (Neumann, 1942). The main factors producing the surface circulation pattern are the wind direction and velocity and the balance between the precipitation plus runoff and evaporation. The main runoff enters the northern Black Sea and evaporation occurs predominantly in the southeastern region. The circulation pattern (Fig. 15.63), is characterized by two large cyclonic eddies. Between these eddies and along the periphery, smaller eddies and counter currents form. Current velocities can reach very high values (40 cm s^{-1}), and this main current field causes the isopycnals of the surface layer to have a convex shape. Near the southern coast of the Black Sea a moderate current of about 20–30 cm s^{-1} flows to the east and turns to the northeast off Eregli, where it increases to 40–50 cm s^{-1} near the coast.

North of Kerempe Bay this current divides into two branches. The main current at first follows its previous direction, but then turns to the north and northwest with some changes in velocity. Having passed Kerempe Bay the second current branch maintains an eastward flow before broadening somewhat off Sinop. Finally, it takes a southeasterly direction after being augmented by an additional current from the northwest. The current velocities are about 20–30 cm s^{-1}. Up to about 37°E the centre of the current lies near the coast but later the current divides into two branches, one travelling north-eastwards and the other flowing eastwards, parallel to the coast. Both currents then deflect to the north. The region bounded by these two branches is one of variable currents. Counter currents and smaller eddies can also be observed. Counter currents exist along the coast of Anatolia, especially in the Bights of Kelkin-Eregli, and Sinop-Bafr-Samsum, and along the coast eastwards to Batumi; they are very dependent on the local wind conditions.

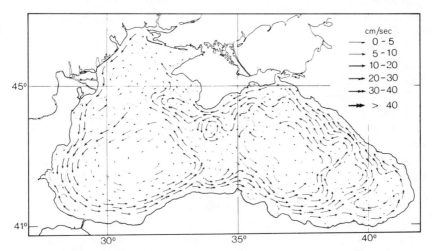

Fig. 15.63. The surface currents in the Black Sea (Neumann, 1942).

In the southeastern part of the Black Sea the current pattern is very complex. Eddies of various sizes may be present, and maximum current velocities of up to 110 cm s^{-1} have been measured and these eddies are thought to give rise to local areas of upwelling. Along the Caucasian coast, the main current direction is to the northwest and between Sochi and Novorossiysk average velocities of 35–50 cm s^{-1} have been measured although considerable variations have been reported.

South of the Straits of Kerch, the current becomes weaker and more unstable and splits into several branches and eddies. Off the southeastern coast of the Crimean Peninsula it turns to the southwest and at about 35°E it divides into two branches, one flowing to the south and the other to the west. The westward branch may reach velocities up to about 40 cm s^{-1}. This branch is augmented by the eddy current from the south and it turns to flow at 90° to the edge of the large shelf area. At 27°E the current turns through almost 90° and flows to the northwest after a sharp turn off Cape

Tarkhankut. Only a small part flows along the 200 m depth contour to the southwest with velocities of about 60–65 cm s^{-1}. This branch later joins the waters from the Dunarea River and completes the circulation pattern.

The currents in the northwest part of the Black Sea over the shelf are highly dependent upon the prevailing winds and on the runoff of the rivers Dnieper and Bug. The fresh water from these two rivers fills the Dnieper Bight, flows along the coast, and is united with that of the Danube. The mean current velocity is 15 cm s^{-1}, but it sometimes reaches 20 cm s^{-1}. Subsequently it increases and reaches 20–30 cm s^{-1} off Constanta. Off Cape Sabla the coastal branch joins branches from the Crimean Peninsula and reaches a velocity of 25–35 cm s^{-1}. A more or less well developed countercurrent with an average velocity of 20 cm s^{-1} occurs along the coast between the Bosporus and the Bight of Burgas. Within the large eddies in the eastern and western Black Sea, the currents are weak and badly developed.

Current systems near the estuaries of the large rivers usually have highly individual patterns. At first the run-off flows at right angles to the coast but then turns gradually to the right. Frequently, small eddies develop which turn the surface current toward the coast. The current regime is best developed in spring because of the increased continental runoff, but in summer and winter the system becomes less pronounced. The influence of wind on the density current is greatest on the shelf. In the Black Sea proper, the current is displaced by 30° to the right of the wind direction. (The angle of displacement is much less than this in shallow water.) The influence of the drift current on the main circulation pattern should not be underestimated.

15.4.5.2. Deep currents

The halocline at a depth of about 70–100 m marks the boundary between surface and deep water. At the centres of the large eddies it is at a depth of ~70 m, but it sinks to a depth of ~100 m in regions where there are high coastal currents. It is not broken down by winter convectional processes. Since the stable halocline prevents contact between deep water and surface waters, the water below it (i.e. at depths greater than 100–200 m) is anoxic and the only renewal, apart from a very restricted downward diffusion of surface water arises from the inflow of Mediterranean water through the Bosporus. On the basis of the Mediterranean water inflow complete renewal of the deep water of the Black Sea in this way has been calculated to take about 2,500 years.

The current velocity remains at its surface value down to a depth of ~20 m below which it decreases as the depth increases to ~200 m. The north-east current flows with a velocity of ~15–20 cm s^{-1} even at a depth of 100 m. In the southeastern corner of the Black Sea, extremely steep velocity gradi-

TABLE 15.17

Mean Surface Water Temperatures off the Coast of the Black Sea (°C)

	Jan.	Feb.	March	April	May	June	July	Aug.	Sept.	Oct.	Nov.	Dec.	Mean annual range
Dnieper Liman	0·3	0·6	2·7	9·0	17·1	21·4	23·0	22·2	17·8	12·5	5·9	1·6	22·7
Northwest Part	1·4	1·3	3·1	8·5	15·8	19·5	21·0	21·6	18·1	14·2	8·2	3·7	20·3
Crimea	4·7	4·6	6·2	9·5	14·7	19·3	21·6	22·4	19·4	15·8	10·7	7·0	17·8
Sevastopol	7·3	5·9	7·6	10·4	14·9	19·7	22·7	23·2	20·7	16·4	12·1	8·8	17·3
Caucasus	7·5	7·4	8·9	11·5	16·1	20·7	24·1	24·1	22·0	18·3	13·5	9·7	17·0
Straits of Kertsch	2·0	2·0	3·9	8·4	15·7	20·9	23·4	23·3	19·3	14·7	9·1	5·5	21·4
Azov Sea	0·3	0·3	1·8	8·4	16·6	21·5	23·8	22·8	17·7	11·6	5·4	1·5	23·5

T

ents between current and countercurrent are encountered, and the boundary between the currents may occur at various depths depending on location. The so-called depth of no motion is estimated to be about 150 m in the central part of the Black Sea and at about 250–300 m near the coasts. Very weak bottom currents are generally thought to flow in the opposite direction to the surface currents.

15.4.6. WATER TEMPERATURE

The mean surface water temperature of the Black Sea is generally slightly higher than that of the air in the coastal regions. Table 15.17 gives a compilation of the mean monthly water temperatures in several localities (see also Figs. 15.64a, b, c). Relative to the Black Sea, the Sea of Azov is warmer in summer and colder in winter. The temperature difference which may amount to as much as 4–5°C is the result of the shallowness of the latter sea. Temperatures of the surface water of the Black Sea proper are generally higher in the winter than are those at the coasts; however, in summer they are almost uniform. Coastal temperatures in the winter season are about 5–8°C in the Black Sea, although higher temperatures occur in the south-eastern regions. The mean surface temperature in summer is 23°C. The surface water temperatures off the southern coast of the Crimean Peninsula are often comparatively low in summer and values below 11°C have been observed near Cape Khersonsky. These low values result from upwelling caused by the prevailing winds and the surface temperature may drop by as much as 11·4°C in 11 hours.

15.4.7. SALINITY

Fig. 15.65a depicts the temperature and salinity profiles at four stations in the Black Sea proper. They indicate the occurrence of a cold water layer at a depth of about 50–75 m throughout the whole year and cold water in winter sinks only to that depth. Renewal of the water down to the halocline is caused only by turbulent diffusion or, in very severe winters, by convection. Seasonal fluctuations of temperature are confined to the water column above this cold layer.

The distribution of the surface salinity during the summer seasons is depicted by Fig. 15.65b. The influence of fresh water runoff is evident not only in the northwestern and western parts, but also in the southeastern region to which a few small rivers contribute relatively large amounts of run-off. Near the estuaries of the rivers, the salinity of the surface water may change markedly with season as the amount of run-off varies. During the winter the surface salinity is influenced by the withdrawal of water during

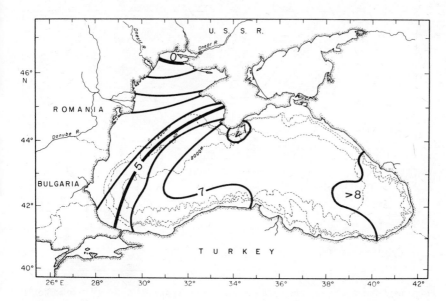

FIG. 15.64 (a). Average surface temperature in the Black Sea in February (°C) (Morskoi Atlas, Vol. II).

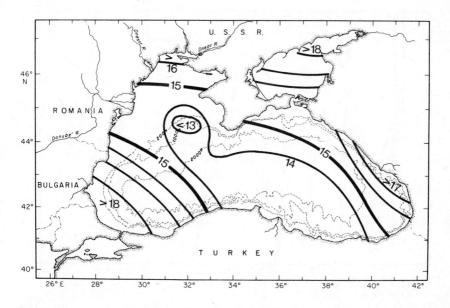

FIG. 15.64 (b). Average surface temperature in the Black Sea in May (°C) (Morskoi Atlas, Vol. II).

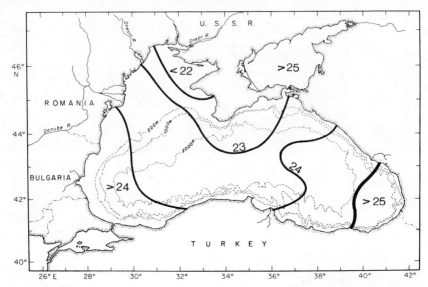

FIG. 15.64 (c). Average surface temperature in the Black Sea in August (°C) (Morskoi Atlas, Vol. II).

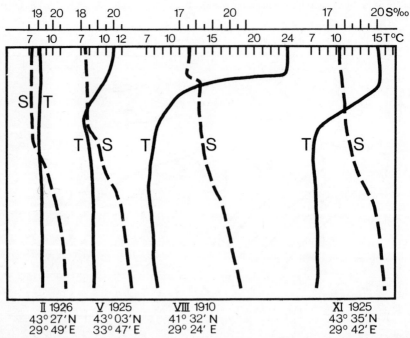

FIG. 15.65 (a). The distribution of temperature and salinity at four stations from the Black Sea (Neumann, 1943).

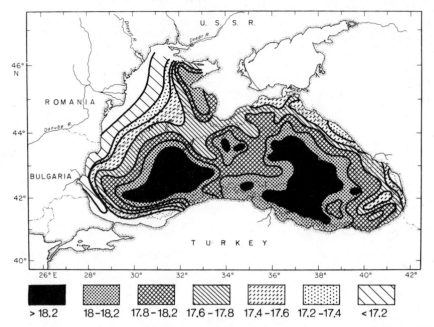

FIG. 15.65 (b). Mean surface salinity (‰) in the Black Sea in summer (July, August, September).

ice formation. The halocline occurs at a depth of about 100 m, and the salinity only increases by 0·5–0·6‰ from a depth of 300 m to the bottom.

Figure 15.66a illustrates the distribution of salinity and temperature over a west–east section in April 1969, and Fig. 15.66b shows similar north–south sections across the western and eastern basins.

15.4.8. DISTRIBUTION OF OXYGEN AND HYDROGEN SULPHIDE

As in the Baltic, a permanent halocline separates the deep water and surface water masses. Because of the cyclonic current system an updoming of the halocline occurs in the central parts of both western and eastern basins; this results also in a corresponding updoming of the boundary between the aerobic and anaerobic water. This boundary lies at a depth of about 250 m at the periphery and at a depth of approximately 150 m in the central parts. The current pattern is a further contributory factor to the slope of the boundary since the current velocities increase considerably towards the coasts, and this results in deeper mixing in these areas.

Because of the comparatively slow inflow of aerated water from the Mediterranean through the Bosporus, the rate of addition of oxygen to the

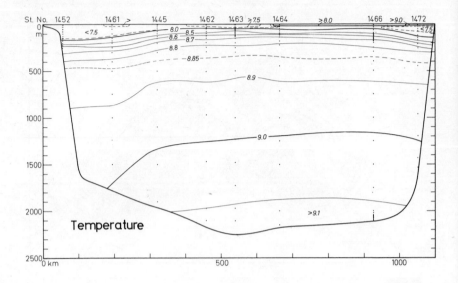

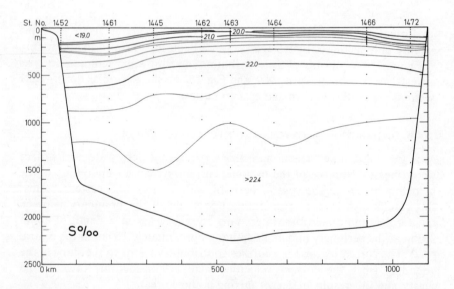

FIG. 15.66 (a). West–east sections of the distribution of temperature (°C) and salinity in April 1969 in the Black Sea (station numbers relate to the Atlantis II cruise).

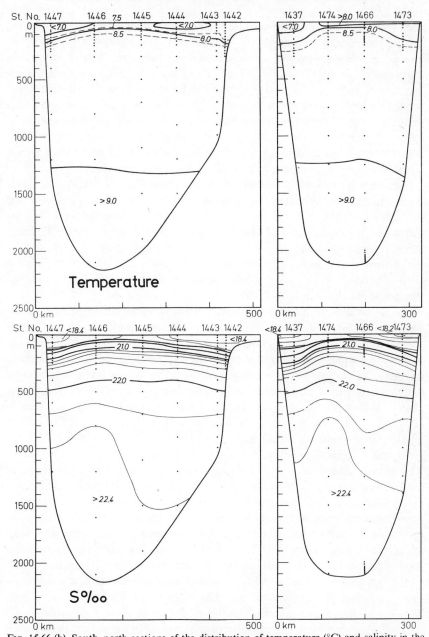

FIG. 15.66 (b). South–north sections of the distribution of temperature (°C) and salinity in the western and eastern Basin of the Black Sea in April 1969 (station numbers relate to the Atlantic II Black Sea cruise).

deep water is inadequate to counterbalance the rate at which it is consumed during the mineralization of organic material. This results in the complete and permanent absence of oxygen below the halocline. Unlike the Baltic, the amount of sub-halocline water is too large to allow any significant renewal by ingress of intrusive water, and for this reason the Black Sea has been anoxic in its deeper layers since the permanent halocline came into existence several thousand years ago. Fig. 15.67 depicts the distribution of oxygen and hydrogen sulphide in April 1969 along a longitudinal and two south–north sections through the Black Sea. As can be seen the shelf area— where it exists—is completely covered with oxygenated water.

Although it has long been believed that intrusion of Mediterranean water is the only means by which the deep water of the Black Sea is renewed this is certainly not the case. On the basis of the average inflow through the Bosporus the residence time of the deep water should be about 2,500 years. However, dating of the deep water by other methods indicates a mean residence time of approximately 800 years. This implies that there must be some vertical mixing which makes a contribution greater than the inflow through the Bosporus.

As can be seen from Fig. 15.68, which shows oxygen profiles for the central parts of the Black Sea and those parts adjacent to the Anatolian coast, there is a striking difference in the structure of the transition layer between the oxic surface water and the anoxic deep water. In the centre of the basin a sharp oxygen gradient marks the boundary between the surface layer and the deep water. The depth of the oxycline is the same as those of the thermo- and haloclines. However, the density increases only from $\sigma_t = 15.24$ at 50 m to $\sigma_t = 16.12$ at 100 m. Nevertheless, this pycnocline forms an effective barrier against vertical mixing. In contrast, along the Anatolian coast the strong currents and complicated eddies result in a deeper mixing of the surface water. The oxygen saturation only decreases slowly below a depth of 120 m and at 200 m the water is still about 20% saturated. The several inversions in the oxygen distribution illustrate the inhomogeneity of the water and suggest the possible occurrence of mixing processes. It seems likely therefore that the predominant part of the exchange process between surface water and deep water takes place along the edge of the Black Sea, especially along the Anatolian coast and the south eastern margin. This assumption is confirmed by the fact that appreciable precipitation of iron and manganese oxides has been observed at stations off the Anatolian coast (Brewer *et al.*, 1969). These precipitates are the result of the oxidation of these metal ions in the aerobic water from the +2 oxidation state (typical of the anoxic deep water) to the readily hydrolysed +3 and +4 oxidation states. In parallel with these oxidation processes, the oxidation of hydrogen sulphide to elemental sulphur

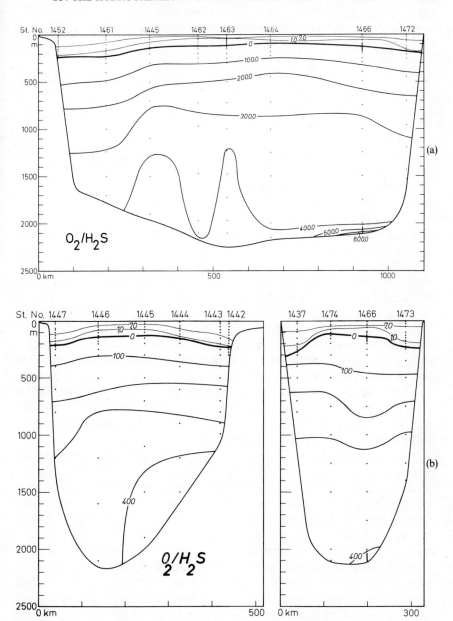

FIG. 15.67. The distribution of oxygen (ml l^{-1}) and hydrogen sulphide (ml l^{-1}) in the Black Sea in March/April 1969, (a) Along a west–east longitudinal section. (b) Along north–south sections in the Western and Eastern Basins. (Grasshoff, Atlantis II data 1969).

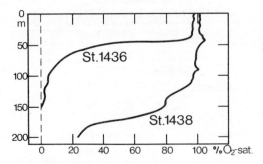

FIG. 15.68. Oxygen profiles in the Black Sea. St. 1436 in the centre of the Eastern Basin, St. 1438 off the Anatolian coast (Grasshoff, Atlantis II cruise 1969).

has also been observed by both turbidity measurements and *in situ* filtration (Sachs, 1969).

On the basis of the salt balance it can be deduced that about 500 km³ of surface water penetrate annually through the halocline into the deep water carrying down about 5×10^6 tons of oxygen and that about 700 km³ of hydrogen sulphide containing deep water are mixed into the surface layers. When account is taken of the 200 km³ of Mediterranean water which intrudes through the Bosporus, the residence time of the deep water can be calculated to be ~ 700 years. This is much more reasonable than the 2500 years estimated on the assumption that the deep water is only renewed by means of the Bosporus inflow only. Carbon-14 dating of the Black Sea deep water indicates a value for the residence time very similar to that calculated from the salt balance. It is likely that most of this vertical exchange occurs along the slope of the southern and the south-eastern coast.

There is a transition zone between the oxygenated surface waters and the anoxic deeper waters (see Figs. 15.69 and 15.67a, b). In this, oxygen is absent and there is no evidence for the formation of hydrogen sulphide. It has been suggested that hydrogen sulphide and oxygen are able to coexist, at least for some time, even though the half life time of hydrogen sulphide in the presence of oxygen under natural conditions (in the presence of bacteria and organic substrates) is known to be only about 30 minutes. Thus, it seems likely that such assertions are often based on analyses of sulphide containing water which has become contaminated with oxygen immediately before analysis. It might be noted that a detailed sampling programme carried out in the transition layer at a number of stations during the Atlantis II cruise gave no evidence for the coexistence of oxygen and hydrogen sulphide. However, even in regions in which it is known that there is strong vertical mixing hydrogen sulphide was never found when oxygen

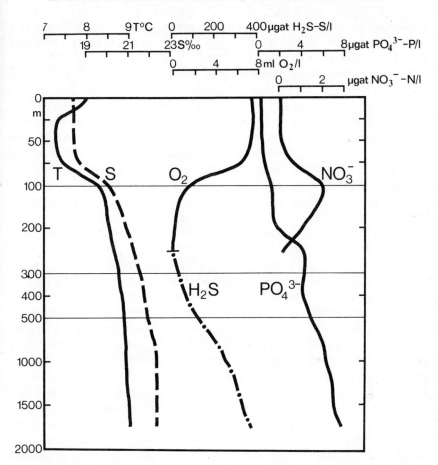

FIG. 15.69. Vertical profiles for temperature, salinity, oxygen, hydrogen sulphide, dissolved phosphate, and nitrate at station Nr. 1461 in the Black Sea, April 1969 (Grasshoff, Atlantis II data).

was present in amounts greater than the analytical error (0.02 ml l^{-1}). These two components could only be present simultaneously in the same body of water if rapid and intensive vertical mixing processes with an amplitude in the order of ten metres and a time constant of less than 30 minutes are occurring.

In the central regions of the Black Sea the oxygen concentration usually decreases suddenly from almost saturation to very low values of less than 1 ml l^{-1} at a depth of 125 to 150 m. This decrease occurs at about the same depth as the permanent pycnocline. Immediately below the pycnocline the dissolved oxygen content decreases exponentially to zero, suggesting that vertical eddy diffusion with only a very small exchange coefficient predominates over most of the Black Sea.

As mentioned above, recent detailed work on the transition layer has shown both oxygen and hydrogen sulphide to be absent. In this layer there is apparently a dynamic equilibrium in which upward diffusing hydrogen sulphide is oxidized by oxygen which diffuses downwards. This concept is supported by the approximately exponential increase of the hydrogen sulphide concentration below the transition layer observed for the central regions of the Black Sea. On the basis of theoretical studies of oxygen distribution by Wyrtki (1962), and Craig and Weiss (1970), Spender and Brewer (1971) have calculated a vertical advective velocity of 0.5 m yr^{-1}, for the entire Black Sea and a vertical eddy diffusion coefficient of 0.014 cm^2 s^{-1}. This latter value is somewhat lower than the generally accepted lower limit for the vertical eddy diffusion coefficient in the oceans (Munk, 1966), and is only 10 times the molecular thermal diffusion coefficient. According to Brewer (1973), this low value may reflect the high stability of the water column which is particularly notable in the central parts of the Black Sea.

The oxygen profiles (Fig. 15.68) show that the vertical mixing processes off the southern and south eatern coast of the Black Sea differ markedly from those which exist in the central parts. The easterly currents (which in some places reach considerable velocities—see Fig. 15.63), the countercurrents and the horizontal eddies enhance vertical mixing. The permanent halocline lies about 100 m deeper than it does in the centre of the Black Sea, and the salinity and especially the oxygen gradients are much less pronounced along the Anatolian and Caucasian coasts; this implies that these coastal zones play an important role in the renewal of the Black Sea deep water. This conclusion is also supported by the observation that suspended iron and manganese oxides occur in the mixing layer of the Anatolian coast (Spencer and Brewer, 1971), these are believed to have been formed following transport (by turbulent mixing) of anoxic water containing iron (II) and manganese (II) ions, from below the halocline, into the oxic layers. Here the iron and manganese are oxidized to higher oxidation states and precipitate as sparingly soluble oxides and hydroxides.

15.4.9. ALKALINITY AND pH

Figure 15.70 shows that the pH values of the surface water lie in the range 8.2–8.3—similar to that for other seas. Oxygen consumption in the subsurface water is accompanied by carbon dioxide formation and as a result the pH falls to values of about 8.0 at the depth of the permanent halocline.

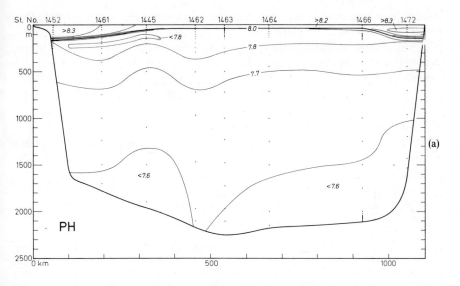

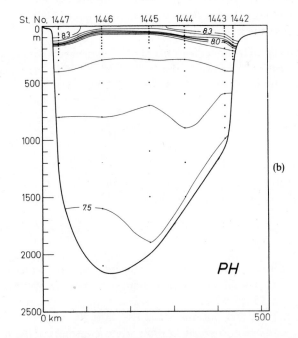

Fig. 15.70. The distribution of pH in the Black Sea. (a) Along a west–east longitudinal section from west to east, (b) Along a north–south section in the western Basin (Grasshoff, Atlantis II data).

The pH 8·0 isoline closely follows the 19·5‰ S isohaline. Below the halocline the pH decreases rapidly to a value of 7·8 and then more slowly to 7·6 at a depth of about 1000 m. Even in the presence of considerable amounts of hydrogen sulphide the pH does not decrease much below 7·5. This contrasts with the low pH values found in the Baltic (< 7·0 during anoxic periods in the deep water). There appear to be three factors contributing to the relatively high-values of the anoxic deep water of the Black Sea:

(i) the amount of oxidizable organic matter reaching the deep water is relatively small. Thus, the average primary production for the Black Sea has been estimated to be about 8·3 mol C m^{-2} yr^{-1} (Sorokin, 1964; Deuser, 1971; Sen Gupta and Jannasch, 1973), but most of the organic matter produced is oxidized both above and just below the halocline, and consequently little CO_2 generation occurs in the deep water. Further, some of the CO_2 is used by chemosynthetic bacteria according to the equation

$$6CO_2 + 24H^+ + 24e^- = (C_6H_{12}O_6) + 6H_2O$$

leading to the production of carbohydrate and to the consumption of CO_2 and hydrogen ions. (Sorokin, 1964; Sen Gupta and Jannasch, 1973; Tuttle and Jannasch, 1973). This process is thought to occur mainly in the anoxic layers just below the halocline;

(ii) The relatively high concentrations of total hydrogen sulphide constitutes a buffering system. Almost all the hydrogen sulphide originates via the reduction of sulphate (Deuser, 1971; see also Chapter 16)

$$(CH_2O)_{106}(NH_3)_{16}H_3PO_4 + 53SO_4^{2-} \rightarrow$$
$$106CO_2 + 53S^{2-} + 16NH_3 + H_3PO_4 + 106 H_2O$$

(Richards, 1965)

Sulphide so formed readily associates with protons to form HS^- and H_2S (for H_2S, $pK'_{s1} = 7·42$ and $pK'_{s2} = 15·4$ at a salinity of 20‰ and a temperature of 10°C (Grasshoff, unpublished)). Over the entire range of possible pH values in sea water the predominant ionic species of the sulphide system will therefore be HS^-;

(iii) The alkalinity of both the surface water and the deep water is unusually high (Fig. 15.71). The alkalinity of the surface water is approximately 3·4 meq l^{-1} (Table 15.18) which is about 2·06 meq l^{-1} greater than that of oceanic water. The contributions of the HS^- and S^{2-} ions to the total alkalinity must be taken into consideration when considering the anoxic deep water, but even so the alkalinity expressed simply as the sum of the equivalents of HCO_3^-, CO_3^{2-} and $B(OH)_4^-$ ions is more than 4·0 meq l^{-1} which corres-

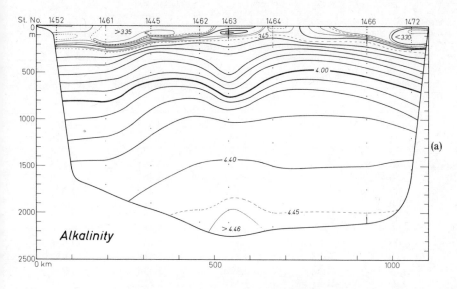

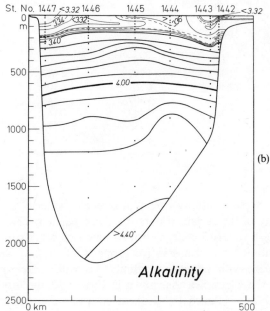

FIG. 15.71. The distribution of the alkalinity (in meq l^{-1}) (corrected for the H_2S alkalinity) in the Black Sea. (a) along a west-east longitudinal section, (b) along a north-south section in the Western Basin (Grasshoff, Atlantis II data).

TABLE 15.18

Alkalinity and Alkalinity Anomaly at St. 1463 in the Central Black Sea (in meq l^{-1})
(Grasshoff, 1969, Atlantis II Data)

Depth	Salinity (‰)	Tot. alk. (corrected for sulphide)	Carbonate alk.	Alk. anomaly
5	18·97	3·385	3·358	2·063
25	18·93	3·396	3·370	2·076
35	18·99	3·397	3·370	2·074
45	20·35	3·374	3·361	1·955
50	—	3·401	3·401	1·958
60	—	3·417	3·417	1·697
70	—	3·412	3·412	1·955
80	21·02	3·390	3·378	1·925
90	—	3·423	3·422	1·959
100	—	3·441	3·441	1·976
115	21·28	3·435	3·416	1·952
125	21·33	3·460	3·436	1·974
130	21·31	3·447	3·423	1·961
150	21·50	3·470	3·441	1·972
200	21·68	3·538	3·487	2·027
250	21·84	3·602	3·541	2·080
300	21·92	3·689	3·606	2·161
500	22·12	3·952	3·808	2·410
1000	22·37	4·292	4·077	2·733
1250	22·38	4·367	4·109	2·807
1500	22·35	4·407	4·148	2·849
1750	22·36	4·443	4·167	2·885
2000	22·39	4·458	4·175	2·898
2150	22·41	4·466	4·163	2·904

ponds to an excess alkalinity over that of ocean water of 2·9 meq 1^{-1}. The high alkalinity of the surface waters is at least partially caused by the large amount of fresh water discharged from a lime-rich catchment area.

If the alkalinity is a conservative property of the sea water the values should decrease with increasing salinity, i.e. with increasing depth. This is not so for the Black Sea. As shown in Fig. 15.70, the total alkalinity increases to values >4·0 meq 1^{-1} in the deep water, even when the values are corrected for the sulphide contributions. Phosphate is not a significant contributory factor as its concentration is only ~0·015 meq 1^{-1}. As the B/Cl ratio in Black Sea deep water is similar to that in ocean water (Kremling, unpublished) this element cannot be responsible for the anomalous alkalinity.

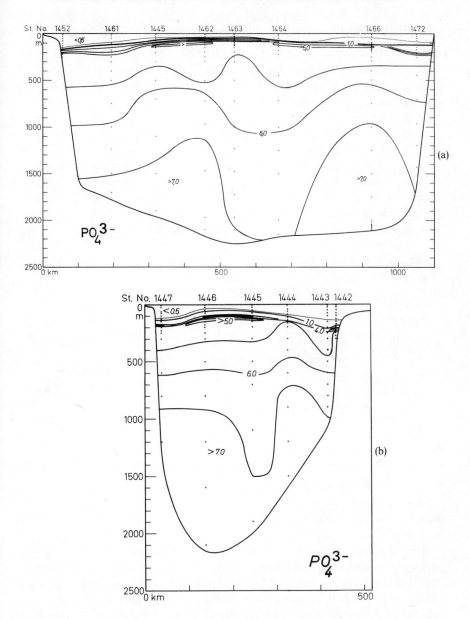

FIG. 15.72. The distribution of dissolved inorganic phosphate in the Black Sea (in μg-at PO_4^{3-}–P l^{-1}). (a) Along a west–east longitudinal section from West to East, (b) Along a north–South section in the Western Basin (Grashoff, Atlantis II data).

15.4.10. NUTRIENTS

15.4.10.1. *Phosphorus*

The permanent halocline clearly separates two distinct water masses as judged by the dissolved inorganic phosphate concentration (Fig. 15.72). In the surface water the concentration of reactive phosphate is comparable to that in other seas. The values are normally low (0.5 µg-at PO_4^{3-}–P l^{-1}) in the euphotic zone following the spring plankton bloom and remain low during the whole summer season. During spring and summer there is a net downward transport of organically bound phosphorus. Each year the upper layer is refertilized with phosphate by convectional overturn of the surface water during the winter. In severe winters the density gradient is reduced sufficiently to permit diffusion of phosphates from below the permanent pycnocline. The upward flux of phosphates from below the pycnocline into the brackish layer may amount to as much as 115000 tons yr^{-1} (Fonselius, 1974; see Fig. 15.73). It is thought that the amount of phosphate which is

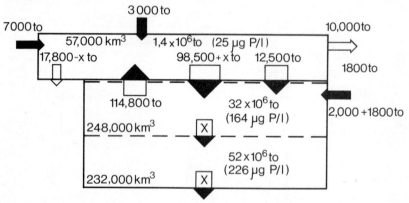

FIG. 15.73. The phosphorus balance in the Black Sea. Fluxes are expressed in tons yr^{-1} (Fonselius, 1973).

deposited as mineral phases on the floor of the deep parts of the Black Sea is very small.

The restrictive diffusional barrier between surface and deep water leads not only to the total permanent depletion of oxygen below the halocline, but also to the accumulation of dissolved phosphate to the extent of ~ 8 µg-at PO_4^{3-}–P l^{-1}. According to Richards (1965), the decomposition of planktonic matter of "normal" composition results in the liberation of 53 moles of sulphide for each mole of phosphate. If it is assumed that all the dissolved phosphate present in the deep water results from decaying organic matter, the average concentration of hydrogen sulphide in the deep water should be about 400 µg-at S^{2-}–S 1^{-1}. This holds only approximately since several side processes also occur (e.g. chemotrophic syntheses of organic material, the fermentation of organic material leading to the formation of e.g. methane; see Chapter 16). Data collected during the Atlantis II cruise in 1969 (Fig. 15.73) show the relationship between hydrogen sulphide and the dissolved phosphate concentration in the deeper layers of the Black Sea. The slope of the curve suggests that most of the phosphate is organogenic. This contrasts with what is found for the Baltic in which large amounts of phosphates dissolve from the sediment under anoxic conditions. There are two principal reasons for this difference:

(i) the pH in the anoxic bottom water of the Black Sea is not low enough to allow the formation of mineralogenic phosphate (see also Fig. 15.70).

(ii) because the Black Sea has been permanently anoxic for several thousand years, all the phosphate should have been leached out of the upper sediments.

15.4.10.2. Nitrogen compounds

It seems likely that the most dominant factor governing primary production in the Black Sea is the availability of nitrogen. As in most seas, the nitrate content of the surface water drops to values of a few tenths of a microgram atom per litre during the spring bloom and remains very low until the onset of the autumnal convectional overturn of the surface layer. Even in the layers above the halocline, but beneath the euphotic zone the nitrate content is not high (Fig. 15.69), values exceeding 2·5 µg-at NO_3^-–N 1^{-1} being exceptional. It is thus clear that this water could not be a major source of nitrate for the surface layer.

Immediately above the hydrogen sulphide-rich region there is a layer in which nitrate behaves as a hydrogen acceptor and is itself reduced to molecular nitrogen by denitrification and thus eliminated from the cycle. On the basis of the box model proposed by Fonselius (p. 574) and assuming an average content of 2 µg-at NO_3^-–N 1^{-1} it can be calculated that ~ 14000 tons of NO_3^-–N are lost annually in this way.

The ammonia content of the surface water is usually low and seldom exceeds 1·0 µg-at NH_4^+–N 1^{-1}. In the deep water below the permanent halocline concentrations of more than 85 µg-at NH_4^+–N 1^{-1} are found. Figure 15.74b shows the relationship between ammonia-nitrogen and hydrogen sulphide in the deep water of the Black Sea on the basis of data collected during the Atlantis II cruise. Even if the analytical method underestimated the ammonia, the slope of the regression line indicates that most of it is organogenic and that it is formed according to a scheme whereby for each atom of organogenic ammonia nitrogen produced 3·3 atoms of sulphide sulphur will be generated from sulphate (Richards, 1965). Some of the sulphur

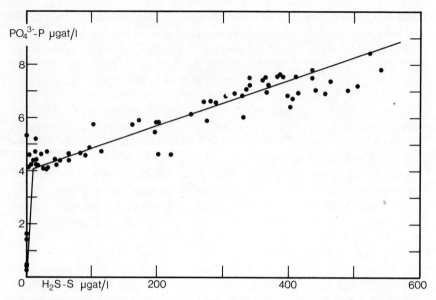

FIG. 15.74 (a). The relationship between the concentrations of dissolved inorganic phosphate and hydrogen sulphide in the Black Sea deep water on the basis of data from the Atlantis II cruise 1969 (Grasshoff, 1969).

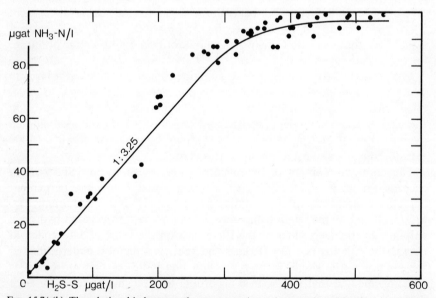

FIG. 15.74 (b). The relationship between the concentrations of ammonia and hydrogen sulphide in the Black Sea deep water on the basis of data from the Atlantis II cruise 1969 (Grasshoff, 1969).

is of course bound as metal sulphides and fermentation and chemosynthesis also take place.

According to Brewer and Murray (1973) there is a net upward flux of ~ 0.3 g m^{-2} yr^{-1} of ammonia-nitrogen to the upper layer and of this 0.075 g are used in biosynthesis. The remainder corresponds to the entry of approximately 90 000 tons of ammonia nitrogen. Some of this ammonia is oxidized via nitrite to nitrate. Significant nitrite maxima can be observed in the oxygenated layers immediately above the halocline.

Although there has been considerable discussion (Sen Gupta, 1969, 1971, 1973; Skopnitsev et al., 1967) about whether hydrogen sulphide and nitrate can coexist, there is little experimental evidence for the existence of nitrate in sulphide containing waters. (The oxidation of ammonia proceeds rapidly in the presence of oxygen particularly if the ammonia content is high). However, it is possible that in areas of strong turbulent mixing sulphide and nitrate might be found together in the boundary layer.

15.4.10.3. *The distribution of silicate*

Because of the large contribution of fresh water it would be expected that the silicate content of the surface waters of the Black Sea will be relatively high, even though all the rivers do not drain rocky catchment areas. According to data collected during the Atlantis II cruise, the water of the productive layers along the coasts contain about 10–20 µg-at reactive silicate l^{-1}, the value increasing in the upper layers towards the central parts (see Fig. 15.75).

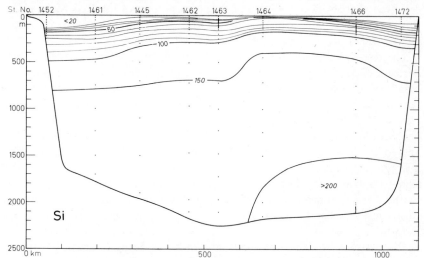

Fig. 15.75. The distribution of the reactive silicate (in µg-at Si l^{-1}) along a east–west longitudinal section of the Black Sea (Grasshoff, Atlantis II data).

With increasing depth the silicate concentration increases to $\sim 100\,\mu$g-at Si 1^{-1} at a depth of 500 m, and to more than 200 µg-at Si 1^{-1} in the bottom water of the eastern basin. A steep silicate concentration gradient is observed at the same depth as the permanent halocline; thus, the updoming of the discontinuity layer is mirrored by the silicate content. The high silicate content of the deep water probably arises from the dissolution of diatom shells as a result of the reductive removal of the Fe(III) coating which is thought to protect them from dissolution in oxic waters.

The distribution of silicate in the surface layer suggests that silicate is not a limiting factor in primary production. Since Mediterranean water is rather poor in silicate (< 10 µg-at Si 1^{-1}) the distribution of silica can be used to study the mechanism of penetration of Mediterranean water over the sill.

15.5. THE RED SEA

15.5.1. GEOGRAPHY AND HYDROGRAPHY

The Red Sea belongs to category (ii) of landlocked seas (see Section 15.1) as it is in a semi-enclosed basin in an arid zone in which evaporation far exceeds precipitation and runoff. Like the other previously discussed enclosed seas, its chemistry is intimately linked with physical and biological processes and especially with topographical features. Morcos (1970) has extensively reviewed the physical and chemical oceanography of the Red Sea, and for this reason, discussion will be confined to the more important aspects which will be illustrated by reference to the author's own work (Grasshoff, 1969).

The Red Sea extends as a rift valley from the narrow land bridge between Africa and Asia in the North to the Straits of Bab el Mandeb. It has a total length of almost 2,200 km and is, on average, about 280 km broad. It is a part of the extensive rift valley system between Asia and Africa. The Sinai Peninsula divides the northern part into the shallow Gulf of Suez and the deep Gulf of Aquaba. Muromtsev (1960) has estimated the area of the Red Sea to be 0.450×10^6 km^2 and its volume to be 0.251×10^6 km^3. The average depth of the Red Sea is 491 m, but it is much deeper along the axial trough in which the depth may exceed 2500 m. The Red Sea is connected to the Mediterranean by the Suez Canal which has no locks; however, this connection is of no practical importance for the exchange of water. In the south, the connection to the Indian Ocean is formed by the narrow and shallow Straits of Bab el Mandeb, in which the deepest part of the sill (to the south east of the Island of Perim) has a depth of only 100 m. This strait limits water exchange between the Red Sea and the Gulf of Aden.

According to Swartz and Arden (1960) the Red Sea was formed by the separation of the two crustal blocks of Africa and Arabia. This separation began during the Lower Eocene and continued throughout the Middle Eocene and Oligocene. The central rift valley is, at present, oceanic crust, and the shallower parts on either side of the trough are down-faulted continental blocks covered initially by carbonate sediments and evaporites and subsequently by coral layers and unconsolidated sediments (Girdler, 1966).

No rivers run into the Red Sea and precipitation over it is extremely small, not exceeding 5 cm yr^{-1}. In contrast, estimates for the annual evaporation rate range from 1·82 m (Privett, 1959) to 2·66 m (Bogdanova, 1966). For the purposes of the present discussion a rate of 2·0 m yr^{-1} will be assumed.

The combination of the very restricted exchange of water with the open sea and the extremely high evaporation rate leads to the production of the dense, highly saline water which forms the deeper layers of the Red Sea. These waters are at all depths very much denser than those at the same depth in the adjacent Gulf of Aden.

According to Thompson (1939) the circulation in the Red Sea is produced by four factors: the tides, the prevailing winds, the evaporation and the differences in the density of the water resulting from seasonal changes in the rate of evaporation. The most important of these factors is the prevailing wind direction, which varies in sympathy with the monsoon periods (Barlow, 1934). From May until September the prevailing wind blows from the north-west. During the rest of the year the more northerly part is subject to wind from the northerly quarter; south of 20°N the wind direction between October and April is mainly south-southeasterly. During the transitional periods winds are weak and variable. The high rate of evaporation in the Red Sea leads to intrusion of water from the Gulf of Aden. During the winter this inflow takes place in the surface layer down to a depth of about 80 m. According to Neumann and MacGill (1961) the inflow during the summer period occurs in the intermediate layers because of the existence of a wind driven surface current flowing towards the Gulf of Aden. At the bottom, a fairly well-defined current transports Red Sea water into the Gulf of Aden. The intensity of the outflow varies seasonally and according to the wind pattern. The deep water circulation results from the sinking of dense highly saline water produced by evaporation. Cooling of the surface water during the winter season contributes further to the increase in density, especially in the northern part of the Red Sea and the Gulf of Aquaba. Conditions are almost isothermal and isohaline at depths greater than 300 m. The source of the deep water can be clearly seen from the longitudinal sections published by Neumann and MacGill (1961); wind induces convergence which leads to sinking of the surface water north of 25°N (Sverdrup et al., 1942; Barlow,

1934). In addition, a transverse water circulation is caused which leads both to an upwelling along the Arabian coast, especially south of 20°N, and to a transverse circulation in summer (Thompson, 1939). This transverse circulation does not affect the deep water in the central trough below 500 m. Using the water exchange in the Straits of Bab el Mandeb to calculate the rate of evaporation, Privett (1959) has concluded that it is likely that the rate of evaporation is greatest during the winter season. His conclusion was based on work by Neumann and McGill who considered that ingoing currents prevail in the Straits of Bab el Mandeb not only during the winter months, but until the middle of June. This deduction by Privett is in conflict with the opinion expressed by Schott (1929).

The water of maximum density is formed in the Gulf of Aquaba during the winter and an overflow across the sill of Tiran transports this dense water into the Red Sea proper.

15.5.2. DISTRIBUTION OF SALINITY AND TEMPERATURE

The southern part of the Red Sea is covered by a relatively thin layer of surface water which has its origin in the Gulf of Aden (see Fig. 15.76). The salinity of this water ranges from 37·5‰ in the southern part to 39·5‰ at about 25°N. The salinity gradient between the surface water and the under-lying Red Sea deep water is greatest in the southern part and least pronounced towards the north. The deep water is almost isohaline with a salinity ranging from 40·0 to slightly above 40·6‰. It shows very little structure; however, such structure as there is is most pronounced during the summer. The pycnocline is most pronounced when the halocline coincides in depth with thermocline. During the summer, the temperature of the surface layer averages 30°C in the southern Red Sea and 24–26°C (see Fig. 15.77) in the northern part. The deep water is again almost isothermal with a temperature of 21·5–21·6°C. In the deeper layers the warmest water occurs at intermediate depths of about 500 to 1 000 m. During the winter months, the temperature of the deep water is almost constant and is, on average, slightly warmer (21·8°C) than in summer because of the convection of dense surface water and its penetration from north to south.

The distribution of salinity in the Straits of Bab el Mandeb is shown in Fig. 15.78. High salinity water ($\sigma_t > 27$) overflows into the Gulf of Aden, and cascades down the southern slope of the sill. As it does so it gradually mixes with less saline water and intrudes into the deep water of the Indian Ocean at a depth appropriate to its density. As a result of the complicated bottom topography of the southern part of the slope, the outflowing water divides into two branches (Fig. 15.79; see also Krause, 1970). By virtue of the high salinity

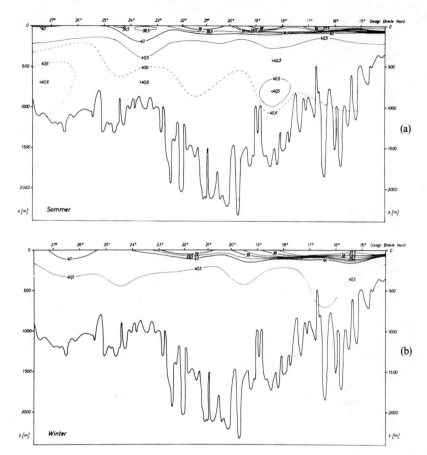

FIG. 15.76. The distribution of salinity (‰) along a longitudinal section of the Red Sea. (a) summer, (b) winter (Siedler, 1968).

and low silicate content of the water these branches can be traced as discrete layers, differing slightly in depth, even south of the equator off the coasts of Somalia and Tanzania. Current measurements (Siedler, 1968) and chemical properties such as the distribution of, e.g. oxygen (Fig. 15.80; Grasshoff, 1969) suggest that the main outflow occurs via the southern branch.

Several deep basins filled with hot brine exist in the trough region of the southern and central Red Sea. The brines have temperatures ranging from 30°C to almost 60°C and are nearly saturated with sodium chloride. Since their discovery a decade ago these waters have been the subject of numerous investigations (see Degens and Ross, 1969 for a review of their chemistry

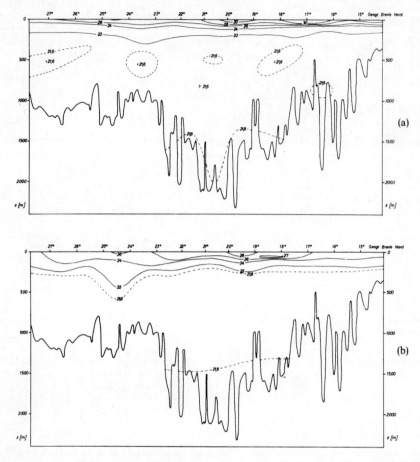

FIG. 15.77. The distribution of temperature (°C) along a longitudinal section of the Red Sea (a) in summer, (b) in winter (Siedler, 1968).

and geochemistry). The basins are potentially of economic importance because their bottom sediments are rich in trace metals such as silver, zinc, cadmium, nickel, copper, manganese, etc., which occur as oxides and/or sulphides. Further exploration of the area in 1972 by the Russian research ship "Valdivia" has led to the discovery of several additional brine areas. The origin of the brine has been the subject of much speculation. However, it now seems to be generally accepted that it has its origin in the penetration of Indian Ocean water into evaporite beds underlying the Red Sea. Flow is induced by the increase in density resulting from dissolution of evaporite minerals, and the brine undergoes geothermal heating before it emerges at

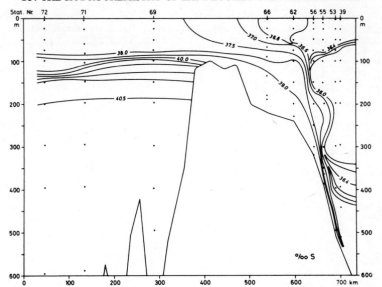

FIG. 15.78. The distribution of salinity in the Straits of Bab el Mandeb (Grasshoff, 1969).

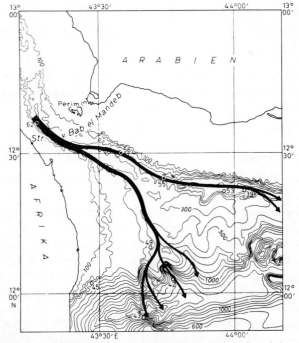

FIG. 15.79. The bottom topography off the Straits of Bab el Mandeb and the resulting outflow at the bottom (Grasshoff, 1969).

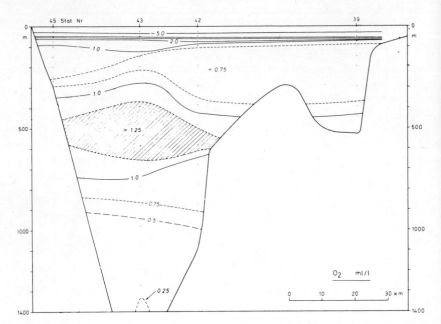

FिG. 15.80. The distribution of oxygen along section C in the Inner Gulf of Aden (Grasshoff, 1969).

depths in excess of 2000 m through the floors of the basins. The brines have no significant influence on the hydrography of the Red Sea because of the very sharp density gradient which separates them from the overlying Red Sea Deep Water.

15.5.3. WATER BALANCE

As has been previously mentioned, unless they are carried out for a complete year direct current measurements (Siedler, 1968) in the Straits of Bab el Mandeb will not lead to realistic estimates of the total exchange of water through the Straits. If measurements are made only for short periods they may be useful in examining the effect of meteoroloical conditions on water flow but will not provide sufficient data to enable the water balance to be assessed. The picture is further complicated by the existence of coastal countercurrents which make the calculation of the water exchange on the basis of current measurements in the middle of the Strait extremely questionable. A more reliable method of calculating the average water exchange is based on the assumption of constant salinity and constant average sea level. Both of these assumptions are strongly supported by data obtained during oceanographic expeditions over the last 50 years. Before the exchange rate can be calculated it is also necessary to know the evaporation rate. No direct measurements of this rate have been made because of the experimental difficulties of measuring it at sea. It is therefore necessary to use assumed values; these range from 182 cm (Privett, 1959) to 266 cm (Bogdanova, 1966) per year. A figure of 200 cm yr^{-1} is probably realistic and has been used by Grasshoff (1969) in developing a model for the water circulation in the Red Sea (Fig. 15.81). In establishing the model it was necessary also to use data on

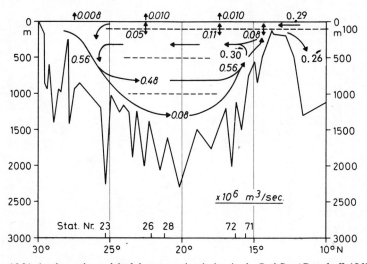

Fig. 15.81. A schematic model of the water circulation in the Red Sea (Grasshoff, 1969).

the axial distribution of salinity and dissolved oxygen, and to make some assumptions about the rate of oxygen consumption in the water. Estimates of mass balance factors derived by various workers have been compiled by Morcos (1970) (Table 15.19).

15.5.4. DISTRIBUTION OF OXYGEN

The dissolved oxygen distribution pattern in the Red Sea shows three distinct layers (Fig. 15.82). The layer above the halocline is saturated, or almost saturated, with respect to oxygen, but the absolute concentrations are comparatively low because of the high salinity and high temperature of the water. A well developed oxygen minimum layer extends from a depth of 300 to about 1000 m. In this, the concentration may be as low as 0·5 ml l^{-1}; however, complete absence of oxygen has never been reported. Because of

TABLE 15.19

Compilation of various estimates of the water budget of the Red Sea (Morcos, 1970) (i = ingoing, u = outgoing undercurrent)

Method	From the principles of conservation of volume and salt			From direct measurement of currents		
Author	Morcos (1970)	Bogdanova (1966)	Grasshoff (1969)	Krümmel (1911)	Vercelli (1925)	Siedler (1968a)
Section				Dumeira	Dumeira	Large Strait
S_i‰	36·7	—	36·6	36·9		
S_u‰	39·7	—	40·5	39·7		
V_i, cm s^{-1}				43 (0–74 m)	$\left\{\begin{array}{l}46·3\ (0{-}50\ m)\\23·1\ (50{-}100\ m)\end{array}\right.$	
V_u cm s^{-1}				43 (74–196 m)	69·4 (100–196 m)	
T_i 10^6 m^3 s^{-1}	0·358	0·730	0·29	0·461	$\left.\begin{array}{l}0·463\\0·116\end{array}\right\}0·579$	0·59
T_u 10^6 m^3 s^{-1}	0·330	0·692	0·26	0·424	0·486	0·43
$T_i - T_u$ 10^6 m^3 s^{-1}	0·028	0·038	0·03	0·037	0·093	0·16
E, 10^6 m^3 s^{-1}	0·0284	0·038	0·028	0·037	0·051	0·028
km^3 yr^{-1}	898	1197	876	1167	1605	876
Area, 10^3 km^2	438	450	438	458·48	458·48	438
E, cm yr^{-1}	205	266	200	254·6	350	200
$T_i - T_u/E$	1	1	1	1	1·8	5·7

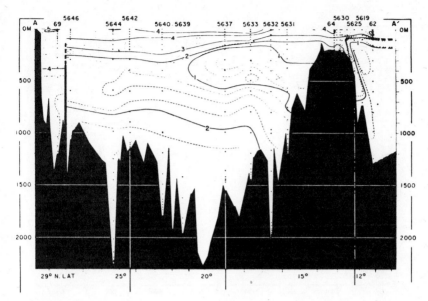

FIG. 15.82. The distribution of dissolved oxygen (ml l⁻¹) in the Red Sea (Neumann and McGill, 1961).

the relatively strong vertical circulation in the north and the small residence time of the deep water in the Red Sea compared to that in the adjacent Indian Ocean. The oxygen concentration in the deep water is significantly higher than that in the intermediate layer. A thin layer of oxygen-depleted water leaves the Red Sea across the sill at Bab el Mandeb (Fig. 15.83a); however, its oxygen content is high in comparison with that of the water lying below the halocline in the Gulf of Aden (Fig. 15.83b). The oxygen content of this core water proves that the latter originates from the upper and not the deeper layers of the Red Sea. The low oxygen content of the Red Sea water noted off the Somalian coast (Dietrich et al., 1967) is a secondary property. Because of its isolation from the surrounding water, this water becomes progressively depleted in oxygen as it moves southwards.

The low oxygen content of the water below the thermo-halocline in the Gulf of Aden indicates that this layer is permanently isolated from direct exchange with the atmosphere and that the thermo-halocline prevails throughout the year.

There is an unusual intrusion of oxygen-depleted water north of the Strait region (Fig. 15.83). This oxygen-poor water is also characterized by its low temperature and high content of nutrients (Grasshoff, 1969). On the basis of these factors and the salinity, it is believed that these intruding water

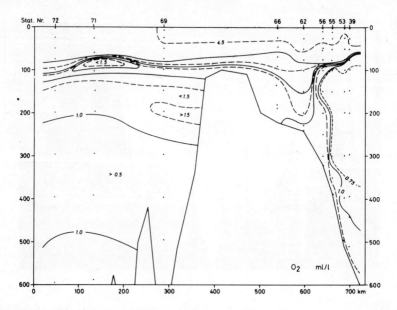

FIG. 15.83a. The distribution of dissolved oxygen (ml l^{-1}) in the Straits of Bab el Mandeb (Grasshoff, 1969).

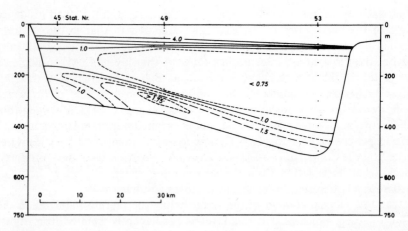

FIG. 15.83b. The distribution of oxygen (ml l^{-1}) along section B in the Inner Gulf of Aden (Grasshoff, 1969).

bodies have originated from depths just below the thermocline in the Gulf of Aden. Grasshoff has suggested that the inflowing current can be divided into two layers, the upper low salinity layer above the thermocline in the

Gulf of Aden, and a deeper thin layer of water depleted in oxygen and rich in nutrients. The latter, although it only flows intermittently over the sill into the Red Sea may, nevertheless, contribute considerably to the nutrient balance.

15.5.5. NUTRIENTS

The surface water of the Red Sea, in common with those of the neighbouring seas, is almost depleted in nutrients. Since most of its surface water originates from the Gulf of Aden, it would not be expected that the Red Sea would be fertilized by this incoming surface water. However, some enrichment of the surface layer does occur as a result of turbulent mixing in the Strait region. The distribution of dissolved phosphate in the Red Sea reflects the general water circulation pattern (see e.g. Fig. 15.84). The intermediate water shows

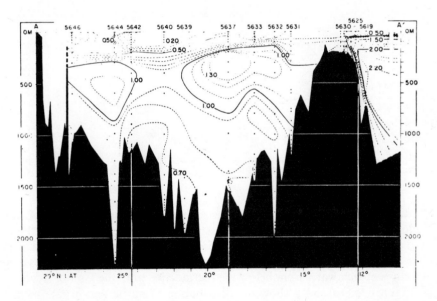

FIG. 15.84. The distribution of total phosphorus (μg-at P l^{-1}) in the Red Sea (Neumann and McGill, 1962).

a phosphate maximum which results from the mineralization of organic matter. Towards the bottom, the phosphate concentration again decreases to low values. This water mass results from the sinking of nutrient-poor-surface water in the northern part of the Red Sea; this then travels southward as the bottom layer. Subsequently, this water is deflected upwards by the

U

rise of the sea floor at the southern end of the Red Sea (see Fig. 15.85) and leads to the upward displacement of the overlying phosphate-rich inter-mediate layer. The intrusion of sub-thermocline water from the Gulf of Aden is clearly demonstrated by the high phosphate content found at Meteor

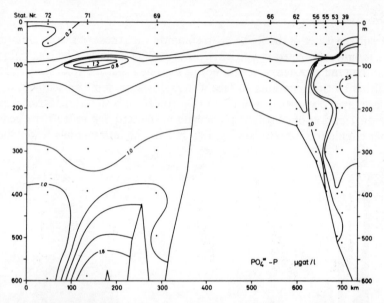

FIG. 15.85. The distribution of dissolved inorganic phosphate (μg-at l^{-1}) in the Straits of Bab el Mandeb (Grasshoff, 1969).

Station 71. On the basis of the average phosphate concentrations of the inflowing and outflowing waters, Grasshoff (1969) has estimated that there is a phosphate–phosphorus deficit of 153,000 tons per annum in the Red Sea. This is probably at least partly counterbalanced by occasional intrusions of nutrient-rich water from below the thermocline from the Gulf of Aden.

Figures 15.86a and b show the distribution of nitrite and nitrate in the region of the Straits of Bab el Mandeb. Nitrate is present at a very low level in the surface water of the Red Sea proper. However, turbulent mixing brings nitrate-rich outflowing water to the surface in the Straits. The unusually high nitrite concentration in the surface waters appears to be the result of its generation from nitrate by plankton in the nutrient-rich water produced by turbulent mixing (Grasshoff, 1969).

A review of the phosphate and nitrate data for the Red Sea suggests that the ratio $\Delta PO_4^{-3} : \Delta NO_3^-$ is $1:20$ and that the concentration of phosphate

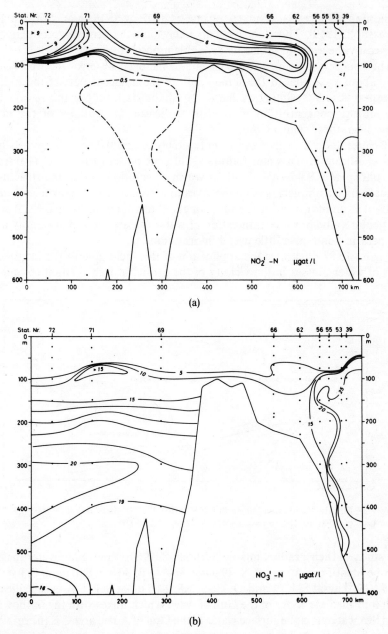

FIG. 15.86. The distribution of nitrite (a) and nitrate (b) in the Straits of Bab el Mandeb in Dec. 1964 (Grasshoff, 1969).

remaining after exhaustion of nitrate is $0.070 \pm 0.25 \, \mu\text{g-at PO}_4^{3-}\text{–P} \, 1^{-1}$ (Grasshoff, 1969). Therefore it seems likely that the primary production is limited by the amount of nitrate available in the productive layers.

The concentration of reactive silicate in the deep water of the Red Sea is only about $\frac{1}{10}$ of that at the same depth in the adjacent part of the Indian Ocean; it is, in fact, similar to that in the Mediterranean. For this reason, the low silicate concentration is a useful indicator for outflow of the deep water from these semienclosed seas.

Because diatom shells dissolve only slowly in aerated water, the distribution of silicate in sea water follows a different pattern from that of nitrate and phosphate. Below the depth at which dissolution of the fine structures of diatoms is complete, silicate behaves in a conservative manner and may be used as an indicator for the spreading and mixing of water masses. It is of particular value in the identification of water masses for which salinity and temperature data give little useful information.

Figure 15.87 illustrates the application of this technique to the identification of water masses in the vicinity of the Straits of Bab el Mandeb and to

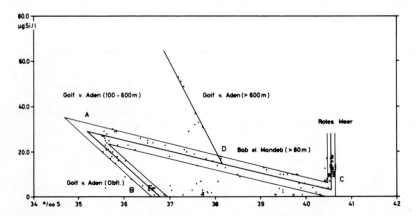

F<small>IG</small>. 15.87. Silicate-salinity diagram from the Red Sea, the Straits of Bab el Mandeb, and the Inner Gulf of Aden (for key to letters see text) (Grasshoff, 1969).

the study of their gradual mixing (Grasshoff, 1969). Four main water masses can be identified: A, $35.1 \, \%_0$S, $29 \, \mu\text{g-at Si} \, 1^{-1}$; B, $36.8 \, \%_0$S, $0 \, \mu\text{g-at Si} \, 1^{-1}$; C, $40.6 \, \%_0$S, $3 \, \mu\text{g-at Si} \, 1^{-1}$; and D $38.16 \, \%_0$S, $14.5 \, \mu\text{g-at Si} \, 1^{-1}$. A is the silicate-rich intermediate water of the Gulf of Aden which mixes with the outflowing Red Sea water. B is the surface water of the Gulf of Aden, and C is the original water from the Red Sea before it is mixed and leaves the Red Sea. The salinity of this water suggests that it originates at a depth below 300 m and, indeed, may be from much deeper. However, its silicate content (Fig. 15.88) clearly demonstrates that its source is in layers shallower than 200 m. The point D marks the composition of the Red Sea water after it has cascaded down the slope into the Gulf of Aden and become separated from the bottom water.

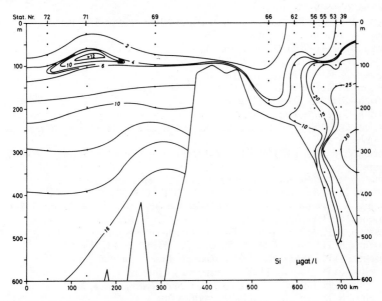

FIG. 15.88. The distribution of dissolved silicate in the Straits of Bab el Mandeb (Grasshoff, 1969

REFERENCES

Ames, L. L. (1959). *Econ. Geol.* **54**, 829.
Andruliewicz, E., Dubrawski, R. and Zmudzinski, L. (1972). 8th Conf. Balt. Oceanographers, Copenhagen.
Arkhangel'skii, A. D. (1928). *BMOIP, odt., geol.* **6**, 77.
Arrhenius, G. (1963). *In* "The Sea" (M. N. Hill, ed.), Interscience, New York.
Barlow, E. W. (1934). *Mar. Obstr.* **11**, 67–8; 110; 150.
Bather, J. M. and Riley, J. P. (1954). *J. Cons. Int. Explor. Mer.* **20**, 145.
Beyer, F. (1970). *In* ICES Coop. Res. Rep. Ser. A No 20.
Bogdanova, A. K. (1966). XX-ème Congress de la C.I.E.S.M.M., 693–696.
Brandt, A. von (1936). *J. Conseil Int. Explor. Mer*, **11**, 314–2.
Brewer, P. G. and Murray, I. W. (1973). *Deep Sea Res.* **20**, 803.
Brewer, P. G., Spencer, D. W. and Sachs, P. L. (1970). *Oceanus,* **15**, 23.
Brogmus, W. (1952). *Kiel. Meeresforsch.* **9**, 15.
Buch, W. (1932). *Havsforskningsinstitutets Skrift* **86**,
Caspers, H. (1957). *In* "Treatise on Marine Ecology and Paleoecology" Memoir, **67**, Vol. 1, Geol. Soc. America, 801.

Cox, R. A., Culkin, F., Greenhalgh, R. and Riley, J. P. (1962). *Nature, Lond.* **193**, 518.
Cox, R. A. and Culkin, F. (1966). *Deep-Sea Res.* **13**, 789.
Craig, H. and Weiss, R. F. (1970). *J. Geophys. Res.* **76**, 5078.
Culkin, F. (1965). *In* "Chemical Oceanography" (J. P. Riley and G. Skinnow, eds.), p. 144. Academic Press London and New York.
Degens, E. T. and Ross, D. A. (1969). "Hot Brines and Recent Heavy Metal Deposits in the Red Sea". Springer-Verlag, New York.
Degens, E. T. and Ross, D. A. (1970). *Naturwissenschaften* **57**, 349.
Deuser, W. G. (1970a). *Science, N.Y.* **168**, 1575.
Deuser, W. G. (1970b). *Oceanus*, **15**, 33.
Deuser, W. G. (1971). *Deep-Sea Res.* **18**, 995.
Dickson, R. R. (1971). *Dtsch. Hydrogr. Z.*, **24**, 97.
Dickson, R. R. (1972). ICES C.M. 1972, C: 10, Hydr. Comm.
Dietrich, G. (1951). *Dtsch. Hydrogr. Z.*, **4**, 129.
Dietrich, G. (1954). Die Küste, **2**, 130.
Dietrich, G., Düing, W., Grasshoff, K. and Koske, P. H. (1967). *Meteor. Forsch. Ergebn.* A, No. 2.
Dietrich, G. and Ulrich, I. (1968).—"Atlas zur Ozeanographie" Bibliogr. Inst. Mannheim.
Ehrhardt, M. (1969). Kieler Meeresforsch. **25**, 71–80.
Fedosov, M. V. and Zaitsev, G. N. (1959). ICES C.M. 1959, No. 66.
Fleming, R. H. (1940). Proc. 6. Pacific Sci. Congr. California, 1939, pp. 535–540.
Fonselius, S. H. (1962). *Fish. Bd Sweden, Ser. Hydr. Rep.* **13**.
Fonselius, S. H. (1966). *Medd. Havsfiskelab. Lysekil* **No. 11**.
Fonselius, S. H. (1967). *Fish. Bd Sweden*, Ser. Hydrogr. **13**, 41 pp.
Fonselius, S. H. (1968). *Medd. Havsfiskelab. Lysekil* **No. 54**.
Fonselius, S. H. (1969a). Thesis, Gothenburg University.
Fonselius, S. H. (1969b). *Fish. Bd Sweden, Ser. Hydrogr. Rep.* **23**.
Fonselius, S. H. (1970). unpublished manuscript.
Fonselius, S. H. (1971). *Medd. Havfiskelab. Lysekil*, No. **144**.
Fonselius, S. H. (1972). *Ambio Special Rep.* No. 1.
Fonselius, S. H. (1973). *In* "The Black Sea" (E. T. Degens and D. A. Ross, eds.). A.A.P.G. Memoir.
Fonselius, S. H. (1974). *In* "The Black Sea, its Geology, Chemistry and Biology". A.A.P.G. Memoir (in press).
Føyn, E. (1970). *ICES Coop. Res. Rep. Ser.* A, No. 20.
Gade, H. G. (1967). *Inst. Water Res. Oslo,* **2**, 1.
Gade, H. G. (1968). *Helg. Wiss. Meeresunters.* **17**, 426.
Gieskes, I. M. and Grasshoff, K. (1969). *Kiel Meeresforsch.* **25**, 105.
Girdler, R. W. (1966). *In* "The World Rift System" (T. N. Irvine, ed.). Geol. Surv. Can. Ottawa, Paper 66 14, pp. 65–77.
Granquist, G. (1952). *Havsforskningsinstitutets Skrift* No. **152**, 1–29.
Grasshoff, K. (1969). *Meteor Forsch. Ergebnisse, Reihe* A, No. 6.
Greenhalgh, R. and Riley, J. P. (1963). *Nature, Lond.* **197**, 171.
Greenhalgh, R. and Riley, J. P. (1971). Cited in Riley and Chester "Introduction to Marine Chemistry", Academic Press, London and New York, p. 81.
Gripenberg, S. (1937). *J. Con. Int. Explor. Mer.* **12**, 293.
Hallberg, R. O., Bågander, L. E., Engvall, A. G. and Schippel, F. A. (1972). *Ambio* **1**, 71.

Hela, I. (1965). *ICES C.M. 1965 Hydr. Comm. Paper* No. 27.
Hela, I. (1966). *Comm. Phys.-Math. Soc. Scient. Fennia* 31.
Hermann, F. und Olsen, V. (1970). *ICES C.M. Hydrogr. Comm.* C 27.
Hesselberg, Th. (1918). *Ann. d. Hydr. u. Mar. meteor.* 118.
Hofman-Bang, O. (1904). "Studie über schwedische Fluss- und Quellwasser". Uppsala, 13.
Hollan, E. (1969a). *Kiel. Meeresforsch.* 25, 19.
Hollan, E. (1969b). *Kiel. Meeresforsch.* 25, 245–267.
Hollan, E. (1973). *Dtsch. Hydrogr. Zeitschr.* 26, 144.
Hupfer, P. (1962). *Veröff. Geophys. Inst. Univ. Leipzig.* 17, 355.
ICES (1970a). *Coop. Res. Rep. Series A.* No. 15.
ICES (1970b). *Coop. Res. Rep. Series A.* No. 20.
Kalle, K. (1943). *Ann. Hydr.* 60, H1.
Knudsen, M. (1899). "Hydrography 2 Luno Copenhagen", 128 pp.
Knudsen, M., Forsch, C. and Sörensen, S. P. L. (1902). *Kiel. Wissen. Meeresunters. Neue Folge* No. 6, 123.
Knudsen, M. (1903). *Publ. Circ. Cons. Expl. Mer.* 5, 3.
Koroleff, F. (1968). *ICES C.M.* C 34.
Koroleff, F. (1974). *9. Conf. Balt. Oceanogr. Kiel Paper* No. 25.
Krause, G. (1970). *Deep-Sea Res.* 17, 385.
Krauskopf, K. B. (1956). *Geochim. Gösmochim. Acta* 10, 1.
Krauss, W. (1963). *Kiel. Meeresforsch.* 29, 119.
Kremling, K. (1968). *ICES Hydrogr. Comm.* C15.
Kremling, K. (1969). *Kiel. Meeresforsch.* 25, 81.
Kremling, K. (1970). *Kiel. Meeresforsch.* 26, 1.
Krümmel, O. (1911). "Handbuch der Oceanographie", Vol. II. Engelhorn, Stuttgart, 766 pp.
Kullenberg, G. (1970). *Tellus,* 22, 3.
Kullenberg, B. and Sen Gupta, R. (1972). Unpublished manuscript Oceanogr. Inst. Rep. No. 6, Gothenburg.
Laking, P. N. (1974). "The Black Sea, A Bibliography." Woods Hole Oceanogr. Inst.
Lisitzin, E. (1967). *Geophysica, Helsingf.* 9, 259.
Magaard, L. (1965). *Kiel. Meeresforsch.* 21, 22.
Majewski, A., Trzosinska, A. and Zmudzinski, L. (1972). 8th Conf. Balt. Oceanographers, Copenhagen.
Marsilli, L. F. (1681). Letter to Queen Kristina of Sweden.
Matthäus, W. (1973). *Beitr. z. Meeresk.* 32, 105.
Merz, A. (1918a). "Bibliographie der Geographie Handbuch. N.F., Festband" (A. Penck, ed.), Stuttgart.
Merz, A. (1921). *Verh. Deutsch. Geogr. Tag* 20.
Merz, A. and Möller, L. (1928). *Veröff. Inst. Meeresforschung Berlin, N.F.A.,* 18, 284 pp.
Mikulski, Z. (1970). *Nordic Hydrology* 4, 216.
Morcos, S. A. (1970). *Oceanog. Mar. Biol. Ann. Rev.* 8, 73.
Morris, A. W. and Riley, J. P. (1966). *Deep Sea Res.* 13, 699.
Morskoi Atlas (1950). Vol. II (Black Sea). U.S.S.R. Admiralty.
Munk, W. H. (1966). *Deep-Sea Res.* 13, 707.
Muromtsev, A. M. (1960). *Dokl. Akad. Nauk. S.S.S.R.* 134, 1443.
Nehring, D. and Rohde, K. H. (1966). *Beitr. z. Meeresk* k 20, 10.

Nehring, D. and Francke, E. (1971). *Fisch. Forsch. wiss. Schrift.* **9**, 35.
Neumann, G. (1942). *Ann. Hydrogr. Marit. Met.* **70**, 265.
Neumann, G. (1943). *Ann. Hydrogr. Marit. Met.* **71**, 1.
Neumann, G. H., Fonselius, S. H. and Wahlmann, L. (1959). *Int. J. Air Pollution* **2**, 132.
Neumann, A. C. and MacGill, D. A. (1961). *Deep Sea Res.* **8**, 223.
Nicolaev, I. I., Krievs, H. K. and Freimane-Apine, S. O. (1964). *Ann. Biol.* **19**, 71.
Park, K. (1968). *ICES Interlab. Rep.* No. **2**.
Privett, P. W. (1959). *Q.J.R. met. Soc.* **85**, 424.
Rankama. K. and Sahama, T. G. (1955). Geochemistry. Chicago Univ. 551–556.
Redfield, A. C., Ketchum, B. H. and Richards, F. A. (1963). *In* "The Sea" (M. N. Hill, ed.), Vol. 2. Interscience, New York.
Richards, F. A. (1965). *In* "Chemical Oceanography (J. P. Riley and G. Skirrow, eds). Vol. I". Academic Press, London and New York.
Richards, F. A. and Vaccaro, R. F. (1956). *Deep Sea Res.* **3**, 214.
Rohde, K. H. (1966). *Beitr. Meeresk.* **19**, 18.
Rohde, K. H. (1967). *Beitr. Meeresk.* **20**, 34.
Rubey, W. W. (1951). *Bull. Geol. Soc. Amer.* **62**, 1111.
Ruppin, E. (1912). *Kiel. Wiss. Meeresunters. N.F.* **14**, 205.
Schott, G. (1929). *Ann. Hydrogr. Berlin.* **57**, 10.
Schulz, B. (1956). *Dtsch. hydrogr. Z. Erg. H.B* (4) No. 1.
Segerstråle, S. G. (1951). *J. Cons. Int. Explor. Mer.* 103.
Sen Gupta, R. (1969). *Tellus,* **21**, 270.
Sen Gupta, R. (1971). *Deep-Sea Res.* **18**, 457.
Sen Gupta, R. (1973). Thesis, Gothenburg University.
Sen Gupta, R. and Jannasch, H. W. (1973). *Int. Rev. ges. Hydrobiol.* **58**, 625.
Shephard, F. P. and Curray, J. R. (1967). *Progr. Oceanogr.* **4**, 283.
Siedler, G. (1968). *Meteor Forsch. Ergebn. Reihe A* No. **4**.
Siedler, G. (1968a). *Kieler Meersf.* **24**, 59.
Sillén, L. G. (1961). *In* "Oceanography" (M. Sears, ed.). Amer. Ass. Adv. Sci. Washington Publ. **67**, p. 549.
Sillén, L. G. (1966). *Tellus,* **18**, 198.
Sjöberg, S., Wulff, F. and P. Wåhlström (1972). *Contrib. Askö Lab. Univ. Stockholm* No. 1.
Skopintsev, B. A. (1959). Intern. Oceanogr. Congress-Preprints. Ed. M. Seas, 953–955, American Ass. for the Advancement of Science, Washington, D.C.
Skopintsev, B. A., Timofeyeva, S. N., Danielenko, A. F. and Sokola, M. V. (1967). *Okeanologiia* **7**, 457.
Sorokin, Y. I. (1964). *J. Cons. Int. Explor. Mer.* **29**, 41.
Soskin, I. M. (1963). "Continuous Changes in the Hydrological Characteristics in the Baltic Sea". Hydrometeorological Press, Leningrad.
Soskin, I. M. and Rosowa, L. V. (1957). *Tr. Gos. Okeanogr. Inst.* **41**, 9.
Spencer, D. W. and Brewer, P. G. (1971). *J. Geophys. Res.* **76**, 5877.
Strakhov, N. M. (1954). *In* "Obrazovanii odadkov v sovremennykh vodoemakh" (D. S. Belyankin, ed.). Izd. An. Moscow, 81.
Svansson, A. (1972). 8. *Conf. Balt. Oceanogr. Copenhagen Paper* No. **26**.
Sverdrup, H. U., Johnson, M. W. and Fleming, R. H. (1942). "The Oceans". Prentice Hall.
Swartz, D. H. and Arden, D. D. (1960). *Bull. Am. Ass. Petrol. Geol.* **44**, 1621–1637.

Thompson, E. F. (1939). *Sci. Rep't. John Murray Exped. 1933–1934*, **2**, No. 4, 105–119.
Thompson, T. G., Jonson, W. R. and Wirth, H. E. (1931). J. Cons. Int. Explor. Mer. **6**, 246.
Trzosinska, A. (1967). *Przegl. Geofiz.* **3/4**, 367.
Trzosinska, A. (1968). *Proc. VI Conf. Baltic Oceanographers.*
Tuttle, J. J. and Jannasch, H. W. (1973). *Mar. Biol.* **20**, 64.
Ullyot, P. and Ilgaz, O. (1946). *Geogr. Rev.* **36**, 44.
Vercelli, F. (1925). *Annali idrogr.* **II**.
Voipio, A. (1957). *Suom. Kemistilehti,* **B30**, 84.
Voipio, A. (1961). *Ann. Acad. Scient. Fenn. Ser. A,* II **106**.
Wattenberg, H. (1949). *Kiel. Meeresforsch.* **6**, 10.
Wittig, H. (1940). *Meeresk. Arbeiten Univ. Kiel,* No. **70**.
Wolf, G. (1973). *Beitr. Meeresk.* **32**, 137.
Wooster, W. S., Lee, A. J. and Dietrich, G. (1969). *Deep-Sea Res.* **16**, 321.
Wyrtki, K. (1954). *Kiel. Meeresforsch.* **10**, 19.
Wyrtki, K. (1962). *Deep-Sea Res.* **9**, 11.
Zarin, E. and Ozolin, J. (1935). *J. Cons Int. Explor. Mer.* **10**, 275.

Appendix

Tables of physical and chemical constants relevant to marine chemistry

TABLE 1

Some physical properties of pure water (after Dorsey, 1940)

Molecular weight	18·0153
Heat of formation	285·89 kJmol^{-1} (at 25°C and 1 atm)
Ionic dissociation constant	10^{-4} M^{-1} (at 25°C and 1 atm)
Heat of ionization	55·71 kJmol^{-1} (at 25°C and 1 atm)
Viscosity	8·949 mP (at 25°C and 1 atm)
Velocity of sound	1496·3 ms^{-1} (at 25°C and 1 atm)
Density	0·9979751 g cm^{-3} (at 25°C and 1 atm)
Freezing point	0°C (at 1 atm)
Boiling point	100°C (at 1 atm)
Isothermal compressibility	45·6 × 10^{-6} atm^{-1} (at 25°C over the range 1–10 atm)
Specific heat at constant volume	4·1786 int.J (g°C)$^{-1}$ (at 25°C and 1 atm)
Thermal conductivity	0·00598 W cm^{-1} °C^{-1} (at 20°C and 1 atm)
Temperature of maximum density	3·98°C (at 1 atm)
Dielectric constant	81·0 (at 1 atm, 17°C, and 60 MHz)
Electrical conductivity	Less than 10^{-8} Ω^{-1} cm^{-1} (at 25°C and 1 atm)

TABLE 2

*Concentrations of the major ions in sea water of various salinities (g kg⁻¹)**

Salinity (‰)	Na⁺	Mg²⁺	Ca²⁺	K⁺	Sr²⁺	B	Cl⁻	SO₄²⁻	Br⁻	F⁻	HCO₃⁻
5	1·539	0·185	0·058	0·057	0·001	0·001	2·763	0·387	0·010	0·0002	0·020
10	3·078	0·370	0·118	0·114	0·002	0·001	5·527	0·775	0·019	0·0004	0·041
15	4·617	0·555	0·177	0·171	0·003	0·002	8·290	1·162	0·029	0·0005	0·061
20	6·156	0·739	0·235	0·228	0·005	0·003	11·054	1·550	0·038	0·0007	0·081
25	7·695	0·924	0·294	0·285	0·006	0·003	13·817	1·937	0·048	0·0009	0·101
30	9·234	1·109	0·353	0·342	0·007	0·004	16·581	2·325	0·058	0·0011	0·122
31	9·542	1·146	0·365	0·353	0·007	0·004	17·133	2·402	0·059	0·0011	0·126
32	9·850	1·183	0·377	0·365	0·007	0·004	17·685	2·480	0·062	0·0012	0·130
33	10·157	1·220	0·388	0·376	0·007	0·004	18·239	2·557	0·063	0·0012	0·134
34	10·465	1·257	0·400	0·388	0·008	0·004	18·791	2·635	0·065	0·0012	0·137
35	10·773	1·294	0·412	0·399	0·008	0·004	19·344	2·712	0·067	0·0013	0·142
36	11·081	1·331	0·424	0·410	0·008	0·005	19·897	2·789	0·069	0·0013	0·146
37	11·389	1·368	0·435	0·422	0·008	0·005	20·449	2·867	0·071	0·0013	0·150
38	11·696	1·405	0·447	0·433	0·009	0·005	21·002	2·944	0·073	0·0014	0·154
39	12·004	1·442	0·459	0·445	0·009	0·005	21·555	3·022	0·075	0·0014	0·158
40	12·312	1·479	0·471	0·456	0·009	0·005	22·107	3·099	0·077	0·0015	0·162
41	12·620	1·516	0·482	0·467	0·009	0·005	22·660	3·177	0·079	0·0015	0·166
42	12·928	1·553	0·494	0·479	0·009	0·005	23·213	3·254	0·081	0·0015	0·170

* Cations concentrations; averages of mean results of Cox and Culkin (1967) and Riley and Tongudai (1967). Sulphate and bromide concentration based on mean values from Morris and Riley (1966).

TABLE 3

Preparation of artificial sea water (S = 35·00‰)

Lyman and Fleming (1940) (g.)		Kalle (1945) (g.)	
NaCl	23·939	NaCl	28·566
$MgCl_2$	5·079	$MgCl_2$	3·887
Na_2SO_4	3·994	$MgSO_4$	1·787
$CaCl_2$	1·123	$CaSO_4$	1·308
KCl	0·667	K_2SO_4	0·832
$NaHCO_3$	0·196	$CaCO_3$	0·124
KBr	0·098	KBr	0·103
H_3BO_3	0·027	$SrSO_4$	0·0288
$SrCl_2$	0·024	H_3BO_3	0·0282
NaF	0·003		
Water to	1 kg	Water to	1 kg

Kester *et al.* (1967)
A. Gravimetric salts $\quad$ g kg^{-1}

NaCl	23·926
Na_2SO_4	4·008
KCl	0·667
$NaHCO_3$	0·196
KBr	0·098
H_3BO_3	0·026
NaF	0·003

B. Volumetric salts (standardized by Mohr method)

	Approx. molarity	Use volume equivalent to
$Mg_2 6H_2O$	1·0 M	1·297 g Mg kg^{-1}
$CaCl_2 2H_2O$	1·0 M	0·406 g Ca kg^{-1}
$SrCl_2 6H_2O$	0·1 M	0·0133 g Sr kg^{-1}

C Water to 1 kg

Note: (i) Allowance must be made for water of crystallization of any of the salts used.
$\quad\;$ (ii) After aeration the pH should lie between 7·9 and 8·3.

TABLE 4

Collected conversion factors

Conversion	Factor	Reciprocal
μg NO_3^- $\longrightarrow \mu$g N	0·2259	4·427
μg NO_2^- $\longrightarrow \mu$g N	0·3045	3·286
μg NH_3 $\longrightarrow \mu$g N	0·8225	1·216
μg NH_4^+ $\longrightarrow \mu$g N	0·7764	1·287
μg PO_4^{3-} $\longrightarrow \mu$g P	0·3261	3·066
μg P_2O_5 $\longrightarrow \mu$g P	0·4364	2·291
μg SiO_2 $\longrightarrow \mu$g Si	0·4675	2·139
μg SiO_4^{4-} $\longrightarrow \mu$g Si	0·3050	3·278
μg N $\longrightarrow \mu$g-at N	0·07138	14·008
μg P $\longrightarrow \mu$g-at. P	0·03228	30·975
μg Si $\longrightarrow \mu$g-at. Si	0·03560	28·09

TABLE 5

Table for conversion of weights of nitrogen, phosphorus and silicon expressed in terms of μg into μg-at.

μg N, P, or Si^{-1}	μg-at Nl^{-1}	μg-at. P l^{-1}	μg-at. Si l^{-1}
1	0·071	0·032	0·036
2	0·143	0·065	0·071
3	0·214	0·097	0·107
4	0·286	0·129	0·142
5	0·357	0·161	0·178
6	0·428	0·194	0·214
7	0·500	0·226	0·249
8	0·571	0·258	0·284
9	0·643	0·291	0·320
10	0·714	0·323	0·356
20	1·428	0·646	0·712
30	2·142	0·968	1·068
40	2·856	1·291	1·424
50	3·569	1·614	1·780
60	4·283	1·937	2·136
70	4·997	2·260	2·492
80	5·711	2·582	2·848
90	6·425	2·905	3·204
100	7·139	3·228	3·560

TABLE 6

Solubility of oxygen (C) in sea water ($cm^3\ dm^{-3}$) with respect to an atmosphere of 20·95% oxygen and 100% relative humidity at a total atmospheric pressure of 760 mm Hg. (UNESCO, 1973)*

T (°C)	Salinity (‰)														
	0	5	10	15	20	25	30	31	32	33	34	35	36	37	38
0	10·22	9·87	9·54	9·22	8·91	8·61	8·32	8·27	8·21	8·16	8·10	8·05	7·99	7·94	7·88
1	9·94	9·60	9·28	8·97	8·68	8·39	8·11	8·05	8·00	7·94	7·89	7·84	7·78	7·73	7·68
2	9·67	9·35	9·04	8·74	8·45	8·17	7·90	7·85	7·79	7·74	7·69	7·64	7·59	7·53	7·48
3	9·41	9·10	8·80	8·51	8·23	7·96	7·70	7·65	7·60	7·55	7·50	7·45	7·40	7·35	7·30
4	9·16	8·86	8·57	8·29	8·02	7·76	7·51	7·46	7·41	7·36	7·31	7·26	7·22	7·17	7·12
5	8·93	8·64	8·36	8·09	7·83	7·57	7·33	7·28	7·23	7·18	7·14	7·09	7·04	7·00	6·95
6	8·70	8·42	8·15	7·89	7·64	7·39	7·15	7·11	7·06	7·01	6·97	6·92	6·88	6·83	6·79
7	8·49	8·22	7·95	7·70	7·45	7·22	6·98	6·94	6·89	6·85	6·81	6·76	6·72	6·67	6·63
8	8·28	8·02	7·76	7·52	7·28	7·05	6·82	6·78	6·74	6·69	6·65	6·61	6·57	6·52	6·48
9	8·08	7·83	7·58	7·34	7·11	6·89	6·67	6·63	6·59	6·54	6·50	6·46	6·42	6·38	6·34
10	7·89	7·64	7·41	7·17	6·95	6·73	6·52	6·48	6·44	6·40	6·36	6·32	6·28	6·24	6·20
11	7·71	7·47	7·24	7·01	6·80	6·58	6·38	6·34	6·30	6·26	6·22	6·18	6·14	6·10	6·07
12	7·53	7·30	7·08	6·86	6·65	6·44	6·24	6·21	6·17	6·13	6·09	6·05	6·01	5·98	5·94
13	7·37	7·14	6·92	6·71	6·50	6·31	6·11	6·07	6·04	6·00	5·96	5·93	5·89	5·85	5·82
14	7·20	6·98	6·77	6·57	6·37	6·17	5·99	5·95	5·91	5·88	5·84	5·80	5·77	5·73	5·70
15	7·05	6·84	6·63	6·43	6·24	6·05	5·87	5·83	5·79	5·76	5·72	5·69	5·65	5·62	5·58
16	6·90	6·69	6·49	6·30	6·11	5·93	5·75	5·71	5·68	5·64	5·61	5·58	5·54	5·51	5·48
17	6·75	6·55	6·36	6·17	5·99	5·81	5·64	5·60	5·57	5·53	5·50	5·47	5·43	5·40	5·37
18	6·61	6·42	6·23	6·05	5·87	5·69	5·53	5·49	5·46	5·43	5·40	5·36	5·33	5·30	5·27
19	6·48	6·29	6·11	5·93	5·75	5·59	5·42	5·39	5·36	5·33	5·29	5·26	5·23	5·20	5·17
20	6·35	6·17	5·99	5·81	5·64	5·48	5·32	5·29	5·26	5·23	5·20	5·17	5·14	5·10	5·07
21	6·23	6·05	5·87	5·70	5·54	5·38	5·22	5·19	5·16	5·13	5·10	5·07	5·04	5·01	4·98
22	6·11	5·93	5·76	5·60	5·44	5·28	5·13	5·10	5·07	5·04	5·01	4·98	4·95	4·92	4·89
23	5·99	5·82	5·65	5·49	5·34	5·18	5·04	5·01	4·98	4·95	4·92	4·89	4·87	4·84	4·81

TABLE 6 cont.

Solubility of oxygen (C) in sea water ($cm^3\ dm^{-3}$) with respect to an atmosphere of 20.95% oxygen and 100% relative humidity at a total atmospheric pressure of 860 mm Hg. (UNESCO, 1973)*

T (°C)	Salinity (‰)														
	0	5	10	15	20	25	30	31	32	33	34	35	36	37	38
24	5.88	5.71	5.55	5.39	5.24	5.09	4.95	4.92	4.89	4.86	4.84	4.81	4.78	4.75	4.73
25	5.77	5.61	5.45	5.30	5.15	5.00	4.86	4.84	4.81	4.78	4.75	4.73	4.70	4.67	4.65
26	5.66	5.51	5.35	5.20	5.06	4.92	4.78	4.75	4.73	4.70	4.67	4.65	4.62	4.59	4.57
27	5.56	5.41	5.26	5.11	4.97	4.83	4.70	4.67	4.65	4.62	4.60	4.57	4.54	4.52	4.49
28	5.46	5.31	5.17	5.03	4.89	4.75	4.62	4.60	4.57	4.55	4.52	4.50	4.47	4.45	4.42
29	5.37	5.22	5.08	4.94	4.81	4.67	4.55	4.52	4.50	4.47	4.45	4.42	4.40	4.37	4.35
30	5.28	5.13	4.99	4.86	4.73	4.60	4.47	4.45	4.43	4.40	4.38	4.35	4.33	4.31	4.28
31	5.19	5.05	4.91	4.78	4.65	4.53	4.40	4.38	4.36	4.33	4.31	4.28	4.26	4.24	4.22
32	5.10	4.96	4.83	4.70	4.58	4.45	4.33	4.31	4.29	4.26	4.24	4.22	4.20	4.17	4.15

* Based on measurements by Carpenter (1966) and Murray and Riley (1969a) fitted by Weiss (1970) to the thermodynamically consistent equation :

$$\ln C = A_1 + A_2(100/T) + A_3 \ln (T/100) + A_4(T/100) + S\%[B_1 + B_2(T/100) + B_3(T/100)^2]$$

where

A_1	A_2	A_3	A_4	B_1	B_2	B_3
-173.4292	249.6339	143.3483	-21.8492	-0.033096	0.014259	-0.0017000

and T and S‰ are the absolute temperature (K) and salinity in parts per mille respectively.

TABLE 7

Solubility of nitrogen in sea water (cm³ dm⁻³) with respect to an atmosphere of 78·084%
nitrogen and 100% relative humidity at a total pressure of 760 mm Hg (Weiss (1970)
from data by Murray and Riley (1969b)).

T(°C)	Salinity ‰								
	0	10	20	30	34	35	36	38	40
−1	—	—	16·28	15·10	14·65	14·54	14·44	14·22	14·01
0	18·42	17·10	15·87	14·73	14·30	14·19	14·09	13·88	13·67
1	17·95	16·67	15·48	14·38	13·96	13·86	13·75	13·55	13·35
2	17·50	16·26	15·11	14·04	13·64	13·54	13·44	13·24	13·05
3	17·07	15·87	14·75	13·72	13·32	13·23	13·13	12·94	12·76
4	16·65	15·49	14·41	13·41	13·03	12·93	12·84	12·66	12·47
5	16·26	15·13	14·09	13·11	12·74	12·65	12·56	12·38	12·21
6	15·88	14·79	13·77	12·83	12·47	12·38	12·29	12·12	11·95
8	15·16	14·14	13·18	12·29	11·95	11·87	11·79	11·62	11·46
10	14·51	13·54	12·64	11·80	11·48	11·40	11·32	11·17	11·01
12	13·90	12·99	12·14	11·34	11·04	10·96	10·89	10·74	10·60
14	13·34	12·48	11·67	10·92	10·63	10·56	10·49	10·35	10·21
16	12·83	12·01	11·24	10·53	10·25	10·19	10·12	9·99	9·86
18	12·35	11·57	10·84	10·16	9·90	9·84	9·77	9·65	9·52
20	11·90	11·16	10·47	9·82	9·57	9·51	9·45	9·33	9·21
22	11·48	10·78	10·12	9·50	9·26	9·21	9·15	9·03	8·92
24	11·09	10·42	9·79	9·20	8·98	8·92	8·87	8·76	8·65
26	10·73	10·09	9·49	8·92	8·71	8·65	8·60	8·50	8·39
28	10·38	9·77	9·20	8·66	8·45	8·40	8·35	8·25	8·15
30	10·06	9·48	8·93	8·41	8·21	8·16	8·12	8·02	7·92
32	9·76	9·20	8·67	8·18	7·99	7·94	7·89	7·80	7·71
34	9·48	8·94	8·43	7·96	7·77	7·73	7·68	7·59	7·51
36	9·21	8·69	8·20	7·75	7·57	7·53	7·48	7·40	7·31
38	8·95	8·46	7·99	7·55	7·38	7·33	7·29	7·21	7·13
40	8·71	8·23	7·78	7·36	7·19	7·15	7·11	7·03	6·95

The solubility at any value of salinity and temperature in the above range can be calculated if the
following constants are substituted in the equation below (Table 6).

A_1	A_2	A_3	A_4	B_1	B_2	B_3
−172·4965	248·4262	143·0738	−21·7120	−0·049781	0·025018	−0·003486

APPENDIX

TABLE 8

Solubility of argon in sea water (cm^{-3} dm^{-3}) with respect to an atmosphere of 0·934 % argon and 100 % relative humidity at a total atmosphere pressure of 760 mm Hg (Weiss (1970) from data by Douglas (1964, 1965)).

T (°C)	0	10	20	30	Salinity ‰ 34	35	36	38	40
−1	——	——	0·4456	0·4156	0·4042	0·4014	0·3986	0·3931	0·3877
0	0·4980	0·4647	0·4337	0·4048	0·3937	0·3910	0·3883	0·3830	0·3777
1	0·4845	0·4524	0·4224	0·3944	0·3837	0·3811	0·3785	0·3733	0·3682
2	0·4715	0·4405	0·4115	0·3845	0·3741	0·3716	0·3691	0·3641	0·3592
3	0·4592	0·4292	0·4012	0·3750	0·3650	0·3625	0·3601	0·3552	0·3505
4	0·4474	0·4184	0·3912	0·3659	0·3562	0·3538	0·3515	0·3468	0·3422
5	0·4360	0·4080	0·3817	0·3572	0·3478	0·3455	0·3432	0·3387	0·3342
6	0·4252	0·3980	0·3726	0·3488	0·3397	0·3375	0·3353	0·3309	0·3265
8	0·4049	0·3794	0·3555	0·3331	0·3246	0·3225	0·3204	0·3162	0·3121
10	0·3861	0·3622	0·3397	0·3186	0·3106	0·3086	0·3066	0·3027	0·2989
12	0·3688	0·3463	0·3251	0·3053	0·2977	0·2958	0·2939	0·2902	0·2866
14	0·3528	0·3316	0·3116	0·2929	0·2857	0·2839	0·2822	0·2787	0·2752
16	0·3380	0·3180	0·2991	0·2814	0·2746	0·2729	0·2712	0·2679	0·2647
18	0·3242	0·3053	0·2875	0·2707	0·2642	0·2626	0·2610	0·2579	0·2548
20	0·3114	0·2935	0·2766	0·2607	0·2546	0·2531	0·2516	0·2486	0·2457
22	0·2995	0·2825	0·2665	0·2514	0·2455	0·2441	0·2427	0·2399	0·2371
24	0·2883	0·2722	0·2570	0·2426	0·2371	0·2357	0·2344	0·2317	0·2291
26	0·2779	0·2626	0·2481	0·2344	0·2292	0·2279	0·2266	0·2241	0·2215
28	0·2681	0·2535	0·2398	0·2268	0·2217	0·2205	0·2193	0·2169	0·2144
30	0·2588	0·2450	0·2319	0·2195	0·2147	0·2136	0·2124	0·2101	0·2078
32	0·2502	0·2370	0·2245	0·2127	0·2081	0·2070	0·2059	0·2037	0·2015
34	0·2420	0·2294	0·2175	0·2062	0·2019	0·2008	0·1997	0·1976	0·1955
36	0·2342	0·2222	0·2109	0·2001	0·1959	0·1949	0·1939	0·1919	0·1899
38	0·2269	0·2154	0·2046	0·1943	0·1903	0·1893	0·1883	0·1864	0·1845
40	0·2199	0·2090	0·1986	0·1888	0·1849	0·1840	0·1831	0·1812	0·1794

The solubility at any value of salinity and temperature in the above range can be calculated if the following constants are substituted in the equation below (Table 6).

A_1	A_2	A_3	A_4	B_1	B_2	B_3
−173·5146	245·4510	141·8222	−21·8020	−0·034474	0·014934	−0·0017729

TABLE 9

Literature citations for solubilities of other gases in sea water

Gas	Reference
Carbon dioxide	Murray and Riley (1971); see also Chapter 9, Table 9.
Helium	Weiss (1971); see also Chapter 8, Table A8.5.
Neon	Weiss (1971); see also Chapter 8, Table A8.4
Krypton	Wood and Caputi (1966); see also Chapter 8, Table 8.5.
Xenon	Wood and Caputi (1966); see also Chapter 8, Table 8.5.
Carbon monoxide	Douglas (1967); see also Chapter 8, Table 8.12.
Hydrogen	Crozier and Yamamoto (1974).

TABLE 10

*The density of artificial sea water as a function of temperature and chlorinity**
(Millero and Lepple, 1973)

Cl (‰)	0°C	5°C	10°C	15°C	20°C	25°C	30°C	35°C	40°C
0	0·999868	0·999992	0·999728	0·999129	0·998234	0·997075	0·995678	0·994063	0·992247
3·42$_6$	1·004944	1·004959	1·004599	1·003921	1·002962	1·001744	1·000295	0·998643	0·996783
6·05$_5$	1·008665	1·008705	1·008292	1·007566	1·006575	1·005335	1·003868	1·002190	1·000307
8·17$_4$	1·011851	1·011731	1·011265	1·010502	1·009472	1·008201	1·006707	1·005013	1·003113
11·69$_5$	1·016982	1·016758	1·016208	1·015368	1·014275	1·012949	1·011407	1·009669	1·007745
13·67$_3$	1·019835	1·019564	1·018970	1·018102	1·016986	1·015641	1·014087	1·012346	1·010406
16·33$_3$	1·023703	1·023352	1·022695	1·021772	1·020611	1·019229	1·017642	1·015866	1·013920
19·05$_6$	1·027648	1·027227	1·026511	1·025538	1·024335	1·022921	1·021311	1·019528	1·017564
21·53$_7$	1·031240	1·030774	1·029989	1·028941	1·027731	1·026307	1·024658	1·022890	1·020925

* These densities are relative to those tabulated by Kell (1967) for pure water assuming the density of pure water is 1·000000 g ml^{-1} at 3·98°C.

TABLE 11

The expansibility of artificial sea water as a function of temperature and chlorinity,
$\alpha \times 10^6$ *(deg.*$^{-1}$*)* *(Millero and Lepple, 1973)*

Cl‰	0°C	5°C	10°C	15°C	20°C	25°C	30°C	35°C	40°C
0·000	−68·1	16·0	87·9	150·7	206·6	257·0	303·1	345·7	385·4
3·426	−46·9	35·5	105·1	165·4	218·7	266·7	310·7	351·8	391·0
6·055	−28·0	49·4	115·2	172·7	224·1	271·0	314·8	356·4	396·7
8·174	−14·8	60·4	124·4	180·5	230·7	276·6	319·4	359·9	398·8
11·695	8·2	79·2	140·2	194·1	242·7	287·2	328·3	366·7	402·6
13·673	18·4	88·1	147·5	199·6	246·6	289·8	330·1	368·6	405·7
16·333	36·1	102·3	159·2	209·4	254·8	296·8	335·8	372·8	407·9
19·056	51·0	115·2	170·2	218·5	262·2	302·4	339·9	375·2	408·8
21·537	61·9	127·6	181·6	227·5	267·9	304·9	340·1	375·0	410·7

$\alpha = -1/d(\partial d/\partial t)$ where d is the density of the sea water.

TABLE 12

The isothermal compressibility of sea water at l atm as a function of salinity and temperature (Lepple and Millero, 1971)

S(‰)	$\beta \times 10^{-6}$ (bar^{-1})								
	0°C	5°C	10°C	15°C	20°C	25°C	30°C	35°C	40°C
0·00	50·886	49·171	47·811	46·736	45·895	45·250	44·774	44·444	44·243
6·14	50·07	48·42	47·10	46·09	45·31	44·71	44·26	43·92	43·75
11·80	49·25	47·70	46·43	45·41	44·66	44·13	43·68	43·34	43·19
14·75	48·84	47·30	46·11	45·15	44·38	43·83	43·43	43·13	43·01
21·01	48·14	46·71	45·59	44·63	43·92	43·45	42·96	42·71	42·63
24·52	47·63	46·25	45·17	44·29	43·61	42·98	42·68	42·33	42·23
29·38	47·01	45·62	44·62	43·74	43·17	42·56	42·24	41·96	41·86
34·25	46·49	45·17	44·15	43·32	42·69	42·18	41·88	41·69	41·55
35·00	46·32	45·03	44·02	43·19	42·58	42·11	41·78	41·49	41·48
39·00	45·84	44·62	43·63	42·80	42·30	41·73	41·53	41·23	41·15

TABLE 13

Observed values for the change in the specific volume of sea water from 0° to T°C at various pressures and salinities. Unit of specific volume = $10^{-6} cm^3 g^{-1}$. (Cox et al., 1970)

	S = 35·00‰						S = 30·50‰				S = 39·50‰			
P, bars absolute →	8·3	201·3	401·2	601·0	800·9	1000·8	8·3	201·3	601·0	1000·9	8·3	201·3	601·0	1000·8
S, ‰ →	35·000	35·004	35·005	35·002	35·002	35·002	30·502	30·504	30·506	30·510	39·503	39·502	39·504	39·507
pH (1 bar, 25°C) →	7·91	7·95	7·94	7·94	8·00	7·96	8·06	7·98	8·03	8·00	8·22	8·18	8·13	8·16
T(°C)														
-2·000	—	—	-277·1	-356·9	-424·3	-480·5	—	—	-341·9	-472·3	—	—	-370·6	-489·4
-1·000	—	-97·5	—	—	—	—	—	-86·9	—	—	—	-107·1	—	—
0·000	0	0	0	0	0	0	0	0	0	0	0	0	0	0
2·000	132·2	224·9	310·03	383·2	444·9	497·6	106·9	204·7	368·6	489·0	155·8	245·2	394·4	504·3
4·000	311·2	489·5	652·0	791·7	910·6	1012·5	262·7	450·5	766·4	998·5	355·8	527·9	815·7	1026·8
6·000	535·0	791·0	1023·0	1225·3	1396·8	1544·7	464·7	734·7	1189·8	1523·0	599·1	846·2	1259·1	1556·1
8·000	801·0	1127·3	1424·4	1683·3	1902·4	2094·3	712·1	1055·7	1637·2	2064·2	883·0	1198·6	1726·2	2117·5
10·000	1107·1	1498·0	1854·4	2163·3	2427·7	2660·1	1000·9	1412·3	2106·9	2623·2	1205·8	1582·7	2216·4	2686·7
12·000	1452·7	1901·6	2312·4	2668·5	2971·6	3243·6	1330·2	1802·6	2603·7	3197·3	1566·5	1999·4	2729·5	3274·6
14·000	1836·3	2336·9	2796·2	3198·3	3535·9	3827·4	1698·8	2226·3	3123·4	3790·3	1962·6	2446·4	3264·0	3872·9
16·000	2255·5	2804·0	3306·4	3745·3	4119·2	4448·9	2104·7	2682·5	3665·3	4398·0	2394·3	2923·1	3818·4	4489·2
18·000	2709·3	3299·8	3840·9	4315·0	4721·5	5075·4	2547·1	3169·6	4228·6	5021·7	2858·7	3428·0	4394·0	5119·5
20·000	3196·3	3823·9	4400·2	4906·4	5341·0	5719·9	3022·8	3685·2	4815·6	5661·0	3355·8	3961·8	4994·4	5764·6
22·000	3717·0	4376·5	4984·1	5516·1	5975·8	6378·6	3533·0	4230·0	5421·9	6315·4	3883·4	4522·1	5610·2	6423·2
24·000	4268·2	4957·5	5591·4	6151·3	6630·1	7051·0	4075·6	4804·1	6049·9	6985·9	4442·8	5108·6	6246·4	7100·4
26·000	4850·1	5564·4	6223·6	6803·5	7303·4	7738·2	4649·9	5403·9	6698·9	7673·4	5031·4	5721·8	6901·7	7789·7
28·000	5461·6	6197·0	6877·1	7472·2	7990·8	8439·8	5255·2	6032·2	7367·9	8374·8	5648·6	6358·8	7576·4	8493·4
30·000	6102·8	6855·3	7554·3	8165·9	8693·8	9159·8	5889·4	6687·1	8056·6	9091·8	6294·8	7021·5	8269·4	9211·1

TABLE 14

Specific gravity and percentage volume reduction of sea water* under pressure (amended from Cox, 1965)

Pressure (db)	Specific gravity	% decrease in volume
0	1·02813	0·000
100	1·02860	0·046
200	1·02908	0·093
500	1·03050	0·231
1,000	1·03285	0·460
2,000	1·03747	0·909
3,000	1·04199	1·349
4,000	1·04640	1·778
5,000	1·05071	2·197
6,000	1·05494	2·609
7,000	1·05908	3·011
8,000	1·06314	3·406
9,000	1·06713	3·794
10,000	1·07104	4·175

* Salinity, 35·00‰; Temperature 0°C.

TABLE 15

Percentage reduction in volume of sea water under a pressure of 1,000 db at various temperatures and salinities. (After Cox, 1965).

S‰	Temperature (°C)			
	0	10	20	30
0	0·500	0·470	0·451	0·440
10	0·486	0·459	0·442	0·432
20	0·474	0·448	0·432	0·423
30	0·462	0·438	0·424	0·415
35	0·457	0·433	0·419	0·411
40	0·450	0·428	0·415	0·407

TABLE 16

Thermal expansion of sea water under pressure $(10^{-6} \, cm^3 \, (^{\circ}C)^{-1})$. *(Bradshaw and Schleicher, 1970)*

Pressure (bars)	Temperature (°C)			
	0	10	20	30
		$S = 30.50\%_0$		
1	39	155	246	324
500	158	229	290	346
1000	240	284	323	362
		$S = 35.00\%_0$		
1	52	162	251	327
500	166	234	293	347
1000	244	286	325	363
		$S = 39.50\%_0$		
1	65	170	256	329
500	174	239	296	348
1000	248	289	326	363

TABLE 17a*

Velocity of sound in sea water†

Pressure (db)	Temperature (°C)						
	0	5	10	15	20	25	30
0	1449·3	1471·0	1490·4	1507·4	1522·1	1534·8	1545·8
1000	1465·8	1487·4	1506·7	1523·7	1538·5	1551·3	1562·5
2000	1482·4	1504·0	1523·2	1540·2	1555·0	1567·9	1579·2
3000	1499·4	1520·7	1538·6	1555·6			
4000	1516·5	1537·7	1555·2	1572·2			
5000	1533·9	1554·8	1571·9	1588·9			
6000	1551·5	1572·1					
7000	1569·3						
8000	1587·3						
9000	1605·4						
10000	1623·5						

* Reproduced by permission of U.S. Navy Oceanographic Office.
† Velocities in m s^{-1}; pressures in decibars above atmosphere. Salinity 35‰. For other salinities see Table 17b.
For detailed tables of the velocity of sound in sea water, see U.S. Naval Oceanographic Office (1962) and Bark *et al.* (1964).

TABLE 17b*

Effect of salinity on sound velocity†

$S‰$	Temperature (°C)						
	0	5	10	15	20	25	30
30	−7·0	−6·7	−6·5	−6·2	−5·9	−5·6	−5·3
32	−4·2	−4·0	−3·9	−3·7	−3·5	−3·4	−3·2
33	−2·8	−2·7	−2·6	−2·5	−2·4	−2·2	−2·1
34	−1·4	−1·3	−1·3	−1·2	−1·2	−1·1	−1·1
35	0	0	0	0	0	0	0
36	1·4	1·3	1·3	1·2	1·2	1·1	1·1
37	2·8	2·7	2·6	2·5	2·4	2·3	2·1
38	4·2	4·1	3·9	3·7	3·6	3·4	3·2
40	7·0	6·8	6·5	6·2	6·0	5·7	5·3

* Reproduced by permission of U.S. Navy Oceanographic Office.

Corrections to be applied to the values in Table 17a for salinities other than 35‰.

TABLE 18

Specific heat of sea water at constant pressure ($J g^{-1} °C^{-1}$) at various salinities and temperatures (Millero et al., 1973).

Salinity, ‰	0°C	5°C	10°C	15°C	20°C	25°C	30°C	35°C	40°C
0	4·2174	4·2019	4·1919	4·1855	4·1816	4·1793	4·1782	4·1779	4·1783
5	4·1812	4·1679	4·1599	4·1553	4·1526	4·1513	4·1510	4·1511	4·1515
10	4·1466	4·1354	4·1292	4·1263	4·1247	4·1242	4·1248	4·1252	4·1256
15	4·1130	4·1038	4·0994	4·0982	4·0975	4·0977	4·0992	4·0999	4·1003
20	4·0804	4·0730	4·0702	4·0706	4·0709	4·0717	4·0740	4·0751	4·0754
25	4·0484	4·0428	4·0417	4·0437	4·0448	4·0462	4·0494	4·0508	4·0509
30	4·0172	4·0132	4·0136	4·0172	4·0190	4·0210	4·0251	4·0268	4·0268
35	3·9865	3·9842	3·9861	3·9912	3·9937	3·9962	4·0011	4·0031	4·0030
40	3·9564	3·9556	3·9590	3·9655	3·9688	3·9718	3·9775	3·9797	3·9795

TABLE 19

The relative partial equivalent heat capacity of sea salt $(cal(eq\,deg)^{-1})$ *(Millero et al., 1973a)*

Salinity	Temperature (°C)						
(‰)	0	5	10	15	20	25	30
0	0	0	0	0	0	ʊ	0
5	2·8	3·1	3·4	3·7	4·0	4·3	4·6
10	4·7	5·0	5·4	5·7	5·9	6·3	6·6
15	6·9	7·1	7·3	7·5	7·7	7·9	8·1
20	9·5	9·4	9·4	9·4	9·3	9·3	9·2
25	12·2	11·9	11·5	11·2	10·8	10·5	10·2
30	15·3	14·6	13·8	13·1	12·3	11·6	10·8
35	18·6	17·4	16·2	15·0	13·8	12·6	11·3
40	22·2	20·5	18·7	17·0	15·2	13·4	11·7

TABLE 20

Thermal conductivity $(K\ in\ 10^{-5}\ W\,cm\,deg^{-1})$ *of sea water* $(S = 34\cdot994‰)$ *as a function of temperature and pressure. (After Castelli et al., 1974)**

Pressure (p)	Temperature (t°C)			
(bars)	1·82	10	20	30
∠00	563	578	594	605
400	570	585	601	613
600	578	592	609	619
800	585	599	615	627
1000	591	606	622	634
1200	596	613	628	641
1400	602	618	634	647

$K = 5\cdot5286 \times 10^{-3} + 3\cdot4025 \times 10^{-7}P + 1\cdot8364 \times 10^{-7}t - 3\cdot3058 \times 10^{-9}t$

* Other data have been published by Caldwell (1974).

Table 21

Freezing point of sea water (T_f) at atmospheric pressure based on the data of Doherty and Kester (1974).

$S\%_{00}$	T_f (°C)	$S\%_{00}$	T_f (°C)	$S\%_{00}$	T_f (°C)
5	-0.275	17	-0.918	29	-1.582
6	-0.328	18	-0.973	30	-1.638
7	-0.381	19	-1.028	31	-1.695
8	-0.434	20	-1.082	32	-1.751
9	-0.487	21	-1.137	33	-1.808
10	-0.541	22	-1.192	34	-1.865
11	-0.594	23	-1.248	35	-1.922
12	-0.648	24	-1.303	36	-1.979
13	-0.702	25	-1.359	37	-2.036
14	-0.756	26	-1.414	38	-2.094
15	-0.810	27	-1.470	39	-2.151
16	-0.864	28	-1.526	40	-2.209

The freezing point at *in situ* pressure is given by
$T_f(°C) = -0.0137 - 0.051990\ S\%_{00} - 0.00007225\ (S\%_{00})^2 - 0.000758z$ where z is the depth in metres.

Table 22

Boiling point elevation of sea water ($S = 35.00\%_{00}$) at various temperatures, (Stoughton and Lietzke, 1967)

Temp. (°C)	30	40	50	60	70	80	90	100
Vap. press. (atm)	0.042	0.073	0.122	0.197	0.309	0.469	0.694	1.003
Elevation of B.P (°C)	0.325	0.350	0.377	0.405	0.433	0.463	0.493	0.524
Temp. (°C)	120	140	160	180	200	220	240	260
Vap. press. (atm)	1.965	3.577	6.119	9.931	15.407	22.99	33.18	46.52
Elevation of B.P (°C)	0.590	0.660	0.735	0.817	0.906	1.003	1.111	1.232

TABLE 23

Osmotic pressure and vapour depression of sea water at 25°C (Robinson, 1954)

	Chlorinity									
	12	13	14	15	16	17	18	19	20	21
Osmotic pressure (atm)	15·51	16·85	18·19	19·55	20·91	22·28	23·366	25·06	26·47	27·89
Vap. press. lowering* × 10^2	1·139	1·237	1·334	1·433	1·532	1·631	1·732	1·832	1·936	2·039

$(p^0 - p)/p^0$ where p and p^0 are the vapour pressures of sea water and pure water respectively ($p^0 = 23.75$ mm at 25°C).

TABLE 24

Surface tension of clean sea water (in $N m^{-1}$) at various salinities and temperatures (from data by Krümmel (1900)* and others (After Fleming and Revelle, 1939)

$S‰$	Temperature (°C)			
	0	10	20	30
0	$75·64 \times 10^{-3}$	$74·20 \times 10^{-3}$	$72·76 \times 10^{-3}$	$71·32 \times 10^{-3}$
10	75·86	74·42	72·98	71·54
20	76·08	74·64	73·20	71·76
30	76·30	74·86	73·42	71·98
35	76·41	74·97	73·53	72·09
40	76·52	75·08	73·64	72·20

Surface tension $(N m^{-1}) = 10^3 (75·64 - 0·144t + 0·0221 S‰)$
* Measurements made on bubbles below the surface, they therefore take no account of the effects of surface contamination which may be very considerable (e.g. see Lumby and Folkard, 1956 and Vol. 2, pp. 233–4.

TABLE 25

The viscosity of sea water (η) at various salinities and temperatures (in centipoises) computed from values for distilled water (η₀) by Korson et al. (1969) using equations developed by Millero (1974)

Salinity ‰	Temperature °C															
	0	2	4	6	8	10	12	14	16	18	20	22	24	26	28	30
0	1·7916	1·6739	1·5681	1·4725	1·3857	1·3069	1·2349	1·1691	1·1087	1·0532	1·0020	0·9547	0·9109	0·8703	0·8326	0·7975
5	1·8049	1·6868	1·5808	1·4849	1·3979	1·3189	1·2466	1·1807	1·1200	1·0644	1·0129	0·9655	0·9215	0·8807	0·8428	0·8076
10	1·8180	1·6995	1·5930	1·4968	1·4095	1·3302	1·2576	1·1913	1·1304	1·0745	1·0228	0·9751	0·9309	0·8900	0·8519	0·8165
15	1·8312	1·7122	1·6054	1·5087	1·4210	1·3412	1·2685	1·2018	1·1407	1·0845	1·0327	0·9847	0·9402	0·8991	0·8608	0·8252
20	1·8445	1·7251	1·6178	1·5208	1·4325	1·3525	1·2794	1·2125	1·1513	1·0945	1·0424	0·9942	0·9495	0·9082	0·8697	0·8339
25	1·8579	1·7380	1·6302	1·5327	1·4442	1·3638	1·2903	1·2231	1·1614	1·1046	1·0522	1·0036	0·9588	0·9172	0·8786	0·8426
30	1·8713	1·7509	1·6427	1·5448	1·4560	1·3751	1·3012	1·2338	1·1717	1·1146	1·0619	1·0132	0·9682	0·9263	0·8875	0·8513
32	1·8767	1·7563	1·6478	1·5497	1·4607	1·3797	1·3057	1·2379	1·1758	1·1186	1·0658	1·0171	0·9719	0·9300	0·8910	0·8547
34	1·8823	1·7643	1·6528	1·5545	1·4652	1·3843	1·3101	1·2423	1·1800	1·1227	1·0698	1·0210	0·9757	0·9336	0·8945	0·8582
36	1·8876	1·7696	1·6578	1·5594	1·4701	1·3888	1·3146	1·2465	1·1841	1·1267	1·0737	1·0248	0·9793	0·9372	0·8981	0·8617
38	1·8932	1·7752	1·6630	1·5644	1·4748	1·3934	1·3189	1·2508	1·1883	1·1308	1·0778	1·0286	0·9831	0·9409	0·9017	0·8651
40	1·8986	1·7805	1·6680	1·5692	1·4795	1·3980	1·3233	1·2551	1·1925	1·1348	1·0817	1·0325	0·9869	0·9446	0·9053	0·8686
42	1·9041	1·7861	1·6732	1·5741	1·4842	1·4026	1·3278	1·2595	1·1967	1·1389	1·0857	1·0363	0·9906	0·9483	0·9089	0·8721

Viscosity of pure water η_t at temperature $t°C$ is given by $\log \dfrac{\eta_t}{\eta_{20}} = \dfrac{1\cdot1709(20 - t) - 0\cdot001827(t - 20)^2}{t + 89\cdot93}$ where η_{20} is the viscosity at 20°C.

Viscosity of sea water calculated from ratio $\dfrac{\eta}{\eta_0} = 1 + ACl_v^{\ddagger} + BCl_v$.

Where Cl_v is the volume chlorinity ($Cl_v = Cl‰ \times$ density) and $A = 0\cdot000366, 0\cdot001403$ and $B = 0\cdot002756, 0\cdot003416$ at 5° and 25°C; constants at other temperatures obtained by linear interpolation or extrapolation.

According to Matthäus (1972) the change in dynamic viscosity ($\Delta\eta_p$, centipoises) produced by increase in pressure (P, $kg\,cm^{-2}$) at temperature, $T°C$) can be calculated from the expression

$$\Delta\eta_p = -1\cdot7913 \times 10^{-4}\,P + 9\cdot5182 \times 10^{-8}\,P^2 + P(1\cdot3550 \times 10^{-5}\,T - 2\cdot5853 \times 10^{-7}\,T^2 - P^2(6\cdot0833 \times 10^{-9}\,T - 1\cdot1652 \times 10^{-10}\,T^2)$$

* The assistance of Miss J. Wolfe with the computations is gratefully acknowledged.

TABLE 26

Relative viscosity of Standard Sea Water ($S = 35.00‰$) at various temperatures and pressures. (Stanley and Batten, 1969)

Pressure, kg cm^{-2}	η_p/η_1 at $-0.024°C$	η_p/η_1 at $2.219°C$	η_p/η_1 at $6.003°C$	η_p/η_1 at $10.013°C$	η_p/η_1 at $15.018°C$	η_p/η_1 at $20.013°C$	η_p/η_1 at $29.953°C$
176	0.9828	0.9852	0.9891	0.9914	0.9949	0.9977	0.9997
352	0.9709	0.9742	0.9814	0.9876	0.9926	0.9972	0.0001
527	0.9620	0.9670	0.9766	0.9843	0.9900	0.9978	1.0031
703	0.9560	0.9626	0.9735	0.9821	0.9915	0.9998	1.0071
878	0.9533	0.9598	0.9733	0.9836	0.9932	1.0040	1.0131
1055	0.9526	0.9600	0.9750	0.9874	0.9964	1.0070	1.0179
1230	0.9533	0.9637	0.9767	0.9902	1.0014	1.0110	1.0244
1406	0.9559	0.9673	0.9821	0.9961	1.0073	1.0166	1.0313

Where η_p/η_1 is the ratio of the viscosity at pressure p (kg cm^{-2}) relative to that at 1 atm.

TABLE 27

Specific conductivity of sea water* (Weyl (1964). From data by Thomas et al., (1934)

S‰	Temperature (°C)					
	25	20	15	10	5	0
10	17·345	15·628	13·967	12·361	10·816	9·341
20	32·188	29·027	25·967	23·010	20·166	17·456
30	46·213	41·713	37·351	33·137	29·090	25·238
31	47·584	42·954	38·467	34·131	29·968	26·005
32	48·951	44·192	39·579	35·122	30·843	26·771
33	50·314	45·426	40·688	36·110	31·716	27·535
34	51·671	46·656	41·794	37·096	32·588	28·298
35	53·025	47·882	42·896	38·080	33·457	29·060
36	54·374	49·105	43·996	39·061	34·325	29·820
37	55·719	50·325	45·093	40·039	35·190	30·579
38	57·061	51·541	46·187	41·016	36·055	31·337
39	58·398	52·754	47·278	41·990	36·917	32·094

* Conductivity in millimho cm^{-1}.

TABLE 28

Effect of pressure on the conductivity of sea water* (after Bradshaw and Schleicher, 1965)

Temp.	Pressure (db)	S‰			Temp.	S‰		
		31	35	39		31	35	39
0°C	1,000	1·599	1·556	1·512	15°C	1·032	1·008	0·985
	2,000	3·089	3·006	2·922		1·996	1·951	1·906
	3,000	4·475	4·345	4·233		2·895	2·830	2·764
	4,000	5·759	5·603	5·448		3·731	3·646	3·562
	5,000	6·944	6·757	6·569		4·506	4·403	4·301
	6,000	8·034	7·817	7·599		5·221	5·102	4·984
	7,000	9·031	8·787	8·543		5·879	5·745	5·612
	8,000	9·939	9·670	9·401		6·481	6·334	6·187
	9,000	10·761	10·469	10·178		7·031	6·871	6·711
	10,000	11·499	11·188	10·877		7·529	7·358	7·187
5°C	1,000	1·368	1·333	1·298	20°C	0·907	0·888	0·868
	2,000	2·646	2·578	2·510		1·755	1·718	1·680
	3,000	3·835	3·737	3·639		2·546	2·492	2·438
	4,000	4·939	4·813	4·686		3·282	3·212	3·142
	5,000	5·960	5·807	5·655		3·964	3·879	3·795
	6,000	6·901	6·724	6·547		4·594	4·496	4·399
	7,000	7·764	7·565	7·366		5·174	5·064	4·954
	8,000	8·552	8·333	8·114		5·706	5·585	5·464
	9,000	9·269	9·031	8·794		6·192	6·060	5·929
	10,000	9·915	9·661	9·408		6·633	6·492	6·351
10°C	1,000	1·183	1·154	1·125	25°C	0·799	0·783	0·767
	2,000	2·287	2·232	2·177		1·547	1·516	1·485
	3,000	3·317	3·237	3·157		2·245	2·200	2·156
	4,000	4·273	4·170	4·067		2·895	2·837	2·780
	5,000	5·159	5·034	4·910		3·498	3·429	3·359
	6,000	5·976	5·832	5·688		4·056	3·976	3·896
	7,000	6·728	6·565	6·402		4·571	4·481	4·390
	8,000	7·415	7·236	7·057		5·045	4·945	4·845
	9,000	8·041	7·847	7·652		5·478	5·369	5·261
	10,000	8·608	8·400	8·192		5·872	5·756	5·640

* Percentage increase compared with the conductivity at one atmosphere.

W

TABLE 29

Conductivity ratio of sea water at 15°C (R_{15}) and 20°C (R_{20}) relative to sea water of salinity 35·000‰. (From data in UNESCO, 1966).

$S‰$	15°C	20°C	$S‰$	15°C	20°C
29·50	0·85795	0·8583	36·00	1·02545	1·0254
30·00	0·87101	0·8714	36·50	1·03814	1·0380
30·50	0·88404	0·8844	37·00	1·05079	1·0506
31·00	0·89705	0·8973	37·50	1·06341	1·0632
31·50	0·91002	0·9103	38·00	1·07601	1·0758
32·00	0·92296	0·9232	38·50	1·08858	1·0883
32·50	0·93588	0·9361	39·00	1·10112	1·1008
33·00	0·94876	0·9489	39·50	1·11364	1·1133
33·50	0·96160	0·9617	40·00	1·12613	1·1257
34·00	0·97444	0·9745	40·50	1·13849	1·1381
34·50	0·98724	0·9873	41·00	1·15103	1·1505
35·00	1·00000	1·0000	41·50	1·16344	1·1629
35·50	1·01275	1·0127	42·00	1·17583	1·1752

For 15°C $S‰ = -0.08996 + 28.29720\,R_{15} + 12.80832\,R_{15}^2 - 10.678969\,R_{15}^3 + 5.98624\,R_{15}^4 - 1.32311\,R_{15}^5.$

TABLE 30

Correction values ($\times 10^4$) to be applied to conductivity ratios measured at temperatures differing from 20°C to correct them to ratios at 20°C (to be used only in conjunction with 20°C ratios in Table 29). (After UNESCO, 1966)

Measured ratio	Temperature (°C)								
	10	12	14	16	18	20	22	24	26
0·85	80	62	45	29	14	0	−14	−26	−38
0·90	56	44	32	21	10	0	−9	−18	−27
0·95	29	23	17	11	5	0	−5	−10	−14
1·00	0	0	0	0	0	0	0	0	0
1·05	−33	−25	−19	−12	−6	0	5	11	15
1·10	−69	−54	−39	−25	−12	0	11	22	32
1·15	−109	−85	−62	−40	−19	0	18	35	50

TABLE 31

Light absorption of typical sea waters. Extinction for 10 cm path length. (After Clarke and James, 1939)

Sample	Wavelength Å							
	3600	4000	5000	5200	6000	7000	7500	8000
Pure water	0·001	0·001	0·002	0·002	0·010	0·025	0·115	0·086
Artificial sea water	0·011	0·003	0·005	0·007	0·010	0·025	0·115	0·086
Ocean water, unfiltered	0·012	0·009	0·007	0·008	0·011	0·025	0·115	0·086
Continental slope waters, unfiltered	0·052	0·030	0·011	0·010	0·012	0·035	0·130	0·088
Continental slope waters, filtered	0·016	0·010	0·005	0·005	0·012	0·030	0·115	0·086
Inshore water unfiltered	0·055	0·042	0·028	0·026	0·035	0·052	0·140	0·100
Inshore water, filtered	0·015	0·010	0·005	0·005	0·010	0·025	0·110	0·086

TABLE 32

Differences between the extinctions of sea waters and pure water*. (From data by Clarke and James, 1939)

Sample	Wavelength Å							
	3600	4000	5000	5200	6000	7000	7500	8000
Artificial sea water	0·010	0·002	0·003	0·005	nil	nil	nil	nil
Ocean water unfiltered	0·011	0·008	0·005	0·006	0·001	nil	nil	nil
Continental slope water, unfiltered	0·051	0·029	0·009	0·008	0·002	0·010	0·015	0·002
Continental slope water, filtered	0·015	0·009	0·003	0·003	0·002	0·005	nil	nil
Inshore water, unfiltered	0·054	0·041	0·026	0·024	0·025	0·027	0·025	0·015
Inshore water, filtered	0·014	0·009	0·003	0·003	nil	nil	nil	nil

* $E_{SW(10 cm)} - E_{PW(10 cm)}$

Note: The values given in Tables 31 and 32 for unfiltered inshore waters should be taken as no more than a rough indication, since actual values vary widely with time and location.

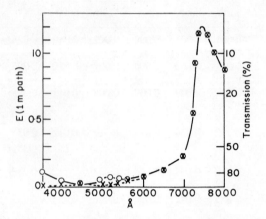

FIG. 1. Extinction (1 m path) against wavelength. Solid line—filtered ocean water. Broken line—pure water. (After Clarke and James, 1939).

Refractive index differences (Δn) for sea water at a wavelength of 589·3 nm at various temperatures and salinities. ($\Delta n = (n - 1\cdot30000) \cdot 10^5$). (Matthäus, 1974)

T[°C]	\multicolumn{21}{c}{S[‰]}																				
	0	2	4	6	8	10	12	14	16	18	20	22	24	26	28	30	32	34	36	38	40
0	3402	3441	3481	3520	3559	3598	3637	3677	3716	3755	3794	3833	3873	3912	3951	3990	4029	4069	4108	4147	4186
1	3400	3439	3478	3517	3556	3595	3634	3674	3713	3752	3791	3830	3869	3908	3947	3986	4025	4064	4103	4142	4181
2	3398	3437	3476	3515	3553	3592	3631	3670	3709	3748	3787	3826	3865	3904	3942	3981	4020	4059	4098	4137	4176
3	3395	3434	3473	3511	3550	3589	3628	3666	3705	3744	3783	3821	3860	3899	3938	3976	4015	4054	4093	4131	4170
4	3392	3431	3469	3508	3547	3585	3624	3662	3701	3740	3778	3817	3855	3894	3933	3971	4010	4048	4087	4126	4164
5	3389	3427	3466	3504	3543	3581	3620	3658	3697	3735	3773	3812	3850	3889	3927	3966	4004	4043	4081	4120	4158
6	3385	3424	3462	3500	3538	3577	3615	3653	3692	3730	3768	3807	3845	3883	3922	3960	3998	4037	4075	4113	4152
7	3381	3419	3458	3496	3534	3572	3610	3648	3687	3725	3763	3801	3839	3878	3916	3954	3992	4030	4068	4107	4145
8	3377	3415	3453	3491	3529	3567	3605	3643	3681	3719	3757	3795	3833	3871	3909	3948	3986	4024	4062	4100	4138
9	3372	3410	3448	3486	3524	3562	3600	3638	3675	3713	3751	3789	3827	3865	3903	3941	3979	4017	4055	4093	4130
10	3367	3405	3443	3481	3518	3556	3594	3632	3669	3707	3745	3783	3821	3858	3896	3934	3972	4010	4047	4085	4123
11	3362	3399	3437	3475	3512	3550	3588	3625	3663	3701	3738	3776	3814	3851	3889	3927	3964	4002	4040	4077	4115
12	3356	3394	3431	3469	3506	3544	3581	3619	3656	3694	3732	3769	3807	3844	3882	3919	3957	3994	4032	4069	4107
13	3350	3387	3425	3462	3500	3537	3575	3612	3649	3687	3724	3762	3799	3837	3874	3911	3949	3986	4024	4061	4098
14	3344	3381	3418	3456	3493	3530	3568	3605	3642	3679	3717	3754	3791	3829	3866	3903	3941	3978	4015	4053	4090
15	3337	3374	3411	3449	3486	3523	3560	3597	3635	3672	3709	3746	3783	3821	3858	3895	3932	3969	4006	4044	4081
16	3330	3367	3404	3441	3478	3515	3552	3590	3627	3664	3701	3738	3775	3812	3849	3886	3923	3960	3997	4035	4072
17	3323	3360	3397	3434	3470	3507	3544	3581	3618	3655	3692	3729	3766	3803	3840	3877	3914	3951	3988	4025	4062
18	3315	3352	3389	3425	3462	3499	3536	3573	3610	3647	3684	3720	3757	3794	3831	3868	3905	3942	3979	4016	4052
19	3307	3344	3380	3417	3454	3491	3527	3564	3601	3638	3675	3711	3748	3785	3822	3858	3895	3932	3969	4006	4042
20	3298	3335	3372	3408	3445	3482	3518	3555	3592	3629	3665	3702	3739	3775	3812	3849	3885	3922	3959	3995	4032
21	3290	3326	3363	3399	3436	3473	3509	3546	3582	3619	3656	3692	3729	3765	3802	3838	3875	3912	3948	3985	4021
22	3281	3317	3354	3390	3427	3463	3500	3536	3573	3609	3646	3682	3719	3755	3792	3828	3865	3901	3938	3974	4011
23	3271	3308	3344	3380	3417	3453	3490	3526	3562	3599	3635	3672	3708	3745	3781	3817	3854	3890	3927	3963	3999
24	3261	3298	3334	3370	3407	3443	3479	3516	3552	3588	3625	3661	3697	3734	3770	3806	3843	3879	3915	3952	3988
25	3251	3288	3324	3360	3396	3433	3469	3505	3541	3578	3614	3650	3686	3723	3759	3795	3831	3868	3904	3940	3976
26	3241	3277	3313	3349	3386	3422	3458	3494	3530	3566	3603	3639	3675	3711	3747	3783	3820	3856	3892	3928	3964
27	3230	3266	3302	3338	3375	3411	3447	3483	3519	3555	3591	3627	3663	3699	3736	3772	3808	3844	3880	3916	3952
28	3219	3255	3291	3327	3363	3399	3435	3471	3507	3543	3579	3615	3651	3687	3723	3759	3796	3832	3868	3904	3940
29	3208	3244	3279	3315	3351	3387	3423	3459	3495	3531	3567	3603	3639	3675	3711	3747	3783	3819	3855	3891	3927
30	3196	3232	3268	3303	3339	3375	3411	3447	3483	3519	3555	3591	3627	3662	3698	3734	3770	3806	3842	3878	3914

TABLE 34

Refractive index differences (Δn) for sea water of salinity 35·00‰ at various temperatures and wavelengths ($\Delta n = (n - 1·30000) \cdot 10^5$). (After Matthäus, 1974)

T[°C]	\multicolumn{15}{c}{Wavelength (nm)}
	404·7	435·8	457·9	467·8	480·0	488·0	501·7	508·5	514·5	546·1	577·0	579·1	589·3	632·8	643·8
0	5099	4840	4684	4621	4549	4504	4433	4400	4372	4240	4130	4124	4091	3961	3929
1	5094	4835	4679	4616	4544	4500	4428	4395	4367	4235	4126	4119	4086	3956	3925
2	5089	4830	4674	4611	4529	4495	4423	4390	4362	4230	4121	4114	4081	3951	3920
3	5084	4825	4669	4606	4534	4489	4418	4385	4357	4225	4115	4109	4076	3946	3914
4	5078	4819	4664	4601	4528	4484	4412	4379	4351	4219	4110	4103	4070	3941	3909
5	5072	4814	4658	4595	4522	4478	4407	4374	4345	4213	4104	4097	4065	3935	3903
6	5066	4807	4652	4589	4516	4472	4400	4367	4339	4207	4098	4091	4058	3929	3897
7	5060	4801	4645	4582	4510	4465	4394	4361	4333	4201	4091	4085	4052	3922	3890
8	5053	4794	4639	4576	4503	4459	4387	4354	4326	4194	4085	4078	4045	3916	3884
9	5046	4787	4632	4569	4496	4452	4380	4347	4319	4187	4078	4071	4038	3909	3877
10	5039	4780	4624	4561	4489	4444	4373	4340	4312	4180	4071	4064	4031	3901	3869
11	5031	4773	4617	4554	4481	4437	4366	4332	4304	4172	4063	4056	4023	3894	3862
12	5023	4765	4609	4546	4473	4429	4358	4325	4297	4164	4055	4048	4016	3886	3854
13	5015	4757	4601	4538	4465	4421	4350	4317	4288	4156	4047	4040	4008	3878	3846
14	5007	4748	4592	4529	4457	4412	4341	4308	4280	4148	4039	4032	3999	3869	3837
15	4998	4740	4584	4521	4448	4404	4333	4300	4271	4139	4030	4023	3991	3861	3829
16	4989	4731	4575	4512	4439	4395	4324	4291	4262	4130	4021	4014	3982	3852	3820
17	4980	4721	4566	4503	4430	4386	4314	4281	4253	4121	4012	4005	3972	3843	3811
18	4971	4712	4556	4493	4421	4376	4305	4272	4244	4111	4002	3995	3963	3833	3801
19	4961	4702	4546	4483	4411	4366	4295	4262	4234	4102	3993	3986	3953	3823	3791
20	4951	4692	4536	4473	4401	4356	4285	4252	4224	4092	3982	3976	3943	3813	3781

TABLE 34 cont.

T[°C]	0,4047	0,4358	0,4579	0,4678	0,4800	0,4880	0,5017	0,5085	0,5145	0,5461	0,5770	0,5791	0,5893	0,6328	0,6438
21	4940	4682	4526	4463	4390	4346	4275	4242	4214	4081	3972	3965	3933	3803	3771
22	4930	4671	4515	4452	4380	4335	4264	4231	4203	4071	3961	3955	3922	3792	3760
23	4919	4660	4504	4441	4369	4324	4253	4220	4192	4060	3951	3944	3911	3781	3749
24	4908	4649	4493	4430	4358	4313	4242	4209	4181	4048	3939	3932	3900	3770	3738
25	4896	4637	4482	4419	4346	4302	4230	4197	4169	4037	3928	3921	3888	3759	3727
26	4884	4626	4470	4407	4334	4290	4219	4186	4157	4025	3916	3909	3877	3747	3715
27	4872	4614	4458	4395	4322	4278	4207	4174	4145	4013	3904	3897	3865	3735	3703
28	4860	4601	4445	4382	4310	4265	4194	4161	4133	4001	3892	3885	3852	3722	3690
29	4847	4589	4433	4370	4297	4253	4182	4149	4120	3988	3879	3872	3840	3710	3678
30	4834	4576	4420	4357	4284	4240	4169	4136	4108	3975	3866	3859	3827	3697	3665

λ[nm]

TABLE 35

Absolute refractive index of sea water ($S = 35\cdot00‰$) as a function of temperature, pressure and wavelength. (Stanley, 1971)

Pressure	Temperature (°C)						
	0·03	5·03	10·03	15·02	20·00	24·99	29·98
			6328 Å				
Atm.	1·34015	1·33977	1·33935	1·33899	1·33850	1·33795	1·33737
352 kg cm²	1·34539	1·34487	1·34431	1·34388	1·34331	1·34270	1·34207
703 kg cm⁻²	1·35025	1·34962	1·34896	1·34844	1·34780	1·34713	1·34647
1055 kg cm⁻²	1·35481	1·35403	1·35380	1·35269	1·35200	1·35129	1·35059
1406 kg cm⁻²	—	1·35813	1·35738	1·35668	1·35592	1·35519	1·35443
			5017 Å				
Atm	1·34455	1·34455	1·34422	1·34379	1·34327	1·34272	1·34215
352 kg cm⁻²	1·35008	1·34969	1·34924	1·34873	1·34813	1·34757	1·34694
703 kg cm⁻²	1·35507	1·35450	1·35394	1·35333	1·35269	1·35208	1·35137
1055 kg cm⁻²	1·35953	1·35891	1·35834	1·35764	1·35695	1·35632	1·35561
1406 kg cm⁻²	—	1·36314	1·36241	1·36166	1·36095	1·36019	1·35946

TABLE 36

Velocity of light (λ = 589·3 nm) in sea water at 1 atm (km s⁻¹) (Sager, 1974)

$S‰$	Temperature (°C)								
	0	5	10	15	20	25	30	35	40
0	224·732	224·749	224·785	224·837	224·904	224·985	225·080	225·185	225·305
2·5	224·650	224·668	224·705	224·759	224·827	224·909	225·004	225·110	225·230
5·0	224·567	224·588	224·626	224·681	224·749	224·832	224·928	225·035	225·156
7·5	224·485	224·507	224·547	224·603	224·672	224·756	224·852	224·960	225·081
10·0	224·402	224·426	224·468	224·524	224·595	224·679	224·776	224·885	225·006
12·5	224·319	224·346	224·388	224·446	224·518	224·603	224·700	224·810	224·931
15·0	224·236	224·265	224·309	224·368	224·441	224·527	224·625	224·735	224·857
17·5	224·154	224·185	224·230	224·290	224·364	224·450	224·549	224·660	224·782
20·0	224·072	224·104	224·151	224·212	224·287	224·374	224·473	224·585	224·707
22·5	223·990	224·024	224·072	224·134	224·210	224·297	224·398	224·510	224·633
25·0	223·907	223·943	223·994	224·057	224·133	224·221	224·322	224·435	224·559
27·5	223·825	223·863	223·915	223·979	224·056	224·145	224·247	224·360	224·485
30·0	223·743	223·783	223·836	223·901	223·979	224·069	224·171	224·285	224·411
32·5	223·661	223·703	223·758	223·823	223·903	223·993	224·096	224·211	224·336
35·0	223·579	223·623	223·679	223·746	223·826	223·917	224·020	224·136	224·262
37·5	223·498	223·543	223·600	223·669	223·749	223·841	223·945	224·061	224·188
40·0	223·416	223·463	223·521	223·591	223·673	223·765	223·870	223·986	224·114

REFERENCES

Bark, L. S., Ganson, P. P. and Meister, N. A. (1964), "Tables of the Velocity of Sound in Sea Water" Pergamon, Oxford.
Bradshaw, A. and Schleicher, K. E. (1965). *Deep-Sea Res.* 12, 151.
Bradshaw, A. and Schleicher, K. E. (1970). *Deep-Sea Res.* 16, 691.
Caldwell, D. T. (1974). *Deep-Sea Res.* 21, 131.
Carpenter, J. H. (1966). *Limnol. Oceanogr.* 11, 264.
Castelli, V. J., Stanley, E. M. and Fischer, E. C. (1974). *Deep-Sea Res.* 21, 311.
Clarke, G. L. and James, H. R. (1939). *J. Opt. Soc. Amer.* 29, 43.
Cox, R. A. (1965). *In* "Chemical Oceanography" (J. P. Riley and G. Skirrow, eds), Vol. I. Academic Press, London.
Cox R. A. and Culkin, F. (1967). *Deep-Sea Res.* 13, 789.
Cox, R. A., McCartney, M. J. and Culkin, F. (1970). *Deep-Sea Res.* 17, 679.
Crozier, T. E. and Yamamoto, S. (1974). *J. Chem. Eng. Data,* 19, 242.
Doherty, B. T. and Kester, D. R. (1974), *J. Mar. Res.* 32, 285
Dorsey, N. E. (1940). "Properties of Ordinary Water-substance". Reinhold, New York.
Douglas, E. (1964). *J. Phys. Chem.* 68, 169.
Douglas, E. (1965). *J. Phys. Chem.* 69, 2608.
Douglas, E. (1967). *J. Phys. Chem.* 71, 1931.
Fleming, R. H. and Revelle, R. R. (1939). "Recent Marine Sediments" (N. Trask ed.). Amer. Soc. Petrol. Geol., Tulsa, Oklahoma.
Kalle, K. (1945). In "Probleme der Kosmischen Physik" 2nd Edn., Vol. 23. Leipzig.
Korson, L., Drost-Hansen, W. and Millero, F. J. (1969). *J. Phys. Chem.* 73, 34.
Kester, D. R., Duedall, I. W., Connor, D. N. and Pytkowicz, R. M. (1967). *Limnol. Oceanogr.* 12, 176.
Krümmel, O. (1900). *Wiss. Meeresuntersuch.* 5, 9.
Lepple, F. K. and Millero, F. J. (1971). *Deep-Sea Res.* 18, 1233.
Lumby, J. R. and Folkard, A. R. (1956). *Bull. Inst. Océanogr. Monaco,* 1080, 1.
Lyman, J. and Fleming, R. H. (1940). *J. Mar. Res.* 3, 134.
Matthäus, W. (1974). *Beitr. Meeresk.* 29, 93.
Matthäus, W. (1974). *Beitr. Meeresk.* 33, 73.
Millero, F. J. (1974). *In* "The Sea" (E. D. Goldberg ed.), Vol. 5, Interscience, New York.
Millero, F. J. and Lepple, F. K. (1973). *Mar. Chem.* 1, 89.
Millero, F. J., Hansen, L. D. and Hoff, E. V. (1973b). *J. Mar. Res.* 31, 21.
Millero, F. J., Perron G. and Desnoyers, J. E. (1973). *J. Geophys. Res.* 78, 4499.
Morris, A. W. and Riley, J. P. (1966). *Deep-Sea Res.* 13, 689.
Murray, C. N. and Riley, J. P. (1969a). *Deep-Sea Res.* 16, 311.
Murray, C. N. and Riley, J. P. (1969b). *Deep-Sea Res.* 16, 297.
Murray, C. N. and Riley, J. P. (1971). *Deep-Sea Res.* 18, 533.
Riley, J. P. and Tongudai, M. (1967). *Chem. Geol.* 2, 263.
Robinson, R. A. (1954). *J. Mar. Biol. Ass. U.K.* 33, 449.
Sager, G. (1974). *Beitr. Meeresk.* 33, 68.
Stanley, E. M. (1971). *Deep-Sea Res.* 18, 833.
Stanley, E. M. and Batten, R. C. (1969). *J. Geophys. Res.* 74, 3415.
Stoughton, R. W. and Lietzke, M. H. (1967). *J. Chem. Eng. Data,* 12, 101.
Thomas, B. D., Thompson, T. G. and Utterback, C. L. (1934). *J. Cons. Int. Explor. Mer,* 9, 28.

UNESCO (1966). International Oceanographic Tables, Vol. 1. National Institute of Oceanography, Wormley, Surrey, England.

UNESCO (1973). International Oceanographic Tables. Vol. 2. National Institute of Oceanography, Wormley, Surrey, England.

U.S. Naval Oceanogr. Office (1962) Tables of sound speed in sea water. Publ. SP58 U.S. Naval Oceanographic Office, Washington, U.S.A.

U.S. Navy (1961). Tables for the velocity of sound in sea water. Bureau of ships reference NObsr 81564 S-7001-0307, Washington, U.S.A.

Weiss, R. F. (1970). *Deep-Sea Res.* **17**, 721.

Weiss, R. F. (1971). *J. Chem. Eng. Data,* **16**, 235.

Weyl, P. (1964). *Limnol. Oceanogr.* **9**, 75.

Wilson, W. D. (1960). *J. Acoust. Soc. Amer.* **32**, 1357.

Wood, D. and Caputi, R. (1966). "Technical Report No. 988" U.S. Naval Radiological Defense Laboratory, San Francisco.

Subject Index

(Numbers in bold type indicate the page on which a subject is treated most fully.)

633